Springer Atmospheric Sciences

More information about this series at http://www.springer.com/series/10176

Alexander V. Ryzhkov • Dusan S. Zrnic

Radar Polarimetry for Weather Observations

Springer

Alexander V. Ryzhkov
Cooperative Institute for Mesoscale
Meteorological Studies
The University of Oklahoma
Norman, OK, USA

Dusan S. Zrnic
National Severe Storms Laboratory
National Oceanic and Atmospheric
Administration
Norman, OK, USA

ISSN 2194-5217 ISSN 2194-5225 (electronic)
ISBN 978-3-030-05092-4 ISBN 978-3-030-05093-1 (eBook)
https://doi.org/10.1007/978-3-030-05093-1

Library of Congress Control Number: 2019930082

This Springer imprint is published by the registered company Springer Nature Switzerland AG
The registered company address is: Gewerbestrasse 11, 6330 Cham, Switzerland

Preface

Prior to the advent of dual-polarization radars, direct interpretation of cloud and precipitation bulk characteristics from radar reflectivity was limited to very few unambiguously defined cases, such as the bright band or large hail. Even then, knowledge of the physical conditions, like the height above ground of enhanced reflectivity areas, was needed to properly interpret and quantify the observations. Polarimetry transformed the way meteorologists look at and interpret the bulk properties of clouds and precipitation. It brought a dramatic change to direct interpretation and quantitative assessment of these properties, so much so that the polarimetric weather radar has transitioned from a scientific instrument to operational use. This is exemplified by the network of the US weather surveillance radars (WSR-88D) which have been upgraded to dual-polarization. Several other countries did or are doing similar upgrade of their national weather radar networks. Moreover, some of the weather radar manufacturers do not even offer single-polarization radars for weather surveillance anymore. The additional information dual-polarization radar provides to forecasters is primarily used for quantitative precipitation estimation (QPE), classification of radar returns, discrimination between meteorological and nonmeteorological scatterers, and severe weather warnings. Identification of meteorological scatterers provides an added positive impact on QPE. Potential improvements of QPE were used as principal justifications for the introduction of the Doppler capability (in the 1980s) and dual-polarization upgrade of operational weather radars. Lately, interest is rising within the numerical weather prediction (NWP) community to incorporate polarimetric data into NWP models either via assimilation or through improvement of their microphysical parameterization.

Although there are numerous scientific papers and reports about weather radar polarimetry and its applications, books on the subject are few. We hope that this monograph adds variety and some material not compiled elsewhere. It is meant for practicing radar meteorologists, hydrologists, cloud physicists, and modelers who are interested in the bulk properties of hydrometeors and quantification of these with the goals to improve precipitation measurements, understanding of precipitation processes, or model forecasts. We have made a deliberate attempt to tightly connect the microphysical processes responsible for the development and evolution of the clouds' bulk physical properties to the polarimetric variables. The book contains instructions on how to simulate realistic polarimetric variables. It also demonstrates that the polarimetric variables from all but the precipitation containing large (Mie) scatterers can be adequately related to bulk precipitation physics using simple closed-form solutions. It also addresses the problem of determining the polarimetric variables from the output of NWP models.

To make the book self-contained, we included fundamental topics in polarimetry such as polarimetric variables, polarimetric radar, and scattering. We hope that the practical aspects, references, and instructions contained herein will be beneficial to those entering this fascinating field as well as to those needing quick answers concerning practical applications of weather radar polarimetry. Much of the material in this book came from the research the authors did at the National Severe Storms Laboratory and the Cooperative Institute for Mesoscale Meteorological Studies. Polarimetric work at the National Severe Storms Laboratory started under the directorship of Dr. E. Kessler and continued under Dr. R. Maddox, Dr. J. Kimpel, and Dr. S. Koch. Crucial were two NSSL radars, first Cimarron and then KOUN, to test various design/engineering aspects of dual polarimetry. These radars also provided valuable data for analysis and interpretation.

The able engineering team responsible for implementation of dual-polarization consisted of D. Sirmans, A. Zahrai, J. Carter, M. Schmidt, and R. Wahkinney. To them and other NSSL support staff, we extend sincere thanks. The National Research Council postdoctoral associates at NSSL, Dr. M. Sachidananda and Dr. N. Balakrishnan, made pioneering contributions at the time when these were most needed. Collaboration with Dr. J. Straka from the University of Oklahoma brought rigor to polarimetric classification of echoes and introduced the subject to the community. Significant contributions to Chaps. 7 and 8 are from Dr. M. Kumjian's works. Discussions with Drs. R. Doviak (NSSL), G. Zhang, V. Melnikov (University of Oklahoma), S. Matrosov (University of Colorado in Boulder), and V. Bringi (Colorado State University) were always illuminating.

We express our deep gratitude to our international partners and colleagues, A. Khain, M. Pinsky, C. Simmer, S. Troemel, K.-E. Kim, D.-I. Lee, G. Lee, R. Kaltenboeck, D. Hudak, S. Boodoo, and many others for their fruitful collaboration and exchange of ideas and data. We also acknowledge help from our colleagues at NSSL/CIMMS, D. Forsyth, T. Schuur, J. Krause, P. Zhang, L. Borowska, H. Reeves, K. Ortega, in the pursuit of our research. The partnership with Drs. R. Palmer and T.-Y. Yu from the Advanced Radar Research Center and their generosity in sharing the OU-PRIME polarimetric data are greatly appreciated.

Last but not least, our students J. Conway, B. Gordon, M. Loney, P, Schlatter, M. Askelson, S. Bachmann, S. Giangrande, S. Ganson, H.-S. Park, J.-Y. Gu, J. Picca, J. Snyder, J. Carlin, P. Bukovcic, D. Mirkovic, E. Griffin, and A. Murphy were sounding boards for testing concepts and ideas, as well as for generating new ones.

Funding for dual-polarization work, although sparse, was sufficient to keep us hungry for more; thank you OAR/NOAA. Token support was provided by the FAA,

NSF, and NASA. Dr. J. Rasmussen, Head of OAR in the 1990s, funded specifically dual-polarization work. Dr. E. Friday, Director of the National Weather Service, arranged the transfer of the WSR-88D (KOUN) from NWS to NSSL. This was a tremendous help to NSSL's research and facilitated the upgrade of the WSR-88D network to dual-polarization.

Norman, OK, USA Alexander V. Ryzhkov
Norman, OK, USA Dusan S. Zrnic

Contents

Abbreviations

α	Canting angle
α	Ratio A/K_{DP}
β	Antenna elevation angle
β	Ratio A_{DP}/K_{DP}
δ	Backscatter differential phase
ε	Dielectric constant
ε_w	Dielectric constant of water
ε_i	Dielectric constant of solid ice
ε_s	Dielectric constant of snow
η	Reflectivity (cross section per unit volume)
η_o	120π (Ω), free space impedance
θ	Angular distance from the beam axis
θ_e	Elevation angle
θ_1	One-way beamwidth between half-power points
λ	Electromagnetic wavelength
Λ	Slope of the exponential raindrop size distribution
Λ_s	Slope of the exponential size distribution of snow/ice
μ	Shape parameter of the Gamma distribution
ρ_a	Mass density of air
ρ_i	Density of solid ice
ρ_{hv}	Correlation coefficient between horizontally and vertically polarized return signals
$\rho_{xh,xv}$	Correlation coefficients between cross-polar (*x*) and copolar (*h*,*v*) components of the returned signal
ρ_s	Density of snow
ρ_w	Density of water
σ	Width of canting angle distribution
σ_b	Backscattering cross section
σ_e	Extinction or attenuation cross section
σ_v	Doppler spectrum width (m s^{-1})
σ_{vn}	Normalized spectrum width ($4\sigma_v T_s/\lambda$)

σ_θ^2	Second central moment of the two-way antenna radiation pattern
τ	Pulse width
τ_s	Range time delay
ϕ	Azimuth
Φ_{DP}	Differential phase
Ψ	Angle between the axis of rotation of scatterer and the direction of wave propagation
ω	Angular frequency
A_{DP}	Specific differential attenuation (dB km^{-1})
$A_{h,v}$	Specific attenuation at orthogonal polarizations (dB km^{-1})
A_i	Aggregation value for i-th radar echo class
$A_1 - A_5$	Angular moments of particle orientation distributions
c	Speed of light (3×10^8 m s^{-1})
$c_{h,v}$	Speeds of EM waves (polarization H or V) in anisotropic medium
C	Capacitance of particle
CDR	Circular depolarization ratio (dB)
C_{dr}	Circular depolarization ratio (linear units)
D_a	Diameter of the antenna system
D_e	Diameter of an equivalent volume spherical raindrop
D_o	Median volume diameter of particle
D_m	Mean volume (or mass-weighted) diameter of particle
D_v	Diffusivity of water vapor
e	Partial pressure of water vapor
$\mathbf{e}_h$, $\mathbf{e}_v$	Unit vectors in the h and v direction, abbreviated with $\mathbf{e}_{h,v}$
E	Electric field intensity
$\mathrm{E}_{h,v}$	Components of the complex $\vec{\mathbf{E}}$ vector along the $\mathbf{e}_{h,v}$ directions
$E_{h,v}$	Phasor representation of the h and v components
E_w	Saturated vapor pressure with respect to water
E_s	Saturated vapor pressure with respect to ice
$\boldsymbol{E}$	Phasor matrix representation: $\boldsymbol{E} = [E_h, E_v]^T$
$\vec{\mathbf{E}}$	Electric field vector (complex) containing time variation
$\vec{\boldsymbol{E}}$	Phasor vector representation: $\vec{\boldsymbol{E}} = E_h\mathbf{e}_h + E_v\mathbf{e}_v$
f	Frequency
f_d	Doppler frequency shift
f_{rim}	Degree of riming
f_w	Mass water fraction
$f^2(\theta,\phi)$	Normalized one-way power gain of radiation pattern
F	Ventilation coefficient
g	Gravitational constant (9.81 m s^{-2})
g	Antenna gain
$I(\mathbf{r},\mathbf{r}_1)$	Illumination function
$I(t)$	In-phase component of the phasor signal
IWC	Ice water content (g m^{-3})

k	Electromagnetic wave number ($2\pi/\lambda$) in vacuum
K	Thermal conductivity of air
$k_{h,v}$	Complex wave number in atmosphere with scatterers
K_{DP}	Specific differential phase (deg km^{-1})
K_w	$(\varepsilon_w - 1)/(\varepsilon_w + 2)$
K_i	$(\varepsilon_i - 1)/(\varepsilon_i + 2)$
$l_{h,v}$	One-way propagation loss due to scatter and absorption (≥ 1)
ln	Natural logarithm
log	Logarithm to base 10
$\boldsymbol{L}$	Matrix of losses
$L_{a,b}$	Shape factors of spheroidal particle
$l_{h,v}$	Loss factors at orthogonal polarizations (≥ 1)
l_r	Range weighting function loss factor (≥ 1)
L_f	Latent heat of fusion (melting)
L_r	10 log l_{r} (dB)
L_v	Latent heat of vaporization
LDR	Linear depolarization ratio (dB)
L_{dr}	Linear depolarization ratio (linear units)
LWC	Liquid water content (g m^{-3})
m	Mass of particle
M	Number of signal samples (or sample pairs) along sample time axis
M_{I}	Number of independent samples
$n_{h,v}$	Refractive index (complex) of atmosphere with hydrometeors
$\Delta n^{(\mathrm{air})}$	Contribution to refractive index by air
$\Delta n_{h,v}^{(\mathrm{scat})}$	Contribution to refractive index by scatterers (hydrometeors)
$N_{h,v}$	White noise power in the orthogonal receiver channels
$N(D_e)$	Drop size distribution (m^{-3} mm^{-1})
N_{Re}	Reynolds number
N_0	Intercept parameter of the exponential and gamma size distribution of raindrops
N_{0s}	Intercept parameter of the exponential size distribution of ice/snow
N_w	Intercept parameter of the normalized size distribution of raindrops
N_0^*	Intercept parameter of the normalized size distribution of ice/snow
N_t	Total concentration of particles (m^{-3})
P	Atmospheric pressure
P^r	Received signal power
P^t	Peak transmitted power
$P^{(i)}(V_j)$	Membership function of the variable "j" and class "i"
PIA	Two-way path-integrated attenuation (dB)
PIA	Two-way path-integrated attenuation in linear scale
Q_w	Total water content
$Q(t)$	Quadrature-phase component of the complex signal
r	Range to scatterer
r_a	Unambiguous range

r_6	6-dB range width of resolution volume
$\mathbf{r_o}$	Vector range to the resolution volume V_6 center
r_m	Aspect ratio of melting graupel/hail
r_w	Axis ratio of raindrops
R	Rain rate (mm h^{-1})
RH	Relative humidity of air (%)
$R_{\mathrm{hh,vv}}(T_s)$	Autocorrelations of weather signal
R_{hv}	Cross-correlation of weather signals of H and V polarization
R_v	Gas constant for water vapor
$s^{(0)}_{\mathrm{hh,\,vv}}$	Forward-scattering coefficient of a scatterer
s_{mn}	Backscattering coefficient of a scatterer, incident polarization is n (H or V), backscattered is m (H or V)
$s_{a,b}$	Backscattering coefficients of a spheroidal scatterer: subscript a is for incident polarization parallel to the rotation axis and subscript b for incident polarization perpendicular to this axis
$\vec{\mathbf{S}}$	Power density of the electromagnetic wave
$\mathbf{S}$	Scattering matrix
S	Snow water equivalent rate (mm h^{-1})
S_w	Vapor saturation ratio with respect to water
S_i	Vapor saturation ratio with respect to ice
SNR	Signal-to-noise ratio (dB)
snr	Signal-to-noise ratio (linear scale)
$\mathbf{T}$	Transmission matrix
T_s	Pulse repetition time
T	Temperature in C^{o}
v	Doppler velocity (m s^{-1})
V_6	Resolution volume size
$V_{\mathrm{hh,vv}}$	Voltage complex representations (contains $j2\pi f$)
$V_{\mathrm{hh,vv}}$	Voltage phasor representations
V_t	Terminal velocity of hydrometeor (m s^{-1})
$W(r)$	Range weighting function
$Z_{\mathrm{H,V}}$	Reflectivity factors for horizontal and vertical polarizations (dBZ)
$Z_{h,v}$	Reflectivity factors for horizontal and vertical polarizations (mm^6 m^{-3})
Z_{dr}	Differential reflectivity in linear units (Z_h/Z_v)
Z_{DR}	Differential reflectivity (dB)
Z_{DP}	Reflectivity difference ($Z_h - Z_v$) in mm^6 m^{-3}

Chapter 1
Polarization, Scattering, and Propagation of Electromagnetic Waves

Understanding polarimetric measurements requires basic knowledge of interaction between electromagnetic (EM) waves and hydrometeors like raindrops, snowflakes, hailstones, or graupel. These particles extract energy from the EM waves and scatter it in all directions and hence are named scatterers. Some, like raindrops, are preferentially oriented and, therefore, interact differently with EM fields than more randomly oriented scatterers like snow aggregates. The number of scatterers interacting with propagating wave is very large and so is the number reflecting the wave. The impact of propagation and reflection on the wave is quantified by the weather radar. Radar measurements are examined with the aim to characterize the bulk properties of the involved scatterers (average concentration, size, shape, orientation, and phase composition). From the bulk properties, the type of hydrometeors in the cloud can be inferred and the amount of precipitation estimated. Telling by remote means the type and amount of precipitation on the ground has been a longstanding goal of meteorologists. The polarimetric radar has promise of achieving this goal.

Fundamental to the scatterers' bulk properties are the scattering characteristics of a single hydrometeor. These are introduced here and quantified in terms of scatterer's physical properties. The concept is then applied to determine propagation effects through the atmosphere filled with a collection of hydrometeors. These effects manifest themselves as attenuation and phase shift of the EM wave. Useful formulas are derived anticipating applications for measurements of bulk precipitation properties.

1.1 Polarization State of Electromagnetic Wave

Coupled fields of electric and magnetic force propagating in space are called electromagnetic (EM) waves. A packet of these waves has spatial and temporal variation of the electric and magnetic fields. Most fields generated by humans for

A. V. Ryzhkov, D. S. Zrnic, *Radar Polarimetry for Weather Observations*,
Springer Atmospheric Sciences, https://doi.org/10.1007/978-3-030-05093-1_1

transmitting information (communication, entertainment, etc.) and remote sensing with radar have sinusoidal variations. Natural sources of EM propagating fields emanating from stars, sun, or lightning discharge do not have simple sinusoidal variations.

At far distance from a source, the vectors of electric and magnetic fields lie in a plane perpendicular to the propagation direction **k** and are perpendicular to each other. This plane is called polarization plane and is defined with two orthogonal directions **h** (horizontal) and **v** (vertical). It is convenient and physically meaningful to designate the horizontal direction as parallel to the locally flat earth surface. Polarization refers to the orientation of the electric field in the polarization plane. The electric field vector of a plane electromagnetic wave traveling in the **r** direction can be expressed in complex notation as a sum of horizontally and vertically polarized components $\vec{\mathbf{E}}_h$ and $\vec{\mathbf{E}}_v$ (Fig. 1.1)

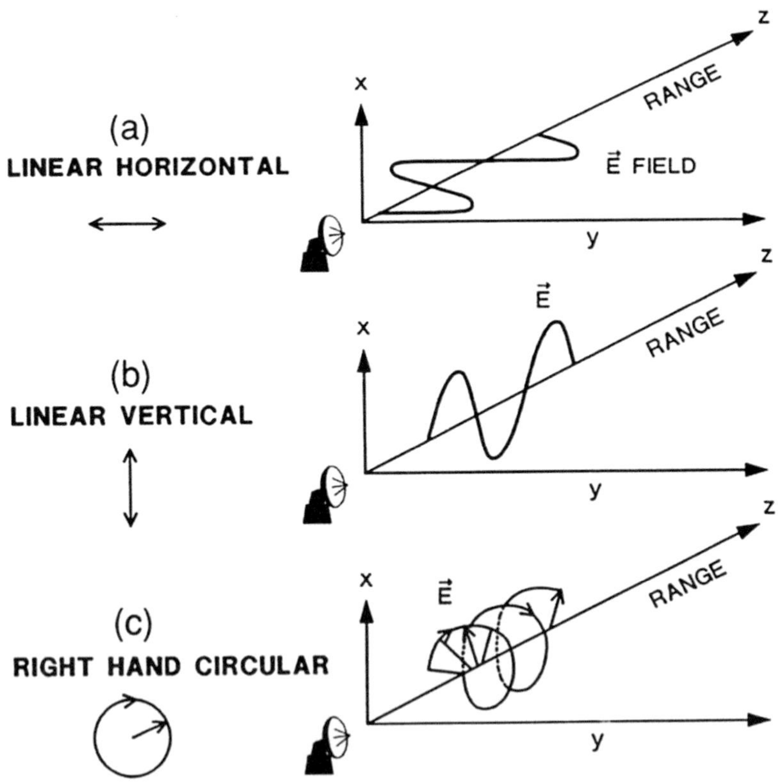

Fig. 1.1 The spatial dependence of the electric field vector for (**a**) horizontally, (**b**) vertically, and (**c**) circularly (right-handed) polarized waves. From Doviak and Zrnic (2006)

$$\vec{\mathbf{E}}(r) = \underset{h}{\vec{\mathbf{E}}}(r) + \underset{v}{\vec{\mathbf{E}}}(r) = \mid E_h(0) \mid \exp[j(2\pi ft - k_h r - \varphi_h)]\mathbf{e}_h + \mid E_v(0) \mid \exp[j(2\pi ft - k_v r - \varphi_v)]\mathbf{e}_v, \left(\mathrm{V\,m^{-1}}\right) \tag{1.1}$$

where $\mathbf{e}_h$ and $\mathbf{e}_v$ are the unit vectors corresponding to the horizontal and vertical polarizations, $|E_h(0)|$ and $|E_v(0)|$ are the amplitudes of horizontally and vertically polarized components of the electric field at $r = 0$, $t = 0$, t is time (s), f is frequency (Hz), $\varphi_{h,v}$ are initial phases (i.e., $E_{h,v}(0) = |E_{h,v}(0)|\exp(-j\varphi_{h,v})$), and $k_{h,v}$ are wave numbers defined as

$$k_{h,v} = 2\pi n_{h,v} f/c \left(\mathrm{m^{-1}}\right). \tag{1.2}$$

In (1.2), c is the speed of light in vacuum and n_h and n_v are the refractive indices of the propagation medium with respect to horizontal and vertical polarizations. In the atmosphere, the refractive indices are complex numbers; the real parts quantify phase shift in (1.1) and imaginary parts quantify attenuation. In vacuum, $n_{h,v} = 1$ and

$$k_{h,v} = k = 2\pi f/c = 2\pi/\lambda \left(\mathrm{m^{-1}}\right), \tag{1.3}$$

where λ is the wavelength. It is understood that the actual fields are real parts of Eq. (1.1) obtained by replacing the exponents with cosines of the arguments (without the imaginary unit j).

In a homogeneous medium having refractive indexes $n_{h,v}$, the EM wave propagates at the speed

$$c_{h,v} = c/\mathrm{Re}(n_{h,v}) \left(\mathrm{m\,s^{-1}}\right). \tag{1.4}$$

The refractive indices are related to the dielectric constants of the medium at orthogonal polarizations ε_h and ε_v as $n_{h,v} = \sqrt{\varepsilon_{h,v}}$. The purposeful distinction between refractive index for horizontally and vertically polarized waves is to quantify propagation of both in the atmosphere filled with nonspherical oriented hydrometeors. At microwave frequencies, the pair of refractive index values $n_{h,v}$ is

$$n_{h,v} = 1 + \Delta n^{(\mathrm{air})} + \Delta n_{h,v}^{(\mathrm{scat})}. \tag{1.5}$$

In (1.5), the component $\Delta n^{(\mathrm{air})}$ represents the contribution of atmospheric gases causing refraction or bending of the wave propagation path (Doviak and Zrnic 2006) and attenuation of the propagating EM field (Ulaby et al. 1981). The real part of the refractive index due to atmospheric gases is

$$\mathrm{Re}\left(\Delta n^{(\mathrm{air})}\right) = \frac{77.6}{T}\left(P + \frac{4810e}{T}\right)10^{-6}, \tag{1.6}$$

where P is the atmospheric pressure (in millibars), e is the partial pressure of water vapor (in millibars), and T is the temperature (in °K). The imaginary part of $\Delta n^{(\mathrm{air})}$ quantifies gaseous absorption. The term $\Delta n_{h,v}{}^{(\mathrm{scat})}$ in (1.5) describes the contribution of atmospheric scatterers (hydrometeors, biota, and others). The contributions at the horizontal and vertical polarizations (H, V) differ if the scatterers are nonspherical and oriented. Hydrometeors such as raindrops or ice crystals falling under the influence of gravity are oriented by drag forces; therefore, generally $\Delta n_h{}^{(\mathrm{scat})} \neq \Delta n_v{}^{(\mathrm{scat})}$.

To explicitly express the wave dependence on attenuation and phase shift represented by the terms $[\mathrm{Im}(k_{h,v})r]$ and $[\mathrm{Re}(k_{h,v})r]$, Eq. (1.1) is rewritten as

$$\mathrm{Re}\left[\vec{\mathbf{E}}\ (r)\right] = \mid E_h(r) \mid \cos\left(2\pi ft - \mathrm{Re}(k_h)r - \varphi_h\right)\mathbf{e}_h + \mid E_v(r) \mid \cos\left(2\pi ft - \mathrm{Re}(k_v)r - \varphi_v\right)\mathbf{e}_v, \tag{1.7}$$

where

$$\mid \mathrm{E}_{h,v}(r) \mid = \mid \mathrm{E}_{h,v}(0) \mid \exp[\mathrm{Im}(k_{h,v})r], \quad \mathrm{Im}(k_{h,v}) < 0. \tag{1.8}$$

Equation (1.7) quantifies how the real part of the wave number (or refractive index) affects the phase of the propagating wave. Through (1.8), the imaginary part quantifies attenuation by the medium filled with atmospheric gases and hydrometeors.

At any given range r, the vectors of horizontally and vertically polarized components of the wave oscillate with frequency f along directions defined by the unit vectors $\mathbf{e}_h$ and $\mathbf{e}_v$. The direction of the composite vector $\mathrm{Re}\left[\vec{\mathbf{E}}\ (r)\right]$ depends on the phase difference, $\Delta\varphi$, between the orthogonal components of the wave:

$$\Delta\varphi = \mathrm{Re}(k_h - k_v)r + \varphi_h - \varphi_v. \tag{1.9}$$

If $\Delta\varphi$ is an integer multiple of π, then the horizontally and vertically polarized components of the electric field are in phase or out of phase and the vector $\mathrm{Re}\left(\vec{\mathbf{E}}\right)$ oscillates along a fixed line, i.e., the electromagnetic wave is linearly polarized (Fig. 1.2a). Otherwise, the instantaneous direction of the vector $\mathrm{Re}\left(\vec{\mathbf{E}}\right)$ changes in time and its tip describes an ellipse; hence, that polarization state is called elliptical (Fig. 1.2b).

If $|E_h(r)| = |E_v(r)|$ and $\Delta\varphi = \pm\pi/2$, the ellipse becomes a circle and the polarization is circular. The electric vector $\mathrm{Re}\left(\vec{\mathbf{E}}\right)$ rotates in the clockwise direction (when viewed in the direction of propagation) if $\Delta\varphi = \pi/2$ (Fig. 1.2c) and in the counterclockwise direction if $\Delta\varphi = -\pi/2$ (Fig. 1.2d). The corresponding polarization states are, respectively, called right-hand circular and left-hand circular. This convention of viewing the vector in the direction of propagation is adopted here. Viewing reflected waves from the propagation direction is standard in the optical community and a

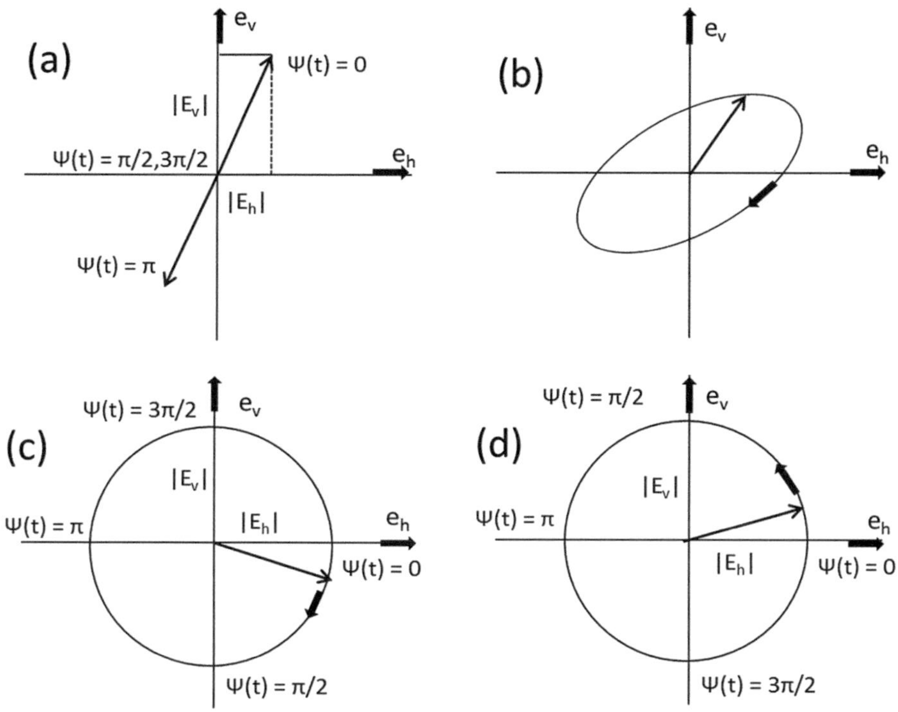

Fig. 1.2 Vector of electric field and trajectories of its tip for (**a**) linear polarization, (**b**) elliptical polarization, (**c**) right-hand circular polarization, and (**d**) left-hand circular polarizations; $\psi(t) = 2\pi ft - \Delta\varphi$ (see (1.7) and (1.9))

comprehensive discussion of the two conventions is in the book by Bringi and Chandrasekar (2001). Note the phase difference (1.9) depends on r if $k_h \neq k_v$. Therefore, the polarization state of the wave changes as it propagates in an anisotropic medium.

It is a standard practice to separate the time-varying factor $\exp(j2\pi ft)$ in the expression (1.1) for complex electric vector

$$\vec{\mathbf{E}}(r) = E_h \exp(j2\pi ft)\mathbf{e}_h + E_v \exp(j2\pi ft)\mathbf{e}_v \left(\mathrm{V\,m^{-1}}\right) \tag{1.10}$$

and analyze the pair of complex amplitudes $E_{h,v}$ referred to as phasors. In matrix notation, the phasor pair is expressed as

$$\boldsymbol{E} = \begin{bmatrix} E_h \\ E_v \end{bmatrix} = \begin{bmatrix} E_h(0)\exp(-jk_h r - j\varphi_h) \\ E_v(0)\exp(-jk_v r - j\varphi_v) \end{bmatrix} \left(\mathrm{V\,m^{-1}}\right). \tag{1.11}$$

In the sequel concerning (1.11) and similar equations, capital italics denote the phasors, capital letters indicate complex quantities containing the time-varying factor $\exp(j2\pi ft)$, and unit vectors (i.e., $\mathbf{e}_h$ or $\mathbf{e}_v$) are explicitly added to clarify the physical meaning behind some derivations.

A propagating wave carries power in the direction of propagation, and the *instantaneous* power density $\left|\vec{\mathbf{S}}\right|$ is the magnitude of the vector product of electric and magnetic fields (Popovic 1971):

$$\left|\vec{\mathbf{S}}\right| = \left|\vec{\mathbf{E}} \times \vec{\mathbf{H}}\right| = \left|\vec{\mathbf{E}}\right|^2 / \eta_0 \left(\mathrm{W\,m^{-2}}\right). \tag{1.12}$$

This definition is valid for plane waves propagating in free space where $\eta_0 = 377\ \Omega$ (ohm) is the free space impedance; it relates the electric and magnetic fields via $\left|\vec{\mathbf{H}}\right| = \left|\vec{\mathbf{E}}\right| / \eta_0$. At a given range r, the instantaneous power density varies rapidly between zero and its maximum. It is the average power density $\overline{|\vec{\mathbf{S}}|}$ over one cycle $T_c = 1/f$ of the wave that matters and can be quantified by standard measuring instruments. In vacuum,

$$\begin{aligned} \overline{|\vec{\mathbf{S}}|} &= \frac{1}{\eta_0 T_c} \int_0^{T_c} |\mathrm{Re}(\mathbf{E})|^2 dt \\ &= \frac{1}{\eta_0 T_c} \int_0^{T_c} \left[E_h^2 \cos^2(2\pi ft - kr - \varphi_h) + E_v^2 \cos^2(2\pi ft - kr - \varphi_v)\right] dt \end{aligned} \tag{1.13}$$

and simplifies to

$$\overline{|\vec{\mathbf{S}}|} = \frac{E_h^2 + E_v^2}{2\eta_0} = \frac{|\mathbf{E}|^2}{2\eta_0} \left(\mathrm{W\,m^{-2}}\right). \tag{1.14}$$

It is evident in (1.14) and intuitively satisfying that the circularly polarized wave (i.e., $E_h = E_v$ and phase difference 90°) carries twice the power of its constituent horizontal and vertical projections.

1.2 Scattering by a Single Particle

Next we consider a plane wave impinging on a scatterer. The wave induces oscillating currents within the scatterer. These currents may not be collinear with the incident electric field direction and produce secondary radiation in all directions including opposite to the incident wave. Although the following exposition concerns a continuous wave, it is applicable to pulsed sinusoidal waves typically emitted by weather surveillance radars; thereby the leading edge of the wave causes transients in the scattered field. These transients decay very fast (in few cycles) compared to the duration of the pulse (hundreds of cycles). Therefore, the steady state is quickly established and its temporal variation expressed by exp($j2\pi ft$) is implicitly assumed throughout the book.

The strength of electric field produced by the scatterer is inversely proportional to the distance r as can be deduced from the following physical considerations. Assume the scatterer intercepts electromagnetic wave with a power density expressed by (1.14). Some of the intercepted energy is reradiated back toward the radar, some in other directions, and some is absorbed and dissipated inside the scatterer. The total power of the scattered radiation (in a lossless medium) passing through a sphere centered on the scatterer doesn't depend on the sphere radius r; hence the power density is inversely proportional to r^2. Therefore, the electric field corresponding to the scattered radiation is inversely proportional to r. Relevant to radar measurements is the portion radiating back and its relation to the backscatter cross section of a scatterer (Doviak and Zrnic 2006, Sect. 3.2).

For quantifying effects of propagation and reflections from an ensemble of hydrometeors, consideration of scattering in several directions is needed. Hence we next present the matrix form of the scattering equation but assume propagation in vacuum. The derived results are valid for other isotropic and nonattenuating media (only the wave number must be modified). Extension to anisotropic media (e.g., precipitation) is made in Chap. 2. With this simplification, we isolate the scatterer's properties from propagation effects (quantified in Sect. 1.3).

The scattering matrix **S** relates two orthogonal components of the scattered field $\boldsymbol{E}^s$ to the orthogonal components of the incident field $\boldsymbol{E}^i$ and it is defined by (e.g., Bohren and Huffman 1983)

$$\boldsymbol{E}^s = \frac{e^{-jkr}}{r}\mathbf{S}\boldsymbol{E}^i \left(\mathrm{V\,m^{-1}}\right), \tag{1.15}$$

where r is the distance between the particle and the measurement location of $\boldsymbol{E}^s$ (very far from the scatterer), and the intervening attenuation is negligible (propagation medium is vacuum). Equation (1.15) in expanded form is written as

$$\begin{bmatrix} E_h^s \\ E_v^s \end{bmatrix} = \frac{e^{-jkr}}{r}\begin{bmatrix} s_{\mathrm{hh}}(\mathbf{k}_i,\mathbf{k}_s) & s_{\mathrm{hv}}(\mathbf{k}_i,\mathbf{k}_s) \\ s_{\mathrm{vh}}(\mathbf{k}_i,\mathbf{k}_s) & s_{\mathrm{vv}}(\mathbf{k}_i,\mathbf{k}_s) \end{bmatrix}\begin{bmatrix} E_h^i \\ E_v^i \end{bmatrix}, \tag{1.16}$$

to explicitly identify the complex amplitudes (phasors) $E_h^{i,s}$ and $E_v^{i,s}$ of horizontally and vertically polarized components of the incident and scattered fields in the coordinate system associated with the unit vector $\mathbf{k}_i$ corresponding to the direction of the propagating incident wave (Fig. 1.3).

If $\mathbf{k}_s = \mathbf{k}_i$ the matrix **S** is called the forward-scattering matrix ($\mathbf{S}^{(0)}$), and it is called the backscattering matrix ($\mathbf{S}^{(\pi)}$) if $\mathbf{k}_s = -\mathbf{k}_i$. The superscript indicates the angle between vectors $\mathbf{k}_s$ and $\mathbf{k}_i$. Generally $\mathbf{S}^{(0)} \neq \mathbf{S}^{(\pi)}$. For radar measurements, the

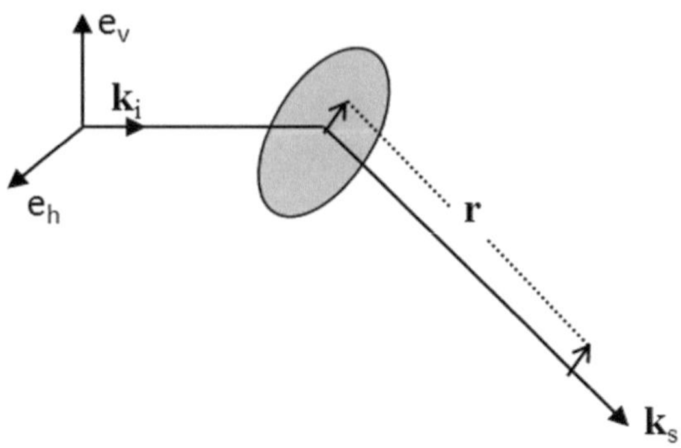

Fig. 1.3 Scattering geometry including coordinate systems and propagation vectors. Superscript *i* indicates incident field and *s* stands for scattered field

backscattering matrix is of primary interest because it relates the properties of the transmitted radiation and backscattered radiation (carrying information about hydrometeors) at the radar location. Henceforth the superscript will be dropped from the backscattering matrix, so that

$$\mathbf{S}=\begin{bmatrix} s_{\mathrm{hh}} & s_{\mathrm{hv}} \\ s_{\mathrm{vh}} & s_{\mathrm{vv}} \end{bmatrix} (\mathrm{m}), \tag{1.17}$$

and, for short, it will be called the scattering matrix.

The term s_{mn} is called backscattering (or backscatter) coefficient. Its second subscript (n) indicates the polarization (h or v) of the incident field transmitted by the radar; its first subscript refers to the polarization of the backscattered field. The coefficient represents intrinsic (inherent) electromagnetic properties of the scatterer causing the radar return. It is complex and its magnitude quantifies the portion of the incident field reflected back whereas its phase indicates the shift upon reflection (backscattering) with respect to the phase of the incident field. Determining the value of s_{mn} for various scatterers can be quite complicated and is a discipline in itself. Nonetheless, certain types of scatterers (spheroids) at wavelengths large compared to scatterers' dimensions offer closed-form solutions (Chap. 5).

The matrix description of scattering is needed because many scatterers in the atmosphere change polarization of the incident EM radiation. Therefore the scattered field $\boldsymbol{E}^s$ contains two orthogonal components (horizontally and vertically polarized) even if the incident wave $\boldsymbol{E}^i$ contains only one. The off-diagonal elements of the scattering matrix $\mathbf{S}$ are equal, i.e., $s_{\mathrm{hv}} = s_{\mathrm{vh}}$. The equality stems from reciprocity principle applicable to passive media. For our purpose, the principle can be phrased as follows. If horizontally polarized field illuminates a scatterer and vertically polarized reflected field is measured, the result is the same as would be obtained if vertically polarized fields were illuminating the scatterer and horizontally polarized fields were measured. Throughout the book, s_{hv} always equals $s_{\mathrm{vh}},$ but for proper association with the transmitted and received polarization, both symbols are often used.

Computations of the elements of the scattering matrix for hydrometeors of complex shapes are generally complicated (particularly for scatterers comparable or larger than the wavelength) and require evaluation of the distributions of the fields inside the scatterer. This, even for ellipsoids, is not trivial. Significant simplification occurs if scatterers are spheroids. These shapes accurately approximate raindrops and graupel and are quite adequate for snowflakes, ice crystals, small hailstones, or insects. In this book, spheroids are considered for modeling scatter of most hydrometeors.

Scattering properties of a spheroid can be represented by a combination of two crossed dipoles oriented along the axis a and b of the projected spheroid onto the plane of polarization. The angle between the projection of the rotation axis a on the plane of polarization and the direction of the vertical electric field is called the canting angle and will be denoted with α. The dipole model is applicable to small arbitrarily shaped hydrometeors and the spheroid approximation (oblate or prolate) is used to model the flat or elongated ones.

If the direction of the incident electric field $\underset{i}{\vec{\mathbf{E}}}$ coincides with the direction of one of the dipoles, then only that particular dipole is excited to produce secondary radiation in the same "copolar" direction as the incident field. This happens, for example, if a raindrop is illuminated by horizontally polarized waves. Raindrops have approximately oblate shapes and their rotation axis a is vertically oriented (see Chap. 4). In this case, polarization of the backscattered wave is also horizontal (Fig. 1.4a). For this specific geometry, the backscatter coefficient $s_{\mathrm{hh}} = s_b$ is called the backscattering amplitude along the axis b, and in the absence of attenuation, the relation (1.15) between backscattered (indicated with superscript b) and incident field (phasor representation) becomes

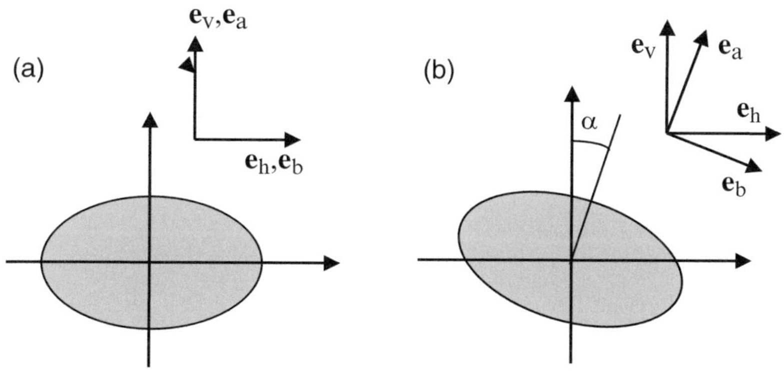

Fig. 1.4 Cross sections in the plane of polarization of (**a**) noncanted and (**b**) canted hydrometeor. Non-depolarizing orientation is on the left and the depolarizing one is on the right

$$E_h^b \mathbf{e}_h = \frac{e^{-jkr}}{r} s_b E_h^i \mathbf{e}_h. \tag{1.18}$$

The same raindrop will produce vertically polarized backscattered wave if the polarization of incident wave is vertical, thus

$$E_v^b \mathbf{e}_v = \frac{e^{-jkr}}{r} s_a E_v^i \mathbf{e}_v, \tag{1.19}$$

where $s_a = s_{vv}$ is the backscattering amplitude along the vertical (rotation) axis a.

If the raindrop is canted in the plane of polarization, then the principal axes of the canted raindrop are not vertical or horizontal (Fig. 1.4b). Both dipoles are excited and each generates secondary radiation along its axis. The vector of horizontally polarized incident field can be represented as a sum of two components aligned with the two principle axes a and b

$$E_h^i \mathbf{e}_h = E_h^i \cos\alpha\, \mathbf{e}_b + E_h^i \sin\alpha\, \mathbf{e}_a = E_b^i\, \mathbf{e}_b + E_a^i\, \mathbf{e}_a, \tag{1.20}$$

where $\mathbf{e}_a$ and $\mathbf{e}_b$ are the unit vectors directed along a and b axes. These unit vectors are related to the unit vectors $\mathbf{e}_h$ and $\mathbf{e}_v$ via the rotation transformation as

$$\begin{bmatrix} \mathbf{e}_a \\ \mathbf{e}_b \end{bmatrix} = \begin{bmatrix} \sin\alpha & \cos\alpha \\ \cos\alpha & -\sin\alpha \end{bmatrix} \begin{bmatrix} \mathbf{e}_h \\ \mathbf{e}_v \end{bmatrix}. \tag{1.21}$$

The backscatter fields from the two diploes are proportional to the product of the incident field components (E_a^i and E_b^i) along their axes and backscattering amplitudes s_a and s_b. Consider a horizontally polarized incident field with magnitude E_h^i. Then

$$\begin{aligned} \vec{E}^b &= \frac{e^{-jkr}}{r}\left(s_a E_a^i \mathbf{e}_a + s_b E_b^i \mathbf{e}_b\right) \\ &= \frac{e^{-jkr}}{r}\left[E_h^i\left(s_a \sin^2\alpha + s_b \cos^2\alpha\right)\mathbf{e}_h + E_h^i (s_a - s_b)\sin\alpha\cos\alpha\, \mathbf{e}_v\right], \end{aligned} \tag{1.22}$$

and the backscattered field $\vec{E}^b$ has both horizontal and vertical components if $s_a \neq s_b$. For vertically polarized incident field, similar derivation yields

$$\vec{E}^b = \frac{e^{-jkr}}{r}\left[E_v^i (s_a - s_b)\sin\alpha\cos\alpha\;\, \mathbf{e}_h + E_v^i\left(s_b \sin^2\alpha + s_a \cos^2\alpha\right)\mathbf{e}_v\right]. \tag{1.23}$$

Compacting (1.22) and (1.23) into a matrix equation produces the following expression for the backscattering matrix of a single spheroidal particle whose axis of rotation is in the plane of polarization:

$$\mathbf{S} = \begin{bmatrix} s_a \sin^2\alpha + s_b \cos^2\alpha & (s_a - s_b)\sin\alpha\cos\alpha \\ (s_a - s_b)\sin\alpha\cos\alpha & s_b \sin^2\alpha + s_a \cos^2\alpha \end{bmatrix} (\mathrm{m}). \tag{1.24}$$

The expression (1.24) applies to a spheroid whose axis of rotation lies in the plane of polarization of the incident wave, therefore the angle ψ between the axis of rotation and the direction of wave propagation is $\pi/2$. If $\psi \neq \pi/2$ (Fig. 1.5), the backscattering amplitudes s_a and s_b in (1.19) should be replaced with the amplitudes s'_{hh} and s'_{vv} that are functions of the angle ψ (Holt 1984). It was shown by Holt and Shepherd (1979) that such dependencies on ψ are simple if scatterers are much smaller than the wavelength. Then the following "backscatter rule" applies

$$s'_{\mathrm{hh}} = s_b \ (\mathrm{m}) \tag{1.25}$$

$$s'_{\mathrm{vv}} = s_a \sin^2\psi + s_b \cos^2\psi \,(\mathrm{m}). \tag{1.26}$$

To grasp the physical meaning behind (1.26), assume that there is no canting in the polarization plane ($\alpha = 0$), the direction of wave propagation $\mathbf{k}$ is in the horizontal plane ($\beta = 0$), and the symmetry axis of oblate spheroid $\mathbf{N}$ rotates from $\psi = \pi/2$ to $\psi = 0$ in the vertical plane x–z (Fig. 1.5). Then the projection of the spheroid onto the polarization plane x–y varies as shown in Fig. 1.6. It is obvious that the horizontal dimension of the projection (b) (proportional to the radar return at horizontal polarization) remains the same, while its vertical projection increases from a (at $\psi = \pi/2$) to b (at $\psi = 0$). In other words, s_{hh} remains equal to s_b, whereas s_{vv} changes from s_a to s_b as reflected in Eq. (1.26).

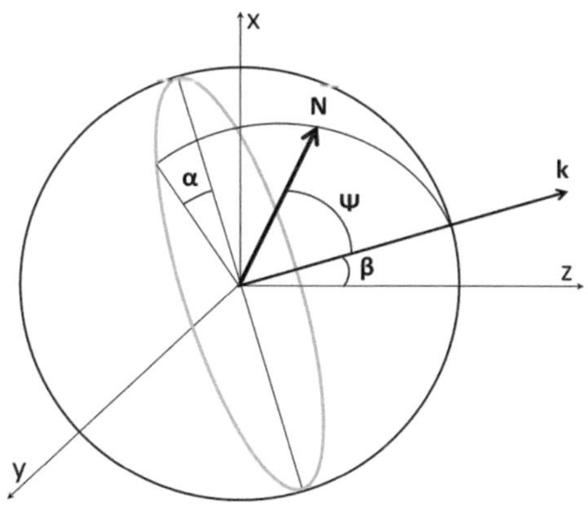

Fig. 1.5 Scattering geometry. Direction **N** denotes orientation of the symmetry axis of the particle, **k** represents the direction of wave propagation; it is perpendicular to the polarization plane and lies in the x, z plane. The polarization plane is depicted by the grey ellipse. The x axis is true vertical and y z is horizontal direction. Canting angle α is the angle between the projections of vector **N** and true vertical x on the polarization plane, ψ is the angle between **N** and **k**, and antenna elevation angle is β

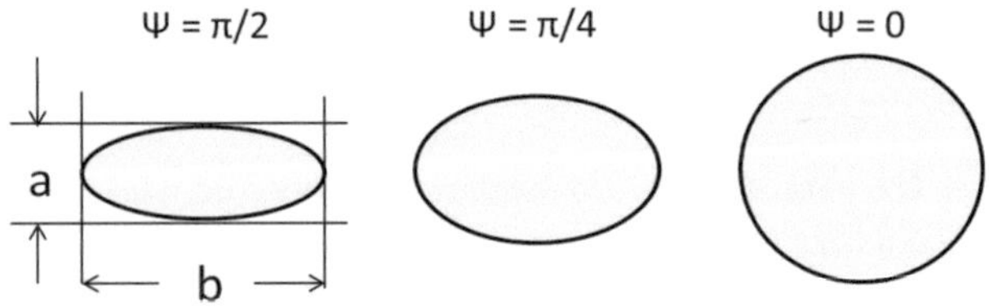

Fig. 1.6 Spheroid projection onto the polarization plane for different values of ψ

Holt and Shepherd (1979) indicate that the backscatter rule applies for oblate raindrops up to 35 GHz provided $\psi > 80°$. After substituting the scattering amplitude s_a in (1.24) with s'_{vv} from (1.26), we obtain the expression for the matrix **S** in the general case of spheroid orientation determined by two angles, α and ψ.

$$\mathbf{S} = \begin{bmatrix} (s_a - s_b)\sin^2\psi\sin^2\alpha + s_b & (s_a - s_b)\sin^2\psi\sin\alpha\cos\alpha \\ (s_a - s_b)\sin^2\psi\sin\alpha\cos\alpha & (s_a - s_b)\sin^2\psi\cos^2\alpha + s_b \end{bmatrix}. \tag{1.27}$$

This relatively simple expression for the scattering matrix **S** is sufficient for interpretation of polarimetric properties of most hydrometeors at the 10, 5, and 3 cm wavelengths (designated with letters *S*, *C*, and *X*). It has been used extensively for modeling backscatter (e.g., Holt 1984; Ryzhkov 2001) and is similarly applied throughout this book. For more rigorous modeling of scattering wherein the backscatter rule does not hold, the reader is referred to Vivekanandan et al. (1991).

The elements of the backscatter matrix **S** are directly related to backscatter cross sections commonly used to quantify reflections from objects. The backscatter cross section of a scatterer is the area which if multiplied with the incoming power density would produce the same backscatter power density as an isotropic scatterer. To include polarization into this definition, let's assume the incident wave has polarization indexed by *n* (stands for either *h* or *v*) and consider the backscattered field E_n^b. According to (1.12), (1.13), and (1.14), the power density at the distance *r* from the scatterer in the backscatter direction ($\mathbf{k}_s = -\mathbf{k}_i$) is

$$\overline{|\vec{\mathbf{S}}|}_{\mathrm{mn}} = \frac{|E_n^i|^2 |s_{\mathrm{mn}}|^2}{2r^2\eta_o}\ \left(\mathrm{W\,m^{-2}}\right). \tag{1.28}$$

If $m = n$, the indicated power density ($\overline{|\vec{\mathbf{S}}|}_{\mathrm{hh}}$ or $\overline{|\vec{\mathbf{S}}|}_{\mathrm{vv}}$) is of the copolar component; otherwise ($\overline{|\vec{\mathbf{S}}|}_{\mathrm{hv}}$ or $\overline{|\vec{\mathbf{S}}|}_{\mathrm{vh}}$) it is of the cross-polar component. The cross-section definition σ_{mn} applicable to dual-polarization radars is the area that intercepts the incoming power density of the field E_n^i so that the backscattered power density at the antenna location corresponding to polarization *m* would be the same if the scatterers were radiating equally in all directions. This means

$$\overline{|\vec{\mathbf{S}}|}_{\mathrm{mn}} = \frac{|E_n^i|^2 \sigma_{\mathrm{mn}}}{8\pi r^2\eta_o}. \tag{1.29}$$

Equating the power densities (1.28) and (1.29) yields

$$\sigma_{\mathrm{mn}} = 4\pi|s_{\mathrm{mn}}|^2\,(\mathrm{m}^2). \tag{1.30}$$

1.3 Propagation Effects

Electromagnetic wave propagating through the atmosphere filled with hydrometeors acquires additional phase shift compared to what it would have in vacuum. Moreover, its amplitude decreases due to the presence of various gas molecules and hydrometeors. Each of the scatterers extracts a puny amount of energy from the wave and reradiates part of it in all directions while absorbing the rest. At microwave frequencies, the absorbed part is dominant (Bohren and Huffman 1983) except for very large hail and attenuation is quantified with the imaginary part of the effective refractive index $n_{h,v}$. In interaction with propagating EM fields, scatterers retard the wave and reduce its propagation speed. This increases the phase shift in proportion to the real part of $n_{h,v}$. In the absence of hydrometeors, the atmosphere is isotropic with respect to polarization (i.e., $n_h = n_v$), therefore, phase shifts and attenuation of the propagating wave are independent of polarization. This is not so if the atmosphere is filled with nonspherical oriented scatterers like raindrops or snowflakes because then $n_h \neq n_v$.

Each scatterer contributes its share to the effective refractive indexes $n_{h,v}$ (1.4) according to its forward-scattering amplitude. The contributions are linear and additive and n_h and n_v depend on the hydrometeor type, concentration, shape, orientation, and wavelength λ. These parameters can often be described with the distribution function $N(\boldsymbol{X})$ of the scatterer physical properties (size, shape, orientation, dielectric constant, etc.) characterized by vector $\boldsymbol{X}$. The function $N(\boldsymbol{X})$ is defined to produce the concentration of particles within the unit volume N_T (in m^{-3}):

$$N_T = \int N(\boldsymbol{X})d\boldsymbol{X}\,(\mathrm{m}^{-3}). \tag{1.31}$$

The particles contribute to the average forward-scattering amplitude $< s^{(0)}_{\mathrm{hh,vv}} >$ per unit volume

$$< s^{(0)}_{\mathrm{hh,vv}} >= \int N(\boldsymbol{X})s^{(0)}_{\mathrm{hh,vv}}(\boldsymbol{X})d\boldsymbol{X}\,(\mathrm{m\,m}^{-3}), \tag{1.32}$$

where the angular brackets indicate probabilistic (ensemble) average of the particles' forward-scattering coefficients at horizontal and vertical polarizations.

Complex amplitudes (phasors) of the horizontally and vertically polarized waves defined in (1.11) are special solutions of the following two coupled differential equations (Oguchi 1983; Bringi and Chandrasekar 2001; Zhang 2016):

$$\frac{d}{dr}\begin{bmatrix} E_h \\ E_v \end{bmatrix} = \mathbf{B}\begin{bmatrix} E_h \\ E_v \end{bmatrix} = \begin{bmatrix} B_{\mathrm{hh}} & B_{\mathrm{hv}} \\ B_{\mathrm{vh}} & B_{\mathrm{vv}} \end{bmatrix}\begin{bmatrix} E_h \\ E_v \end{bmatrix}. \tag{1.33}$$

Under influence of gravity, hydrometeors acquire horizontal orientation and, depending on the species, can wobble (graupel, hail), tumble (hail), or oscillate (drops). If, on the average, the horizontal (or vertical) orientation is maintained, there is no coupling, and the matrix **B** is diagonal with terms

$$B_{\mathrm{hh}} = -jkn_h, \quad B_{\mathrm{vv}} = -jkn_v, \quad B_{\mathrm{hv}} = B_{\mathrm{vh}} = 0, \tag{1.34}$$

and the horizontally and vertically polarized waves propagate independently.

If the average orientation of particles is neither horizontal nor vertical (i.e., the mean canting angle is different from 0 or $\pi/2$), then the matrix **B** has nonzero off-diagonal terms, propagation changes the polarization, and the waves at the two orthogonal polarizations are not independent. This change is commonly called "depolarization on propagation" and might be misleading as the prefix (de) usually means decrease in the suffix. The confusion arises because there is a term "degree of polarization" quantifying the polarized component (i.e., h or v) versus nonpolarized component characterized by random change in polarization (Mott 1992). The dominant orientation of hydrometeors is horizontal and has motivated the choice of the pair of horizontal and vertical polarizations for probing precipitation for two reasons. First, as implied by the discussion, the two waves would not exchange energy between each other and thus obfuscate the information about precipitation being probed. Second, the horizontally and vertically polarized components of the wave are aligned with the shortest and longest dimensions of hydrometeors, and, therefore, the contrast in the strength of reflections at these two polarizations is largest.

A canted nonspherical hydrometeor produces secondary radiation of both polarizations in the backward direction if illuminated with either horizontally or vertically polarized incident wave. Secondary radiation occurs also in the forward direction and the propagating wave changes its original polarization state by acquiring cross-polar component. It can be shown that the general form of the matrix **B** is (Oguchi 1983; Bringi and Chandrasekar 2001)

$$\mathbf{B} = \begin{bmatrix} -jk' + p_{\mathrm{hh}} & p_{\mathrm{hv}} \\ p_{\mathrm{vh}} & -jk' + p_{\mathrm{vv}} \end{bmatrix}, \tag{1.35}$$

where

$$\mathbf{P} = \begin{bmatrix} p_{\mathrm{hh}} & p_{\mathrm{hv}} \\ p_{\mathrm{vh}} & p_{\mathrm{vv}} \end{bmatrix} = -j\lambda < \mathbf{S}^{(0)} > = -j\lambda \begin{bmatrix} < s_{\mathrm{hh}}^{(0)} > & < s_{\mathrm{hv}}^{(0)} > \\ < s_{\mathrm{vh}}^{(0)} > & < s_{\mathrm{vv}}^{(0)} > \end{bmatrix}. \tag{1.36}$$

In (1.35) $k' = k(1 + \Delta n^{(\mathrm{air})})$ where $\Delta n^{(\mathrm{air})}$ is defined in (1.6). Because $\Delta n^{(\mathrm{air})}$ is independent of polarization, none of the polarimetric variables is affected by $\Delta n^{(\mathrm{air})}$.

Therefore, distinction between k and k' is unnecessary unless attenuation or absolute phase shift due to propagation is considered.

The solution of (1.33) applicable to the nondiagonal matrix B is (Oguchi 1983)

$$\begin{bmatrix} E_h(r) \\ E_v(r) \end{bmatrix} = \begin{bmatrix} T_{\mathrm{hh}} & T_{\mathrm{hv}} \\ T_{\mathrm{vh}} & T_{\mathrm{vv}} \end{bmatrix} \begin{bmatrix} E_h(0) \\ E_v(0) \end{bmatrix}, \tag{1.37}$$

where $\boldsymbol{T}$ is the transmission matrix; its elements for a *uniform propagation path* are

$$\begin{aligned} T_{\mathrm{hh}} &= e^{-j\lambda_1 r} \cos^2\alpha + e^{-j\lambda_2 r} \sin^2\alpha \\ T_{\mathrm{vv}} &= e^{-j\lambda_1 r} \sin^2\alpha + e^{-j\lambda_2 r} \cos^2\alpha \\ T_{\mathrm{hv}} &= T_{\mathrm{vh}} = \left(e^{-j\lambda_1 r} - e^{-j\lambda_2 r}\right) \sin\alpha \cos\alpha. \end{aligned} \tag{1.38}$$

In (1.38),

$$\begin{aligned} \lambda_1 &= k + \frac{j}{2}(p_{\mathrm{hh}} + p_{\mathrm{vv}} + \gamma) \\ \lambda_2 &= k + \frac{j}{2}(p_{\mathrm{hh}} + p_{\mathrm{vv}} - \gamma) \\ \gamma &= \sqrt{(p_{\mathrm{hh}} - p_{\mathrm{vv}})^2 + 4p_{\mathrm{hv}}p_{\mathrm{vh}}} \\ \alpha &= \frac{1}{2}\tan^{-1}\left(\frac{2p_{\mathrm{hv}}}{p_{\mathrm{hh}} - p_{\mathrm{vv}}}\right). \end{aligned} \tag{1.39}$$

The angle α indicates the mean canting angle of particles along the uniform propagation path (Fig. 1.4b).

The amplitudes of the elements of the transmission matrix characterize attenuation losses. In addition to the matrix $\mathbf{T}$, it is convenient to introduce the matrix of losses $\mathbf{L}$ as

$$\mathbf{L} = \begin{bmatrix} l_h^{-1} & l_{hv}^{-1} \\ l_{vh}^{-1} & l_v^{-1} \end{bmatrix} = \begin{bmatrix} |T_{\mathrm{hh}}|^2 & |T_{\mathrm{hv}}|^2 \\ |T_{\mathrm{vh}}|^2 & |T_{\mathrm{vv}}|^2 \end{bmatrix}, \tag{1.40}$$

and the loss terms ls are larger than 1.

In the absence of depolarization on propagation ($p_{\mathrm{hv}} = p_{\mathrm{vh}} = 0$, $\alpha = 0$), the expressions for λ_1 and λ_2 in (1.39) simplify to

$$\begin{aligned} \lambda_1 &= k + jp_{\mathrm{hh}} = k + \lambda < s_{\mathrm{hh}}^{(0)} > \\ \lambda_2 &= k + jp_{\mathrm{vv}} = k + \lambda < s_{\mathrm{vv}}^{(0)} > \end{aligned} \tag{1.41}$$

and

$$E_h(r) = E_h(0)\exp\left[-j\left(k + \lambda < s_{\mathrm{hh}}^{(0)} >\right)r\right] = E_h(0)[l_h(r)]^{-1/2}\exp[-j\Phi_h(r)]$$
$$E_v(r) = E_v(0)\exp\left[-j\left(k + \lambda < s_{\mathrm{vv}}^{(0)} >\right)r\right] = E_v(0)[l_v(r)]^{-1/2}\exp[-j\Phi_v(r)] \tag{1.42}$$

where

$$l_{h,v}^{-1}(r) = \exp\left\{\left[2\mathrm{Im}(n^{(\mathrm{air})}) + \frac{\lambda^2}{\pi}\mathrm{Im}(< s_{\mathrm{hh,\,vv}}^{(0)} >)\right]kr\right\}, \tag{1.43}$$

$$\Phi_{h,v} = \left[1 + \mathrm{Re}\left(n^{(air)}\right) + \frac{\lambda^2}{2\pi}\mathrm{Re}\left(< s_{\mathrm{hh,\,vv}}^{(0)} >\right)\right]kr, \tag{1.44}$$

and the **T** matrix is

$$\mathbf{T} = \begin{bmatrix} e^{-jk_h r} l_h^{-1/2} & e^{-jk_v r} l_{hv}^{-1/2} \\ e^{-jk_h r} l_{vh}^{-1/2} & e^{-jk_v r} l_v^{-1/2} \end{bmatrix}. \tag{1.45}$$

It follows from (1.5), (1.43), and (1.44) that the ensemble average of refractive index of the medium associated with hydrometeors is

$$n_{h,v}^{(\mathrm{scat})} = \frac{\lambda^2}{2\pi} < s_{\mathrm{hh,\,vv}}^{(0)} > . \tag{1.46}$$

Logarithms of attenuation factors $l_{\mathrm{hh,vv}}$ are commonly used and can be represented as products of specific attenuation $A_{h,v}$ at orthogonal polarizations and distance r

$$10\log(l_{h,v}) = A_{h,v} r\,(\mathrm{dB}), \tag{1.47}$$

where

$$A_{h,v} = -8.686\left[k\mathrm{Im}\left(\Delta n^{(air)}\right) + \lambda\mathrm{Im}\left(< s_{\mathrm{hh,\,vv}}^{(0)} >\right)\right]\left(\mathrm{dB\,m}^{-1}\right). \tag{1.48}$$

Specific attenuation $A_{h,v}$ is often expressed in dB km^{-1}. Note that in (1.1) the selected convention for complex representation of time dependence is exp($j2\pi ft$), i.e., the same as in Doviak and Zrnic (2006). In this convention, the imaginary parts of the refractive index of clear air and of hydrometeors are negative making $A_{h,v}$ in (1.48) positive.

Attenuation in atmospheric gases is determined by the term $\mathrm{Im}(\Delta n^{(air)})$ and is described in more detail in the Appendix A. At microwave frequencies, attenuation in atmospheric gases is much smaller than attenuation in precipitation, i.e., the specific attenuation is primarily determined by the second term in (1.43). For nonspherical but preferentially oriented hydrometeors, A_h differs from A_v and this difference is called specific differential attenuation A_{DP}, explicitly

$$A_{\mathrm{DP}} = A_h - A_v = -8.686\,\lambda\,\mathrm{Im}(< s_{\mathrm{hh}}^{(0)} > - < s_{\mathrm{vv}}^{(0)} >). \tag{1.49}$$

The contribution of atmospheric gases to the phases $\Phi_{h,v}$ is much larger than the one of hydrometeors. However, the difference $\Phi_h - \Phi_v$ is entirely determined by anisotropic hydrometeors and can be represented as

$$\Phi_h(r) - \Phi_v(r) = \frac{1}{2}\Phi_{\mathrm{DP}}(r) = K_{\mathrm{DP}}r, \tag{1.50}$$

where

$$\begin{aligned} K_{\mathrm{DP}} &= \mathrm{Re}(k_h - k_v) = \mathrm{Re}\left(\Delta n_h^{(\mathrm{scat})} - \Delta n_v^{(\mathrm{scat})}\right)k \\ &= \lambda\mathrm{Re}\left(< s_{\mathrm{hh}}^{(0)} > - < s_{\mathrm{vv}}^{(0)} >\right), \end{aligned} \tag{1.51}$$

and k_h and k_v are defined by

$$k_{h,v} = k + \lambda < s_{\mathrm{hh,vv}}^{(0)} > . \tag{1.52}$$

In radar application, the difference $\Phi_h - \Phi_v$ multiplied by 2 is called differential phase upon propagation Φ_{DP}. Multiplication by 2 is needed because the wave traverses the propagation path twice, once on the way from the radar to the scatterers and the second time back from the scatterers to the radar.

The derived equations apply to homogenous media (uniform propagation paths). Extension to inhomogeneous media can be made by replacing the product $k_{h,v}r$ with $\int_0^r k_{h,v}(r)dr = \Phi_{h,v}(r)$.

References

Bohren, C., & Huffman, D. (1983). *Absorption and scattering of light by small particles*. Hoboken, NJ: Wiley. 530pp.

Bringi, V., & Chandrasekar, V. (2001). *Polarimetric Doppler weather radar. Principles and applications* (p. 636). Cambridge, UK: Cambridge University Press.

Doviak, R., & Zrnic, D. (2006). *Doppler radar and weather observations* (2nd ed.). Reprint, Mineola, NY: Dover. 562pp.

Holt, A. (1984). Some factors affecting the remote sensing of rain by polarization diversity radar in the 3- to 35-GHz frequency range. *Radio Science, 19*, 1399–1412.

Holt, A., & Shepherd, J. (1979). Electromagnetic scattering by dielectric spheroids in the forward and backward directions. *Journal of Physics A: Mathematical and General, 12*, 159–166.

Mott, H. (1992). *Antennas for radar and communications: A polarimetric approach*. Hoboken, NJ: Wiley. 521pp.

Oguchi, T. (1983). Electromagnetic wave propagation and scattering in rain and other hydrometeors. *Proceedings of the IEEE, 71*, 1029–1078.

Popovic, B. D. (1971). *Introductory engineering electromagnetics*. Menlo Park, CA: Addison-Wesley. 634pp.

Ryzhkov, A. V. (2001). Interpretation of polarimetric radar covariance matrix for meteorological scatterers. Theoretical analysis. *Journal of Atmospheric and Oceanic Technology, 18*, 315–328.

Ulaby, F., Moore, R., & Fung, A. (1981). *Microwave remote sensing. Active and passive* (Vol. 1). Boston: Artech House. 456pp.

Vivekanandan, J., Adams, W., & Bringi, V. (1991). Rigorous approach to polarimetric radar modeling of hydrometeor orientation distributions. *Journal of Applied Meteorology, 30*, 1053–1063.

Zhang, G. (2016). *Weather radar polarimetry*. Boca Raton, FL: CRC Press. 304pp.

Chapter 2
Polarimetric Doppler Radar

Single-polarization weather radars transmit and receive linearly (most often horizontally) polarized waves. Propagation can change the polarization of the wave. Linearly polarized wave can become elliptically polarized if the directions of electric field and the average of the projections of hydrometeor orientation on the plane of polarization are not aligned or are not orthogonal. Furthermore, change of polarization can occur upon reflection (backscattering) from hydrometeors if these have preferential orientation not aligned or not orthogonal to the incident fields. Both these changes convey information about precipitation, but with pulsed radars, backscattering properties are easier to resolve in range. This is because the propagation effect accumulates with distance, whereas the backscattering effects are confined within the pulse volume depth. Conventional radars transmitting horizontally polarized waves are designed to receive only the horizontally polarized component of the return. Hence, some information about propagation and backscattering properties of the medium is lost. These properties relate to the scatterers' size, number, orientation, shape, and refractive index.

Early dual-polarization weather radars could receive simultaneously both orthogonally polarized components of the reflected electromagnetic wave. That is, if horizontally polarized wave was transmitted, both horizontally and vertically polarized components were received (Browne and Robinson 1952; Shupiatsky and Morgunov 1963). Another type of polarimetric radar applied to observations of hydrometeors transmits circularly polarized waves (right-hand or left-hand) and receives both right-hand and left-hand circular components of the return wave (McCormick and Hendry 1975).

One issue these dual-polarization radars had is detection of the relatively weak orthogonal cross-polar component. If horizontally or vertically polarized radiation is transmitted, the power of the cross-polar component (i.e., with vertical or horizontal polarizations) is 20–30 dB lower than the power of the copolar component (i.e., horizontal or vertical) of the reflected wave from meteorological scatterers. Therefore, the cross-polar component can be completely overwhelmed by noise if the copolar component is weak to begin with. In contrast, the cross-polar component is

A. V. Ryzhkov, D. S. Zrnic, *Radar Polarimetry for Weather Observations*,
Springer Atmospheric Sciences, https://doi.org/10.1007/978-3-030-05093-1_2

much stronger than the copolar component if a circularly polarized wave is transmitted. Then, the copolar component can be buried in noise. Propagation through rain at low elevations transforms circularly polarized wave into elliptically polarized by affecting differently its constituent *H* and *V* components (Sect. 1.3). Because it reduces the propagation effects on the polarization state of the wave, the *H*/*V* polarization basis has become the choice for meteorological applications.

Seliga and Bringi (1976) suggested comparing two copolar components of radar signal to infer scatter type and compute rain rate. They introduced differential reflectivity to characterize the difference between the magnitudes of the copolar radar returns at horizontal and vertical polarizations. Both horizontally and/or vertically polarized copolar components are much stronger than the corresponding cross-polar components and, therefore, are less affected by noise.

Using orthogonally polarized waves for remote sensing of clouds and precipitation is equivalent to probing hydrometeor media along two orthogonal directions in the plane of polarization. This is because the radar signal at horizontal polarization is primarily determined by the horizontal dimension of hydrometeors (for hydrometeors small compared to wavelength), whereas the signal at vertical polarization is mainly affected by the vertical dimension. Thus, information about the shape and orientation of atmospheric particles can be inferred. The difference between the two orthogonally polarized fields at microwave frequencies is determined not only by the shape and orientation of the scatterers but also by their size and refractive index (Chap. 4). Because the refractive index depends on the hydrometeor's density and water content, these properties are also influencing the measurements and can be inferred under certain conditions.

In this chapter, basic weather radar principles are presented and polarimetric and scattering properties of a single hydrometeor are related to the transmitted and received radar signals.

2.1 Pulsed Doppler Radar

We start with a quick overview of weather radar principle independent of polarization. Essential components of the transmitter and receiver of the pulsed Doppler weather radar are drawn in Fig. 2.1.

Components in the block "frequency and timing generator" make the radar tick. This is where timing, pulse duration, and two frequencies are generated: the local oscillator frequency f_{lo} and the coherent oscillator frequency f_{if}. The subscript *if* stands for intermediate frequency (typically tens of MHz). The local oscillator frequency is in the microwave range which is much higher than the intermediate frequency. Use of intermediate rather than a single frequency is for practical reasons (explained below). The two frequencies are synchronized and/or controlled so that there is no drift between them. The frequency and timing generator as well as the digital receiver are controlled by a computer. This offers great flexibility for generating various pulse lengths, pulse repetition times (PRTs), filter parameters, and so on.

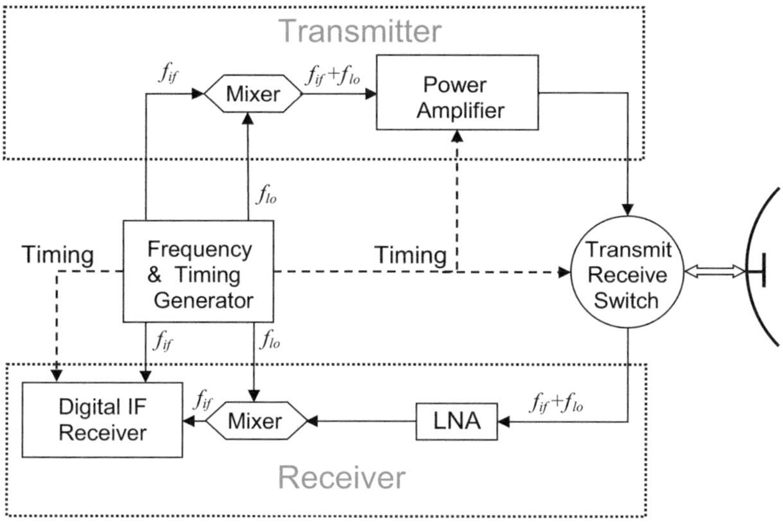

Fig. 2.1 Typical block diagram of the transmitter and receiver (without signal processing part)

For transmission, first the two sinusoids are applied to a nonlinear device (mixer) which produces the signal with a frequency equal to the sum of the two. This process of mixing (up converting) can be expressed formally with the symbol * as

$$\begin{aligned}\cos\left(2\pi f_{lo}t\right) * \cos\left(2\pi f_{if}t - \varphi^t\right) &= \cos\left[2\pi\left(f_{lo} + f_{if}\right)t - \varphi^t\right] \\ &= \cos\left(2\pi ft - \varphi^t\right).\end{aligned} \tag{2.1}$$

where φ^t is an unknown but constant phase (in the coherent radar); it indicates that the phase is preserved and imbedded in the transmitted pulsed sinusoid. Here we assume this phase is referenced to the point in the antenna where the pulsed sinusoid is transformed into the radiating electric field. It is also the point where the incoming fields are transformed into microwave signals. We refer to this point as a reference plane. Although inaccessible to measurement, we evaluate the phase differences between various signals with respect to the reference plane.

The signal (2.1) is modulated to a duration τ (Fig. 2.2) in front of the power amplifier (implied in Fig. 2.1 but not explicitly shown), amplified, and channeled through the transmit/receive (T/R) circuit to the antenna. The timing diagram of the process (Fig. 2.2) consists of a timing spike (or clock), pulse envelope and sequence, continuous signal, and the pulsed sinusoid. The time scale is for conceptual purpose and does not represent relations typical in weather surveillance radars. At frequency of 10 GHz (wavelength 3 cm), the sinusoid period is 0.1 ns, and at 3 GHz (10 cm wavelength), it is 0.33 ns. A typical pulse length τ is 1 μs, and pulse repetition time T_s is 1 ms.

Fig. 2.2 Timing (clock) signal, pulse sequence (envelope), local oscillator signal (continuous), and pulsed (transmitted) waveform

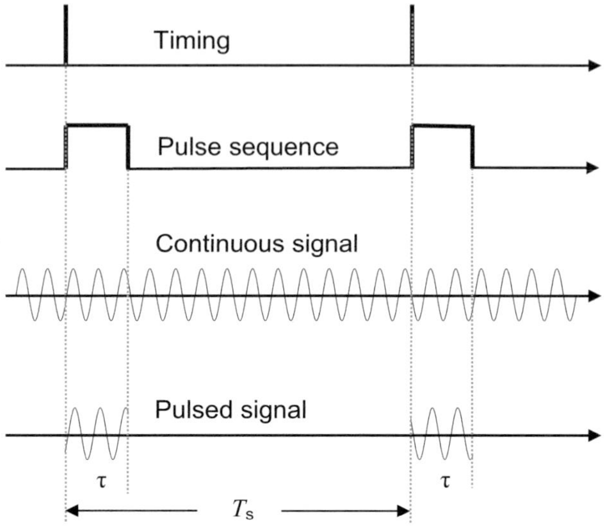

The mathematical expression of the pulsed sinusoid is $\cos(2\pi ft - \varphi^t)\, p(t)$ where

$$\begin{aligned} p(t) &= 1 \quad \text{if} \quad t < \tau \\ p(t) &= 0 \quad \text{if} \quad t > \tau. \end{aligned} \tag{2.2}$$

During transmission (over τ s), the T/R switch (Fig. 2.1) disconnects the receiver to protect its very sensitive components from the huge transmitted power. Immediately after transmission, the T/R circuit switches the power coming back from the antenna to the receiver. The reflected (backscattered) microwave radiation off a single scatterer is intercepted by the antenna and passed via the T/R switch to the low-noise amplifiers (LNA). Upon amplification, the frequency of the signal is shifted down by the mixer to the intermediate value f_{if}. Shifting the frequency from the carrier ($f = f_{lo} + f_{if}$) to f_{if} reduces noise by devices which would occur if direct removal of f were to be made. The signal is then converted to digital values and the intermediate frequency is removed. This process is formalized and simplified to convey the main message next.

The pulsed signal accumulates phase shift on the way between the generator and the output of the antenna; this phase plus the phase of the generator is φ^t (at the reference plane). The signal's phase is further shifted through propagation, backscattering (this includes Doppler shift), and passage through the receiver which produces additional phase shift φ^r. Therefore, the cumulative phase shift at the point before conversion to digital value is $\varphi^t + \varphi^c$. For example, if the polarization of the transmitted and received signal is horizontal, then $\varphi^c = \arg(V_{\text{hh}}) + \varphi^r$ (see Eq. (2.10)) and we represent the signal as

$$x(t) = A\cos\left(2\pi f_{if}t - \varphi^t - \varphi^c\right)p(t - 2r_o/c_h), \tag{2.3}$$

where the range to scatterer is r_o, c_h is the propagation speed in the medium (for H polarization, Eq. (1.4)), and A is the amplitude. The pulse function p indicates that $x(t)$ is a delayed pulsed sinusoid.

The signal (2.3) at the intermediate frequency is sampled (at time instances t_i) and digitized in the digital receiver (Fig. 2.3). The mathematical operation (indicated in Fig. 2.3) consists of multiplication with digital values of the intermediate frequency generator $\cos(2\pi f_{if}\, t_i)$ (upper branch in Fig. 2.3). These values after multiplication with $A\cos(2\pi f_{if}\, t_i - \varphi^t - \varphi^c)$ produce two "sinusoids": one has argument equal to the sum and the other is equal to the difference of arguments of the constituent components. Therefore, the sum sinusoid has frequency $2f_{if}$ and it is filtered by the digital low-pass filter (LPF). The difference sinusoid is kept and is called the in-phase component I. Similar process produces the quadrature component Q after multiplication by $\sin(2\pi f_{if}t_i)$ (lower branch in Fig. 2.3). Two components are needed to unambiguously measure the phase $\varphi^t + \varphi^c$ over a 360° interval. Therefore, the phasor

$$I + jQ = A\cos\left(-\varphi^t - \varphi^c\right) + jA\sin\left(-\varphi^t - \varphi^c\right) \tag{2.4}$$

is not just a convenient symbol facilitating manipulations of sinusoidal signals, it is a pair of real physical signals represented with a complex number often referred to as phasor. This complex physical signal is also called *complex envelope* because it consists of the real envelope of the high frequency sinusoid and it contains its phase as well.

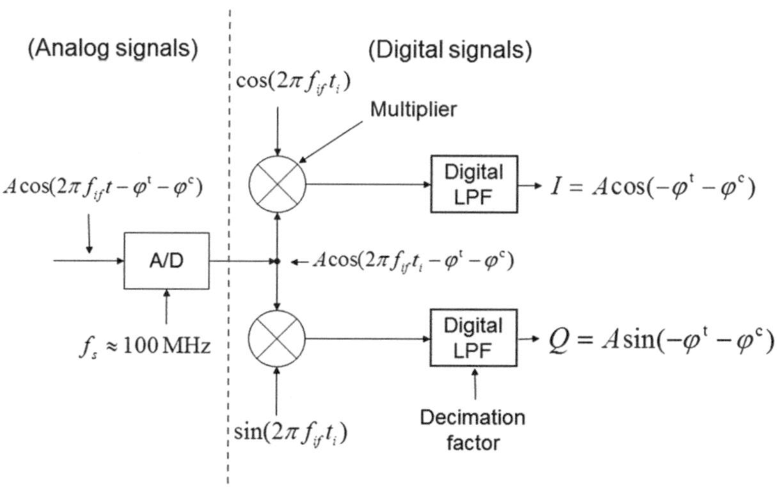

Fig. 2.3 Conceptual diagram of the digital receiver and down converter indicating the essential operations. The dashed vertical line indicates where digital processing begins. Analog to digital conversion is abbreviated with A/D, LPF stands for low-pass filter, and f_s is the sampling frequency. Decimation means increase of the samples spacing from $1/f_s \approx 0.01$ µs to few µs

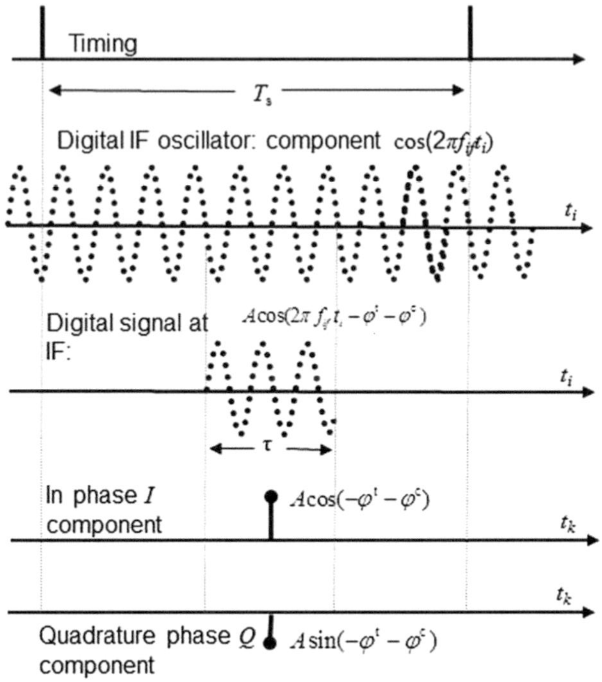

Fig. 2.4 Conceptual timing diagram of processes in the digital receiver. The return signal is assumed to be a sinusoidal pulse such as would be produced by a single point scatterer

Time relations between the digital oscillator cycles, the sampled returned sinusoidal pulse, and the I, Q pair are shown in Fig. 2.4. Only one pair is shown per pulse duration; it is obtained if all the products of multiplication (contained within the pulse duration) are averaged in the filtering process. This is routinely done with digital matched filters.

2.2 Polarimetric Doppler Radar

Here we discuss dual-polarization radar for probing weather with horizontally and vertically polarized electric fields. The way these two components are transmitted and received has vital bearing on the type of measurement that can be made. Consider a parabolic antenna reflector illuminated by a feedhorn. The mechanical orientation of the feed can be made to transmit the waves with linear polarization of arbitrary orientation. A device (orthomode coupler) with two inputs connected to the feedhorn is needed to produce an H field if the power is supplied to the H port via a waveguide or V field if it is supplied to the V port (Fig. 2.5). Early polarimetric measurements were made by transmitting wave with one polarization (H) and receiving both H and V (Browne and Robinson 1952; Shupiatsky and Morgunov 1963).

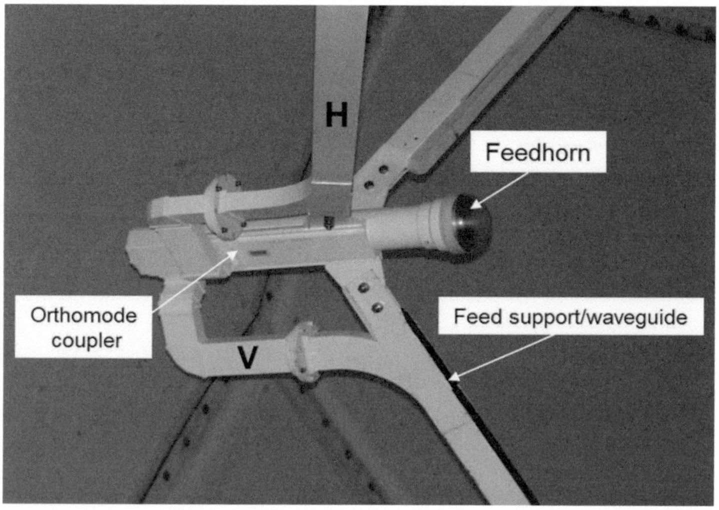

Fig. 2.5 Dual-polarization horn assembly; the horn is protected with a semispherical cover. The feedhorn extends into the orthomode coupler to which the waveguides for horizontal (H) and vertical (V) polarizations are connected. The assembly is inside a radome of the research WSR-88D in Norman OK (designated as KOUN)

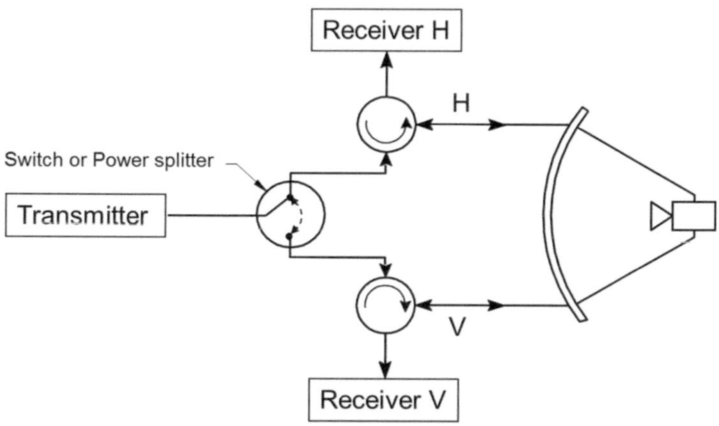

Fig. 2.6 Simplified block diagram of the dual-polarization radar which can simultaneously transmit and receive horizontally and vertically polarized waves (if a power splitter replaces the switch) or transmit H polarization and receive both (indicated by the position of the switch)

A simplified block diagram of the dual-polarization radar consists of a single transmitter and two receivers (Fig. 2.6). The arrangement concisely describes the following three polarimetric modes of operation (Fig. 2.7). The first is a fully polarimetric mode whereby complete backscattering information about scatterers is collected; it entails rapid (pulse to pulse) switching of transmitter power alternately to the H and V antenna ports and receiving signals at both polarizations (Fig. 2.7a).

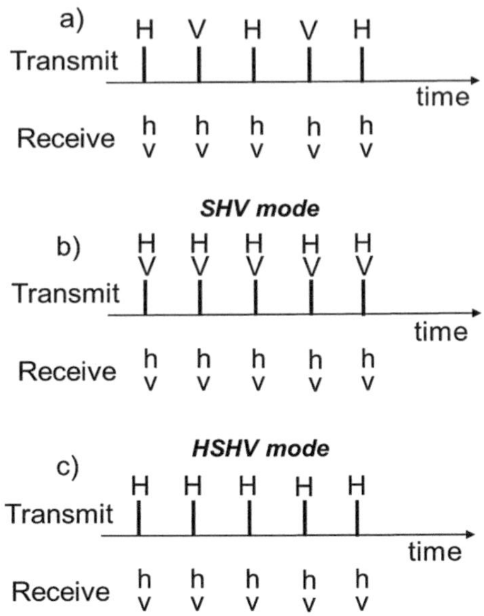

Fig. 2.7 Different schemes for transmission/reception: (**a**) alternate transmission and reception of *H*- and *V*-waves, (**b**) simultaneous transmission and reception of *H*- and *V*-waves, and (**c**) transmission of *H* and reception of both *H*- and *V*-waves

The second is a simultaneous HV (SHV) mode whereby the two components are summed up at the antenna to produce a polarization state that depends on the magnitudes and phase difference between the two orthogonally polarized components of the transmitted wave (Fig. 2.7b). The third polarimetric mode uses transmission of *H* and reception of both *H* and *V* simultaneously (HSHV mode) or transmission of *V* and reception of both (VSVH mode) (Fig. 2.7c).

The fully polarimetric mode has been explored in research environments (Brunkow et al. 2000; Lutz et al. 1997; Schroth et al. 1988) and some of the implications are discussed in Sect. 3.3. Crucial is switching transmission between *H* and *V* with a high-power microwave switch and then simultaneous reception and processing by two receivers as indicated in the timing diagram (Fig. 2.7a). Radars with one receiver have also been used where after *H* transmission the voltage samples of a horizontally polarized component of the backscattered wave are processed (*h* in Fig. 2.7a) followed by *V* transmission and (*v*) reception (Zahrai and Zrnic 1993). In the single-receiver switched mode, it is possible to obtain the complete polarimetric information by suitably connecting the receiver to the *H* or *V* receiving ports (Zrnic 1991). In the switching modes, Doppler and polarimetric variables are somewhat compromised (Sachidananda and Zrnic 1989).

The preferred mode for operations is the SHV mode (Fig. 2.7b). In this mode the switch in Fig. 2.6 is replaced with the power splitter to equally supply the antenna *H* and *V* ports. Two parallel receivers process the return to get the most significant polarimetric information. This mode has been implemented on the polarimetric WSR-88Ds (Doviak et al. 2000).

The HSHV (Fig. 2.7c) mode (and its dual VSVH) is available on some radars; all transmitted power is channeled to the *H* antenna port, i.e., only horizontally polarized wave is transmitted. Two receivers sample copolar (i.e., horizontally polarized)

and cross-polar (vertically polarized components) of the radar return. In this mode, some of the first polarimetric measurements have been made (Browne and Robinson 1952; Shupiatsky and Morgunov 1963). Non-polarimetric aspects of the Weather Surveillance Radar-88 Doppler (WSR-88D of the National Weather Service) are explained by Zrnic (2012).

2.3 Relations Between Fields and Voltages

The electromagnetic radiation is focused in one direction by an antenna usually consisting of a parabolic dish and a horn assembly (Fig. 2.5). Its characteristics are gain g and beamwidth θ_1. Antenna gain g is defined as the ratio of the power density in the direction of maximum radiation to the power density that an isotropic radiator would produce if supplied by the same input power. It is customary to express this ratio in decibel (dB) units. Perfectly isotropic radiator is an intellectual construct not realizable but extremely handy for defining relations between transmitted power and returned power from scatterers. By definition, a fictitious isotropic radiator has a unit gain ($g = 1$ or 0 dB) because it transmits equal power in all directions. The spatial dependence of radiation power on direction describes the antenna pattern; it is a function of the aperture size, wavelength, and distribution of excitation currents. Most of the energy is contained in a single main lobe of the pattern, whereas a small portion is radiating in all directions and forms a myriad of minute sidelobes. Narrow "pencil beams" are common on weather radar antennas. The one-way beamwidth θ_1 (3-dB width) is defined as the angular width of the main lobe between the points where the radiated power is one half of the value at the maximum. An excellent approximation of the main lobe part of the one-way antenna power pattern is the Gaussian function

$$f^2(\theta, \phi) = \exp\left[-\left(\theta^2 + \phi^2\right)/4\sigma_\theta^2\right], \tag{2.5}$$

where the angle coordinates θ, ϕ are with respect to the beam axis and σ_θ^2 is the second central moment of the two-way antenna pattern function $f^4(\theta, \phi)$ (Doviak and Zrnic 2006). The 3-dB beamwidth θ_1 and σ_θ^2 are related by

$$\sigma_\theta^2 = \theta_1^2/16 \ln 2. \tag{2.6}$$

For quantitative precipitation measurements, a relation between the transmitted power and the backscattering cross sections (1.30) is needed. In pulsed radars, the transmitter generates a pulsed sinusoidal voltage which is applied to the antenna port. The voltage is transformed by the antenna into a proportional transmitted EM field. The antenna assembly (feed shown in Fig. 2.5) is designed so that it can produce a desired polarization (e.g., linear H or V, or circular). Herein we are primarily concerned with linear orthogonal polarizations. The field at large distance from the antenna is carried as a plane wave continually (and gradually) changing as it propagates through precipitation. Precipitation also continually backscatters the wave and changes it on the way back to the antenna. The antenna intercepts the wave and linearly transforms it into signals and these appear as voltages at the receiver's output. Therefore, the transmitter and receiver voltages are linearly and uniquely related (in amplitude and phase) to the transmitted and received electric fields. That way the information about scatterers carried by the electric fields is transferred to the receiver voltages. However, the received fields alone, without comparison to the transmitted fields (their amplitude, phase, time of transmission, etc.), carry very little information. Thus a comparison between the two is an extremely important component of quantitative weather observation.

Here we derive relations between the transmitted and received voltages and scatterer properties quantified by the scattering coefficients. To start, assume the transmitter power supplied to the antenna is P_h^t and the polarization of radiated field is horizontal "*h*" (the following reasoning holds also for the "*v*" polarization). The atmosphere is filled with scatterers and we consider a single one located at range *r* (Fig. 2.8). The incident power density $\overline{|\vec{\mathbf{S}}^i\,|}_{\text{hh}}$ on this scatterer (Fig. 2.8, see also Eq. (1.14)) produced by a narrow-beam antenna having the pattern $f^2(\theta, \varphi)$ and gain *g* is

$$\overline{|\vec{\mathbf{S}}^i(r,\theta,\phi)\,|}_{\text{hh}} = \frac{P_h^t}{4\pi l_h(r) r^2} g f^2(\theta,\phi). \tag{2.7}$$

The fraction in (2.7) represents uniform power density at *r* which an isotropic antenna would produce but is reduced by the attenuation factor $l_h(r)$. The modifier

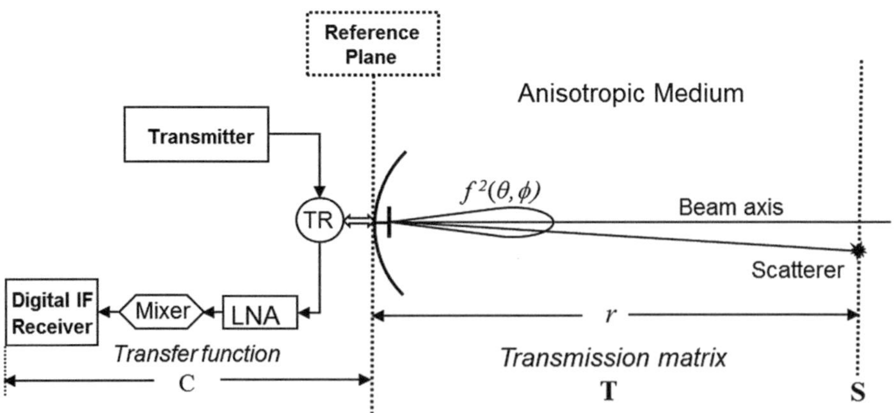

Fig. 2.8 Simplified diagram of transmission and reception illustrating one transmitter and receiver of a dual-polarization radar. The "reference plane" separates the radar system from the environment. Backscattering matrix of the scatterer is **S**. TR is the transmitter/receiver switch and LNA is the low-noise amplifier

$gf^2(\theta, \phi)$ quantifies the spatial gain distribution of the beam with the maximum (g) on the beam axis (the scatterer in Fig. 2.8 is off-beam axis where the power density is smaller).

Recall the power density is a product of electric and magnetic field intensities (1.12). Therefore the incident field phasor E_h^i from (1.11) with propagation effects from (1.42) included can be expressed via power density and the intrinsic impedance of air η_0 (see Eq. (1.14)) as follows:

$$E_h^i = \left(2\overline{|\vec{\mathbf{S}}^i|_{\mathrm{hh}}\eta_0}\right)^{1/2} \exp\left[-j\Phi_h(r) - j\varphi_h^t\right] = \frac{\left(P_h^t\right)^{1/2} g^{1/2} f(\theta,\phi) \eta_0^{1/2}}{(2\pi)^{1/2} l_h^{1/2}(r) r} \exp\left[-j\Phi_h(r) - j\varphi_h^t\right], \tag{2.8}$$

where φ_h^t is the initial phase of the transmitted wave at the antenna and the phase $\Phi_h(r)$ from (1.42) is equal to $\int_0^r \mathrm{Re}[k_h(r)]dr$. Note that the wavenumber's dependence on polarization affects the phase of the wave and, therefore, the phase difference between the H and V fields. This difference is a useful polarimetric variable—differential phase (Chap. 5)—and, for this reason, we distinguish between $\mathrm{Re}(k_h)$ and $\mathrm{Re}(k_v)$ in computations of phases (see Eq. (1.51)). The attenuation factor $l_h(r)$ is included in (2.8). The difference of wavelengths at orthogonal polarizations has negligible effect on the amplitude of the fields and the antenna gain. In computations of these, it is ignored, so $\lambda = \lambda_h = \lambda_v$.

Equation (2.8) is written for incident horizontally (H) polarized field. The hydrometeor backscatters a copolar (H) and cross-polar (V) components of the electric field. According to (1.15), these two components at the antenna are

$$E_{\mathrm{hh}}^s = \frac{\left(P_h^t\right)^{1/2} g^{1/2} f(\theta,\phi) \eta_0^{1/2}}{(2\pi)^{1/2} l_h(r) r^2} \exp\left(-j2\Phi_h(r) - j\varphi_h^t\right) s_{\mathrm{hh}} \tag{2.9a}$$

$$E_{\mathrm{vh}}^s = \frac{\left(P_h^t\right)^{1/2} g^{1/2} f(\theta,\phi) \eta_0^{1/2}}{(2\pi)^{1/2} l_h^{1/2}(r) l_v^{1/2}(r) r^2} \exp\left(-j\Phi_h(r) - j\Phi_v(r) - \varphi_h^t\right) s_{\mathrm{vh}}. \tag{2.9b}$$

It is explicit in (2.9) that the wave carrying a cross-polar component of the field from the scatterer to the radar experiences different attenuation (l_v) and phase shift (Φ_v) compared to the copolar component. Two analogous equations describe the received fields if the transmitted field has vertical polarization. These are obtained by relabeling the subscripts (h for v and v for h) in (2.9).

The power of the backscattered signal with polarization "h" at the antenna output P_h^r (Fig. 2.8) is a product of the power density of the backscattered radiation and the effective area of the antenna A_e (proportional to its physical size) determined by (Doviak and Zrnic 2006, Sect. 3.4)

$$A_e = \frac{g\lambda^2}{4\pi} f^2(\theta, \phi), \tag{2.10}$$

where the minute dependence of the wavelength λ on polarization is ignored.

Hence, combining (2.9a), (2.10), (1.28), and (1.30), we get

$$P_{\mathrm{hh}}^r = \frac{\left|E_{\mathrm{hh}}^s\right|^2}{2\eta_0} A_e = \frac{P_h^t g^2 \lambda^2 f^4(\theta, \phi)}{(4\pi)^2 l_h^2(r) r^4} |s_{\mathrm{hh}}|^2 = \frac{P_h^t g^2 \lambda^2 f^4(\theta, \phi)}{(4\pi)^3 l_h^2(r) r^4} \sigma_{\mathrm{hh}}. \tag{2.11a}$$

A similar equation is valid for the cross-polar power P_{vh}^r, i.e., the one originating from σ_{vh}:

$$P_{\mathrm{vh}}^r = \frac{\left|E_{\mathrm{vh}}^s\right|^2}{2\eta_0} A_e = \frac{P_h^t g^2 \lambda^2 f^4(\theta, \phi)}{(4\pi)^2 l_h(r) l_v(r) r^4} |s_{\mathrm{vh}}|^2 = \frac{P_h^t g^2 \lambda^2 f^4(\theta, \phi)}{(4\pi)^3 l_h(r) l_v(r) r^4} \sigma_{\mathrm{vh}}. \tag{2.11b}$$

For vertically polarized transmitted fields, the two received powers are described with the same relations except the subscripts in (2.11) are interchanged. In reciprocal media like precipitation $\sigma_{\mathrm{vh}} = \sigma_{\mathrm{hv}}$, therefore if $P_h^t = P_v^t$, the received powers are equal: $P_{\mathrm{hv}}^r = P_{\mathrm{vh}}^r$.

Equations (2.11a) and (2.11b) are the *polarimetric radar equations* for an individual scatterer and are valid for propagation through precipitation with attenuation. They don't contain information about phases of transmitted and received waves but are sufficient to quantify the polarization dependencies of the received powers of the radar return signal.

Voltages contain both amplitudes and phases of the transmitted and received signals and are important carriers of additional information available through polarimetric measurements. The equations for voltages come directly from (2.11). For simplicity, we assume the relation between transmitter power and its voltage is the same at both polarizations and there is no coupling between the *H*- and *V*-signals in the receiver and transmitter.

The equations relating the voltages V_h^r and V_v^r at the output of the digital receiver corresponding to a copolar and cross-polar components of the radar signal and the voltage of the transmitted signal at horizontal polarization V_h^t are

$$V_h^r = C_h e^{-j2k_h r} s_{\mathrm{hh}} V_h^t, \tag{2.12a}$$

$$V_v^r = C_v e^{-jk_v r} e^{-jk_h r} s_{\mathrm{vh}} V_h^t, \tag{2.12b}$$

where

$$C_h = \frac{g\lambda f^2(\theta,\phi)}{4\pi l_h(r)r^2}\exp(-j\varphi_h^r) \tag{2.13a}$$

and

$$C_v = \frac{g\lambda f^2(\theta,\phi)}{4\pi[l_h(r)l_v(r)]^{1/2}r^2}\exp(-j\varphi_v^r) \tag{2.13b}$$

In (2.12), $V_h^t = (P_h^t)^{1/2}\exp[j(2\pi ft - \varphi_h^t)]$ is the complex representation where the varying phase term $2\pi ft$ is explicitly included, whereas in the phasor

$$V_h^t = (P_h^t)^{1/2}\exp(-j\varphi_h^t) = I_h + jQ_h \tag{2.13c}$$

this term is implied (but its time origin is lost). In most expressions, we use the phasor notation. Further, it is implicit that the voltage is applied to a 1 Ω resistor. The transmitted phase φ_h^t includes all the delays and shifts from the physical location where the voltage is generated up to the antenna, i.e., "reference plane," where the field has polarization "*h*." The "reference plane" (Fig. 2.8) is a fictitious plane separating the radar effects from the sought-after medium effects. In (2.13), φ_h^r and φ_v^r are the phase shifts generated in the two receivers corresponding to horizontal and vertical polarizations.

From the comparison of Eqs. (2.12) and (2.13), it is evident that $|V_h^r|^2 = P_{\text{hh}}^r$ and $|V_v^r|^2 = P_{\text{vh}}^r$. In practice, calibration of the magnitudes is essential and done with great care to obtain quantitative measurements of precipitation. This is not so with the phase shifts φ_h^r and φ_v^r introduced by the receivers; the absolute values are not needed. What matters are the phase differences because these can be related to the properties of backscattering hydrometeors (including velocity) as well as the properties of precipitation along propagation path.

Equations (2.12a) and (2.12b) and a similar pair for the case where the transmitted wave polarization is vertical can be written explicitly in a matrix form

$$\begin{pmatrix} V_h^r \\ V_v^r \end{pmatrix} = \begin{pmatrix} C_h & 0 \\ 0 & C_v \end{pmatrix}\mathbf{T}\begin{pmatrix} s_{\text{hh}} & s_{\text{hv}} \\ s_{\text{vh}} & s_{\text{vv}} \end{pmatrix}\mathbf{T}\begin{pmatrix} V_h^t \\ V_v^t \end{pmatrix} = \mathbf{CTSTV}^t. \tag{2.14}$$

The transmission matrix **T** (Chap. 1) describes the effects of attenuation and propagation. By setting $V_h^t \neq 0$ and $V_v^t = 0$, the two linear Eqs. (2.12a and 2.12b) are obtained and two more equations are generated by setting $V_h^t = 0$ and $V_v^t \neq 0$. Eq. (2.14) is valid for waves propagating in precipitation and reflecting from a single scatterer.

If propagation medium is void of hydrometeors, then the magnitude of the diagonal terms in the matrix **T** is 1 and the off-diagonal terms are 0. If the hydrometeors along the propagation path are nonspherical and have their mean orientations either horizontal or vertical, then the diagonal elements of the matrix **T** are different (and generally less than one), while the off-diagonal elements are again zeroes. The off-diagonal terms would be different from zero if the mean orientation of hydrometeors along the propagation path is different from horizontal or vertical.

2.4 Doppler Shift and Differential Phase

The phases of received signals (corresponding to either polarization) contain important information about velocity of the reflecting scatterer and its microphysical properties (phase, shape, orientation, and size). The phases are also affected by microphysical properties of the scatterers filling the atmosphere along the propagation path between the point of reflection and the radar. The two effects can be separated if microphysical properties gradually change along the propagation path.

Let's assume that the propagation path is filled with nonspherical hydrometeors and their characteristics vary along the path but there is no change of polarization on propagation (Fig. 2.8). Then the **T** matrix in (2.14; see also 1.37) is

$$\mathbf{T} = \begin{pmatrix} \exp\left(-j\int_0^r k_h dr\right) & 0 \\ 0 & \exp\left(-j\int_0^r k_v dr\right) \end{pmatrix}. \tag{2.15}$$

If a horizontally polarized wave (proportional to the complex voltage V_h^t) is transmitted, then (2.14) can be written as

$$\begin{pmatrix} V_h^r \\ V_v^r \end{pmatrix} = \begin{pmatrix} V_{\mathrm{hh}} \\ V_{\mathrm{vh}} \end{pmatrix} = \mathbf{CTST}\begin{pmatrix} V_h^t \\ 0 \end{pmatrix}, \tag{2.16}$$

and the voltage corresponding to the horizontally polarized (copolar) component of the backscattered wave is

$$V_h^r = V_{\mathrm{hh}} = |C_h| \exp\left(-j2\int_0^r k_h dr - j\varphi_h^r\right) s_{\mathrm{hh}} V_h^t, \tag{2.17a}$$

whereas the voltage corresponding to the vertically polarized (cross-polar) component of the backscattered wave is

$$V_v^r = V_{\mathrm{vh}} = |C_v| \exp\left(-j\int_0^r (k_h + k_v) dr - j\varphi_v^r\right) s_{\mathrm{vh}} V_h^t. \tag{2.17b}$$

The notation with two subscripts is to identify the polarization of the transmitted field (second index) and the returned field (first index). Similar pair of relations holds if a vertically polarized wave is transmitted (i.e., $V_h^t = 0$ and $V_v^t \neq 0$ in (2.14)).

In the SHV mode, $V_h^t \neq 0$ and $V_v^t \neq 0$ in (2.14) and the complex voltages in the receivers (one for H the other for V polarizations) are

$$V_h^r = C_h\left[\exp\left(-j2\int_0^r k_h dr\right)s_{\text{hh}}V_h^t + \exp\left(-j\int_0^r (k_h + k_v)dr\right)s_{\text{hv}}V_v^t\right] \quad (2.18\text{a})$$

$$V_v^r = C_v\left[\exp\left(-j2\int_0^r k_v dr\right)s_{\text{vv}}V_v^t + \exp\left(-j\int_0^r (k_h + k_v)dr\right)s_{\text{hv}}V_h^t\right]. \quad (2.18\text{b})$$

If the canting angle α of the hydrometeor is small, then the magnitude of the off-diagonal terms of the scattering matrix **S** is much smaller than either of the diagonal terms (i.e., $|s_{\text{hv}}| << |s_{\text{hh}}|, |s_{\text{vv}}|$, Eq. (1.24)). Then for most practical applications, the terms containing s_{hv} in (2.18) can be neglected and

$$\begin{aligned} V_h^r &\approx C_h \exp\left(-j2\textstyle\int_0^r k_h dr\right)s_{\text{hh}}V_h^t = V_{\text{hh}} \\ V_v^r &\approx C_v \exp\left(-j2\textstyle\int_0^r k_v dr\right)s_{\text{vv}}V_v^t = V_{\text{vv}} \end{aligned} \quad (2.19)$$

indicating there is no coupling between H- and V-waves. This means the voltage in the receiver for horizontally polarized field is proportional to the transmitted voltage generating the horizontally polarized field. The same relation holds for the vertical polarization.

The case of significant coupling for which the cross-polarization terms can't be dismissed is considered in Sect. 6.9. Otherwise, we assume that (2.19) is valid. Distinction between the phases φ_h^t and $\varphi_v^t$of the transmitter voltages $V_h^t$and V_v^t is important in the SHV mode because their difference specifies the polarization of the composite transmitted field.

The phases of the voltages in the two receivers are sums of the phases contributed by each individual term in (2.19), that is

$$\begin{aligned} \arg[V_{\text{hh}}(t)] &= 2\pi ft - 2\text{Re}\textstyle\int_0^r k_h dr - \delta_{\text{hh}} - \varphi_h^t - \varphi_h^r = \Phi_{\text{hh}}(t) \\ \arg[V_{\text{vv}}(t)] &= 2\pi ft - 2\text{Re}\textstyle\int_0^r k_v dr - \delta_{\text{vv}} - \varphi_v^t - \varphi_v^r = \Phi_{\text{vv}}(t) \end{aligned} \quad (2.20)$$

In (2.20),

$$\delta_{\text{hh}} = -\arg(s_{\text{hh}}), \quad \delta_{\text{vv}} = -\arg(s_{\text{vv}}) \quad (2.21)$$

are the phase shifts of the backscattered wave with respect to the incident wave caused by the scatterer (Sect. 1.2). These backscatter phase shifts depend on the wavelength, size, composition, shape, and orientation of the scatterer. Suffice to say that scatterers very small compared to wavelength cause negligible backscatter phase shifts (Chap. 5, Sect. 5.5).

To better convey the physical meaning of the phases and their differences, we resort to a simple graphical representation of the pulse propagation and reflection. In the SHV mode (Fig. 2.7b), the pulsed polarimetric radar transmits simultaneously continuous series of pulsed electromagnetic waves at horizontal and vertical polarizations with duration τ and envelop $p(t)$ separated by the pulse repetition time T_s (Fig. 2.2).

The propagation speed of H- and V-waves in clear air is c/n where $n = [1 + \mathrm{Re}(\Delta n^{(air)})]$ as defined by (1.4). Hydrometeors filling the atmosphere decrease the propagation speed to $c/[1 + \mathrm{Re}(\Delta n^{(air)} + \Delta n^{(scat)})]$ (see (1.5)). If the hydrometeors are nonspherical and oriented, like large raindrops (having horizontal dimension larger than vertical), the refractive index for horizontally polarized radiation is larger than for the vertically polarized radiation. Therefore, the propagation speed of the H-wave ($c_h = c/\mathrm{Re}(n_h)$) is lower than the speed of the V-wave ($c_v = c/\mathrm{Re}(n_v)$, i.e., $c_h < c_v$, and the spatial length of the H-pulse ($c_h\tau$) is shorter than the length of the V-pulse ($c_v\tau$) (Fig. 2.9).

Although the H-wave and V-wave pulses are transmitted simultaneously in the SHV mode, the V-pulse reaches the scatterer located at the distance r_1 from the radar ahead of the H-pulse (Fig. 2.10). A small fraction of the incident V-radiation is reflected and propagated back to the radar. The front end of the reflected pulse arrives back at the radar at time $t = 2r_1/c_v$. The next V-pulse transmitted at time T_s and reflected by the same scatterer is received by the radar at $t = T_s + 2r_2/c_v$ (where $r_2 \neq r_1$) if the scatterer moves in the radial direction with respect to the radar. Similarly, first and second reflected H-pulses reach the radar at the times $2r_1/c_h$ and $T_s + 2r_2/c_h$ respectively (see Fig. 2.10).

The radar receiver samples the reflected signals at both polarizations at times τ_s, $\tau_s + T_s$, $\tau_s + 2T_s$, ... (Fig. 2.10). The H- and V-signals at τ_s and $\tau_s + T_s$ are characterized by their amplitude (i.e., $|V_{hh}|$ or $|V_{vv}|$) or magnitude of sinusoidal oscillations in Fig. 2.10 and phase (i.e., location of the sinusoid with respect to τ_s and $\tau_s + T_s$).

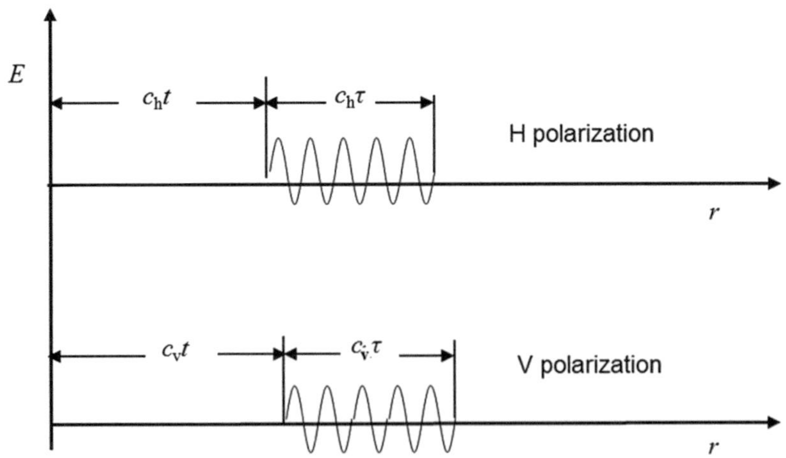

Fig. 2.9 Instantaneous snapshot of propagating pulses at horizontal and vertical polarizations at time t after the back edges of the pulses are transmitted by the radar. The range is r and the effects are exaggerated for visual clarity, and attenuation within the pulse is ignored

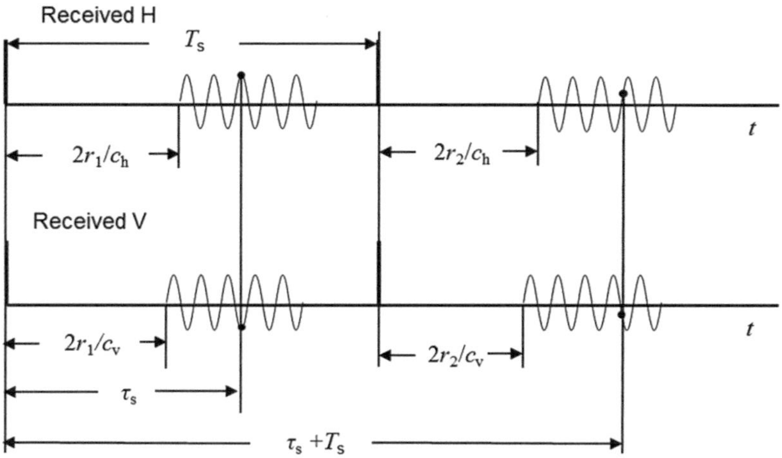

Fig. 2.10 Conceptual plot illustrating transmission and reception of two consecutive pulses at horizontal and vertical polarizations reflected from a single scatterer

Let's consider the sequence of two reflected pulses at horizontal polarization. If the scatterer does not move in the radial direction, then the difference between the phases of the two consecutive reflected pulses taken at times τ_s and $\tau_s + T_s$ is equal to

$$\Delta\Phi_{\mathrm{hh}}(T_s) = \Phi_{\mathrm{hh}}(\tau_s + T_s) - \Phi_{\mathrm{hh}}(\tau_s) = 2\pi f T_s \tag{2.22}$$

as follows from (2.20). The phase difference is slightly different if the scatterer moves in the direction toward or away from the radar (i.e., $r_1 \neq r_2$):

$$\begin{aligned}\Delta\Phi_{\mathrm{hh}}(T_s) &\approx 2\pi f T_s - 2\mathrm{Re}[k_h(r_1)](r_2 - r_1) = 2\pi f T_s - 2\mathrm{Re}[k_h(r_1)]vT_s \\ &= 2\pi(f + f_d)T_s,\end{aligned} \tag{2.23}$$

where

$$f_d = -\frac{2v}{\lambda_h} \tag{2.24}$$

is the Doppler frequency and v is the Doppler velocity.

In the previous example, we compared the phases of two successive pulses at horizontal polarization. Let's compare the phases of the *H*- and *V*-waves at coincident times in Fig. 2.10 (i.e., either τ_s or $\tau_s + T_s$). The phases are different and the difference is not related to the movement of the scatterer. Instead, it is caused by the difference in the time of arrival of the reflected *H*- and *V*-pulses. This phase difference is called the differential phase Φ_{DP} and is expressed as

$$\Phi_{\mathrm{DP}}(r) = \varphi_h^c - \varphi_v^c = -\arg(V_{\mathrm{hh}}) + \arg(V_{\mathrm{vv}}) = \delta(r) + \Phi_{\mathrm{DP}}^{(\mathrm{pro})}(r) + \Phi_{\mathrm{DP}}^{(\mathrm{sys})}, \quad (2.25)$$

where

$$\delta = \delta_{\mathrm{hh}} - \delta_{\mathrm{vv}} \quad (2.26)$$

is the *backscatter differential phase*,

$$\Phi_{\mathrm{DP}}^{(\mathrm{pro})}(r) = 2\int_0^r \mathrm{Re}(k_h - k_v)dr \quad (2.27)$$

is the *differential phase due to propagation*, and

$$\Phi_{\mathrm{DP}}^{(\mathrm{sys})} = \varphi_h^t + \varphi_h^r - \varphi_v^t - \varphi_v^r \quad (2.28)$$

is the *system differential phase*.

Clearly, the total system differential phase is a sum of three terms: (1) backscatter differential phase δ attributed to backscattering from a particular hydrometeor, (2) propagation differential phase $\Phi_{\mathrm{DP}}^{(\mathrm{pro})}(r)$ caused by cumulative difference between propagation speeds of the horizontally and vertically polarized waves in the anisotropic medium, and (3) system differential phase $\Phi_{\mathrm{DP}}^{(\mathrm{sys})}$ attributed to the radar hardware. The backscatter differential phase depends on the scattering properties of the scatterer which reflects incident electromagnetic wave, whereas propagation differential phase is the phase which is accumulated over the propagation path and depends on the scattering properties of atmospheric scatterers along the propagation path.

A simple graphical illustration of the Doppler and differential phases is presented in Fig. 2.11 where phasors V_{hh} and V_{vv} are depicted. Note, both Doppler phase and differential phase are independent of the amplitudes of reflected signals, i.e., the changes of the phasor lengths in Fig. 2.11 do not affect $\Delta\Phi_d$ and Φ_{DP}. Therefore, phase measurements are independent of radar calibration, attenuation, and partial beam blockage all of which impact amplitudes. Although the directions of the *H*- and *V*-vectors change from pulse to pulse due to the Doppler effect, the average difference between the directions of the two vectors (i.e., differential phase) is preserved.

2.5 Measurements of Single-Particle Scattering

The hydrometeors filling the radar resolution volume collectively contribute to the backscattered fields and, therefore, the derived radar variables. To quantify the bulk contribution, one can start with a single scatterer and compute its backscattering coefficients. Then assuming some distribution of hydrometeors and superposing linearly the backscattering coefficients, one can compute the polarimetric variables.

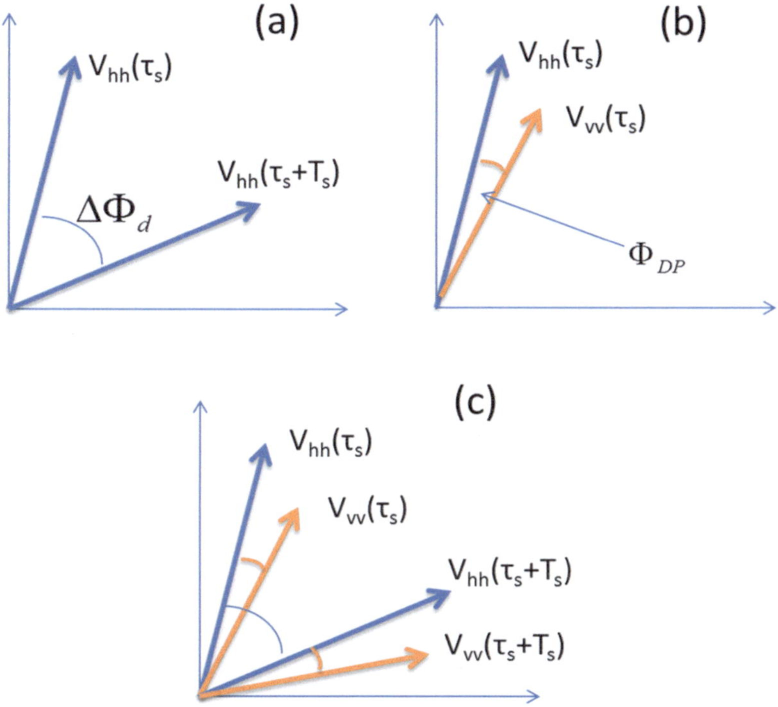

Fig. 2.11 (**a**) Phasors $V_{\text{hh}}(\tau_s)$ and $V_{\text{hh}}(\tau_s + \text{T}_s)$ of two successive samples of received signals at horizontal polarization. The Doppler phase shift $\Delta\Phi_d$ is represented by the angle between these two phasors. (**b**) Phasors $V_{\text{hh}}(\tau_s)$ and $V_{\text{vv}}(\tau_s)$ of the radar signals at orthogonal polarizations received simultaneously. The differential phase Φ_{DP} is represented by the angle between the two phasors. (**c**) Phasors of two successive sample pairs of received signals at horizontal and vertical polarizations

By comparing with measurements and changing the assumed distribution, one can find a good match between the computations and observations. This is the principal reason for computations of the scattering coefficients. Direct radar measurement of the scattering characteristics of a single scatterer is possible in a laboratory setup, and even there it may be challenging.

The scattering coefficients (1.17) contain all the information dual-polarization measurements can provide about the properties of a single scatterer, and, as indicated in (2.17), these are linearly related to the voltages (at the digital receiver output). Measuring scattering characteristics of a single hydrometeor are not easy, but it is instructive to consider the concept to determine what can be measured. For example, to obtain the absolute phase of a coefficient, an equivalent number of wavelengths along the round trip path (including the radar system) would need to be known and removed from the voltage (I and Q). Differential measurements can provide phase differences and are one of the strengths polarimetry brings to bear. Because there are three distinct scattering coefficients, there would be three phase differences. These are expressed as arguments of products such as $s_{\text{hh}}^* s_{\text{vv}}$. If the system differential phase

(2.28) is removed from $\arg\left(V_{\mathrm{hh}}^{*}V_{\mathrm{vv}}\right)$, the backscatter differential phase (2.26) can be measured. It is very convenient for accounting (enumeration) to list all the products combinations of the three coefficients in the following matrix (Zrnic 1991)

$$\begin{bmatrix} |s_{\mathrm{hh}}|^2 & s_{\mathrm{hh}}^{*}s_{\mathrm{hv}} & s_{\mathrm{hh}}^{*}s_{\mathrm{vv}} \\ & |s_{\mathrm{hv}}|^2 & s_{\mathrm{hv}}^{*}s_{\mathrm{vv}} \\ & & |s_{\mathrm{vv}}|^2 \end{bmatrix}. \tag{2.29}$$

The matrix is symmetric and the lower triangular half is omitted. The diagonal terms are proportional to cross sections and these in turn are proportional to powers (magnitude squared of measured voltages). Cross sections relate to the size, shape, orientation, and composition of the scatterer. There are nine independent real quantities in (2.29): the three cross sections and three complex cross products off the diagonal. This is all the backscattering information a single scatterer probed by weather radar returns. A spherical drop has $s_{\mathrm{hh}} = s_{\mathrm{vv}}$, $s_{\mathrm{hv}} = 0$, and the matrix (2.29) is characterized by only one nonzero term ($s_{\mathrm{hh}} = s_{\mathrm{vv}}$), whereas nonspherical raindrops are characterized by two different nonzero terms (s_{hh} and s_{vv}). If drops are canted, then $s_{\mathrm{hv}} \neq 0$ (Chap. 1).

Precipitation is a collection of scatterers and its polarimetric properties can be condensed into (2.29). Moreover, each hydrometeor within the radar resolution volume (roughly defined by the beamwidth and pulse length) contributes linearly to the voltage from which bulk properties of radar quantities (like average cross sections per m^3) can be inferred. These radar quantities are mapped into bulk precipitation properties like amount of liquid water in the cloud, rain rate, hail size, and so on. Relations between the voltages and a collection of backscattering hydrometeors are derived in Sect. 3.3.

For reference, we explicitly expand the voltage Eq. (2.19) for a single scatterer "*i*" assuming a transmitted voltage $\left(P_h^t\right)^{1/2}$ and including its phase as follows:

$$V_{\mathrm{hh}}(i) = \frac{\left(P_h^t\right)^{1/2} g\lambda f^2(\theta_i, \phi_i)}{4\pi r_i^2 l_h(r_i)} s_{\mathrm{hh}}^{(i)} \exp\left[-j2\int_0^{r_i} \mathrm{Re}(k_h)dr - j\varphi_h^t - j\varphi_h^r\right], \tag{2.30}$$

where $V_{\mathrm{hh}}(i)$ is the voltage at the digital receiver output. It is assumed the transmitted pulse is rectangular and no noise filter is applied. Equation of the same form (after replacing subscript "*h*" with "*v*" everywhere) is valid for the $V_{\mathrm{vv}}(i)$. The expression for the cross-polar voltage $V_{\mathrm{vh}}(i)$ has the geometric mean of cumulative attenuations in the denominator and the cumulative phase due to propagation is $\int_0^{r_i} \mathrm{Re}(k_h)dr + \int_0^{r_i} \mathrm{Re}(k_v)dr$.

An example of the voltage phasors measured at the output of the digital receiver is shown in Fig. 2.12. The radar is a research WSR-88D and the returned pulse is from a radio tower at about 29.35 km in range (Fig. 2.13). The pulse length is 250 m and the radar return is sampled at 50 m spacing. The tower produces a very stable return (no Doppler shift). The phase in the middle of the pulse can be obtained

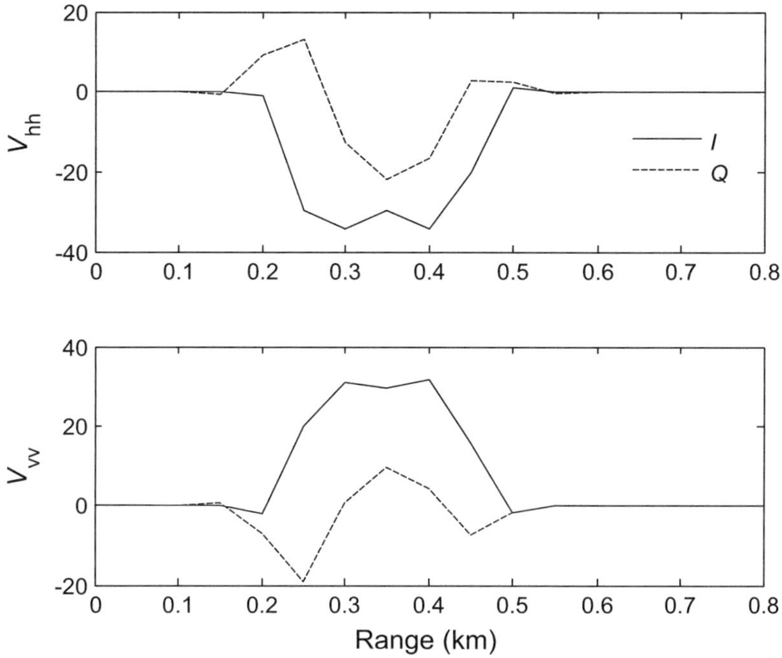

Fig. 2.12 Voltages V_{hh} and V_{vv} at the outputs of digital receivers for *H* and *V* polarization. The range 0 is at 29 km from the radar. Nominal pulse length is 250 m and the samples are spaced 50 m apart

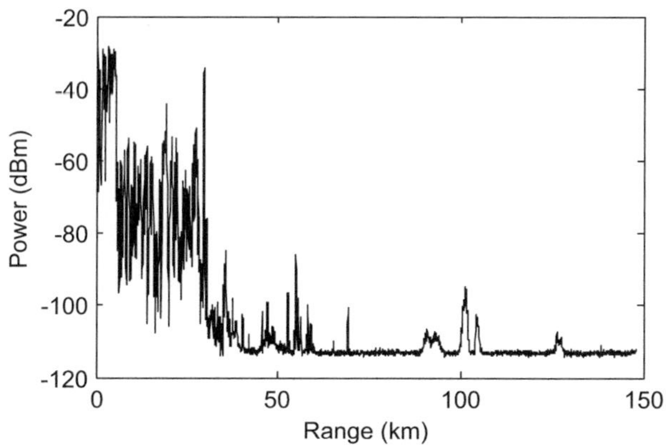

Fig. 2.13 Power in dBm (decibel with respect to 1 mW) at horizontal polarization. The noise level is at −113 dBm and the tower that generated the *I* and *Q* signals in Fig. 2.12 has reflected the sharp signal at 29.35 km

from the *I* and *Q* values. Since $V_{hh} = -29 - j22$ and $V_{vv} = 29 + j10$ in relative units, then the voltage phases at orthogonal polarizations are $\varphi_h^t + \varphi_h^r + \delta_{hh} = -143^\circ$ and $\varphi_v^t + \varphi_v^r + \delta_{vv} = 19^\circ$ where the two backscatter phases are unknown.

References

Browne, I. C., & Robinson, N. P. (1952). Cross polarization of the radar melting band. *Nature, 170*, 1078–1079.

Brunkow, D., Bringi, V., Kennedy, P., Rutledge, S., Chandrasekar, V., Mueller, E., & Bowie, R. (2000). A description of the CSU-CHILL National Radar Facility. *Journal of Atmospheric and Oceanic Technology, 17*, 1596–1608.

Doviak, R., & Zrnic, D. (2006). *Doppler radar and weather observations* (2nd ed.). Reprint, Mineola, NY: Dover Publications, 562 pp.

Doviak, R. J., Bringi, V., Ryzhkov, A., Zahrai, A., & Zrnic, D. S. (2000). Considerations for polarimetric upgrades of operational WSR-88D radars. *Journal of Atmospheric and Oceanic Technology, 17*, 257–278.

Lutz, J., Rilling, R., Wilson, J., Weckwerth, T., & Vivekanandan, J., (1997). S-Pol after three operational deployments, technical performances, siting experiences and some data examples. In *Preprints, 28th Conference on Radar Meteorology* (pp. 286–287). Austin, TX: American Meteorology Society.

McCormick, G., & Hendry, A. (1975). Principles for the radar determination of the polarization properties of precipitation. *Radio Science, 10*, 421–434.

Sachidananda, M., & Zrnic, D. S. (1989). Efficient processing of alternately polarized Radar Echoes. *Journal of Atmospheric and Oceanic Technology, 6*, 173–181.

Schroth, A. C., Chandra, M. S., & Mesichner, P. F. (1988). A *C*-band coherent polarimetric radar for propagation and cloud physics research. *Journal of Atmospheric and Oceanic Technology, 5*, 803–822.

Seliga, T., & Bringi, V. (1976). Potential use of radar differential reflectivity measurements at orthogonal polarizations for measuring precipitation. *Journal of Applied Meteorology, 15*, 69–76.

Shupiatsky, A. B., & Morgunov S. P. (1963). The application of polarization methods to radar studies of clouds and precipitation. In *Tr. Vsesoiuzn. Nauch. Meteor. Soveshch., Leningrad* (pp. 295–305.) (in Russian).

Zahrai, A., & Zrnic, D. S. (1993). The 10-cm-wavelength polarimetric weather radar at NOAA's National Severe Storms Laboratory. *Journal of Atmospheric and Oceanic Technology, 10*, 649–662.

Zrnic, D. S. (1991). Complete polarimetric and Doppler measurements with a single receiver radar. *Journal of Atmospheric and Oceanic Technology, 8*, 159–165.

Zrnic, D. S. (2012). Doppler radar for USA weather surveillance. In J. Bech (Ed.), *Doppler radar observations*. Rijeka, Croatia: IntechOpen www.intechopen.com.

Chapter 3
Scattering by Ensemble of Hydrometeors: Polarimetric Perspective

Scatterers collectively contribute to the returned signals. The volume from which most of the contribution emanates is called the resolution volume or sample volume. In this chapter, we explain the range dimension of this volume and the range dependence of the weight given to the scatterers. Less attention is devoted to the lateral dimension defined by the beamwidth because it is more intuitive to readers. We combine the returns from multiple scatterers into an amalgamation to compute the stable statistical parameters of the returned signals. These are the second-order moments or correlations and powers of the various signals (H and V). We define the second-order moments and polarimetric variables available in different modes of operations (SHV, HSHV, and VSVH; Chap. 2). The effects of orientation distributions of particles on the polarimetric variables are extensively discussed.

3.1 Range Weighting Function

The spatial extent of the propagating rectangular pulse is $c\tau$ and the function $p(t - r/c)$ describes its spatial form (time-space envelope). At a fixed time instant, it indicates the distance and extent of the illuminated region in space. As time progresses, the illumination (pulse) moves forward at the propagation speed. While propagating, the pulse initiates reflections and at a fixed instant all scatterers between its leading and trailing edges are scattering. The ones at the front just begin while the ones at the rear stop scattering. Therefore, the extent of scattering is over the pulse length $c\tau$. However, the contiguous region in space from which backscattered fields return to the radar *at the same time* has depth $c\tau/2$; this is the inherent range resolution of the pulse. The receiver further processes the return and thus changes the effective depth of the probing pulse (see Doviak and Zrnic 2006, Sect. 4.4). A heuristic explanation of the concept follows.

Consider a region filled with scatterers and assume attenuation is negligible (at least within the pulse length). Let the transmitted pulse envelope $p(t)$ be

A. V. Ryzhkov, D. S. Zrnic, *Radar Polarimetry for Weather Observations*,
Springer Atmospheric Sciences, https://doi.org/10.1007/978-3-030-05093-1_3

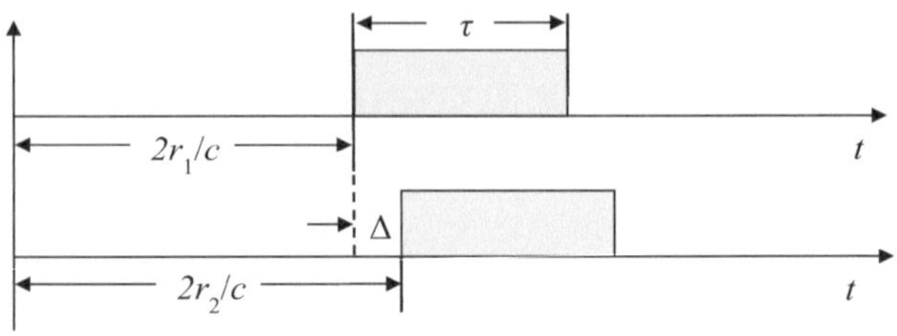

Fig. 3.1 Illustration of the range resolution corresponding to a rectangular pulse of duration τ

rectangular. The backscatter signal from the scatterer at the range r_1 is recorded by the radar receiver within a time interval from $2r_1/c$ to $2r_1/c + \tau$ (Fig. 3.1, top). The corresponding radar return from the scatterer at the range r_2 is received between $2r_2/c$ and $2r_2/c + \tau$ (Fig. 3.1, bottom). The radar signals from the two scatterers are received *simultaneously* (i.e., shaded rectangles in Fig. 3.1 overlap) only if $2r_2/c - 2r_1/c < \tau$ or $r_2 - r_1 < c\tau/2$.

Therefore if $r_2 - r_1 < c\tau/2$, then all the scatterers between these ranges contribute to the returns at the same time and can't be resolved. This means that the range weighting (of powers) function is determined by the pulse envelope function $p(t)$ compressed by the factor of 2

$$|W(r - r_0)|^2 = \left|p\left[\frac{2(r - r_0)}{c}\right]\right|^2, \tag{3.1}$$

where r_0 is the distance to the center of the radar resolution volume.

An additional subtlety is the shape of the transmitted pulse envelope $p(t)$ which by choice is not rectangular. By design, pulse envelopes have tight bandwidth and carrier frequencies are separated to prevent transmitted radiation from interfering with other receivers. The frequency content (distribution) of the rectangular-pulsed sinusoid is given by its Fourier transform, $\sin(f - f_c)/(f - f_c)$, and the distribution of power is the square of this function; f_c is the carrier (transmitted) frequency. The excessive frequency spread of this waveform is caused by the sharp edges of the rectangular pulse; hence pulses are shaped to constrain their bandwidth. In some modern radars (WSR-88D included), the envelope of the pulse is measured, while the pulse is being transmitted. A small part of it is diverted to the receiver (Fig. 2.3) where it is sampled and many digital I and Q values are generated. The phases of these digital samples represent the reference for comparison with returned pulses. On radars with unknown phase of transmitted pulse, these measurements are made routinely to obtain Doppler information. On fully coherent radars, adjusting the phase in the receiver can improve the quality of data if there are variations in phases of the transmitted pulses.

The returned pulse is filtered to reduce white noise entering the receiver and noise generated by the receiver components. Filters are designed to pass most of the signal variation (bandwidth) and reduce as much noise as possible. Samples of the transmitted pulse $p(t_m)$ are needed to compute the weights $h(t_i)$ of the digital filter. A good filter will pass sufficient amount of signal with little smearing while reducing the noise. Once chosen for a specific pulse shape and duration, the filter weights are kept constant. The representation of the filter output is the convolution

$$y(t_i) = \sum_{i=1}^{k} h(t_i - t_m)p(t_m); \tag{3.2}$$

therefore the amplitude range weighting function is

$$W(r_i) = y(-ct_i/2). \tag{3.3}$$

This is a mirror image (flipped about the origin) of $y(t_i)$, and r_0 is defined as the distance to its peak.

The digital operation offers flexibility to choose the filter coefficients h. The sampled $p(t_m)$ in (3.2) is complex (I and Q components); therefore two identical digital operations are executed. In some application, the weights h can be complex to alter phases of the signal. A perfectly matched filter has the amplitude weighting function $h(t)$ equal to the mirror image of the transmitted pulse shape. In convolution (3.2), the p is flipped and slid through h while summing the products and the maximum occurs when the two shapes exactly overlap—this also minimizes white noise. A matched filter for the rectangular pulse is a rectangle. Thus the running product of the two, $W(r)$, has a triangular shape, the magnitude square has a cusp-like shape, and its area (integral) is 2/3 of the rectangle (see text after Eq. (3.12)).

From (3.2) and (3.3) it follows that the range weighting function can be obtained if the pulse is passed through the digital receiver and the result is reversed in range. Radar return from a localized object like a radio tower (Fig. 2.13) replicates the transmitted pulse very well. Therefore, the magnitude squared of the return at the output of the digital receiver (after reversing the axis) depicts the range weighting function. The example in Fig. 3.2 is from WSR-88D (KOUN) and the range weighting function is plotted in dB scale (the corresponding I, Q pair is plotted in Fig. 2.12). The nominal pulse width ($c\tau/2$) is 240 m and the samples are (for this particular example) spaced 50 m apart. In operations, the samples are spaced at 250 m. Low-pass filter has spread the pulse and, therefore, the range weighting function is wider and the resolution is worse than a nominal pulse width. It is important to distinguish resolution from sample spacing. The former refers to region in space over which the range weighting function has significant influence. The sample spacing (often referred to as gate spacing) separates locations from which volume data are taken. The sample spacing can be smaller or larger than the width of $|W(r)|^2$. In the former case, the information in adjacent samples is redundant but can be used to reduce error in estimates (Torres and Zrnic 2003; Ivic et al. 2003). Typically, the spacing is chosen to match the width of the rectangular pulse $p(t)$.

The $10\log(|W|^2)$ range dependence in Fig. 3.2 is not symmetric and the envelope is almost 400 m wide albeit the weights corresponding to this width are about 60 dB

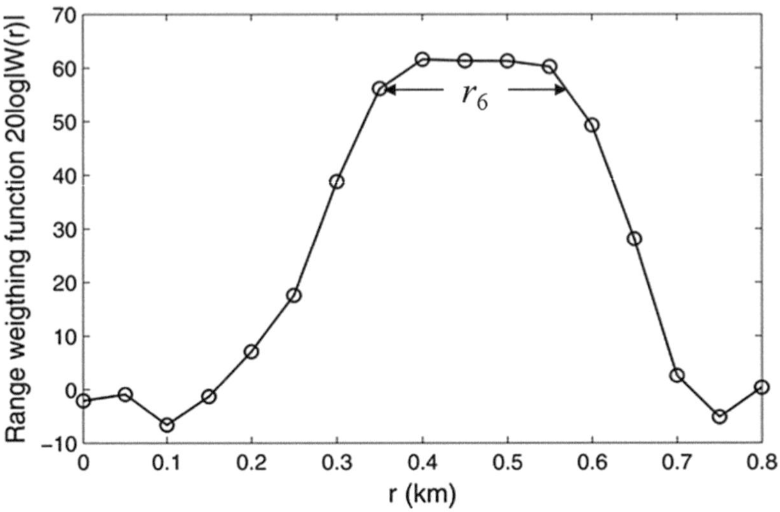

Fig. 3.2 Range weighting function of the research WSR-88D (KOUN). The scale is in dB. The function was obtained by inverting the envelope of the pulse returned from a radio tower. Sample spacing at the digital receiver output is 50 m, and the samples are indicated with circles. A nominal pulse width $c\tau/2$ is 240 m

below the peak value. With r_6 we define the width of $|W(r)|^2$ where its value is within 6 dB of the peak (Fig. 3.2). In this example $r_6 = 240$ m. Sample spacing of 250 m (as on WSR-88D) is well matched to r_6. Nonetheless, if strongly reflecting scatterers (50 dB larger than some weak precipitation) are at the pulse skirts (say, 40 dB down from the peak), they could overwhelm the weak returns within r_6 and appear in two or three contiguous range samples.

The volume in space from which scatterers contribute to the voltages is between two concentric spheres centered on the radar and separated by the depth of the range weighting function. Contribution by a scatterer depends on its position within this volume and is weighted by the product $f^4(\theta_i, \phi_i)|W(r_0,r_i)|^2$ where "i" specifies the location of the scatterer in range and with respect to the beam axis. The location of the resolution volume center is r_0 and it is implicit that the beam is stationary (pointing in a fixed direction). Following Doviak and Zrnic (2006), we use the symbol V_6 to define the volume where the values of $f^4(\theta_i, \phi_i)|W(r_0,r_i)|^2$ are within ¼ of the maximum (from 6 dB below the peak to the peak). The antennas are designed to have narrow beams and low sidelobes, therefore, most of the radar returned signal comes from the contribution of scatterers illuminated by the main lobe. Scatterers from everywhere else also contribute but are weighted much less than those in the resolution volume V_6.

3.2 Powers and Correlations: Ensemble of Scatterers

In this section, powers and correlations of various radar signals originating from collection of scatterers are derived. These are stable statistical quantities from which polarimetric variables are estimated.

3.2.1 Powers

To quantify the radar return in the presence of many scatterers, the range weighting function is introduced into the voltage radar equation for a single scatterer (2.30) and contributions of N_s scatterers are summed up. The composite voltage (phasor) at the digital receiver's output is

$$V_{\mathrm{hh}} = \sum_{i=1}^{\mathrm{N}_s} \frac{\left(P_h^t\right)^{1/2} g\lambda f^2(\theta_i, \phi_i)}{4\pi r_i^2 l_h(r_i)} W_i s_{\mathrm{hh}}^{(i)} \exp\left\{-j\left[2\mathrm{Re}(k_h) r_i + \varphi_h^t + \varphi_h^r\right]\right\}. \qquad (3.4)$$

In a compact form

$$V_{\mathrm{hh}} = B \sum_{i=1}^{\mathrm{N}_s} F_i G_i s_{\mathrm{hh}}^{(i)} \exp[-j2\mathrm{Re}(k_h) r_i] \qquad (3.5\mathrm{a})$$

where

$$B = \frac{\left(P_h^t\right)^{1/2} g\lambda}{4\pi} \exp\left(-j\varphi_h^{(\mathrm{sys})}\right), \quad \varphi_h^{(\mathrm{sys})} = \varphi_h^t + \varphi_h^r;$$

$$F_i = \frac{1}{r_i^2 l_h(r_i)}, \quad \text{and } G_i = f^2(\theta_i, \phi_i) W_i. \qquad (3.5\mathrm{b})$$

Because the phase $2\mathrm{Re}(k_h)r_i$ changes dramatically at the spatial scale of the radar wavelength, the average sum of the contributions of different scatterers within the radar resolution volume (much larger than the wavelength) to the complex voltage V_{hh} is close to zero. This means that averages of I and Q are close to zero. This is not so with power measurements.

Magnitude squared (proportional to power) of (3.4) can be expressed as a double summation (product of V_{hh} in (3.4) and its conjugate):

$$|V_{\mathrm{hh}}|^2 = |B|^2 \sum_{k=i=1}^{N_s} F_i^2 |G_i|^2 \left| s_{\mathrm{hh}}^{(i)} \right|^2$$
$$+ |B|^2 \sum_{i \neq k}^{N_s} F_i^2 G_i^* G_k s_{\mathrm{hh}}^{*(i)} s_{\mathrm{hh}}^{(k)} \exp\{j2[\mathrm{Re}(k_h)(r_i - r_k)]\}. \qquad (3.6)$$

Separation of the double summation (3.6) into a single summation (first term) and double (second term) is done to segregate the interaction of individual scatterer's contributions. In the single sum, each term corresponds to a specific scatterer, whereas in the double sum, each term corresponds to a pair of scatterers. The single sum term is sought after because it contains "permanent" information about scatterer physical properties. It is independent of the scatterers' relative positions and represents the mean power (scaled by appropriate unit conversion) usually referred to the power at the antenna. The double sum consists of product pairs and depends on relative distances between scatterers in each pair. Each pair enters into the double sum twice (interchange "*i*" with "*k*") and the sum of such two pairs is real (cosine multiplied by an amplitude factor). Therefore, the double sum is real positive or negative but the total of (3.6) cannot be smaller than zero.

Expression (3.6) is valid for one spatial configuration of scatterers, a snapshot over the length of the transmitted pulse (few μs) during which scatterers have very little time to rearrange (Fig. 3.3). Change in position is caused by turbulence and other variations of the wind field. Another measurement T_s seconds later (long enough for the scatterers relative positions to change) would produce the same first sum and a very different double sum. Averaging over many sequential measurements reduces the fluctuating double sum (associated with interaction of scatterers) to a very small value, and the contribution by individual scatterers (i.e., first term in (3.6)) becomes dominant. If the number of scatterers is large (about ten suffices), the distribution of the power (3.6) is exponential (Doviak and Zrnic 2006).

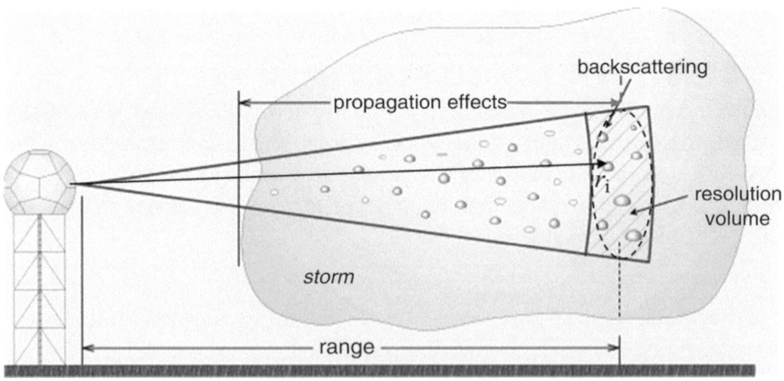

Fig. 3.3 Scattering by an ensemble of hydrometeors illustrating propagation, backscattering, and resolution volume V_6 (its vertical cross section is indicated with dashed contour)

Assume the voltage V_{hh} in (3.4) is measured (at the same time delay) after each transmitted pulse for a total of M measurements. Let $V_{hh}(m)$ represent the mth sample and the power estimate is obtained by averaging the power of each sample. Denote this estimate with $\hat{P}$, and let it be given by the average of (3.6) assuming a unit conversion from voltage to power (1 ohm resistor). Therefore

$$\hat{P}_{hh} = \overline{|V_{hh}|^2} = \frac{1}{M}\sum_{m=1}^{M}|V_{hh}(m)|^2, \tag{3.7a}$$

and by analogy

$$\hat{P}_{vv} = \overline{|V_{vv}|^2} = \frac{1}{M}\sum_{m=1}^{M}|V_{vv}(m)|^2 \tag{3.7b}$$

$$\hat{P}_{hv} = \overline{|V_{hv}|^2} = \frac{1}{M}\sum_{m=1}^{M}|V_{hv}(m)|^2 \tag{3.7c}$$

where the overbar denotes the time average.

Next we postulate that the scattering process is stationary (it is not a function of time) and that spatial and temporal averages in the limit of a very large number of trials are equal. That is as $M \to \infty$, $\hat{P}_{hh}$ from one cloud (single member of ensemble) is the same as would be obtained by averaging M simultaneous measurements each from a different cloud. Such clouds are indistinguishable hence can be described by distribution functions like (1.31).

Denote the spatial (ensemble) expected value with $<P_{hh}>$ and the temporal expectation as $\lim\limits_{M\to\infty}\left(\hat{P}_{hh}\right)$. The purpose of averaging is to closely approximate the ensemble mean value (considered to be true) by a temporal averaging of a fairly large number of returns. Formally we write

$$< P_{hh} >= |B|^2 < \sum_{k=i=1}^{N_s} F_i^2|G_i|^2 s_{hh}^{*(i)} s_{hh}^{(i)} > \tag{3.8}$$

because ensemble average of the second term in (3.6) is 0.

To proceed further, consider the distribution function of scatterers' properties $N(\boldsymbol{X})$ per unit volume (Chap. 1, Eq. (1.31)), where the vector $\boldsymbol{X}$ terms represent such properties (e.g., equivalent volume diameter D, axis ratio of spheroids, and so on). Assume $N(\boldsymbol{X})$ does not change inside the radar resolution volume. Then the summation can be replaced with integration over the resolution volume of the ensemble average of $s_{hh}^{*(i)} s_{hh}^{(i)}$. Denote this ensemble average per unit resolution volume with $<|s_{hh}|^2>$

$$< |s_{\text{hh}}|^2 >= \int N(\boldsymbol{X})|s_{\text{hh}}|^2 d\boldsymbol{X}. \tag{3.9}$$

It suffices to assume (3.9) is constant within the resolution volume but can be different from volume to volume. It may change in a deterministic way within the resolution volume if precipitation has a "deterministic" profile in height (evolves from top to bottom of cloud). To model change one can impose a spatial dependence on $<|s_{\text{hh}}|^2>$ and integrate (3.9), but there is no way to retrieve arbitrary spatial distribution from the integrated value (i.e., the forward problem is unique; the inverse is not).

One more simplification is needed before integration over the resolution volume can be accomplished; it is to replace the r_i in r_i^4 and $l_h^2(r_i)$ with the value at the center of the resolution volume r_0. This is reasonable at centimeter wavelengths and far from the radar where the depth of the resolution volume is small compared to r_0 and attenuation within the resolution volume is negligible. Introducing (3.9) into (3.8) and integrating over the resolution volume produces

$$< P_{\text{hh}} >= \frac{P_h^t g^2 \lambda^2}{16\pi^2 r_0^2 l_h^2(r_0)} < |s_{\text{hh}}(r_0)|^2 > \int_{r_0-\Delta_1}^{r_0+\Delta_2} \int_0^{\pi} \int_0^{2\pi} |W(r)|^2 f^4(\theta,\phi) r^2 \sin\theta d\theta d\phi dr \tag{3.10}$$

which represents the expected power from scatterers centered at range r_0 and extending over the interval $r_0 - \Delta_1$ to $r_0 + \Delta_2$ where $|W(r)| \neq 0$. Assuming a Gaussian antenna pattern and a rectangular range weighting function, the integral reduces to

$$r_0^2 \Delta \int_0^{\pi} \int_0^{2\pi} f^4(\theta,\phi) \sin\theta d\theta d\phi = r_0^2 \frac{\pi c \tau \theta_1^2}{16 \ln 2}, \quad \text{where } \Delta = \Delta_1 + \Delta_2 = c\tau/2. \tag{3.11}$$

Then the weather radar equation for a *rectangular pulse* and no filtering by the receiver (digital or analogue) can be obtained by substituting (3.11) in (3.10). The equation needs to include the *actual transmitted pulse shape and receiver filter effect* (Doviak and Zrnic 2006); these cause small but nontrivial loss, l_r (r stands for this range weighting function loss). The loss is

$$l_r = \frac{c\tau}{2} / \int_r |W(r)|^2 dr, \tag{3.12}$$

and for a rectangular pulse and a matched filter (with impulse response equal to the pulse shape), the $l_r = 1.5$ (or 1.76 in dB) (Doviak and Zrnic 2006).

With this modification, the weather radar equation can be written as

$$< P_{\rm hh} >= \frac{P_h^t g^2 \lambda^2 c\tau\theta_1^2}{2^8 \pi r^2 l_h^2(r) l_r \ln(2)} < |s_{\rm hh}|^2 > , \tag{3.13}$$

where r is the range to the resolution volume (or sampling volume) center and $<|s_{\rm hh}|^2>$ is the ensemble (probabilistic) average representative of the scatterers at that range. This radar equation can be expressed in terms of the backscatter cross section using the relation (1.30).

Equations for the ensemble average of the other two powers (3.7b and 3.7c) have the same form, and appropriate change of subscripts in (3.13) produces either one. A minor deviation is the expression for $<P_{\rm hv}>$ that has the product $l_h l_v$ in the denominator.

3.2.2 Correlations

Two additional significant measurements are routinely made with Doppler polarimetric radars. These are the correlations

$$\hat{R}_{\rm hv} = \overline{V_{\rm hh}^* V_{\rm vv}} = \frac{1}{M} \sum_{m=1}^{M} V_{\rm hh}^*(mT_s) V_{\rm vv}(mT_s) \tag{3.14a}$$

and

$$\hat{R}_{\rm hh}(T_s) = \overline{V_{\rm hh}^*(mT_s) V_{\rm hh}[(m+1)T]_s} = \frac{1}{M} \sum_{m=1}^{M-1} V_{\rm hh}^*(mT_s) V_{\rm hh}[(m+1)T_s]. \tag{3.14b}$$

The derivation applied to obtain the powers holds for $\hat{R}_{\rm hv}$, but particular attention must be given to the ensemble average of the scattering coefficient products. Ensemble average of the cross products (scatterer pairs) vanishes, but the self-product term has the proportionality:

$$\hat{R}_{\rm hv} \propto s_{\rm hh}^{*(i)} s_{\rm vv}^{(i)} \exp\left\{ j \left[\Phi_{\rm DP}^{(\rm pro)}(r_i) + \Phi_{\rm DP}^{(\rm sys)} \right] \right\}.$$

The $\Phi_{\rm DP}^{(\rm pro)}(r_i)$ term accumulates slowly with range and in the ensemble average can be assumed constant within the resolution volume (same as the $< s_{\rm hh}^{*(i)} s_{\rm vv}^{(i)} >$ term), so r_i is replaced with the value at the center of V_6 and the ensemble average of the correlation becomes

$$< R_{\text{hv}} >= \frac{\sqrt{P_h^t P_v^t} g^2 \lambda^2 c\tau\theta_1^2}{2^8 \pi r^2 l_h(r) l_v(r) l_r \ln(2)} < s_{\text{hh}}^* s_{\text{vv}} > \exp\left\{ j \left[\Phi_{\text{DP}}^{(\text{pro})}(r) + \Phi_{\text{DP}}^{(\text{sys})} \right] \right\}, \quad (3.15)$$

where $l_h l_v$ indicates a product of amplitude attenuation factors of H and V fields. A special case when Φ_{DP} and $<s_{\text{hh}}{}^* s_{\text{vv}}>$ vary within the resolution volume will be considered in Chap. 6 (Sect. 6.8). It is assumed that both receivers are calibrated; hence no differential gain is present in (3.15), but even if it were, it would not matter because the normalized form (division by $<P_{\text{hh}}>^{1/2} <P_{\text{vv}}>^{1/2}$) is used to characterize scatterers. The $< s_{\text{hh}}^* s_{\text{vv}} >$ term is influenced by hydrometeors' wobbling, rotation, oscillation, melting, and evolution of size distribution (e.g., via coalescence or breakup of particles).

Ensemble average of the autocorrelation $<R_{\text{hh}}(T_s)>$ involves similar derivation and summation over single scatterer contribution (single sum term in (3.6)):

$$\begin{aligned} < R_{\text{hh}}(T_s) > &= < V_{\text{hh}}^*(mT_s) V_{\text{hh}}[(m+1)T_s] > \\ &\approx |B|^2 \sum_i F_i^2 |G_i|^2 \left| s_{\text{hh}}^{(i)} \right|^2 \exp[-j2\text{Re}(k_h) v_i T_s], \end{aligned} \quad (3.16)$$

where v_i is the radial velocity of the ith scatterer. Assuming Gaussian distribution of radial velocities within the radar resolution volume,

$$p(v_i) = \frac{1}{\sqrt{2\pi}\sigma_v} \exp\left[-\frac{(v_i - < v >)^2}{2\sigma_v^2} \right] \quad (3.17)$$

and replacing summation in (3.16) by integration yields

$$< R_{\text{hh}}(T_s) >= < P_{\text{hh}} > \rho(T_s) \exp[-j2\text{Re}(k_h) < v > T_s]. \quad (3.18)$$

In (3.18),

$$\rho(T_s) = \exp\left\{ -2[\text{Re}(k_h)]^2 \sigma_v^2 T_s^2 \right\} = \exp\left(-8\pi^2 \sigma_v^2 T_s^2 / \lambda^2 \right). \quad (3.19)$$

The Doppler velocity is obtained from the phase of $<R_{\text{hh}}(T_s)>$ as

$$< v >= -\frac{\lambda}{4\pi T_s} \arg[< R_{\text{hh}}(T_s) >], \quad (3.20)$$

and the Doppler spectrum width σ_v can be determined from the ratio of the magnitude of $<R_{\text{hh}}(T_s)>$ and $<P_{\text{hh}}>$ (Doviak and Zrnic 2006; Melnikov and Doviak 2002; Zhang 2016):

$$\sigma_v = -\frac{\lambda}{2\sqrt{2}\pi T_s} \ln\left(\frac{|< R_{\text{hh}}(T_s) >|}{< P_{\text{hh}} >}\right). \tag{3.21}$$

Assume that the receiver and transmitter are well calibrated, i.e., the systematic offsets in gains, transmitted powers, etc. are accounted for. Then the system constant appearing in the expressions of various variables is

$$C_1 = \frac{P^t g^2 \lambda^2 c\tau \theta_1^2}{2^8 \pi l_r \ln(2)}. \tag{3.22}$$

With this, a succinct summary of the equations for powers and correlations (ensemble averages) in dual-polarization measurements follows.

3.2.2.1 Simultaneous SHV Mode (No Coupling on Propagation Path)

$$< P_{\text{hh}}(r_0) >= \frac{C_1}{r_0^2 l_h^2(r_0)} < |s_{\text{hh}}(r_0)|^2 >, \tag{3.23}$$

$$< P_{\text{vv}}(r_0) >= \frac{C_1}{r_0^2 l_v^2(r_0)} < |s_{\text{vv}}(r_0)|^2 >, \tag{3.24}$$

$$\begin{aligned} < R_{\text{hv}}(r_0) >= & \frac{C_1}{r_0^2 l_h(r_0) l_v(r_0)} < s_{\text{hh}}^*(r_0) s_{\text{vv}}(r_0) > \\ & \exp\left\{ j\left(\Phi_{\text{DP}}^{(\text{pro})}(r_0) + \Phi_{\text{DP}}^{(\text{sys})}(r_0)\right)\right\}. \end{aligned} \tag{3.25}$$

3.2.2.2 H Transmitted and Simultaneous H and V Received, HSHV Mode of Operations

In addition to the powers (3.23) and (3.24), this mode provides

$$< P_{\text{hv}}(r_0) >= \frac{C_1}{r_0^2 l_h(r_0) l_v(r_0)} < |s_{\text{hv}}(r_0)|^2 >, \tag{3.26}$$

$$\begin{aligned} < R_{\text{xh}}(r_0) >= & \frac{C_1}{r_0^2 l_h(r_0) l_v(r_0)} < s_{\text{hh}}^*(r_0) s_{\text{vh}}(r_0) > \\ & \exp\left[j\Phi_{\text{DP}}^{(\text{pro})}(r_0)/2 + j\Phi_{\text{DP}}^{(\text{rec})}\right], \end{aligned} \tag{3.27}$$

where the subscript x indicates the cross-polar component correlated with V_{hh} (as in Ryzhkov 2001), and $\Phi_{\text{DP}}^{(\text{rec})} = \varphi_h^r - \varphi_v^r$ is the system differential phase on reception.

3.2.2.3 V Transmitted and Simultaneous V and H Received, VSVH

In addition to the (3.23) and (3.26) terms, this mode provides

$$< R_{\mathrm{xv}}(r_0) >= \frac{C_1}{r_0^2 l_h(r_0) l_v(r_0)} < s_{\mathrm{vv}}^*(r_0) s_{\mathrm{vh}}(r_0) > \exp\left[-j\Phi_{\mathrm{DP}}(r_0)/2 - j\Phi_{\mathrm{DP}}^{(\mathrm{rec})}\right], \tag{3.28}$$

where $s_{\mathrm{vh}}(r_0) = s_{\mathrm{hv}}(r_0)$ is assumed.

The range (r_0) locates the center of the resolution volume and will be implied in subsequent sections. The ensemble average is the true value intrinsic to the V_6 and *per unit volume*. The other variables l_h (r_0), $l_v(r_0)$, and $\Phi_{\mathrm{DP}}(r_0)$ exhibit cumulative and generally nonlinear dependence on the medium up to the range r_0.

All information dual-polarization radar can provide about intrinsic properties of the medium, i.e., combinations of $< s_{\mathrm{mn}}^* s_{\mathrm{kl}} >$ are contained in the six equations (3.23)–(3.28). For accounting purposes, these can be put in a 3 × 3 covariance matrix (Zrnic 1991) by bracketing each term in (2.29) with $< >$. A formal use of this matrix postulating power conservation has been proposed by Tragl (1990), but not often practiced. In the measured powers and correlations, the intrinsic properties are coupled with propagation effects and need to be removed for quantifying the former.

3.3 Polarimetric Variables: Definitions

Polarimetric variables are related to the ensemble averages of various products of the backscattering matrix elements. Which polarimetric variable can be measured depends on the hardware configuration and these are discussed here.

3.3.1 Simultaneous Transmission/Reception (SHV) Mode

This mode is operational on radars belonging to several national meteorological services and is the one we concentrate on. Hence, we define the polarimetric variables measurable in it. The fundamental measurements are listed in Eqs. (3.23)–(3.25) and the intrinsic quantities contained in these can be conveniently visualized as vertexes in the triangular matrix (2.29). Purposeful combination of the three can reveal more about the precipitation than is possible from each alone. The list of useful single and combined polarimetric variables follows.

3.3.1.1 Equivalent Reflectivity Factor

We start with (3.23) and express $<|s_{hh}|^2>$ via the radar equivalent reflectivity factor Z_{hh}

$$Z_{hh} = \frac{4\lambda^4}{\pi^4 |K_w|^2} < |s_{hh}|^2 > , \tag{3.29}$$

where $K_w = \frac{\varepsilon_w - 1}{\varepsilon_w + 2}$ and ε_w is the dielectric constant of water (Chap. 4) so that at microwave frequencies (wavelengths 0.01–0.1 m) $|K_w|^2$ is between 0.91 and 0.93. Combining (3.22), (3.23), and (3.29) yields a standard form of the weather radar equation (Doviak and Zrnic 2006):

$$P_{hh} = \left(\frac{\pi^3 P_h^t g^2 c\tau\theta_1^2}{2^{10} \ln 2 \lambda^2 l_r} \right) \frac{|K_w|^2}{r_0^2 l_h^2} Z_{hh}. \tag{3.30}$$

In (3.30), the term within the brackets is an absolute constant and the $|K_w|^2 = 0.93$ is used regardless of what types of scatterers (rain, snow, insects, and other) are in the radar resolution volume V_6. Usually in real-time computations, attenuation due to propagation through rain is ignored and the ensuing attenuated reflectivity is

$$\bar{Z}_{hh} = Z_{hh}/l_h^2. \tag{3.31}$$

Correcting $\bar{Z}_{hh}$ for attenuation is not trivial and the methods for such correction are discussed in Chap. 6. Methods for attenuation correction relying on the differential phase are particularly efficient. Throughout the remainder of this monograph, the subscript on reflectivity at H polarization is omitted unless there is a reason to contrast it with V polarization.

3.3.1.2 Differential Reflectivity

Differential reflectivity estimate (Seliga and Bringi 1976) is

$$\bar{Z}_{dr} = \frac{P_{hh}}{P_{vv}} = \frac{l_v^2}{l_h^2} \frac{< |s_{hh}|^2 >}{< |s_{vv}|^2 >} = \frac{l_v^2}{l_h^2} Z_{dr}. \tag{3.32}$$

As in (3.31), the designation in (3.32) separates the measured differential reflectivity ($\bar{Z}_{dr}$) affected by differential attenuation (quantified by the ratio l_v^2/l_h^2) from the intrinsic (true) value (Z_{dr}). The true value in dB units is

$$Z_{\mathrm{DR}} = 10\log\left(\frac{< |s_{\mathrm{hh}}|^2 >}{< |s_{\mathrm{vv}}|^2 >}\right). \tag{3.33}$$

Thus in logarithmic (dB) units (3.32) becomes

$$\bar{Z}_{\mathrm{DR}} = 10\log\left(\bar{Z}_{\mathrm{dr}}\right) = Z_{\mathrm{DR}} - 20\log(l_h/l_v). \tag{3.34}$$

We emphasize that the attenuation factors are cumulative functions of range not necessarily linearly related to r. Concerning notation throughout this book, the uppercase subscript designates logarithmic units and the lowercase linear units.

Through the ratio (3.32), the concentration (total number) of particles per unit volume is eliminated from Z_{dr}, therefore this variable relates to the remaining bulk precipitation properties. This is seen from the definition of the two intrinsic parameters of scatterers $<|s_{\mathrm{hh}}|^2>$ and $<|s_{\mathrm{vv}}|^2>$. Each is proportional to the same concentration (number per unit volume) because it refers to the same set of particles. Thus Z_{dr} conveys information about the average shape, orientation, and composition of particles in the resolution volume.

3.3.1.3 Cross-Correlation Coefficient

The complex cross-correlation coefficient ρ_{hv} (Sachidananda and Zrnic 1985) combines the three second-order temporal moments (measured in the SHV mode) into a single value:

$$\bar{\rho}_{\mathrm{hv}} = \frac{< R_{\mathrm{hv}} >}{(< P_{\mathrm{hh}} >< P_{\mathrm{vv}} >)^{1/2}}. \tag{3.35}$$

Following (3.23)–(3.25),

$$\bar{\rho}_{\mathrm{hv}} = \frac{< s_{\mathrm{hh}}^* s_{\mathrm{vv}} >}{\left(< |s_{\mathrm{hh}}|^2 >< |s_{\mathrm{vv}}|^2 >\right)^{1/2}} \exp\left[j\left(\Phi_{\mathrm{DP}}^{(\mathrm{pro})} + \Phi_{\mathrm{DP}}^{(\mathrm{sys})}\right)\right] \tag{3.36}$$

Similarly to differential reflectivity, we distinguish the measured value of the cross-correlation coefficient from its intrinsic value defined as

$$\rho_{\mathrm{hv}} = \frac{< s_{\mathrm{hh}}^* s_{\mathrm{vv}} >}{\left(< |s_{\mathrm{hh}}|^2 >< |s_{\mathrm{vv}}|^2 >\right)^{1/2}}. \tag{3.37}$$

In (3.37), the argument of ρ_{hv} is the backscatter differential phase δ (2.26):

$$\delta = \arg < s_{\text{hh}}^* s_{\text{vv}} > . \tag{3.38}$$

Analogous to Z_{dr}, ρ_{hv} is independent of concentration because it cancels in (3.37). It is primarily influenced by the hydrometeors' shape and canting angle distributions (Sachidananda and Zrnic 1985). The backscatter differential phase (3.38) is insignificant at sizes small compared to wavelength but can be substantial in hail (Balakrishnan and Zrnic 1990) or wet snow (Chap. 5).

3.3.2 *HSHV and VSVH Modes*

In the HSHV mode, two independent measurements (unavailable in the SHV mode) can be made. From the P_{vh} the cross-polar reflectivity factor Z_{vh} can be computed but is not used. Rather it is the ratio $\bar{L}_{\text{drh}} = P_{\text{vh}}/P_{\text{hh}} = \bar{Z}_{\text{vh}}/\bar{Z}_{\text{hh}}$ (intrinsic $L_{\text{drh}} = P_{\text{vh}}/P_{\text{hh}} = Z_{\text{vh}}/Z_{\text{hh}}$) called a linear depolarization ratio that is often examined. Linear depolarization ratio is usually expressed in logarithmic units: $LDR(\text{dB}) = 10\log(\bar{L}_{\text{drh}})$. The $\left(V_{\text{hh}}^* V_{\text{vh}}\right)$ covariance produces the correlation coefficient ρ_{xh} (Zrnic 1991; Ryzhkov 2001). For completeness we list these as

$$L_{\text{drh}} = P_{\text{vh}}/P_{\text{hh}} = \left(\frac{< |s_{\text{vh}}|^2 >}{< |s_{\text{hh}}|^2 >}\right) l_h/l_v = L_{\text{drh}} l_h/l_v \tag{3.39}$$

and

$$\rho_{\text{xh}} = \frac{< s_{\text{hh}}^* s_{\text{vh}} >}{\left(< |s_{\text{hh}}|^2 >< |s_{\text{vh}}|^2 >\right)^{1/2}} \exp\left[j\left(\Phi_{\text{DP}}^{(\text{pro})}/2 + \Phi_{\text{DP}}^{(\text{rec})}\right)\right]. \tag{3.40}$$

This intrinsic ρ_{xh} is obtained directly from the corresponding correlation of voltages and from powers because in its computation, the attenuation factors cancel.

In the VSVH mode, well-calibrated radar has $P_{\text{hv}} = P_{\text{vh}}$; therefore the linear depolarization ratio $L_{\text{drv}} = L_{\text{drh}} Z_{\text{dr}}$ is redundant. The only additional information is in the co-cross-polar correlation coefficient (Ryzhkov et al. 2002):

$$\rho_{\text{xv}} = \frac{< s_{\text{vv}}^* s_{\text{hv}} >}{\left(< |s_{\text{vv}}|^2 >< |s_{\text{vh}}|^2 >\right)^{1/2}} \exp\left[-j\left(\Phi_{\text{DP}}^{(\text{pro})}/2 + \Phi_{\text{DP}}^{(\text{rec})}\right)\right]. \tag{3.41}$$

All possible measurements of intrinsic properties can be compacted and conveniently arranged in the following matrix (derived from the covariance matrix (2.29)):

$$\begin{bmatrix} Z_{\rm hh} & \rho_{\rm xh} & \rho_{\rm hv} \\ & Z_{\rm hv} & \rho_{\rm xv} \\ & & Z_{\rm vv} \end{bmatrix}. \tag{3.42}$$

In (3.42), there are three real power quantities and three complex correlation coefficients for a total of nine quantities measurable with polarimetric radar. Depending on scatterer types, some of these are redundant. Ways to make measurements of all these have been devised (Zrnic 1991) and those tested use sequential transmission of H and V waves with simultaneous reception of the two orthogonally polarized components of the radar return (Lutz et al. 1997; Brunkow et al. 2000). In such sequential transmissions, the differential phase and Doppler phase shift are coupled and need to be separated (Sachidananda and Zrnic 1989). In the SHV mode, the variables on the vertices of the large triangle in (3.42) are measured, while in the HSHV (VSVH) modes, the variables in the upper (lower) small triangles are available.

3.4 Effects of Particle Orientations

The backscatter polarimetric variables Z, $Z_{\rm DR}$, LDR, and $\rho_{\rm hv}$ depend on the scattering properties of the ensemble of reflecting hydrometeors via second-order moments $<|s_{\rm hh}|^2>$, $<|s_{\rm vv}|^2>$, $<|s_{\rm hv}|^2>$, and $<s_{\rm hh}^* s_{\rm vv}>$, whereas the propagation parameters $K_{\rm DP}$, A_h, and $A_{\rm DP}$ are functions of first-order moments of the forward-scattering matrix $<s_{\rm hh}^{(0)}>$ and $<s_{\rm vv}^{(0)}>$. These moments are determined by the distributions of hydrometeors' sizes, shapes, orientations, and phase composition. The dependencies of the scattering matrix **S** elements on particle orientations are explicitly described by (1.27) for the cases of backward and forward scattering. The canting angle α and angle ψ (Fig. 1.5) determine scatter properties which can be described by distribution functions. Hence ensemble averages of second-order moments provide the bulk property as would be observed with the radar. Assuming distributions of particle orientations and particle size, shapes, and dielectric constants are independent and multiple scattering is insignificant, the expressions for the second-order moments are (Ryzhkov 2001)

$$\begin{aligned}
&<|s_{\rm hh}|^2>=<|s_b|^2> -2{\rm Re}<s_b^*(s_b-s_a)>A_2+<|s_b-s_a|^2>A_4\\
&<|s_{\rm vv}|^2>=<|s_b|^2> -2{\rm Re}<s_b^*(s_b-s_a)>A_1+<|s_b-s_a|^2>A_3\\
&<|s_{\rm hv}|^2>=<|s_b-s_a|^2>A_5\\
&<s_{\rm hh}^* s_{\rm vv}>=<|s_b|^2>+<|s_b-s_a|^2>A_5-<s_b^*(s_b-s_a)>A_1\\
&\qquad -<s_b(s_b^*-s_a^*)>A_2
\end{aligned} \tag{3.43}$$

and for the first-order moments of the forward scattering matrix

$$\begin{aligned} &< s_{\mathrm{hh}}^{(0)} >=< s_b^{(0)} > - < s_b^{(0)} - s_a^{(0)} > A_2 \\ &< s_{\mathrm{vv}}^{(0)} >=< s_b^{(0)} > - < s_b^{(0)} - s_a^{(0)} > A_1 \end{aligned} \tag{3.44}$$

where angular moments A_1–A_5 are defined by

$$\begin{aligned} &A_1 =< \sin^2\psi \cos^2\alpha >, \quad A_2 =< \sin^2\psi \sin^2\alpha >, \quad A_3 =< \sin^4\psi \cos^4\alpha >, \\ &A_4 =< \sin^4\psi \sin^4\alpha >, \quad A_5 =< \sin^4\psi \cos^2\alpha \sin^2\alpha >. \end{aligned} \tag{3.45}$$

In (3.43)–(3.45), the ensemble averages are over particle sizes, shapes, and dielectric constants and $s_{a,b} = s_{a,b}^{(\pi)}$ and $s_{a,b}^{(0)}$ are backward and forward scattering amplitudes along the symmetry axis of spheroid a and its transverse axis b (Chap. 1).

Simple formulas expressing angular moments exist for three special cases: (a) completely random orientation of hydrometeors, (b) random orientation in the horizontal plane, and (c) two-dimensional axisymmetric Gaussian distribution. In the latter case, the mean orientation of the hydrometeors is in the direction $(\overline{\psi}, \overline{\alpha})$, and the width of the angular distribution is determined by a dispersion parameter σ. If the angle $\overline{\psi}$ is not very close to zero and the width of the angular distribution is not very large, then axisymmetric Gaussian distribution of orientations can be approximated by

$$p(\psi, \alpha) = \frac{1}{2\pi\sigma\sigma_\alpha} \exp\left[-\frac{(\psi - \overline{\psi})^2}{2\sigma^2} - \frac{(\alpha - \overline{\alpha})^2}{2\sigma_\alpha^2} \right], \tag{3.46}$$

where the parameters σ and $\sigma_\alpha = \sigma / \sin\overline{\psi}$ define the width of the angular distribution along ψ and α directions (see Fig. 1.5). This type of angular distribution represents a wide class of atmospheric scatterers that can be modeled as oblate spheroids, including raindrops, snowflakes, graupel, and hail. The expressions for angular moments in all three cases are in the Appendix B.

Next we consider three special cases of particle orientations and their impact on polarimetric radar variables.

3.4.1 *Completely Random Orientation of Hydrometeors*

From the Appendix B, assuming $\theta_v = 0$ in (B.4), $A_1 = A_2 = 1/3$, $A_3 = A_4 = 1/5$, and $A_5 = 1/15$. Therefore, from (3.43) and (3.45)

$$< |s_{\mathrm{hh}}|^2 >=< |s_{\mathrm{vv}}|^2 >, \quad < s_{\mathrm{hh}}^{(0)} >=< s_{\mathrm{vv}}^{(0)} >, \quad \mathrm{Im} < s_{\mathrm{hh}}^* s_{\mathrm{vv}} >= 0 \tag{3.47}$$

and it follows from (3.33), (3.38), (1.49), and (1.51) that $Z_{\mathrm{DR}}=0$ dB, $\delta=0$, $A_{\mathrm{DP}}=0$, and $K_{\mathrm{DP}}=0$. Regardless of the sizes and shapes of hydrometeors, differential reflectivity, backscatter differential phase, specific differential attenuation, and specific differential phase for the ensemble of randomly oriented particles are equal to zero. Nonetheless, such orientation depolarizes significantly the backscattered EM waves creating substantial cross-polar component of the radar return (P_{vh}) and therefore also high linear depolarization ratio (L_{dr}). In this case, from (3.43) and basic definitions, the variables ρ_{hv} and L_{dr} are related by

$$\rho_{\mathrm{hv}} = 1 - 2L_{\mathrm{dr}}. \tag{3.48}$$

Propagation through the medium filled with randomly oriented scatterers does not alter the polarization state of EM waves.

3.4.2 Noncanted Hydrometeors (α = 0, σ = 0)

If all particles are equally oriented and the axis of rotation is vertical, the mean canting angle $\overline{\alpha}$ is zero. The angular moments are given by the formulas (B.7) in Appendix B, where the parameter $r_\sigma = \exp(-2\sigma^2)$ is equal to 1 because $\sigma = 0$ and $A_1 = A_3 = 1$, $A_2 = A_4 = A_5 = 0$. In this arrangement,

$$\begin{aligned} &< |s_{\mathrm{hh}}|^2 > = < |s_b|^2 >, \quad < |s_{\mathrm{vv}}|^2 > = < |s_a|^2 >, \quad < |s_{\mathrm{hv}}|^2 > = 0, \\ &< s_{\mathrm{hh}}^{(0)} > = < s_b^{(0)} >, \quad < s_{\mathrm{vv}}^{(0)} > = < s_a^{(0)} > \end{aligned} \tag{3.49}$$

and

$$\begin{aligned} &Z_{\mathrm{dr}} = \frac{< |s_b|^2 >}{< |s_a|^2 >}, \quad K_{\mathrm{DP}} = \lambda \mathrm{Re}\left(< s_b^{(0)} > - < s_a^{(0)} >\right), \\ &A_{\mathrm{DP}} = -8.686\lambda \mathrm{Im}\left(< s_b^{(0)} > - < s_a^{(0)} >\right), \end{aligned} \tag{3.50}$$

and the linear depolarization ratio L_{dr} is zero (i.e., there is no depolarization by backscattering).

3.4.3 Gaussian Distribution of Particle Orientations (α = 0, σ ≠ 0)

Here we examine the impact of the width of the canting angle distribution σ on the polarimetric variables. For the sake of simplicity, we assume all scatterers are oblate spheroids and have the same size, shape, and dielectric constant; therefore

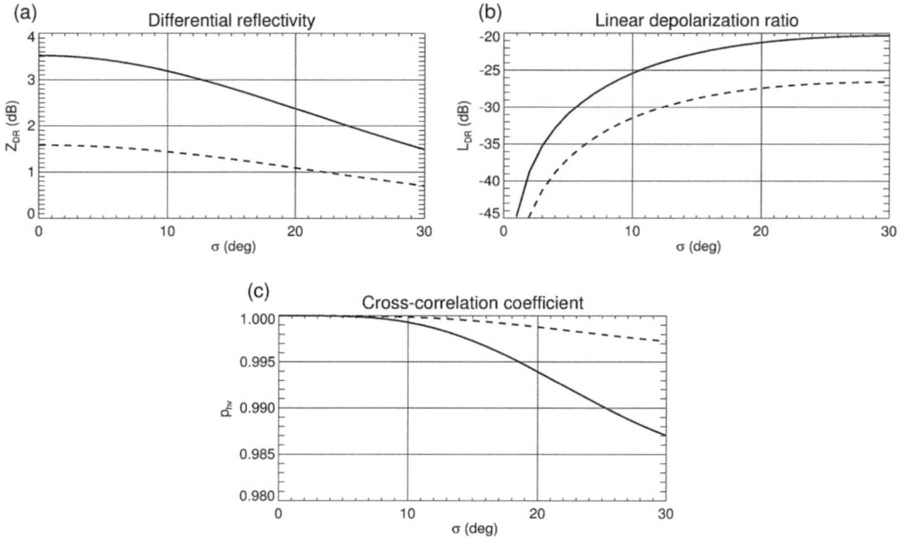

Fig. 3.4 Dependencies of (**a**) Z_{DR}, (**b**) L_{DR}, and (**c**) ρ_{hv} on the width of the canting angle distributions σ for $s_b/s_a = 1.5$ (solid curve) and for $s_b/s_a = 1.2$ (dashed curve)

$$\frac{< |s_b|^2 >}{< |s_a|^2 >} = \left(\frac{s_b}{s_a}\right)^2. \tag{3.51}$$

We also assume the ratio of the scattering amplitudes s_b/s_a is real. The dependencies of Z_{DR}, L_{DR}, and ρ_{hv} on σ for $s_b/s_a = 1.2$ and 1.5 are computed using Eqs. (3.43) and (B.9) and the corresponding plots are in Fig. 3.4. It is evident the increase in the randomness of the scatterers' orientations causes decrease of Z_{DR} and ρ_{hv} and increase in *LDR*. The dependencies on the degree of common alignment are more pronounced for more oblate hydrometeors (at a given refractive index), i.e., for larger ratio s_b/s_a. This general fact is important in interpretation of polarimetric data.

Further analysis of (3.44) shows that the polarimetric propagation variables K_{DP} and A_{DP} depend linearly on the difference of the angular moments $(A_1 - A_2)$ as

$$\begin{aligned} K_{DP} &= \lambda \mathrm{Re}\left(< s_b^{(0)} > - < s_a^{(0)} >\right)(A_1 - A_2) \\ A_{DP} &= -8.686\lambda \mathrm{Im}\left(< s_b^{(0)} > - < s_a^{(0)} >\right)(A_1 - A_2), \end{aligned} \tag{3.52}$$

where

$$A_1 - A_2 = \frac{1}{2} r_\sigma (1 + r_\sigma) \tag{3.53}$$

and

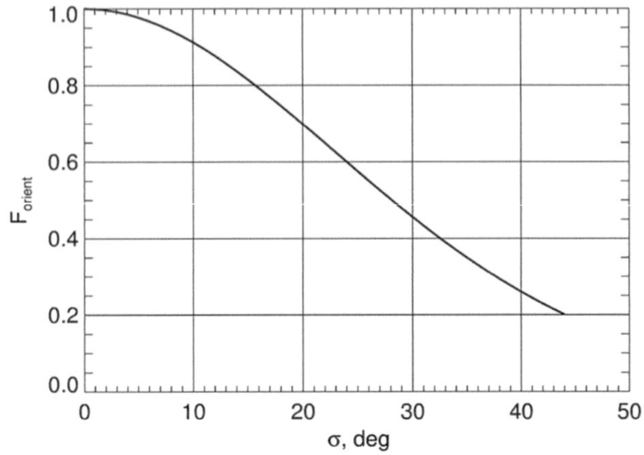

Fig. 3.5 Orientation factor F_{orient} as a function of σ

$$r_\sigma = \exp\left(-2\sigma^2\right) \tag{3.54}$$

(see Appendix B).

Equation (3.53) is valid for relatively narrow angular distributions with $\sigma < 40$–$50°$. The dependence of the particle orientation factor $F_{\text{orient}} = A_1 - A_2$ on the width of the canting angle distribution σ is plotted in Fig. 3.5.

References

Balakrishnan, N., & Zrnic, D. (1990). Use of polarization to characterize precipitation and discriminate large hail. *Journal of the Atmospheric Sciences, 47*, 1525–1540.

Brunkow, D., Bringi, V., Kennedy, P., Rutledge, S., Chandrasekar, V., Mueller, E., et al. (2000). A description of the CSU-CHILL National Radar Facility. *Journal of Atmospheric and Oceanic Technology, 17*, 1596–1608.

Doviak, R., & Zrnic, D. (2006). *Doppler radar and weather observations* (2nd ed.). Reprint, Mineola, NY: Dover Publications, Inc., 562pp.

Ivic, R. I., Zrnic, D. S., & Torres, S. M. (2003). Whitening in range to improve weather radar spectral moment estimates. Part II: Experimental evaluation. *Journal of Atmospheric and Oceanic Technology, 20*, 1449–1459.

Lutz, J., Rilling, B., Wilson, J., Weckwerth, T., & Vivekanandan, J. (1997). S-Pol after three operational deployments, technical performances, siting experiences and some data examples. In *Preprints, 28th Conference on Radar Metoerology* (pp. 286–287). Austin, TX: American Meteor Society.

Melnikov, V., & Doviak, R. (2002). Spectrum widths from echo power differences reveal meteorological features. *Journal of Atmospheric and Oceanic Technology, 19*, 1793–1810.

Ryzhkov, A. (2001). Interpretation of polarimetric radar covariance matrix for meteorological scatterers: Theoretical analysis. *Journal of Atmospheric and Oceanic Technology, 18*, 315–328.

Ryzhkov, A., Zrnic, D., Hubbert, J., Bringi, V., Vivekanandan, J., & Brandes, E. (2002). Polarimetric radar observations and interpretation of co-cross-polar correlation coefficients. *Journal of Atmospheric and Oceanic Technology, 19*, 340–354.

Sachidananda, M., & Zrnic, D. (1985). Z_{DR} measurement consideration for a fast scan capability radar. *Radio Science, 20*, 907–922.

Sachidananda, M., & Zrnic, D. (1989). Efficient processing of alternately polarized radar echoes. *Journal of Atmospheric and Oceanic Technology, 6* , 173–181.

Seliga, T., & Bringi, V. (1976). Potential use of radar differential reflectivity measurements at orthogonal polarizations for measuring precipitation. *Journal of Applied Meteorology, 15* , 69–76.

Torres, S., & Zrnic, D. (2003). Whitening in range to improve weather radar spectral moment estimates. Part I: Formulation and simulation. *Journal of Atmospheric and Oceanic Technology, 20* , 1443–1448.

Tragl, K. (1990). Polarimetric radar backscattering from reciprocal random targets. *IEEE Transactions on Geoscience and Remote Sensing, 28*, 856–864.

Zhang, G. (2016). *Weather radar polarimetry* (Vol. 304). Boca Raton, FL: CRC Press.

Zrnic, D. (1991). Complete polarimetric and Doppler measurements with a single receiver radar. *Journal of Atmospheric and Oceanic Technology, 8* , 159–165.

Chapter 4
Microphysical and Dielectric Properties of Hydrometeors

Microphysical properties of hydrometeors such as size, shape, orientation, and phase composition, and distribution of these properties over an ensemble of particles determine polarimetric radar variables. In this chapter, an overview of microphysical properties of different hydrometeor types is provided. Different forms of size distributions (SD) of raindrops and ice particles are discussed, and the statistics of the key parameters of SD such as liquid or ice water content, mean volume diameter, and normalized concentration are presented. The chapter contains basic information about density of atmospheric particles and their axis ratios and orientations. Special attention is given to dielectric properties of hydrometeors including basic formulas for dielectric constant of fresh water, solid ice, and dry/wet snow, graupel, and hail.

4.1 Size Distributions

4.1.1 Raindrop Size Distributions

Liquid hydrometeors in the atmosphere can be in the form of cloud droplets with sizes typically below 0.05 mm or raindrops with sizes up to 8–9 mm. Raindrops of maximal size usually originate from melting hail and easily break up either spontaneously or by collisions with other raindrops. The raindrop size distributions are commonly modeled by a few analytical forms with the exponential one introduced by Marshall and Palmer (1948) being the simplest

$$N(D) = N_0 \exp(-\Lambda D), \tag{4.1}$$

where the slope Λ is determined by rain rate R

A. V. Ryzhkov, D. S. Zrnic, *Radar Polarimetry for Weather Observations*, Springer Atmospheric Sciences, https://doi.org/10.1007/978-3-030-05093-1_4

$$\Lambda = 4.1R^{-0.21} \tag{4.2}$$

and $N_0 = 8000\ \mathrm{m}^{-3}\ \mathrm{mm}^{-1}$. In (4.1), R is in mm h^{-1}, Λ is in mm^{-1}, and D is the equivolume diameter of the raindrop, i.e., spherical drop with the same volume or mass as nonspherical raindrop. The Marshall-Palmer DSD is parametrized by only one variable, rain rate R. A more general form of DSD is described by the three-parameter gamma distribution (Ulbrich 1983):

$$N(D) = N_0 D^{\mu} \exp(-\Lambda D) = N_0 D^{\mu} \exp[-(3.67 + \mu)D/D_0], \tag{4.3}$$

with the median volume diameter D_0 defined as

$$\frac{\pi \rho_w}{6} \int_0^{D_0} D^3 N(D) dD = \frac{LWC}{2}, \tag{4.4}$$

i.e., D_0 divides total liquid water content LWC into two equal parts. Zhang et al. (2001) suggested that the parameters Λ and μ in the gamma distribution are interdependent and introduced a "constrained gamma distribution":

$$N(D) = N_0 D^{\mu} \exp[-\Lambda(\mu) D]. \tag{4.5}$$

Although the factors Λ and μ are usually well correlated, the $\Lambda(\mu)$ dependencies may differ in different climates and regions (Cao et al. 2008).

Testud et al. (2001) and Illingworth and Blackman (2002) introduced the concept of a "normalized" distribution to describe raindrop spectra according to which the DSD concentration $N(D)$ normalized by a certain parameter N_w is an almost unique function of a "normalized" diameter which is the ratio of diameter D and the mean volume diameter D_m:

$$N(D)/N_w = F(D/D_m), \tag{4.6}$$

where

$$N_w = \frac{4^4}{\pi \rho_w} \frac{LWC}{D_m^4}, \tag{4.7}$$

$$\mathrm{LWC} = \frac{\pi \rho_w}{6} \int D^3 N(D) dD \tag{4.8}$$

is the liquid water content, and

$$D_m = \frac{\int D^4 N(D) dD}{\int D^3 N(D) dD}, \tag{4.9}$$

is the mean volume (or mass-weighted) diameter which is very close to the median volume diameter D_0:

$$D_m = \frac{4 + \mu}{3.67 + \mu} D_0, \tag{4.10}$$

for a gamma distribution (Ulbrich 1983; Bringi et al. 2003).

The stability of the shape of the normalized distribution (or function F) implies that only two parameters, N_w and D_m, are often needed to describe the DSD. In the case of exponential distribution (4.1),

$$N_w = N_0, \quad D_m = \frac{4}{\Lambda}, \tag{4.11}$$

and

$$N(D)/N_w = \exp(-4D/D_m). \tag{4.12}$$

Therefore, N_w may be interpreted as the intercept parameter of an exponential DSD with the same LWC and D_m as the actual one. For the gamma distribution (4.3),

$$N_w = N_0 D_m^{\mu} \frac{\Gamma(4+\mu)}{\Gamma(4)} \frac{4^4}{(4+\mu)^{4+\mu}}, \quad D_m = \frac{4+\mu}{\Lambda}, \tag{4.13}$$

so that

$$N(D)/N_w = f(\mu) \left(\frac{D}{D_m}\right)^{\mu} \exp\left[-(4+\mu)\frac{D}{D_m}\right], \tag{4.14}$$

where

$$f(\mu) = \frac{\Gamma(4)}{4^4} \frac{(4+\mu)^{(4+\mu)}}{\Gamma(4+\mu)}. \tag{4.15}$$

In (4.13), (4.14), and (4.15), Γ is the gamma function.

The intercept parameter N_w is defined by two physical quantities, LWC and D_m, and is more useful than N_0 which has units dependent on the factor μ (for a gamma distribution). Moreover, N_0 varies 7 orders of magnitude, whereas N_w changes 2–3 orders of magnitude and more adequately reflects the variability of total raindrop concentration. The difference between the conventional and normalized raindrop

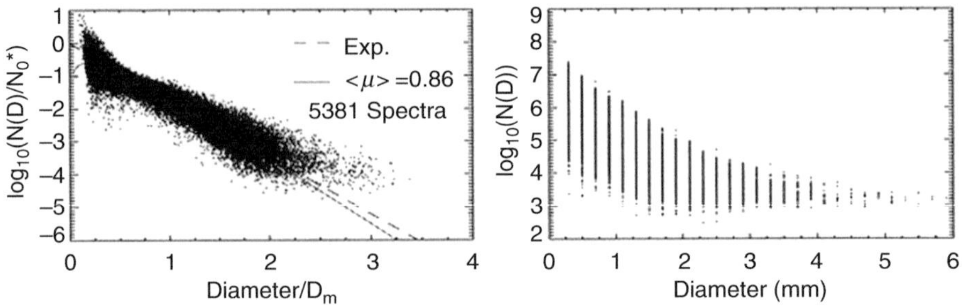

Fig. 4.1 The scatterplots of log[$N(D)/N_w$] vs D/D_m (left panel) and of log[$N(D)$] vs D for a disdrometer dataset obtained in the TOGA COARE field campaign. Adapted from Testud et al. (2001). © American Meteorological Society. Used with permission

size distributions is illustrated in Fig. 4.1 where the scatterplots of log[$N(D)/N_w$] vs D/D_m and of log([$N(D)$)] vs D are compared for a large disdrometer dataset obtained during the TOGA COARE field campaign (Testud et al. 2001). It is obvious that the scatterplot of log[$N((D)/N_w$] vs D/D_m is much tighter.

As will be shown in Chap. 10, one of the important properties of N_w is that the rain rate normalized by N_w (R/N_w) is almost uniquely determined by the reflectivity factor in linear scale or specific differential phase normalized by N_w (Z/N_w, K_{DP}/N_w) as opposed to the nonunique $R(Z)$ dependence.

A recent study by Thurai et al. (2017) shows that the classical exponential or gamma functions (4.1), (4.5), and (4.14) are reasonable fits for size distributions of raindrops exceeding 0.6–0.7 mm but may not adequately represent smaller, drizzle-size drops. They found that the so-called generalized gamma distribution (Lee et al. 2004) provides a closer fit to the size distribution measurements with a resolution of 50 microns across the continuum of raindrop sizes and highlights a drizzle mode of the DSDs.

4.1.2 *Microphysical Factors Affecting Raindrop Size Distributions*

There are a number of microphysical factors affecting precipitation production and the shape of the raindrop size distribution. These include collisions, coalescence, breakup, evaporation, melting, size distribution, density of ice particles above the freezing level, and size sorting caused by vertical air motions and wind shear. Generally, rain can be classified as "warm" if the bulk of precipitation is formed below the freezing level and "cold" if a large proportion of rain is generated from ice above the freezing level. Most of the rain in the tropics is formed via the "warm rain" process facilitated by abundant moisture, low cloud base, high freezing level, and relatively modest updrafts. Under this scenario, excessive water vapor is condensed at lower altitude, and most of the condensed water is converted into precipitation-size drops below the freezing level via collision-coalescence process. Under the "cold rain" scenario which is more typical for midlatitudes, vapor condensation usually starts at higher levels, and the bulk of moisture can be transported by stronger updrafts to the altitudes well above the melting layer so that large amount of ice can be produced either directly from the water vapor via deposition or by freezing of supercooled water droplets. Further growth of snowflakes occurs via aggregation or by accretion of supercooled cloud drops. If aggregation is a dominant process, then the density of snow right above the melting layer is low. If accretion or riming prevails, then higher-density snow or graupel is generated. Large graupel particles or frozen raindrops in the intense updrafts may grow to the size of hailstones. Big snowflakes and graupel/hail particles produce large raindrops after they melt.

Under both scenarios of the "warm" and "cold" rain, larger precipitation particles grow at the expense of smaller particles (via collision-coalescence, accretion, or aggregation). This means that the increase of the median or mean volume diameter would be generally accompanied by the decrease of total concentration of raindrops, i.e., N_w is expected to decrease with increasing D_0 or D_m. However, this is not always the case, especially for heavy tropical "warm" rain. If the processes of raindrop collision-coalescence and breakup balance each other, then DSD attains an equilibrium shape which practically does not change with further increase of total concentration of raindrops. The increase of rain amount is achieved through the increase of raindrop concentration at relatively constant median raindrop diameter (e.g., Zawadzki and De Agostinho Antonio 1988; Hu and Srivastava 1995; Atlas and Ulbrich 2000).

Multiple measurements of DSDs around the globe confirm these general features of raindrop size distributions. One of the largest datasets of DSDs measured by 2D-video disdrometer was collected in Oklahoma (Schuur et al. 2005). It contains 47,144 DSDs measured during a 7-year period. The dependence of the median value of $\log(N_w)$ on D_m for the Oklahoma dataset is shown in Fig. 4.2. Bringi et al. (2003) summarized the $\log(N_w) - D_m$ dependencies obtained from a number of climatic

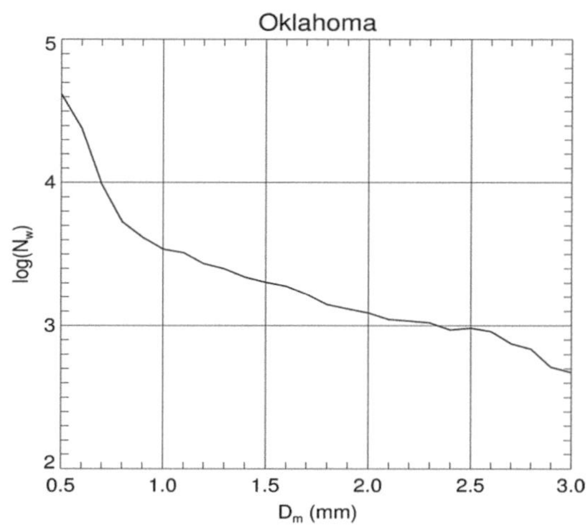

Fig. 4.2 Dependence of the median value of $\log(N_w)$ on the mean volume diameter of raindrop D_m from a large disdrometer dataset in Oklahoma

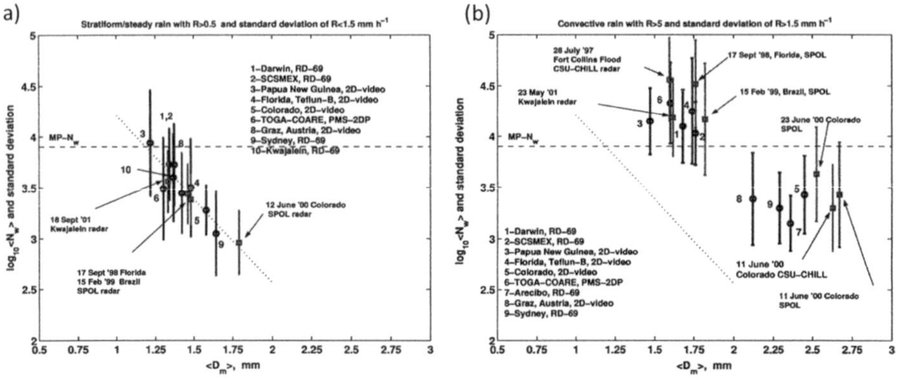

Fig. 4.3 The average value of $\log(N_w)$ (with $\pm 1\sigma$ std dev bars) vs average D_m from disdrometer data and radar retrievals for (**a**) stratiform and (**b**) convective rain. Dotted line is the fit to stratiform rain. Adapted from Bringi et al. (2003). © American Meteorological Society. Used with permission

regimes with a majority of observations taken in tropical rain (Fig. 4.3). The dependencies for stratiform rain are displayed in Fig. 4.3a, whereas the dependencies for convective rain are in Fig. 4.3b. Within each category (i.e., stratiform or convective), the intercept parameter N_w characterizing concentration decreases with increasing D_m. However, for the majority of tropical rain events, $\log(N_w)$ is significantly higher in convective rain than in stratiform rain with the same range of D_m (Bringi et al. 2009; Thompson et al. 2015; Wen et al. 2016). This is an important property of tropical rain where the equilibrium DSD is often attained for moderate-to-heavy rain. Zawadzki and De Agostinho Antonio (1988) found that the shape of DSD in tropical rain in Brazil practically remains unchanged if rain rate changes from 50 to 150 mm/h (Fig. 4.4). In another example from Brazil, Bringi et al. (2002) also showed big difference between the histograms of $\log(N_w)$ for rain rates below 10 mm h^{-1} and above 10 mm/h^{-1}, while the corresponding histograms of D_0 are almost indistinguishable for lower and higher rain rates (Fig. 4.5). In heavy tropical rain, D_m (D_0) stabilizes for very high rain rates. This is illustrated by the scatterplots of D_m (D_0) versus rain rate obtained from disdrometer measurements in Florida and Oklahoma (Fig. 4.6). Indeed, the median or mean volume diameter has average value of about 2 mm for rain rates exceeding 50 mm h^{-1} in both climate regions.

Note that rain events of a tropical type are infrequent in the US Great Plains but produce very heavy rainfall with all attributes characteristic of classical rain in the tropics. However, for the majority of continental storms producing heavy rain, the dominant shape of DSD is quite different from equilibrium which is so common in heavy tropical rain. This is illustrated in Fig. 4.7 where the Oklahoma median DSDs corresponding to different reflectivities are plotted. It is evident that for higher reflectivities of 50–55 dBZ, the raindrop size distributions in continental rain have the shape either close to exponential or gamma distribution with negative value of the shape factor μ. For $Z = 50$–55 dBZ, a slope of continental DSD is much smaller than the one of the equilibrium tropical DSD for raindrops with diameter exceeding

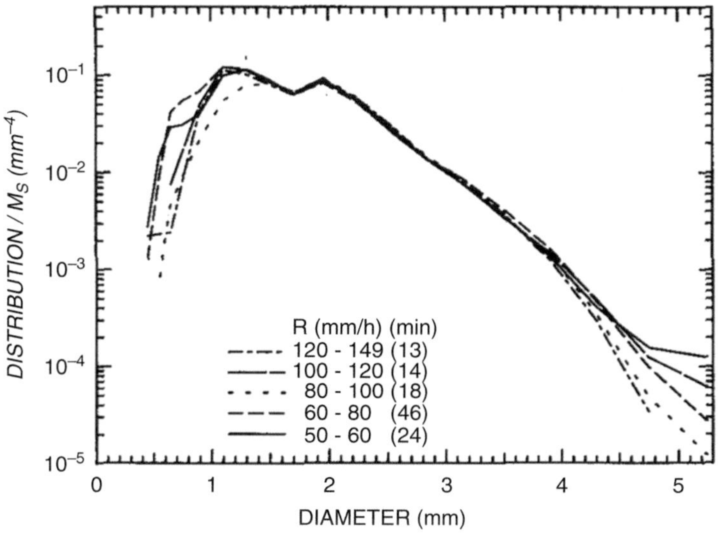

Fig. 4.4 Example of the equilibrium DSDs normalized by liquid water content M_g for different rain rates in tropical rain in Brazil. Adapted from Zawadzki and De Agostinho Antonio (1988). © American Meteorological Society. Used with permission

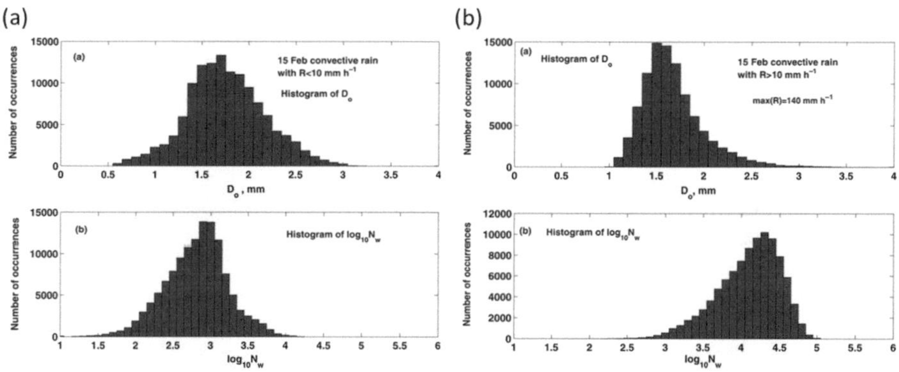

Fig. 4.5 Example of the histograms of D_0 and $\log(N_w)$ in tropical rain with (**a**) rain rate less than 10 mm/h and (**b**) rain rate higher than 10 mm h^{-1}. Adapted from Bringi et al. (2002). © American Meteorological Society. Used with permission

2 mm as comparison of Figs. 4.4 and 4.7 demonstrates. This difference is attributed to the high end of the raindrop spectrum which is formed by melting of large graupel and hail in "cold" continental rain. This process is absent in the maritime tropical environment where "warm" rain is generated in the absence of ice. It is worth noting that the continental DSDs acquire a shape closer to gamma for lower reflectivities and the shape factor μ increases with decreasing Z (Fig. 4.7). Median DSDs associated with low and high $\log(N_w)$ in Oklahoma are dramatically different (Fig. 4.8).

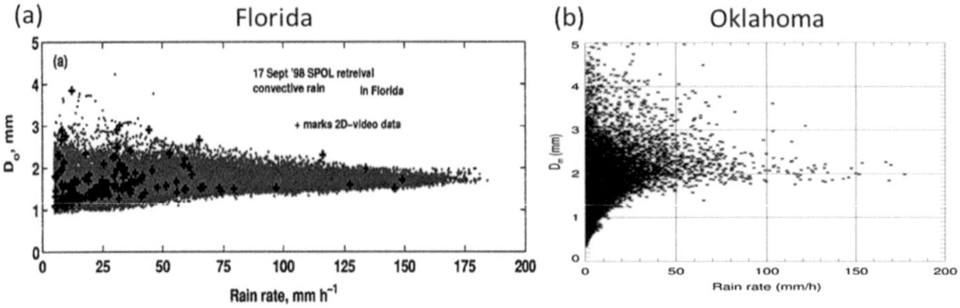

Fig. 4.6 Scatterplots of median and mean volume diameters versus rain rate: (**a**) for rain event in Florida (17 September 1998) and (**b**) for large DSD dataset in Oklahoma. Panel (**a**) is adapted from Bringi et al. (2003). © American Meteorological Society. Used with permission

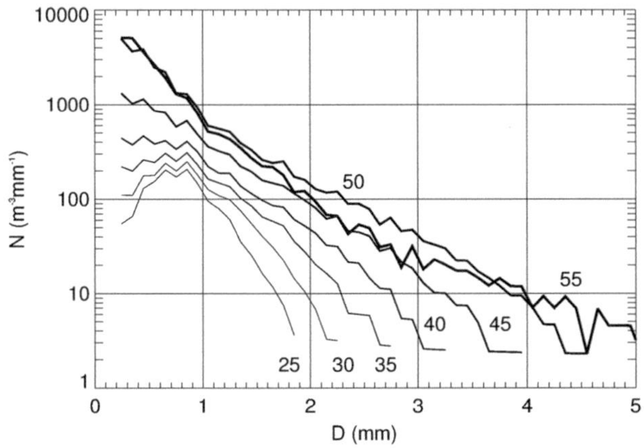

Fig. 4.7 Median DSDs for different reflectivities estimated from large Oklahoma dataset

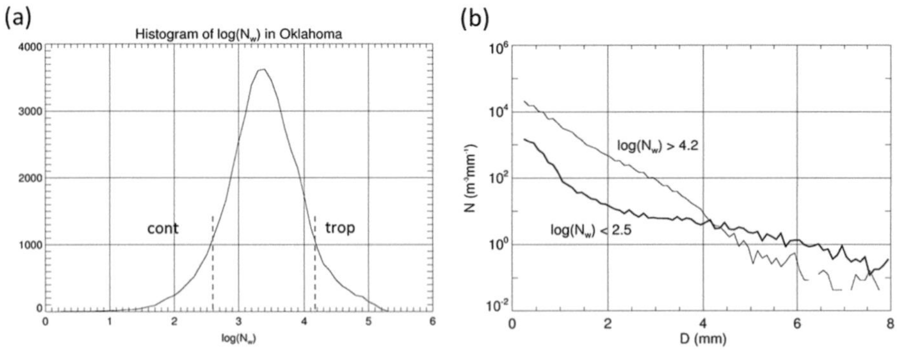

Fig. 4.8 (**a**) Histogram of $\log(N_w)$ and (**b**) median DSDs for "tropical" rain with $\log(N_w) > 4.2$ and for "continental" rain with $\log(N_w) < 2.5$ retrieved from large DSD dataset in Oklahoma

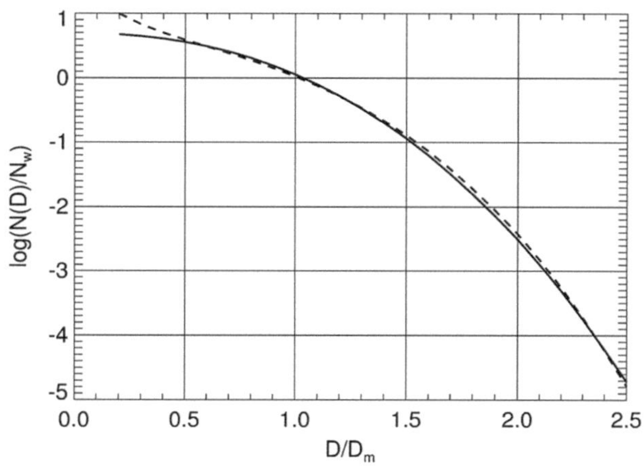

Fig. 4.9 Normalized raindrop size distributions in Korea (solid line) and Oklahoma (dashed line)

Based on the histogram of $\log(N_w)$ for the whole Oklahoma DSD dataset (Fig. 4.8a), we classify the DSDs with $\log(N_w) < 2.5$ as "very continental" rain heavily affected by melting of graupel/hail, whereas the DSDs with $\log(N_w) > 4.2$ are conditionally attributed to "very tropical" rain with little impact of graupel/hail. It is not surprising that the DSD for "continental rain" has much smaller Λ (thus, larger D_m) and intercept than the one for "tropical" rain (Fig. 4.8b). This fundamental difference in DSDs of these rain types has very important implications for polarimetric rainfall estimation.

Recent comparison of the statistics of N_w and D_m from Oklahoma and Korea (Bang and Lee 2017) shows that the median value of $\log(N_w)$ in Korea is 3.65 which is 0.38 higher than the one for Oklahoma, whereas the median diameter D_m is 0.97 mm versus 1.25 mm in Oklahoma. This reflects general difference between the more tropical rain in Korea and the more continental rain in Oklahoma. At the same time, normalized raindrop size distributions are almost indistinguishable (Fig. 4.9) which is another proof of a universal nature of the normalized size distribution defined by Eq. (4.6).

4.1.3 Size Distributions of Ice Crystals and Snowflakes

Ice crystals and snowflakes constitute a wide class of hydrometeors which are very different in their microphysical and polarimetric properties from liquid particles such as cloud droplets and rain. Pristine ice crystals are relatively small, whereas maximal size of heavily aggregated snowflakes can reach 2 or even 3 cm. Ice particles are characterized by a huge variety of shapes, orientations, and phase composition which are strongly affected by various microphysical processes. This makes polarimetric methods particularly attractive for studying such processes.

Similar to raindrops, most of ice species have their size distributions close to exponential or gamma types. In the exponential function

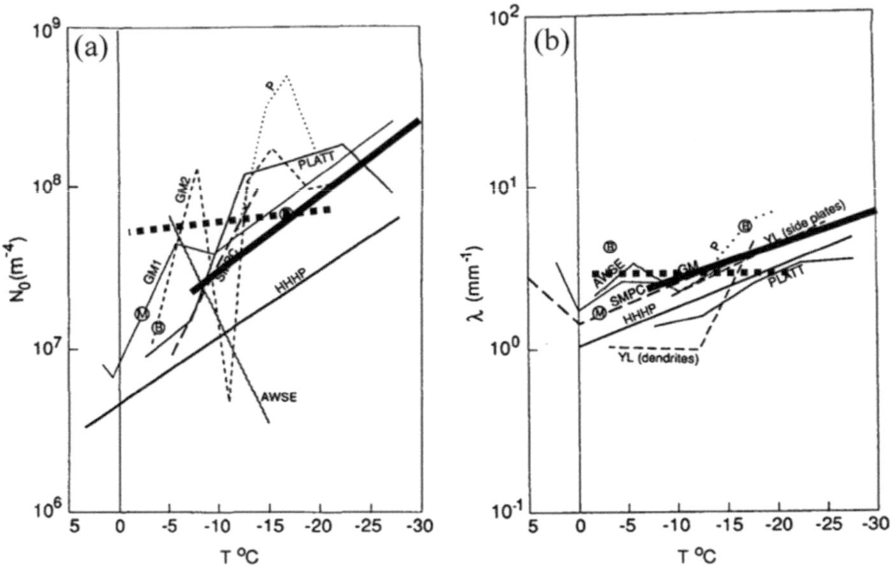

Fig. 4.10 Global ice particle size spectra parameters (**a**) N_{0s} and (**b**) Λ_s as functions of temperature, as summarized in Ryan (1996). The linear fits obtained in the study of Woods et al. (2008) are overlaid with heavy lines. The dashed heavy line is for size spectra with column or needle habits, and the solid heavy line is for size spectra without these habits. From Woods et al. (2008). © American Meteorological Society. Used with permission

$$N(D) = N_{0s}\exp(-\Lambda_s D), \tag{4.16}$$

size D means the equivolume diameter of ice crystal or snowflake, the slope Λ_s usually varies between 1 and 10 mm^{-1}, and the intercept N_{0s} can span several orders of magnitude. As opposed to raindrops, the parameters Λ_s and N_{0s} have pronounced dependence on temperature and both decrease with increasing temperature (Fig. 4.10). Ice particles forming near the top of the clouds grow by deposition, riming, and aggregation as they fall down to the ground; hence their mean volume diameter D_{ms} (inversely proportional to the intercept Λ_s) increases, and total concentration decreases mainly as a result of aggregation.

Delanoe et al. (2005, 2014) showed that a concept of normalized size distribution is applicable to ice and snow as well. In full analogy to Eq. (4.6), a normalized size distribution of ice can be represented as a function of the ratio of equivolume diameter of ice particle D (or its equivalent-melted diameter D_{eq}) and mean volume diameter D_m

$$N\left(D_{\mathrm{eq}}\right)/N_0^* = F\left(D_{\mathrm{eq}}/D_m\right), \tag{4.17}$$

where

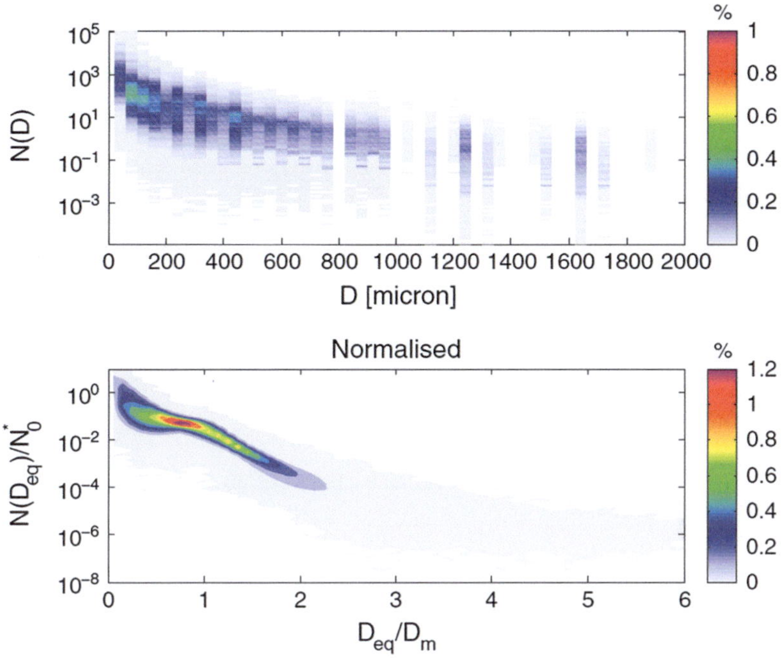

Fig. 4.11 Impact of the normalization approach on the ice particle size distribution for a large dataset obtained from aircraft in situ measurements during six field campaigns. From Delanoe et al. (2014). Copyright 2014; with permission from Elsevier

$$D_{\mathrm{eq}} = (6m/\pi\rho_w)^{1/3}, \tag{4.18}$$

and

$$N_0^* = \frac{4^4}{\pi\rho_w}\frac{IWC}{D_m^4}. \tag{4.19}$$

In (4.18) and (4.19), ρ_w is the density of water, m is the mass of the ice particle, *IWC* is the ice water content, and D_m is the mean equivalent volume diameter of melted snowflakes defined as in (4.9) for raindrops. Delanoe et al. (2014) combined in situ aircraft measurements of size distributions of ice collected in six field campaigns around the world and summarized the results using a conventional and normalization approach (Fig. 4.11). It is very clear that the overwhelming majority of size distributions collapses into a very narrow area in the $N(D_{eq})/N_0^* - D_{eq}/D_m$ plane which tells that the two parameters, N_0^* and D_m, are generally sufficient to describe an arbitrary size distribution of ice. The dependence of N_0^* on temperature shown in Fig. 4.12 is similar to the dependence of N_0 in Fig. 4.10. Its mean can be approximated by the formula (Delanoe et al. 2014):

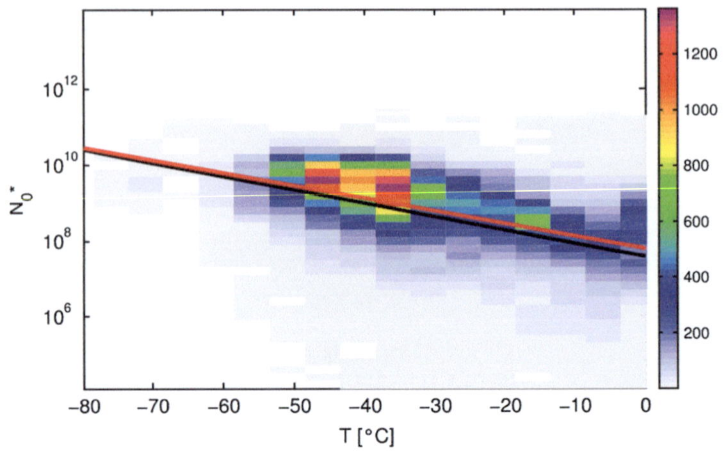

Fig. 4.12 General dependence of N_0^* on temperature. From Delanoe et al. (2014). Copyright 2014; with permission from Elsevier

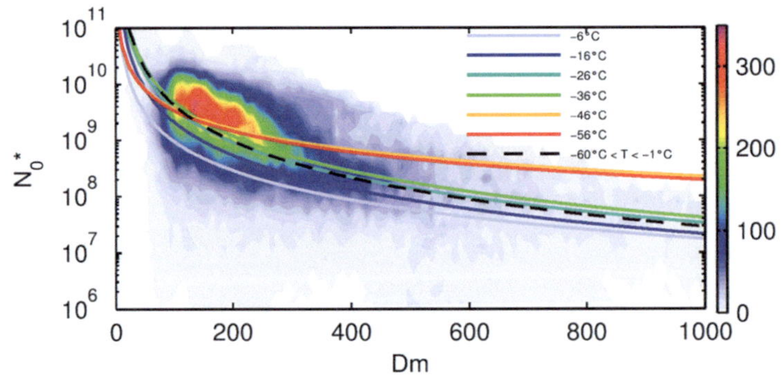

Fig. 4.13 Scatterplot of N_0^* vs D_m for different temperatures. N_0^* is expressed in m^{-4} and D_m is in microns. From Delanoe et al. (2014). Copyright 2014; with permission from Elsevier

$$\log\left(N_0^*\right) = 7.59 - 0.035T, \tag{4.20}$$

where N_0^* is expressed in m^{-4}. A scatterplot of N_0^* vs D_m is displayed in Fig. 4.13. It is evident that for a given mean volume diameter of ice particle, its concentration expressed by N_0^* increases with decreasing temperature.

Ice water content *IWC* and precipitation flux, i.e., snow rate *S*, are very important bulk parameters which characterize ice and snow, and herein we present some general information about their ranges and temperature dependencies in clouds using the results of the recent study by Heymsfield et al. (2017). Ice water content is determined from the size distribution of ice as

$$IWC = \frac{\pi}{6}\int \rho_s(D)D^3N(D)dD, \tag{4.21}$$

where $\rho_s(D)$ is the density of ice particle and *IWC* is expressed in g m^{-3}. A snow rate in mm h^{-1} is

$$S = 6\times 10^{-4}\pi \int \frac{\rho_s(D)}{\rho_w} D^3 V_t(D)N(D)dD, \tag{4.22}$$

where V_t is the terminal velocity of ice particle and ρ_w is the density of water.

Cumulative probability distributions of *IWC* and *S* as functions of temperature from large datasets are plotted in Figs. 4.14 and 4.15. Note that both *IWC* and *S* do not change much with height below the dendritic growth layer (DGL) (−20 to −10 °C) because the process of snow aggregation which is dominant below DGL tends to just redistribute snow mass across a size spectrum without changing the total mass and its flux. A more detailed description of the microphysical processes of snow formation is contained in Chap. 7, and their impact on quantitative radar estimation of ice and snow is discussed in Chaps. 10 and 11.

4.1.4 Size Distributions of Graupel and Hail

Graupel and hail are generated in convective clouds by accretion of supercooled liquid droplets on ice particles or by freezing of supercooled raindrops lofted in updrafts. There is no clear terminological distinction between graupel and hail, but a general consensus is that graupel commonly is smaller than 1 cm and has density lower than that of solid ice (0.92 g cm^{-3}) but higher than 0.4 g cm^{-3}. Hailstones are much larger and may exceed 10 cm. Because only armored research aircrafts can penetrate convective clouds containing graupel and hail, direct measurements of size distributions of graupel and hail aloft are very scarce.

Occasional in situ measurements of size distributions of ice particles aloft in deep convective clouds or hailstorms often reveal a bi-exponential type of particle spectra with different slopes for graupel and hail (Smith et al. 1976; Spahn and Smith 1976) which can be approximated by

$$N(D) = N_{0g}\exp(-\Lambda_g D) + N_{0h}\exp(-\Lambda_h D), \tag{4.23}$$

where subscripts *g* and *h* stand for graupel and hail. The size distributions of graupel and hail of different maximal size represented by (4.23) are shown in Fig. 4.16. Such distributions have been used in a modeling study of melting hail by Ryzhkov et al. (2013) and yielded realistic vertical profiles of polarimetric radar variables in hailstorms. Subsequently, the distributions were used for devising an operational

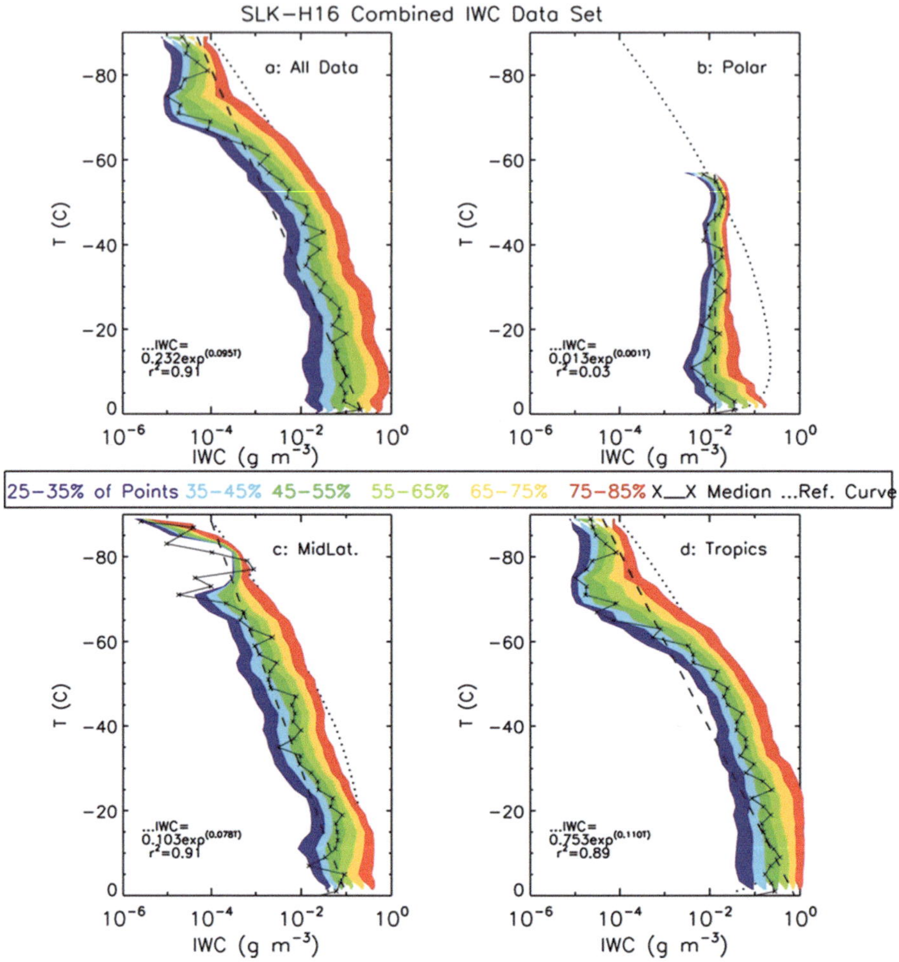

Fig. 4.14 Cumulative probability distributions of ice water content as a function of temperature obtained from many aircraft measurements for different types of clouds. Different percentiles of probability distributions are color coded. Adapted from Heymsfield et al. (2017). © American Meteorological Society. Used with permission

algorithm for hail detection and determination of its size on the WSR-88Ds (Ortega et al. 2016).

Large hailstones may not completely melt as they fall to the surface, and statistics of size distributions of hail observed at the ground are summarized in the papers by Ulbrich and Atlas (1982), Cheng and English (1983), and Cheng et al. (1985) among others. Ulbrich and Atlas (1982) performed a comprehensive analysis of the 1976 National Hail Research Experiment hailpad data, and their results are presented in Fig. 4.17. It is clear that a slope Λ_h of an exponential size distribution of hailstones is much smaller than the one of snowflakes and raindrops. Median and mean volume

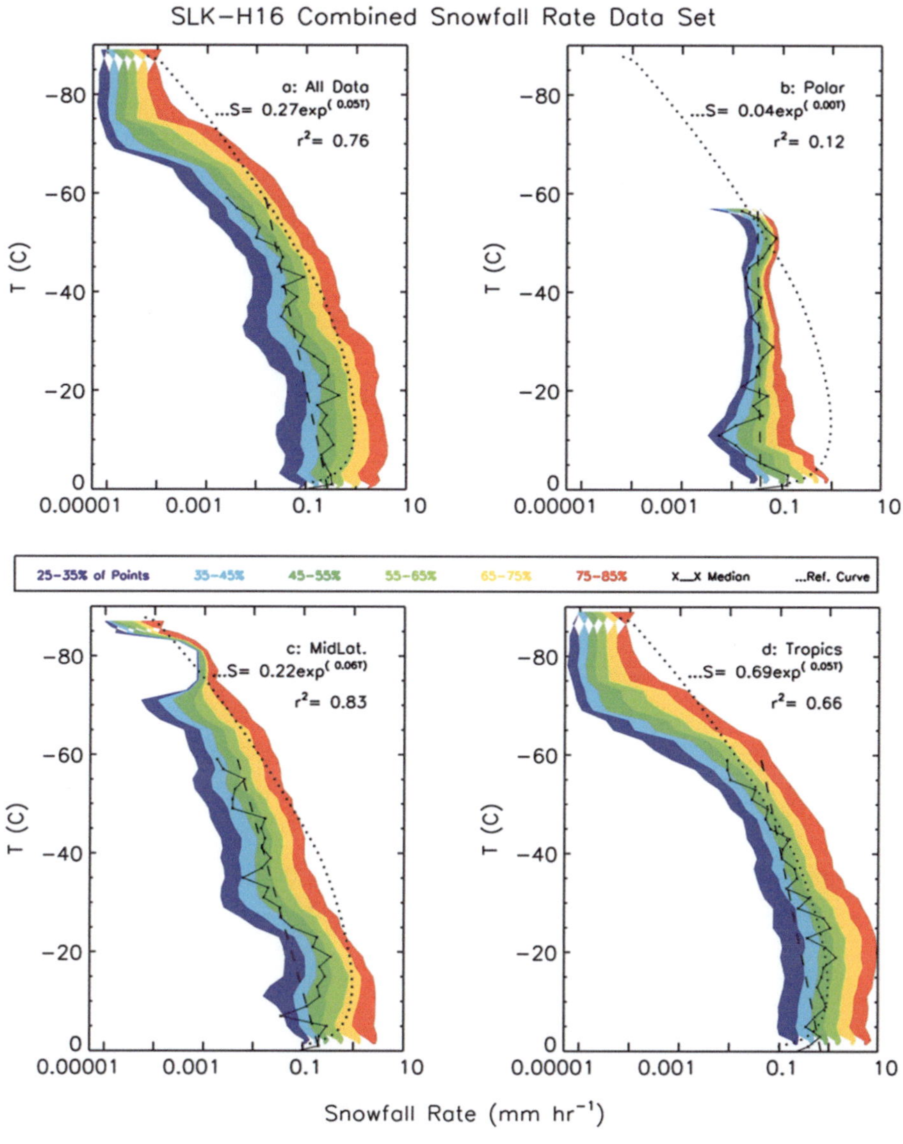

Fig. 4.15 Same as in Fig. 4.14 but for snow water equivalent rate. Adapted from Heymsfield et al. (2017). © American Meteorological Society. Used with permission

diameter of hail (marked as D_0 and D_m, respectively, in Fig. 4.17) commonly vary between 0.6 and 1.8 cm. High correlation between maximal hailstone size $D_{\max}$ and slope Λ_h is also notable so that the distribution of a product of Λ_h and $D_{\max}$ has a sharp maximum at about 8 (Fig. 4.17b). Cheng and English (1983) and Cheng et al.

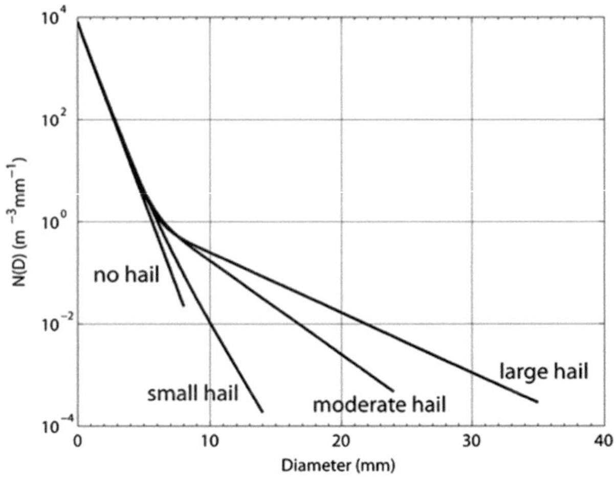

Fig. 4.16 Size distributions of graupel/hail size distributions in the case of no hail, small hail, moderate hail, and large hail used in the modeling study of Ryzhkov et al. (2013). Adapted from Ryzhkov et al. (2013). © American Meteorological Society. Used with permission

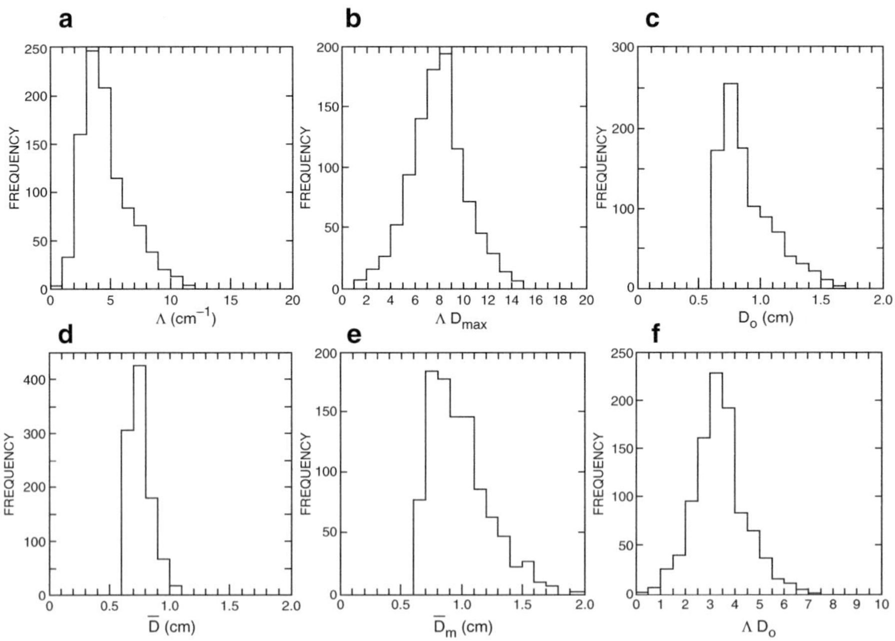

Fig. 4.17 Histograms of hail size distribution parameters for 1976 National Hail Research Experiment hailpads. (**a**) Λ, (**b**) ΛD_{max}, (**c**) D_0, (**d**) $\bar{D}$, (**e**) $\bar{D}_m$, and (**f**) ΛD_0 stand for average, mean volume, and median diameter, respectively. Adapted from Ulbrich and Atlas (1982). © American Meteorological Society. Used with permission

(1985) reported strong dependence of the intercept N_{0h} on the slope Λ_h of the size distribution which is quantified by the relation

$$N_{0h} = (34T_{CB} - 193)\Lambda_h^{4.11}, \tag{4.24}$$

where T_{CB} is the cloud-base temperature in °C, Λ_h is expressed in mm^{-1}, and N_{0h} is in m^{-3} mm^{-1}. The value of intercept increases with the cloud-base temperature.

4.2 Density of Hydrometeors

Density and phase composition of hydrometeors determine their dielectric constant that has very strong impact on their scattering properties. Density of water is 1.0 g cm^{-3}. Density of snow strongly depends on snow habit and the microphysical process determining snow growth. The density of the aggregated or irregular snowflakes is approximately inversely proportional to their diameter. An ensemble mean density $<\rho_s>$ is defined as a ratio of snow mass (measured by weighing gauges) and its volume (measured by disdrometer). Multiple relations between $<\rho_s>$ and mean volume diameter D_0 have been reported for irregular or aggregated snow (e.g., Brown and Francis 1995; Matrosov 1997; Brandes et al. 2007; Szyrmer and Zawadzki 2010; Tiira et al. 2016). For example, the relation from Brandes et al. (2007) is

$$< \rho_s >= 0.178 D_0^{-0.922} \tag{4.25}$$

(see Fig. 4.18).

It can be shown that in the case of exponential size distribution, the dependence $\rho_s(D)$ has the same exponent as the $<\rho_s>(D_0)$ relation but with a slightly different intercept (which is about 18% smaller). Most of the density relations cited in the literature have the exponent close to −1, but the intercept can vary significantly. Matrosov (1997) claims that the intercept is higher for ice particles observed at higher altitudes (using in situ aircraft measurements) compared to the one for snow near the surface (from ground measurements). The intercept of the power-law density-size relation is higher for rimed snow. Zawadzki et al. (2005) use the degree of riming factor, f_{rim}, to account for riming of different intensity so that

$$\rho_s(D) = \alpha_u f_{rim} D^{-\beta} = \alpha D^{-\beta} \tag{4.26}$$

and $f_{rim} = 1$ for unrimed snow. Another measure of riming is the rime mass fraction FR which is the ratio of the rime mass to the snowflake mass so that $f_{rim} = 1/(1 - FR)$ (Moisseev et al. 2017).

Anisotropic pristine crystals usually have higher density, and its dependence on their size is less pronounced than for the aggregated snow. For dendrites,

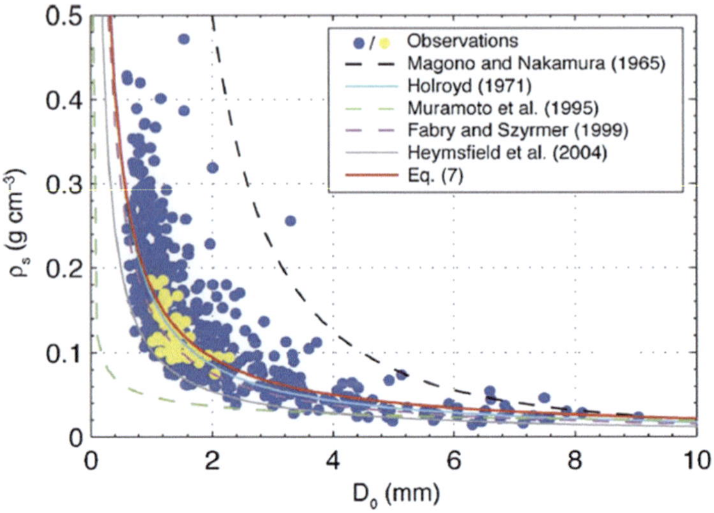

Fig. 4.18 Relationships between bulk snow density and equivolume diameter of snowflake. Equation (4.25) is represented by thick red curve. Adapted from Brandes et al. (2007). © American Meteorological Society. Used with permission

$$\rho(D) = 0.247D^{-0.377} \tag{4.27}$$

(Pruppacher and Klett 1997; Moisseev et al. 2015). In (4.27), D is the equivolume diameter in cm and ρ is in g cm^{-3}. Riming tends to increase density. Moreover, the type of the dependency of ρ on D changes so that ρ usually increases with increasing size as opposed to aggregated snow. Heymsfield and Wright (2014) came up with the relation

$$\rho(D) = 0.18D^{0.33}, \tag{4.28}$$

which is very similar to the theoretical relation of Leinonen and Szyrmer (2015) obtained for heavily rimed snow. In (4.28), D is in cm and ρ is in g cm^{-3}. Once rimed snow reaches the density of 0.2–0.4 g cm^{-3}, it is usually called graupel. Graupel in strong convective updrafts has density varying between 0.2 and 0.9 g cm^{-3}, i.e., closer to the density of solid ice. If graupel forms via riming of snow, its bulk density increases with size. However, if graupel forms by collisions between frozen drops and smaller droplets, its bulk density decreases with increasing graupel size (Khain and Pinsky 2018). The density of hail is usually close to the density of solid ice 0.92 g cm^{-3} if hailstones grow in the wet growth regime (see Chap. 7), but it can be substantially lower for the dry regime growth (Farley 1987; Knight et al. 2008).

4.3 Shapes and Orientations of Hydrometeors

The polarimetric variables are sensitive to the hydrometeor shape and orientation. Hydrometeors are usually modeled as oblate or prolate spheroids, and each is characterized by its equivalent volume (or equivolume) diameter $D = (ab^2)^{1/3}$, where a is a symmetry (or rotation) axis of spheroid and b is its transverse axis. For oblate hydrometeors $a < b$, whereas for prolate hydrometeors $a > b$.

4.3.1 Axis Ratios of Raindrops

Average shape of raindrops and its dependence on size has been investigated in numerous theoretical and observational studies. A thorough overview can be found in Bringi and Chandrasekar (2001, Chap. 7). There is a wide consensus that the relation suggested by Brandes et al. (2002)

$$r_w = 0.9951 + 0.02510D - 0.03644D^2 + 0.005303D^3 - 0.0002492D^4 \quad (4.29)$$

is a good approximation for the dependence of the axis ratio r_w on the equivolume diameter D (mm). It was determined by combining the observations of Pruppacher and Pitter (1971), Chandrasekar et al. (1988), Beard and Kubesh (1991), and Andsager et al. (1999) as indicated in Fig. 4.19. Although different forms of the shape-diameter relation exist for raindrops, (4.29) approximates quite well the average of these r_w—D dependencies. As Fig. 4.19 shows, large drops are more oblate than small ones.

4.3.2 Aspect Ratio of Ice Crystals

Pristine ice crystals have great variety of shapes, and the relation between the smaller and larger dimensions for different ice habits can be approximated by the power-law relation

$$h = cL^d, \quad (4.30)$$

where h is the smaller and L is the larger dimension of the crystal, both expressed in mm. The parameters c and d in (4.30) are listed in Table 4.1 for different types of crystals (Matrosov et al. 1996). The dependencies of the aspect ratios of different crystal habits on their maximal size are displayed in Fig. 4.20. Overall, ice crystals have more extreme shapes than raindrops, particularly dendrites, hexagonal plates, and elementary needles which are most "anisotropic" among the crystal habits.

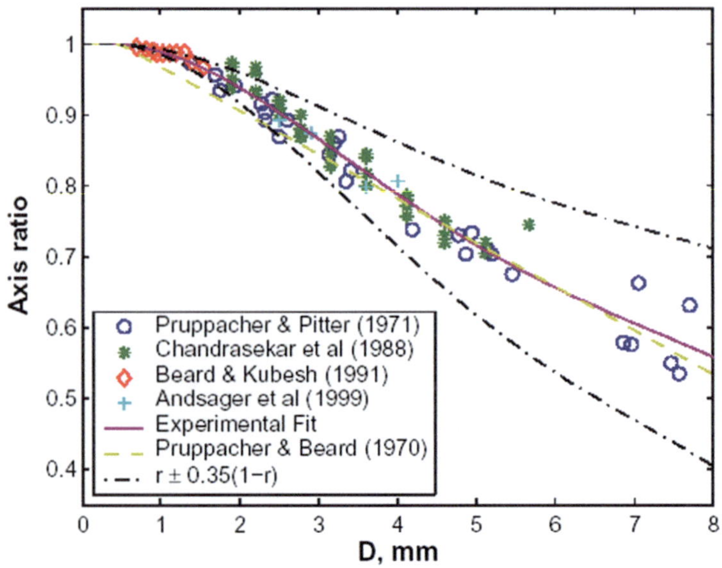

Fig. 4.19 Axis ratio of raindrop versus its equivolume diameter from various sources. The Brandes et al. (2002) approximation (4.29) is indicated by the solid line. From Brandes et al. (2002). © American Meteorological Society. Used with permission

Table 4.1 Parameters *c* and *d* in (4.30) for different types of crystals

Crystal habit	*c*	*d*
1. Dendrites	0.038	0.377
2. Solid thick plate	0.230	0.778
3. Hexagonal plates	0.047	0.474
4. Solid columns ($L/h < 2$)	0.637	0.958
5. Solid columns ($L/h > 2$)	0.308	0.927
6. Hollow columns ($L/h < 2$)	0.541	0.892
7. Hollow columns ($L/h > 2$)	0.309	0.930
8. Long solid columns	0.128	0.437
9. Solid bullets ($L < 0.3$ mm)	0.250	0.786
10. Hollow bullets ($L > 0.3$ mm)	0.185	0.532
11. Elementary needles	0.073	0.611

4.3.3 Aspect Ratios of Dry Aggregated Snowflakes and Dry Graupel and Hail

In the process of aggregation, crystals clump and eventually produce low-density "fluffy" aggregates. The shape of such aggregates tends to be more spherical than the shape of crystals. From the aggregate appearance, it is very difficult to identify the type of crystals the aggregate originated from. Aspect ratio of dry aggregated snowflakes usually varies between 0.5 and 0.7 (Straka et al. 2000; Korolev and

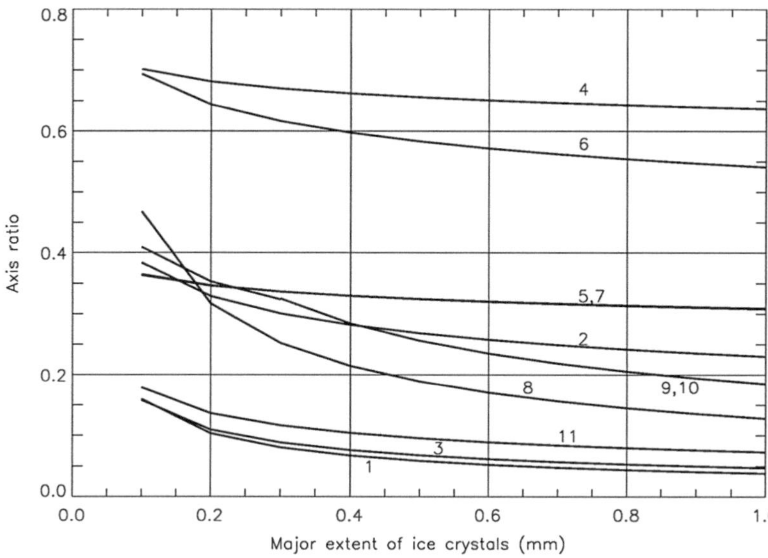

Fig. 4.20 Aspect ratio of ice crystal types as a function of particle major dimension. Numbers indicate crystal habits listed in Table 4.1

Isaac 2003) and is almost independent of the particle size (Fig. 4.21). The aspect ratio generally increases with riming. Garrett et al. (2015) found the median axis ratio of unrimed aggregates equal to 0.60 versus 0.70 for moderately rimed snowflakes and 0.85 for graupel in the observations with a Multi-Angle Snowflake Camera (MASC).

Aspect ratio of dry graupel or hail varies between 0.6 and 0.9 (Straka et al. 2000). For modeling purposes, Ryzhkov et al. (2011) apply the following dependence of the aspect ratio of graupel and hail (r_{gh}) on equivolume diameter D (mm):

$$\begin{aligned} r_{gh} &= 1.0 - 0.02D \quad if \quad D < 10mm \\ r_{gh} &= 0.8 \quad if \quad D > 10mm. \end{aligned} \tag{4.31}$$

4.3.4 Aspect Ratio of Melting Crystals and Snow

The shape and aspect ratio of melting crystals and snow are not as well documented as those of dry snow. It is reasonable to assume that the shape of melting ice particle gradually changes with increasing mass water fraction f_w so that it eventually acquires the shape of the raindrop with the same mass. A linear dependence of the aspect ratio on f_w was suggested in modeling studies by Jung et al. (2008, 2010) and Ryzhkov et al. (2011). Hence, if ice crystal or snowflake melts into raindrop with

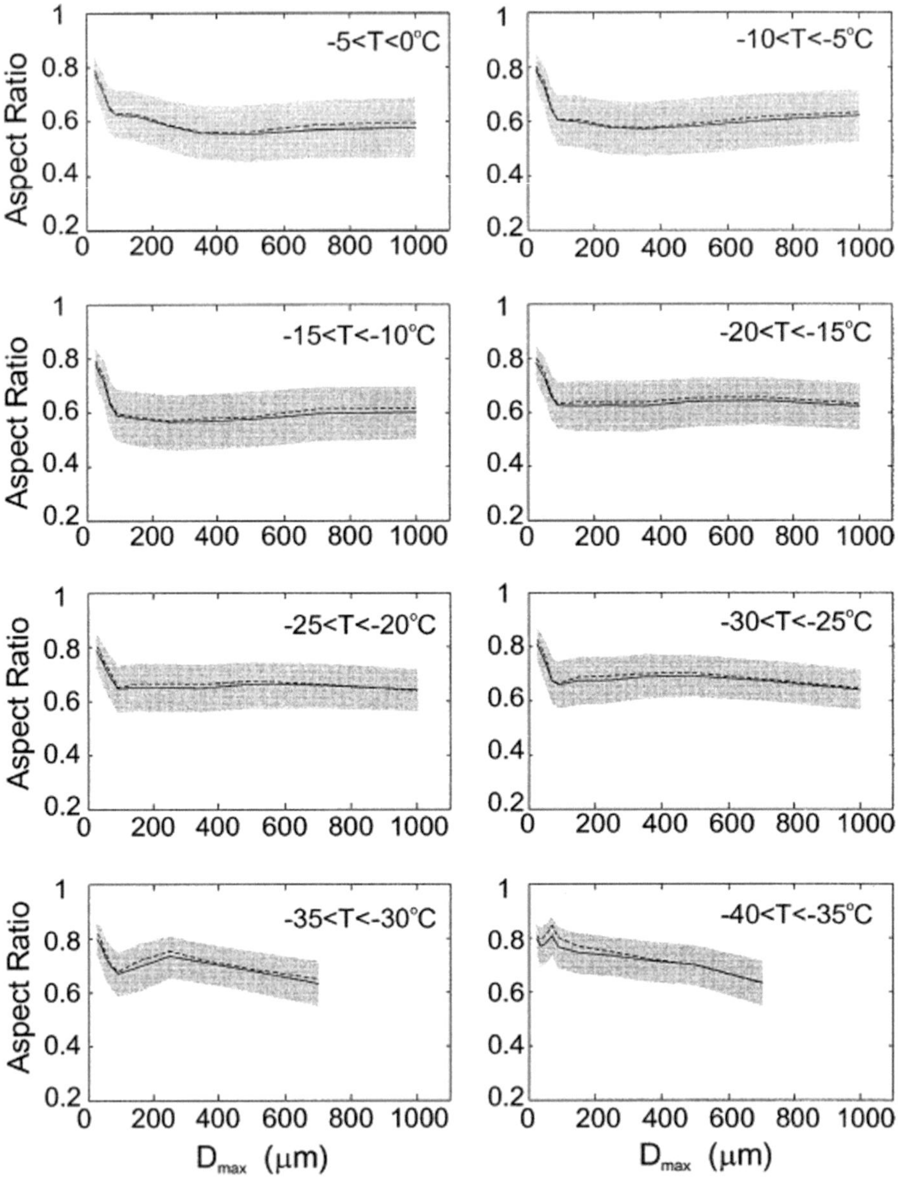

Fig. 4.21 Mean (solid curves) and median (dashed curves) aspect ratio of particles in ice clouds vs particle length. The gray bands indicate the standard deviation. Adapted from Korolev and Isaac (2003). © American Meteorological Society. Used with permission

diameter D, its aspect ratio r_m changes during melting proportionally to the mass water fraction f_w

$$r_m = r_{\mathrm{ds}} + f_w(r_w - r_{\mathrm{ds}}), \tag{4.32}$$

where r_{ds} is the aspect ratio of dry snow and r_w is the axis ratio of raindrop of diameter D which is produced as a result of crystal or snowflake melting. The value of r_w is determined by (4.29).

4.3.5 Aspect Ratio of Melting Graupel and Hail

The aspect ratio of melting graupel and hail was investigated in the experimental study of Rasmussen et al. (1984). Linear approximations to the results from this study suggested by Ryzhkov et al. (2011) are

$$\begin{aligned} r_m &= r_{gh} - 5.0(r_{gh} - 0.8)f_w \quad \text{if} \quad f_w < 0.2 \\ r_m &= 0.88 - 0.40f_w \quad \text{if} \quad 0.2 < f_w < 0.8 \\ r_m &= 2.8 - 4.0r_w + 5.0(r_w - 0.56)f_w \quad \text{if} \quad f_w > 0.8. \end{aligned} \tag{4.33}$$

4.3.6 Orientations of Hydrometeors

All polarimetric variables depend on the angular moments of the distributions of particle orientations. These moments can be approximated with relatively simple analytical expressions in the cases of uniformly distributed random orientations or narrow two-dimensional axisymmetric Gaussian distribution of orientations (Chap. 3). For the sake of simplicity, it is assumed that most of the oblate hydrometeors can be characterized by a two-dimensional Gaussian distribution of orientations (3.46) with zero mean canting angle, i.e., the mean direction of their axis of rotation is vertical. Then the spread of the axis orientation is characterized by a single parameter σ—the width of the canting angle distribution. It is also reasonable to assume that prolate particles (such as columns and needles) are on average randomly oriented in the horizontal plane but exhibit flutter which could be described with a uniform or Gaussian distribution (Zrnic and Ryzhkov 2004; Melnikov and Straka 2013).

Notable exception is the orientation of small crystals under the influence of strong electric fields in thunderstorms. The effects of crystal orientation in electrically charged zones were observed with polarimetric radars having circular (e.g., Hendry and McCormick 1976; Krehbiel et al. 1996) and linear polarizations (e.g., Caylor and Chandrasekar 1996; Metcalf 1997). These studies showed that pristine crystals with low inertia tend to align along the direction of electrostatic field that generally is neither horizontal nor vertical. After lightning strike occurs, the magnitude of electric field abruptly diminishes, and crystals lose their preferred orientation. The high degree of common alignment can be restored if another buildup of charge occurs; otherwise crystals would revert to their typical orientation in the absence of electric activity, i.e., with larger dimension in the horizontal plane. Pristine crystals with nonzero canting angles can produce strong depolarization of electromagnetic waves which can have pronounced impact on the fields of differential reflectivity and differential phase (Ryzhkov and Zrnic 2007); this is discussed in Chap. 6.

The width of the canting angle distribution σ is different for various species of hydrometeors. Raindrops have higher degree of common alignment than ice and mixed-phase hydrometeors. Direct measurements of drop canting using 2D-video disdrometer (Huang et al. 2008) and polarimetric radar retrievals (Ryzhkov et al. 2002) show that in rain the width σ usually varies between 5 and 10°. Wider canting angle distributions, however, might occur in areas of strong turbulence that tends to make raindrop orientations more chaotic.

Information about orientation habits of frozen and mixed-phase hydrometeors is very scarce. Distribution of orientations of ice particles depends on their falling behavior which is primarily a function of the Reynolds number

$$N_{\mathrm{Re}} = V_t D/\nu, \tag{4.34}$$

where V_t is the particle terminal velocity, D is the equivolume diameter, and ν is the kinematic viscosity of the air. In the absence of electrostatic field, symmetrical planar ice crystals with Reynolds number in the range $1 < N_{\mathrm{Re}} < 100$ tend to fall with their major axis oriented preferentially in the horizontal plane with very small flutter. Matrosov et al. (2005) utilized radar measurements in the horizontal-vertical polarization basis and the slant 45–135° polarization basis to estimate such flutter for falling pristine dendrites which were characterized by Reynolds numbers approximately between 40 and 100. The corresponding parameter σ was about $9° \pm 3°$, i.e., very close to the one for rain. Melnikov and Straka (2013) used a different radar retrieval technique based on the combination of Z_{DR} and ρ_{hv} to estimate the value of σ in the dendritic growth layer where Z_{DR} exceeds 4 dB. They found that σ attributed to dendrites or hexagonal plates varies between 2° and 20° and the median is within the interval of 9°–15° consistent with the results of Matrosov et al. (2005).

As crystals start aggregating, their Reynolds number increases in proportion to D. Consequently, eddies at the rear of falling crystals begin to shed, and secondary motions including glide-pitch oscillations and spiraling are introduced at $N_{\mathrm{Re}} > 100$ (Pruppacher and Klett 1997). Therefore, the width of the canting angle distribution of crystals growing via aggregation increases. Hendry et al. (1987) estimate (based on the analysis of data collected by a radar with circular polarization) that the average σ in moderate-to-heavy snow consisting of dry aggregates is about 40°. Similar result was recently obtained by Garrett et al. (2015) from the surface-based measurements with a MASC camera. They also emphasize strong impact of turbulence on the magnitude of σ and report median values equal 40° and 23° for strong and weak turbulence.

Large oblate particles of higher density such as graupel and hail exhibit very complicated falling patterns including tumbling, gyration, and wobbling. These were examined in the theoretical and laboratory studies (e.g., Kry and List 1974), in the analysis of internal structure of hailstones collected on the ground (Knight and Knight 1970), and in the stroboscopic photos obtained near ground (Matson and Huggins 1980). Retrievals from data of circularly polarized radar by Hendry et al. (1976) indicate that the parameter σ is larger than 40° in hail regions of storms.

As far as mixed-phase particles are concerned, it is reasonable to assume that similar to the aspect ratio, the width of the canting angle distribution σ varies linearly with the fractional mass water content f_w (i.e., mass of liquid divided by total mass):

$$\sigma = \sigma_s + f_w(\sigma_r - \sigma_s), \tag{4.35}$$

where σ_r is the width of the canting angle distribution of raindrops, whereas σ_s is the corresponding width of the distribution of frozen hydrometeors. This assumption is consistent with the findings of Hendry et al. (1987) where the spread σ within the melting layer of stratiform precipitation is about 14°–30°. This is between typical value for rain (5°–10°) and dry aggregated snow (about 40°). The dependence (4.35) has been utilized in theoretical studies of Jung et al. (2008, 2010) and Ryzhkov et al. (2011) to convert cloud model outputs into realistically looking fields of polarimetric variables.

4.4 Dielectric Properties of Hydrometeors

The scattering amplitudes s_a and s_b of a hydrometeor (Sect. 1.2) depend on dielectric properties of its constituents, i.e., air, ice, water, or their mixture. Each constituent is characterized by a complex dielectric constant or permittivity $\varepsilon = \mathrm{Re}(\varepsilon) - j\,\mathrm{Im}(\varepsilon)$, where $\mathrm{Re}(\varepsilon)$ and $\mathrm{Im}(\varepsilon)$ are positive if the $\exp(j\omega t)$ convention is adopted (Sect. 1.1). By definition, the refractive index n of the medium is $\varepsilon^{1/2}$.

The dielectric constant of air depends on atmospheric pressure, humidity, temperature, and wavelength of electromagnetic radiation (Eq. (1.6) and Appendix A). At microwave frequencies, refractive index of air is close to 1 (within 0.1%). Variations of the real part of ε (or n) with height cause refraction (bending) of radar rays, whereas imaginary part of ε is responsible for absorption of the microwave radiation (Appendix A).

The spectrum of microwave frequencies is divided into frequency bands defined in Table 4.2. This designation dates back to the World War II when it was meant to conceal the operating frequencies of military radars. The nomenclature remains to this day. Most of weather surveillance radars operate at three frequency bands: S, C, and X. In this monograph, we often utilize terms S, C, and X without specifying concrete microwave frequency to distinguish between long-wave (S band), medium-

Table 4.2 Radar frequency bands and wavelengths

Band	Frequencies (GHz)	Wavelengths (cm)
L	1–2	30–15
S	2–4	15–7.5
C	4–8	7.5–3.75
X	8–12	3.75–2.5
Ku	12–18	2.5–1.67
K	18–27	1.67–1.11
Ka	27–40	1.11–0.75

wave (C band), and short-wave (X band) weather radars. For precise interpretation of scattering and quantitative measurements of precipitation, the exact frequency is often required.

4.4.1 *Dielectric Constant of Fresh Water and Solid Ice*

Dielectric constant of fresh water is a function of wavelength and temperature. The formulas for computing real and imaginary parts of ε from Ray (1972) are presented in Appendix C. The dependencies of real and imaginary parts of ε on the radar wavelength in the microwave range of frequencies for five different temperatures are illustrated in Fig. 4.22. The values of ε at four wavelengths 11.0 cm (S band), 5.45 cm (C band), 3.2 cm (X band), and 0.86 cm (Ka band) at different temperatures are listed in Table 4.3.

The real part of dielectric constant of solid ice with density $\rho_i = 0.92$ gcm^{-3} is equal to 3.17 and is independent of the wavelength and temperature. Its imaginary part is very small (less than 0.01 at all temperatures and wavelengths below 11 cm); hence in all computations in the microwave frequency range, one can assume $\varepsilon_i = 3.17$.

4.4.2 *Dielectric Constant of Dry Snow, Graupel, and Hail*

Dry snow is a mixture of air and solid ice, and its dielectric constant depends on the relative fractions of ice and air, in other words on snow density which can vary over a broad range. According to Maxwell-Garnett (1904), dielectric constant of dry snow is determined by the volume fraction of ice f_{vi} in the mixture with air and dielectric constants of solid ice ε_i and air ε_a:

$$\varepsilon_s = \varepsilon_a \left(1 + \frac{3f_{\text{vi}} \dfrac{\varepsilon_i - \varepsilon_a}{\varepsilon_i + 2\varepsilon_a}}{1 - f_{\text{vi}} \dfrac{\varepsilon_i - \varepsilon_a}{\varepsilon_i + 2\varepsilon_a}} \right). \tag{4.36}$$

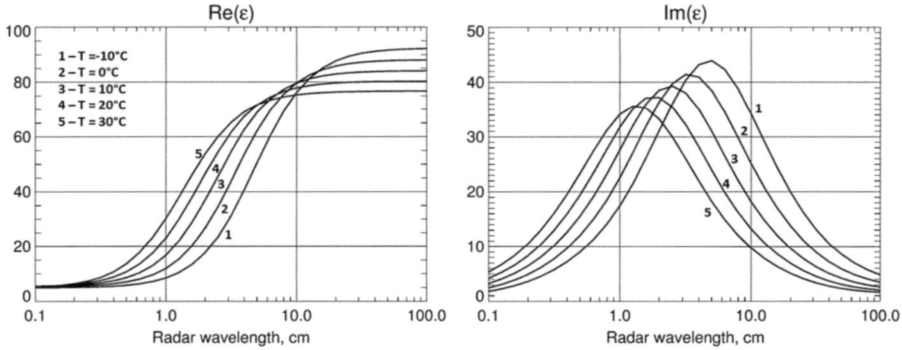

Fig. 4.22 Dependence of the magnitudes of real and imaginary parts of dielectric constant of fresh water on wavelength for five different temperatures

Table 4.3 Dielectric constant of water as a function of wavelength and temperature

Temperature	$\lambda = 11.0$ cm	$\lambda = 5.45$ cm	$\lambda = 3.2$ cm	$\lambda = 0.86$ cm
−10 °C	78.3 – j 32.2	54.3 – j 43.6	31.8 – j 40.6	7.56 – j 15.2
0 °C	81.0 – j 23.3	65.2 – j 37.2	44.5 – j 41.4	10.3 – j 19.8
10 °C	80.4 – j 16.8	70.9 – j 29.4	55.1 – j 37.9	14.1 – j 24.7
20 °C	78.3 – j 12.1	72.8 – j 22.4	62.1 – j 32.0	19.2 – j 29.2
30 °C	75.6 – j 8.91	72.4 – j 17.0	65.5 – j 25.8	25.6 – j 32.4

Ice-volume fraction f_{vi} in a mixture with air can be expressed via densities of snow and ice. Snow density ρ_s is

$$\rho_s = f_{\text{vi}}\rho_i + (1 - f_{\text{vi}})\rho_a, \tag{4.37}$$

where ρ_i and ρ_a are the densities of ice and air, respectively. It follows from (4.37) that

$$f_{\text{vi}} = \frac{\rho_s - \rho_a}{\rho_i - \rho_a} \approx \frac{\rho_s}{\rho_i}. \tag{4.38}$$

Using (4.38) and taking into account that $\varepsilon_a \approx 1$, we rewrite (4.36) as

$$\varepsilon_s = 1 + \frac{3\frac{\rho_s}{\rho_i}\frac{\varepsilon_i - 1}{\varepsilon_i + 2}}{1 - \frac{\rho_s}{\rho_i}\frac{\varepsilon_i - 1}{\varepsilon_i + 2}}. \tag{4.39}$$

For many radar applications, it is more convenient to represent (4.39) in the Debye form

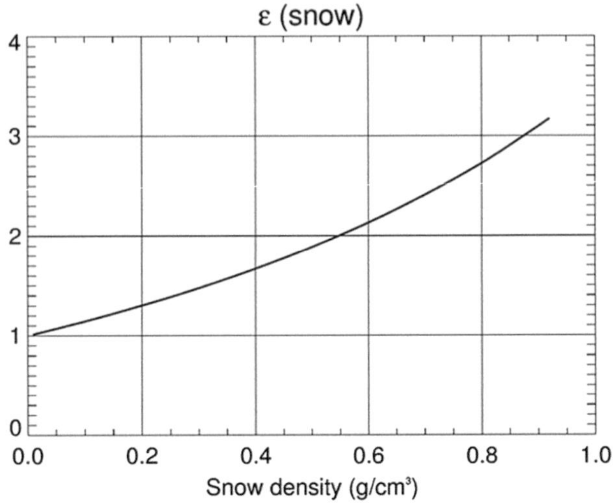

Fig. 4.23 Dependence of the dielectric constant of dry snow on its density

$$\frac{\varepsilon_s - 1}{\varepsilon_s + 2} = \frac{\rho_s}{\rho_i} \frac{\varepsilon_i - 1}{\varepsilon_i + 2}. \tag{4.40}$$

Depending on the snowflake size and degree of riming, the density of snow can vary significantly (Sect. 4.2). Bulk density of pristine crystals is close to that of solid ice (0.92 g cm^{-3}). Once elementary crystals start aggregating, their size increases, and bulk density decreases dramatically and can drop as low as 0.01 g cm^{-3}. According to (4.39) and (4.40), the decrease of bulk snow density from 0.92 g cm^{-3} to 0.01 g cm^{-3} reduces the magnitude of dielectric constant ε_s from 3.17 to 1.01, i.e., the dielectric constant of heavily aggregated dry snow is close to the dielectric constant of pure air (Fig. 4.23).

Equations (4.39) and (4.40) are applicable to dry graupel or hail if the snow density is replaced by graupel or hail density. Graupel and hail grow mainly by accretion/riming of supercoooled cloud droplets, and the density of riming deposit increases with increasing cloud droplet size, ice particle surface temperature, and impact velocity (Macklin 1962; Farley 1987). The density of dry graupel and hail usually varies between 0.2 and 0.92 g cm^{-3}, and the corresponding value of dielectric constant ε_s changes from 1.30 to 3.17.

4.4.3 Dielectric Constant of Wet Snow

Dielectric constant of wet snow depends on snow density and volume fraction of water in the snowflake. Furthermore, the way melted water is distributed within the melting snowflake also affects the value of its dielectric constant. Fabry and Szyrmer (1999) describe at least six different models of such distributions.

Here we present two models and their combination based on two Maxwell-Garnett formulas for the mixture of water and snow. One implies that snow is considered as "matrix," whereas water is treated as "inclusions"

$$\varepsilon_{\mathrm{ws}}^{(1)} = \varepsilon_s \left(1 + \frac{3f_{\mathrm{vw}} \frac{\varepsilon_w - \varepsilon_s}{\varepsilon_w + 2\varepsilon_s}}{1 - f_{\mathrm{vw}} \frac{\varepsilon_w - \varepsilon_s}{\varepsilon_w + 2\varepsilon_s}} \right), \tag{4.41}$$

and the other specifies water as matrix and snow as inclusions.

$$\varepsilon_{\mathrm{ws}}^{(2)} = \varepsilon_w \left[1 + \frac{3(1 - f_{\mathrm{vw}}) \frac{\varepsilon_s - \varepsilon_w}{\varepsilon_s + 2\varepsilon_w}}{1 - (1 - f_{\mathrm{vw}}) \frac{\varepsilon_s - \varepsilon_w}{\varepsilon_s + 2\varepsilon_w}} \right]. \tag{4.42}$$

In (4.41) and (4.42), ε_s is obtained from (4.39), ε_w is the dielectric constant of water, and f_{vw} is the volume fraction of water in a melting snowflake. Equations (4.41) and (4.42) produce different dielectric constants of melting snowflake for the same volume fraction of water. This is a significant source of uncertainty in the computation of scattering amplitudes. A weighted combination of ${\varepsilon_{\mathrm{ws}}}^{(1)}$ and ${\varepsilon_{\mathrm{ws}}}^{(2)}$ has been often used because (4.41) is more accurate if the water fraction is low, whereas (4.42) is preferable if the water fraction is high (Meneghini and Liao 1996; Matrosov 2008). The combination is

$$\varepsilon_{\mathrm{ws}} = \frac{1}{2} \left[(1 + \kappa)\varepsilon_{\mathrm{ws}}^{(1)} + (1 - \kappa)\varepsilon_{\mathrm{ws}}^{(2)} \right], \tag{4.43}$$

where parameter κ is a function of f_{vw}

$$\kappa = \mathrm{Erf}\left(\frac{1 - f_{\mathrm{vw}}}{f_{\mathrm{vw}}} - 0.2 \right) \quad \text{if} \quad f_{\mathrm{vw}} > 0.01 \tag{4.44}$$

and Erf is the error function. Equations (4.43) and (4.44) are similar to those used by Meneghini and Liao (1996) who approximate quite well the "Maxwell Garnett weighted rule" suggested by Matrosov (2008).

More complex models of concentrically melting snowflakes are advocated by Fabry and Szyrmer (1999) and Liao and Meneghini (2005). These models account for the higher melted water fraction at the exterior of the melting snowflake than at its interior and are much more computationally consuming.

4.4.4 Dielectric Constant of Wet Graupel and Hail

Generally, wet hailstones cannot be represented by hydrometeors of uniform dielectric constant because water accumulates at their surface during melting process or growth in the wet regime. If initial density of dry hail (graupel) is less than the density of solid ice (0.92 g cm^{-3}), then water first soaks the particle's interior and fills all air cavities (Phillips et al. 2007). At this stage of melting (or growth), the hydrometeor can be considered as uniformly filled. Once melted water saturates the particle interior, it starts building an exterior shell, thus creating a two-layer structure.

Formula (4.41) is suitable for computing dielectric constant of spongy graupel in the soaking stage. At the later melting stage, when two-layer structure is built, the dielectric constant of water (ε_w) represents the outer shell, and the dielectric constant of the inner core is determined by (4.41).

References

Andsager, K., Beard, K., & Laird, N. (1999). Laboratory measurements of axis ratios for large drops. *Journal of the Atmospheric Sciences, 56*, 2673–2683.

Atlas, D., & Ulbrich, C. (2000). An observationally based conceptual model of warm oceanic convective rain in the tropics. *Journal of Applied Meteorology, 39*, 2165–2181.

Bang, W., & Lee, G.-W. (2017). Characteristics of DSD at two different climates and their implications into radar precipitation estimation. In *Proceedings of 32nd International Union of Radio Science General Assembly and Scientific Symposium* (p. 91).

Beard, K., & Kubesh, R. (1991). Laboratory measurements of small raindrop distortion. Pt 2: Oscillation frequencies and modes. *Journal of the Atmospheric Sciences, 48*, 2245–2264.

Brandes, E., Zhang, G., & Vivekanandan, J. (2002). Experiments in rainfall estimation with polarimetric radar in a subtropical environment. *Journal of Applied Meteorology, 41*, 674–685.

Brandes, E., Ikeda, K., Zhang, G., Schoenhuber, M., & Rasmussen, R. (2007). A statistical and physical description of hydrometeor distributions in Colorado snowstorms using a video-disdrometer. *Journal of Applied Meteorology, 46*, 634–650.

Bringi, V. N., & Chandrasekar, V. (2001). *Polarimetric Doppler weather radar: Principles and applications*. Cambridge, UK: Cambridge University Press 636pp.

Bringi, V., Huang, G.-J., Chandrasekar, V., & Gorgucci, E. (2002). A methodology for estimating the parameters of a gamma raindrop size distribution model from polarimetric radar data: Application to a squall-line event from the TRMM/Brazil campaign. *Journal of Atmospheric and Oceanic Technology, 19*, 633–645.

Bringi, V., Chandrasekar, V., Hubbert, J., Gorgucci, E., Randeu, W., & Schoenhuber, M. (2003). Raindrop size distribution in different climatic regimes from disdrometer and dual-polarized radar analysis. *Journal of the Atmospheric Sciences, 60*, 354–365.

Bringi, V., Williams, C., Thurai, M., & May, P. (2009). Using dual-polarized radar and dual-frequency profiler for DSD characterization: A case study from Darwin, Australia. *Journal of Atmospheric and Oceanic Technology, 26*, 2107–2122.

Brown, P. R. A., & Francis, P. N. (1995). Improved measurements of ice water content in cirrus using a total-water probe. *Journal of Atmospheric and Oceanic Technology, 12*, 410–414.

Cao, Q., Zhang, G., Brandes, E., Schuur, T., Ryzhkov, A., & Ikeda, K. (2008). Analysis of video disdrometer and polarimetric radar data to characterize rain microphysics in Oklahoma. *Journal of Applied Meteorology and Climatology, 47*, 2238–2255.

Caylor, I. J., & Chandrasekar, V. (1996). Time-varying crystal orientation in thunderstorms observed with multiparameter radar. *IEEE Transactions on Geoscience and Remote Sensing, 34*, 847–858.

Chandrasekar, V., Cooper, W., & Bringi, V. (1988). Axis ratios and oscillations of raindrops. *Journal of the Atmospheric Sciences, 45*, 1323–1333.

Cheng, L., & English, M. (1983). A relationship between hailstone concentration and size. *Journal of the Atmospheric Sciences, 40*, 204–213.

Cheng, L., English, M., & Wong, R. (1985). Hailstone size distributions and their relationship to storm thermodynamics. *Journal of Climate and Applied Meteorology, 24*, 1059–1067.

Delanoe, J., Protat, A., Testud, J., Bouniol, D., Heymsfiled, A., Bansemer, A., et al. (2005). Statistical properties of the normalized ice particles size distribution. *Journal of Geophysical Research, 110*, D10201.

Delanoe, J., Heymsfield, A., Protat, A., Bansemer, A., & Hogan, R. (2014). Normalized particle size distribution for remote sensing applications. *Journal of Geophysical Research – Atmospheres, 119*, 4202–4227.

Fabry, F., & Szyrmer, W. (1999). Modeling the melting layer. Part II: Electromagnetic. *Journal of the Atmospheric Sciences, 56*, 3593–3600.

Farley, R. (1987). Numerical modeling of hailstorms and hailstone growth. Pt II. The role of low-density riming growth in hail production. *Journal of Climate and Applied Meteorology, 26*, 234–254.

Garrett, T., Yuter, S., Fallgatter, C., Shkurko, K., Rhodes, S., & Endries, J. (2015). Orientations and aspect ratios of falling snow. *Geophysical Research Letters, 42*. https://doi.org/10.1002/2015GL064040.

Hendry, A., & McCormick, G. C. (1976). Radar observations of alignment of precipitation particles by electrostatic fields in thunderstorms. *Journal of Geophysical Research, 81*, 5353–5357.

Hendry, A., McCormick, G., & Barge, B. (1976). The degree of common orientations of hydrometeors observed by polarization diversity radars. *Journal of Applied Meteorology, 15*, 633–640.

Hendry, A., Antar, Y., & McCormick, G. (1987). On the relationship between the degree of preferred orientation in precipitation and dual-polarization radar echo characteristics. *Radio Science, 22*, 37–50.

Heymsfield, A., & Wright, R. (2014). Graupel and hail terminal velocities: Does a "supercritical" Reynolds number apply? *Journal of the Atmospheric Sciences, 71*, 3392–3403.

Heymsfield, A., Kramer, M., Wood, N., Gettleman, A., Field, P., & Liu, G. (2017). Dependence of ice water content and snowfall rate on temperature, globally: Comparison of in situ observations, satellite active remote sensing retrievals, and global climate model simulations. *Journal of Applied Meteorology and Climatology, 56*, 189–215.

Hu, Z., & Srivastava, R. (1995). Evolution of raindrop size distribution by coalescence, breakup, and evaporation: Theory and observations. *Journal of the Atmospheric Sciences, 52*, 1761–1783.

Huang, G., Bringi, V., & Thurai, M. (2008). Orientation angle distributions of drops after 80 m fall using a 2D-video disdrometer. *Journal of Atmospheric and Oceanic Technology, 25*, 1717–1723.

Illingworth, A., & Blackman, T. (2002). The need to represent raindrop size spectra as normalized gamma distributions for the interpretation of polarization radar observations. *Journal of Applied Meteorology, 41*, 286–297.

Jung, Y., Zhang, G., & Xue, M. (2008). Assimilation of simulated polarimetric radar data for a convective storm using the ensemble Kalman filter. Part I: Observation operators for reflectivity and polarimetric variables. *Monthly Weather Review, 136*, 2228–2245.

Jung, Y., Xue, M., & Zhang, G. (2010). Simulations of polarimetric radar signatures of a supercell storm using a two-moment bulk microphysical scheme. *Journal of Applied Meteorology and Climatology, 49*, 146–163.

Khain, A., & Pinsky, M. (2018). *Physical processes in clouds and cloud modeling*. Cambridge, UK: Cambridge University Press. 626 pp.

Knight, C., & Knight, N. (1970). The falling behavior of hailstones. *Journal of the Atmospheric Sciences, 27*, 672–680.

Knight, C., Schlatter, P., & Schlatter, T. (2008). An unusual hailstorm on 24 June 2006 in Boulder, Colorado. Part II: Low-density growth of hail. *Monthly Weather Review, 136*, 2833–2848.

Korolev, A., & Isaac, G. (2003). Roundness and aspect ratio of particles in ice clouds. *Journal of the Atmospheric Sciences, 60*, 1795–1808.

Krehbiel, P. R., Chen, T., McCrary, S., Rison, W., Gray, G., & Brook, M. (1996). The use of dual-channel circular-polarization radar observations for remotely sensing storm electrification. *Meteorology and Atmospheric Physics, 59*, 65–82.

Kry, P., & List, R. (1974). Angular motions of freely falling spheroidal hailstone models. *Physics of Fluids, 17*, 1093–1102.

Lee, G., Zawadzki, I., Szyrmer, W., Sempere-Torres, D., & Uijlenhoet, R. (2004). A general approach to double-moment normalization of drop-size distributions. *Journal of Applied Meteorology, 43*, 264–281.

Leinonen, J., & Szyrmer, W. (2015). Radar signatures of snowflake riming: A modeling study. *Earth and Space Science, 2*, 346–358.

Liao, L., & Meneghini, R. (2005). On modeling air/spaceborne radar returns in the melting layer. *IEEE Transactions on Geoscience and Remote Sensing, 43*, 2799–2809.

Macklin, W. (1962). The density and structure of ice formed by accretion. *Quarterly Journal of the Royal Meteorological Society, 88*, 30–50.

Marshall, J. S., & Palmer, W. (1948). The distribution of raindrops with size. *Journal of Meteorology, 5*, 165–166.

Matson, R. J., & Huggins, A. W. (1980). The direct measurement of the sizes, shapes, and kinematics of falling hailstones. *Journal of the Atmospheric Sciences, 37*, 1107–1125.

Matrosov, S., Reinking, R., Kropfli, R., & Bartram, B. (1996). Estimation of ice hydrometeor types and shapes from radar polarization measurements. *Journal of Atmospheric and Oceanic Technology, 13*, 85–96.

Matrosov, S. (1997). Variability of microphysical parameters in high-altitude ice clouds: Results of the remote sensing method. *Journal of Applied Meteorology, 36*, 633–648.

Matrosov, S., Reinking, R., & Djalalova, I. (2005). Inferring fall attitudes of pristine dendritic crystals from polarimetic radar data. *Journal of the Atmospheric Sciences, 62*, 241–250.

Matrosov, S. (2008). Assessment of radar signal attenuation caused by the melting hydrometeor layer. *IEEE Transactions on Geoscience and Remote Sensing, 46*, 1039–1047.

Maxwell Garnett, J. C. (1904). Color in metal glasses and in metallic films. *Philosophical Transactions of the Royal Society London A, 203*, 385–420.

Melnikov, V., & Straka, J. (2013). Axis ratios and flutter angles of cloud ice particles: Retrievals from radar data. *Journal of Atmospheric and Oceanic Technology, 30*, 1691–1703.

Meneghini, R., & Liao, L. (1996). Comparisons of cross sections of melting hydrometeors as derived from dielectric mixing formulas and a numerical method. *Journal of Applied Meteorology, 35*, 1658–1670.

Metcalf, J. I. (1997). Temporal and spatial variations of hydrometeor orientation of hydrometeors in thunderstorms. *Journal of Applied Meteorology, 36*, 315–321.

Moisseev, D., Lautaportti, S., Tyynela, J., & Lim, S. (2015). Dual-polarization radar signatures in snowstorms: Role of snowflake aggregation. *Journal of Geophysical Research – Atmospheres, 120*, 12644–12655.

Moisseev, D., von Lerber, A., & Tiira, J. (2017). Quantifying the effect of riming on snowfall using ground-based observations. *Journal of Geophysical Research – Atmospheres, 122*, 4019–4037.

Ortega, K., Krause, J., & Ryzhkov, A. (2016). Polarimetric radar characteristics of melting hail. Part III: Validation of the algorithm for hail size discrimination. *Journal of Applied Meteorology and Climatology, 55*, 829–848.

Phillips, V., Pokrovsky, A., & Khain, A. (2007). The influence of time-depending melting on the dynamics and precipitation production in maritime and continental storm clouds. *Journal of the Atmospheric Sciences, 64*, 338–359.

Pruppacher, H., & Pitter, R. (1971). A semi-empirical determination of the shape of cloud and rain drops. *Journal of the Atmospheric Sciences, 28*, 86–94.

Pruppacher, H., & Klett, J. (1997). *Microphysics of clouds and precipitation* (Vol. 954). Dordrecht, The Netherlands: Kluwer Academic.

Rasmussen, R. M., Levizzani, V., & Pruppacher, H. R. (1984). A wind tunnel study on the melting behavior of atmospheric ice particles. III: Experiment and theory for spherical ice particles of radius >500 μm. *Journal of the Atmospheric Sciences, 41*, 381–388.

Ryan, B. (1996). On the global variation of precipitating layer clouds. *Bulletin of the American Meteorological Society, 77*, 53–70.

Ryzhkov, A. V., Zrnic, D. S., Hubbert, J. C., Bringi, V. N., Vivekanandan, J., & Brandes, E. A. (2002). Polarimetric radar observations and interpretation of co-cross-polar correlation coefficients. *Journal of Atmospheric and Oceanic Technology, 19*, 340–354.

Ryzhkov, A., & Zrnic, D. (2007). Depolarization in ice crystals and its effect on radar polarimetric measurements. *Journal of Atmospheric and Oceanic Technology, 24*, 1256–1267.

Ryzhkov, A., Pinsky, M., Pokrovsky, A., & Khain, A. (2011). Polarimetric radar observation operator for a cloud model with spectral microphysics. *Journal of Applied Meteorology and Climatology, 50*, 873–894.

Ryzhkov, A., Kumjian, M., Ganson, S., & Khain, A. (2013). Polarimetric radar characteristics of melting hail. Pt I: Theoretical simulations using spectral microphysical modeling. *Journal of Applied Meteorology and Climatology, 52*, 2849–2870.

Schuur, T. J., Ryzhkov, A. V., & Clabo, D. R. (2005). Climatological analysis of DSDs in Oklahoma as revealed by 2D-video disdrometer and polarimetric WSR-88D. In Preprints, *32nd Conference on Radar Meteorology*. CD-ROM, 15R.4.

Smith, P., Musil, D., Weber, S., Spahn, J., Johnson, G., & Sand, W. (1976). Raindrop and hailstone distributions inside storms. In *Preprints, International Conference on Cloud Physics* (pp. 252–257). Boulder, CO: American Meteor Society.

Spahn, J., & Smith, P. (1976). Some characteristics of hailstone size distributions inside hailstorms. In *Preprints, 17th Conference on Radar Meteorology* (pp. 187–191). Seattle, WA: American Meteorological Society.

Straka, J. M., Zrnic, D. S., & Ryzhkov, A. V. (2000). Bulk hydrometeor classification and quantification using multiparameter radar data. Synthesis of relations. *Journal of Applied Meteorology, 39*, 1341–1372.

Szyrmer, W., & Zawadzki, I. (2010). Snow studies. Part II: Average relationship between mass of snowflakes and their terminal fall velocity. *Journal of the Atmospheric Sciences, 67*, 3319–3335.

Testud, J., Oury, S., Black, R., Amayenc, P., & Dou, X. (2001). The concept of "normalized" distribution to describe raindrop spectra: A tool for cloud physics and cloud remote sensing. *Journal of Applied Meteorology, 40*, 1118–1140.

Thompson, E., Rutledge, S., Dolan, B., & Thurai, M. (2015). Drop size distributions and radar observations of convective and stratiform rain over the equatorial Indian and west Pacific Oceans. *Journal of the Atmospheric Sciences, 72*, 4091–4125.

Thurai, M., Gatlin, P., Bringi, V., Petersen, W., Kennedy, P., Notaros, B., et al. (2017). Toward completing the raindrop size spectrum: Case studies involving 2D-video disdrometer, droplet spectrometer, and polarimetric radar measurements. *Journal of Applied Meteorology and Climatology, 56*, 877–896.

Tiira, J., Moisseev, D., Lerber, A., Ori, D., Tokay, A., Bliven, L., et al. (2016). Ensemble mean density and its connection to other microphysical properties of falling snow as observed in southern Finland. *Atmospheric Measurement Techniques, 9*, 4825–4841.

Ulbrich, C. (1983). Natural variations in the analytical form of the raindrop size distribution. *Journal of Climate and Applied Meteorology, 22*, 1764–1775.

Ulbrich, C., & Atlas, D. (1982). Hail parameter relations: A comprehensive digest. *Journal of Applied Meteorology, 21*, 22–43.

Wen, L., Zhao, K., Zhang, G., Xue, M., Zhou, B., Liu, S., et al. (2016). Statistical characteristics of raindrop size distributions observed in East China during the Asian summer monsoon season using 2-D video disdrometer and Micro Rain Radar data. *Journal of Geophysical Research – Atmospheres, 121*, 2265–2282.

Woods, C., Stoelinga, T., & Locatelli, D. (2008). Size spectra of snow particles measured in wintertime precipitation in the Pacific Northwest. *Journal of the Atmospheric Sciences, 65*, 189–205.

Zawadzki, I., & De Agostinho Antonio, M. (1988). Equilibrium raindrop size distributions in tropical rain. *Journal of the Atmospheric Sciences, 45*, 3452–3459.

Zawadzki, I., Szyrmer, W., Bell, C., & Fabry, F. (2005). Modeling of the melting layer. Part III: The density effect. *Journal of the Atmospheric Sciences, 62*, 3705–3723.

Zhang, G., Vivekanandan, J., & Brandes, E. (2001). A method for estimating rain rate and drop size distribution from polarimetric radar measurements. *IEEE Transactions on Geoscience and Remote Sensing, 39*(4), 830–840.

Zrnic, D., & Ryzhkov, A. (2004). Polarimetric properties of chaff. *Journal of Atmospheric and Oceanic Technology, 21*, 1017–1024.

Chapter 5
Polarimetric Variables

The polarimetric variables can be either directly measured or estimated using polarimetric radar. These include reflectivity factors Z_H and Z_V at orthogonal polarizations, differential reflectivity Z_{DR}, cross-correlation coefficient ρ_{hv}, total differential phase Φ_{DP}, specific differential phase K_{DP}, backscatter differential phase δ, linear depolarization ratio *LDR*, circular depolarization ratio *CDR*, specific attenuation $A_{h,v}$ of horizontally/vertically polarized waves, and specific differential attenuation A_{DP}. In this chapter, we provide an overview of these variables, their physical meaning, and dependencies on the primary microphysical properties of hydrometeors such as size, concentration, shape, orientation, and physical composition. These dependencies are quantified via relatively simple formulas for small particles using Rayleigh approximation or T-matrix simulations for hydrometeors of larger size. The overview is given for liquid- (raindrops), solid- (ice, snow), and mixed-phase (melting ice/snow) hydrometeors in the S, C, and X microwave frequency bands commonly utilized for operational polarimetric weather radars.

5.1 Scattering Amplitudes and Cross Sections in the Rayleigh Approximation

As can be seen from (1.27) for the scattering matrix **S**, the scattering properties of atmospheric particles are determined by their orientations and scattering amplitudes s_a and s_b. Recall that according to definitions given in Sect. 1.2, s_a is the scattering amplitude if the electric field vector of the incident wave is parallel to the symmetry axis of the hydrometeor, and s_b stands for the scattering amplitude if the electric vector is perpendicular to the symmetry axis. The scattering of the incident microwave radiation which illuminates the particle occurs in all directions, and the scattering amplitude depends on the relative directions of propagation of the incident and scattered wave (see Fig. 1.3). As shown in Chap. 3, the polarimetric variables in

A. V. Ryzhkov, D. S. Zrnic, *Radar Polarimetry for Weather Observations*, Springer Atmospheric Sciences, https://doi.org/10.1007/978-3-030-05093-1_5

the backscattered direction such as Z_H, Z_{DR}, LDR, δ, and $|\rho_{hv}|$ depend on the backward scattering amplitudes $s_{a,b}^{(\pi)}$ corresponding to the case when the directions of propagation of the incident and scattered waves are opposite. Measurements of the forward scattered fields in the same direction as propagation yields Φ_{DP}, K_{DP}, A_H, and A_{DP} which depend on the forward scattering amplitudes $s_{a,b}^{(0)}$. The scattering amplitudes depend on the particle size, shape, and refractive index. The latter one is a function of particle physical properties (phase composition, density, water content, and temperature) and radar wavelength.

If the hydrometeors are relatively small compared to the radar wavelength λ and are modeled as oblate or prolate spheroids, then simple analytical formulas for the forward and backward scattering amplitudes $s_a^{(0,\pi)}$ and $s_b^{(0,\pi)}$ can be obtained in the Rayleigh approximation (Van de Hulst 1981):

$$s_{a,b}^{(0)} = s_{a,b}^{(\pi)} = \frac{\pi^2 D^3}{6\lambda^2} \frac{1}{L_{a,b} + \frac{1}{\varepsilon - 1}}, \tag{5.1}$$

where $D = (ab^2)^{1/3}$ is the equivolume diameter of the particle (a is rotation axis of spheroid), ε is the dielectric constant, and L_a and L_b are the shape parameters defined as

$$L_a = \frac{1+f^2}{f^2}\left(1 - \frac{\arctan f}{f}\right), \quad f = \sqrt{\frac{b^2}{a^2} - 1}, \quad L_b = \frac{1 - L_a}{2}, \tag{5.2}$$

for oblate spheroids ($a < b$) and

$$L_a = \frac{1 - e^2}{e^2}\left(\frac{1}{2e} \ln\left(\frac{1+e}{1-e}\right) - 1\right), \quad e = \sqrt{1 - \frac{b^2}{a^2}}, \quad L_b = \frac{1 - L_a}{2}, \tag{5.3}$$

for prolate spheroids ($a > b$).

The dependencies of shape parameters on the aspect ratio of oblate and prolate spheroids are illustrated in Fig. 5.1. Cross sections of spheroidal particle in the directions along their principle axes a and b are defined as.

$$\sigma_{a,b} = 4\pi \left| s_{a,b}^{(\pi)} \right|^2. \tag{5.4}$$

In the case of spherical particle ($a = b$), $L_a = L_b = 1/3$, and

$$s_a^{(\pi)} = s_b^{(\pi)} = \frac{\pi^2 D^3}{2\lambda^2} \frac{\varepsilon - 1}{\varepsilon + 2}. \tag{5.5}$$

Consequently

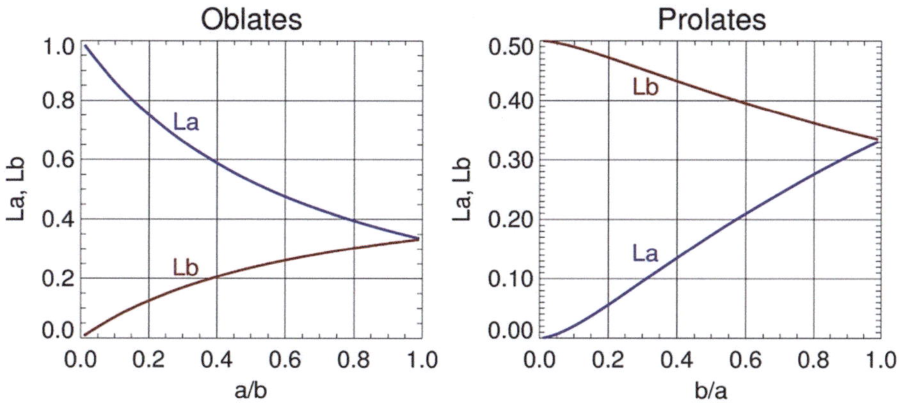

Fig. 5.1 Dependencies of the shape parameters $L_{a,b}$ on the aspect ratio of oblate and prolate spheroids

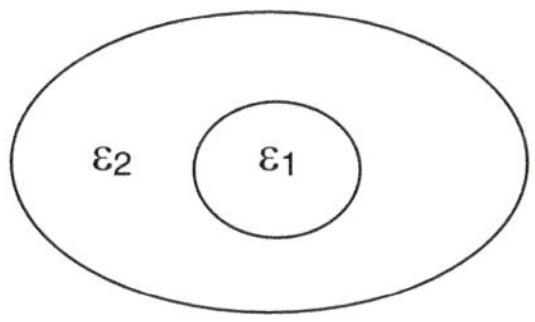

Fig. 5.2 Geometry of a two-layer spheroid

$$\sigma_a = \sigma_b = \frac{\pi^5 D^6}{\lambda^4}\left|\frac{\varepsilon - 1}{\varepsilon + 2}\right|^2. \tag{5.6}$$

Equation (5.1) is valid for uniformly filled spheroids. Some of the mixed-phase hydrometeors such as water-coated melting hailstones or graupel are better represented by two-layer spheroids (Fig. 5.2). In that case, the following more complicated formula for the scattering amplitudes $s_a^{(0,\pi)}$ and $s_b^{(0,\pi)}$ is applicable in the Rayleigh approximation (Bohren and Huffman 1983):

$$s_{a,b}^{(0)} = s_{a,b}^{(\pi)}$$
$$= \frac{\pi^2 D^3}{6\lambda^2} \frac{(\varepsilon_2 - 1)\left[\varepsilon_2 + (\varepsilon_1 - \varepsilon_2)\left(L_{a,b}^{(1)} - \xi L_{a,b}^{(2)}\right)\right] + \xi\varepsilon_2(\varepsilon_1 - \varepsilon_2)}{\left[\varepsilon_2 + (\varepsilon_1 - \varepsilon_2)\left(L_{a,b}^{(1)} - \xi L_{a,b}^{(2)}\right)\right]\left[1 + (\varepsilon_2 - 1)L_{a,b}^{(2)}\right] + \xi L_{a,b}^{(2)}\varepsilon_2(\varepsilon_1 - \varepsilon_2)}, \tag{5.7}$$

where ξ is the volume fraction of the inner spheroid and $L_{a,b}^{(1)}$ and $L_{a,b}^{(2)}$ are the shape parameters for the inner and outer spheroids. In general, $L_{a,b}^{(1)} \neq L_{a,b}^{(2)}$.

Note that in the Rayleigh approximation, there is no difference between the forward and backward scattering amplitudes. This is generally not so for particles

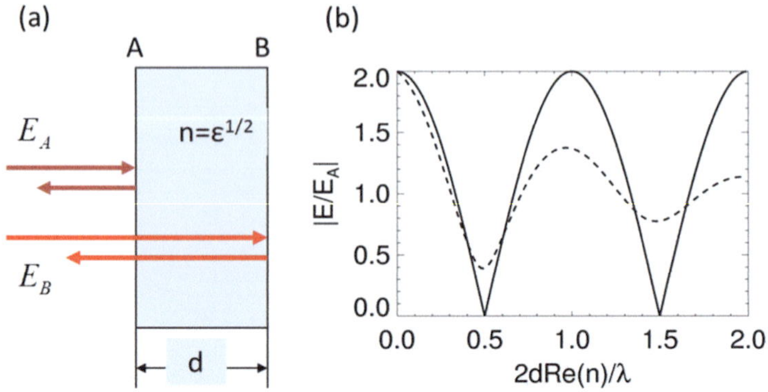

Fig. 5.3 Illustration of the interference concept, (**a**) slab with incident field *EA* and reflected fields off its front and back, and (**b**) ratio of the magnitudes of the reflected and incident field

with sizes comparable to or larger than the radar wavelength. The Rayleigh formulas for the scattering amplitudes are easy to use and instrumental for explaining the physical meaning of all polarimetric variables and their microphysical interpretation.

The Rayleigh approximation is valid if the ratio (or "resonance parameter") $RP = D|\varepsilon|^{1/2}/\lambda$ is less than 0.3–0.4 (Ryzhkov et al. 2011). As formulas (5.1), (5.6), and (5.7) show, the magnitudes of the scattering amplitudes $s_{a,b}$ and cross sections $\sigma_{a,b}$ increase in proportion to D^3 and D^6, respectively, as long as the equivolume diameter D is sufficiently small compared to the wavelength. The size dependence becomes more complex for larger D primarily because of the interference between the incident wave that reflects from the near side of the particle boundary and the wave which penetrates into the particle and is reflected back from the inner rear side of the hydrometeor boundary.

The concept of interference of the two waves is illustrated in the case of a dielectric slab illuminated by a plane electromagnetic wave at the incidence angle 90° (Fig. 5.3). The returned wave is a sum of the waves E_A reflected from boundary A and E_B which penetrates the slab and reflects from boundary B plus the first round trip wave within the slab (B to A to B then through A), etc. For simplicity, consider only the wave E_B. Generally it experiences amplitude loss if the slab is filled with absorbing material and acquires additional phase (compared to E_A) accumulated along the additional propagation path forth and back within the slab. The sum of the two reflected waves is

$$E = E_A + E_B = E_A\left[1 + \exp\left(-\frac{j4\pi dn}{\lambda}\right)\right] = E_A\left[1 + \eta\exp\left(-\frac{j4\pi d\mathrm{Re}(n)}{\lambda}\right)\right], \quad (5.8)$$

where d is the thickness of the slab, n is the refractive index of the medium filling the slab, and attenuation coefficient η is

$$\eta = \exp\left(\frac{4\pi d \mathrm{Im}(n)}{\lambda}\right). \tag{5.9}$$

A normalized amplitude of the sum wave E as a function of the parameter $d\mathrm{Re}(n)/\lambda$ for lossless slab ($\mathrm{Im}(n) = 0$) and a slab with losses ($\mathrm{Im}(n) < 0$) is plotted in Fig. 5.3b. It is obvious that the dependence of $|E|$ on the slab thickness d is nonmonotonic due to interference of the waves E_A and E_B. The resulting reflected signal is amplified if the two waves add in phase and is reduced if they are summed with opposite signs. The depth of such oscillations (or resonances) is larger for low-loss medium whereby almost complete cancellation of the two waves occurs.

Atmospheric hydrometeors are modeled by spheroids rather than dielectric slabs, but the essence of the interference is similar because the hydrometeor equivolume diameter D is roughly analogous to the slab depth d. It is not possible to offer simple close-form solutions of the scattering problem if the "resonance parameter" RP is larger than 0.3–0.4. Instead, "T-matrix" or other numerical methods are utilized for computing the scattering amplitudes (e.g., Waterman 1965; Barber and Yeh 1975; Mishchenko 2000; Bringi and Chandrasekar 2001). In the case of uniformly filled particle with large resonance parameter, the Mishchenko T-matrix code (2000) available on the web can be used. However, the Mishchenko code is not designed to handle two-layer spheroids like water-coated large hailstones. Instead, a two-layer version of the T-matrix code can be utilized (Bringi and Seliga 1977). This version of a code has been used, for example, by Aydin and Zhao (1990), Depue et al. (2007), and Ryzhkov et al. (2011).

Rayleigh approximation provides a very simple and economic way to compute scattering amplitudes for a wide range of practical problems. Moreover, closed-form solution allows straightforward physical interpretation of the results. These make Rayleigh formulas very attractive for converting the output of the numerical weather forecast models to the fields of polarimetric variables which are needed for assimilation of radar data into such models.

The limits of applications of the Rayleigh formulas at S ($\lambda = 10$–11 cm), C ($\lambda = 5$–6 cm), and X ($\lambda = 3.2$ cm) microwave frequencies commonly used for weather radars depend on tolerable errors in computation of the basic variables Z, Z_{DR}, and K_{DP}. Ryzhkov et al. (2011) evaluated these limits by comparing the estimates of Z, Z_{DR}, and K_{DP} from the Rayleigh formulas and T-matrix method. It was assumed that tolerated errors in the Z computation is 1 dB, Z_{DR} is 0.1–0.2 dB (Ryzhkov et al. 2005a), and a fractional error in the K_{DP} estimation is 20%. The accuracy of Z estimate of 1 dB is consistent with practically achievable precision of the Z measurements by weather radars. The 1 dB uncertainty in Z translates into the accuracy of rain rate (R) estimation of about 16% according to the basic NEXRAD $R(Z)$ relation $Z = 300\ R^{1.4}$. The needed accuracy of the K_{DP} computation which would yield the same error in rain rate according to the relation $R(K_{\mathrm{DP}}) = 44.0\ K_{\mathrm{DP}}^{0.822}$ at S band (Ryzhkov et al. 2005b) is about 20%. The accepted requirement to the accuracy of Z_{DR} estimation is 0.1–0.2 dB (Zrnic et al. 2010). Maximal sizes of raindrops and dense graupel/hail (with the density 0.92 g cm^{-3}) for which the Rayleigh formulas are valid at S, C, and X bands are listed in Table 5.1.

Table 5.1 Maximal sizes of raindrop and dense graupel/hail for which the Rayleigh formulas are valid at S, C, and X bands

Frequency band	Rain	Dry solid ice
S	$D < 6.3$ mm	$D < 10$ mm
C	$D < 3.4$ mm	$D < 5.1$ mm
X	$D < 2.3$ mm	$D < 3.0$ mm

5.2 Reflectivity

5.2.1 Rayleigh Formulas for Reflectivity

Reflectivity factor at horizontal polarization is defined as (3.29) (Doviak and Zrnic 2006)

$$Z_h = \frac{4\lambda^4}{\pi^4 |(\varepsilon_w - 1)/(\varepsilon_w + 2)|^2} < |S_{\text{hh}}|^2 > , \tag{5.10}$$

where ε_w is the dielectric constant of water and angular brackets indicate integration over the ensemble of particles. If the scatterers are spherical, then according to (3.43) and (5.5)

$$< |S_{\text{hh}}|^2 >=< \left|s_a^{(\pi)}\right|^2 >=< \left|s_b^{(\pi)}\right|^2 > \quad \text{and}$$

$$Z_h = \frac{< |(\varepsilon - 1)/(\varepsilon + 2)|^2 D^6 >}{|(\varepsilon_w - 1)/(\varepsilon_w + 2)|^2}. \tag{5.11}$$

For raindrops, $\varepsilon = \varepsilon_w$ and (5.11) is simplified to a well-known formula for the reflectivity factor

$$Z_h =< D^6 >= \int D^6 N(D) dD, \tag{5.12}$$

where $N(D)$ is a function describing size distribution of raindrops. If units of D and $N(D)$ in (5.12) are mm and m^{-3} mm^{-1}, respectively, then Z_h is expressed in mm^6 m^{-3}. After normalizing this Z by 1 mm^6 m^{-3} and taking 10 logarithms in base 10, one obtains the reflectivity in dBZ.

In the case of frozen particles whose dielectric constant does not change with size (e.g., graupel or small hail),

$$Z_h = \frac{|(\varepsilon - 1)/(\varepsilon + 2)|^2}{|(\varepsilon_w - 1)/(\varepsilon_w + 2)|^2} \int D^6 N(D) dD. \tag{5.13}$$

Comparing (5.12) and (5.13), it can be concluded that reflectivity factors of raindrops and frozen particles with identical size distributions differ by the ratio in front of integral in (5.13) which depends on the difference between dielectric constants of water and ice. In the case of solid ice with bulk density $\rho_i = 0.92$ g cm^{-3}, this ratio is about 0.19 for dielectric constants of water at different wavelengths and temperatures listed in Table 4.3. This means that the reflectivity factor of spherical solid ice particles is about 7.2 dB less than reflectivity of spherical liquid particles with the same size distribution provided that the particles are small with respect to the wavelength.

Dry snow usually has dielectric constant varying across its size spectrum because snow density decreases with increasing size (Eq. 4.25 and Fig. 4.18). Using (4.39), Eq. (5.13) can be rewritten for snow as

$$Z_h = \frac{|K_i|^2}{|K_w|^2 \rho_i^2} \int \rho_s^2(D) D^6 N(D) dD. \tag{5.14}$$

In (5.14), ρ_s is the density of snow which depends on D, $K_i = (\varepsilon_i - 1)/(\varepsilon_i + 2)$, and ε_i is the dielectric constant of solid ice with density ρ_i. Because ρ_s is almost inversely proportional to diameter of snowflakes (see (4.25)), the reflectivity factor of dry snow is approximately proportional to the fourth moment of size distribution as opposed to Z_h which is proportional to the sixth moment of size distributions for rain and graupel/small hail. This circumstance and because $|K_i|^2/|K_w|^2$ is less than 1 explain why the reflectivity of snow is usually much lower than the one of rain although snowflakes may have larger sizes and higher total concentrations than raindrops.

5.2.2 *Reflectivity of Raindrops*

Using (5.1) and (3.43), the most general formulas for the reflectivity factors at horizontal and vertical polarizations can be written as

$$\begin{aligned} Z_h &= \frac{4\lambda^4}{\pi^4 |K_w|^2} \left[< |s_b|^2 > -2\mathrm{Re} < s_b^* (s_b - s_a) > A_2 + < |s_b - s_a|^2 > A_4 \right], \\ Z_v &= \frac{4\lambda^4}{\pi^4 |K_w|^2} \left[< |s_b|^2 > -2\mathrm{Re} < s_b^* (s_b - s_a) > A_1 + < |s_b - s_a|^2 > A_3 \right], \end{aligned} \tag{5.15}$$

where the angular moments A_1–A_4 are defined by (3.45). Because raindrops have orientations of rotation axis approximately distributed according to the two-dimensional axisymmetric Gaussian distribution with the mean canting angle equal to zero, the angular moments of raindrops can be estimated using formulas (B.7)–(B.9), assuming that the width of the canting angle distribution σ is 10°

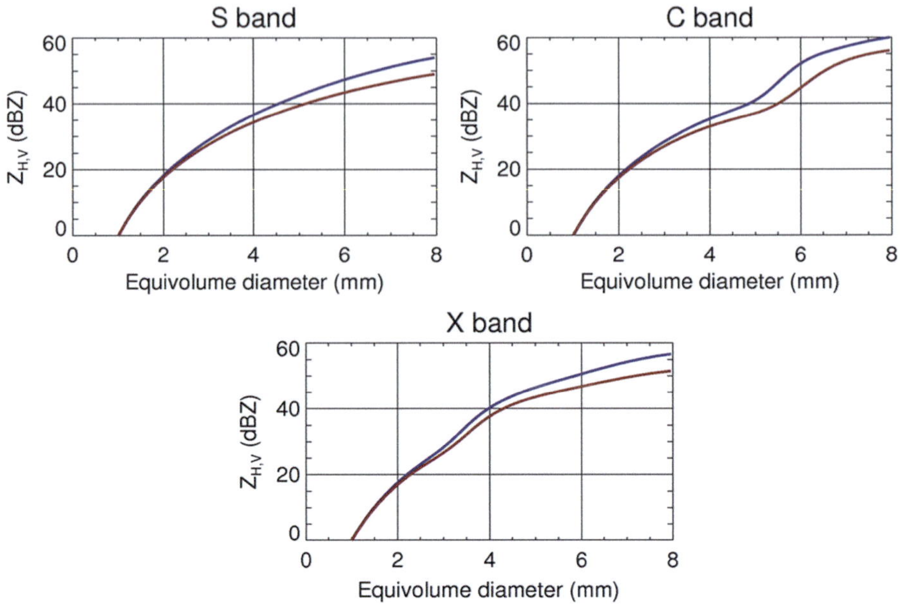

Fig. 5.4 Dependencies of normalized reflectivity factors at orthogonal polarizations on the equivolume raindrop diameter D at S (λ = 11.0 cm), C (λ = 5.45 cm), and X (λ = 3.2 cm) bands. Blue curves denote horizontal polarization and red curves are for vertical polarization

(Huang et al. 2008). The dependencies of normalized reflectivity factors at orthogonal polarizations on the equivolume raindrop diameter D at three radar frequencies are displayed in Fig. 5.4. For any given raindrop size D, the reflectivity factor shown in Fig. 5.4 is computed for the monodispersed size distribution with concentration of one raindrop per cubic meter (i.e., $N(D) = 1\ \mathrm{m}^{-3}$). The normalized reflectivity factor is expressed in logarithmic scale, i.e., $Z_{H,V} = 10\log(Z_{h,v})$, where $Z_{h,v}$ is in $\mathrm{mm}^6\ \mathrm{m}^{-3}$. It is also assumed that raindrops of a given size have variety of orientations described by the Gaussian distribution of canting angles with $\sigma = 10°$ and the shape of raindrop is determined by (4.29).

It is evident from Fig. 5.4 that $Z_H > Z_V$ for raindrops due to their oblate shape. There is practically no difference between reflectivities at the same polarization at all three wavelengths at raindrop sizes less than 2 mm because the Rayleigh approximation is valid. The resonance scattering effects become pronounced at C and X bands for larger raindrop sizes where reflectivity factors exhibit a wiggle and are different at the three wavelengths.

5.2.3 Reflectivity of Frozen Particles

As mentioned earlier, reflectivity of a frozen particle composed of dry ice is lower than reflectivity of a raindrop of the same size due to the difference in dielectric constant. This is illustrated in Fig. 5.5a where the Z_H-D dependencies for raindrops, graupel with density of solid ice ρ_i, and snowflakes with much lower density (4.25) are displayed for S band (red, green, and blue curves, respectively). It is noteworthy that the reflectivity of snowflake with $D > 4$ mm is almost 40 dB lower than that of raindrop with the same size.

Hailstones can be much larger than raindrops and graupel, and the effects of resonance scattering become obvious at sizes exceeding 10 mm where the parameter $RP = D|\varepsilon|^{1/2}/\lambda$ can be of the order of one or larger. The normalized reflectivity of dry graupel/hail at S, C, and X bands in a broad range of sizes from 0 to 40 mm is plotted in Fig. 5.6. The dependence of Z on D is nonmonotonic in the proximity of resonance sizes. It is also important that reflectivity at shorter wavelengths is significantly lower than at S band for $D > 10$ mm and the differences between Z(S), Z(C), and Z(X) can be as high as 20 dB at certain hail sizes. The positive

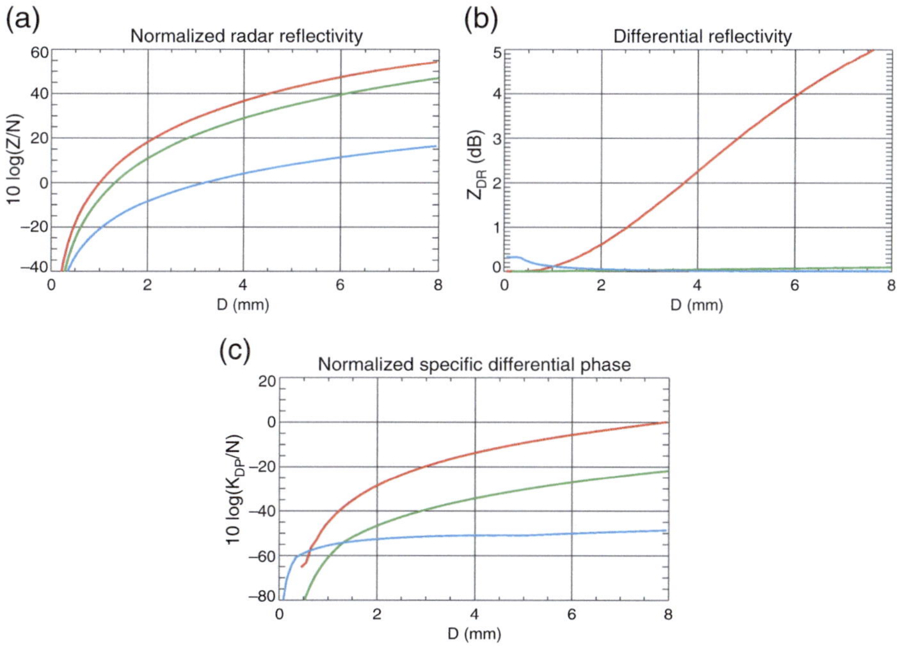

Fig. 5.5 Size dependencies of (**a**) normalized reflectivity, (**b**) differential reflectivity, and (**c**) normalized specific differential phase for raindrops (red curves), dry graupel (green curves), and dry aggregated snow (blue curves) at S band. Aspect ratios of raindrops r_w and graupel r_{gh} are determined from (4.29) and (4.31). For snow, aspect ratio is equal to 0.6. The width of canting angle distribution σ is equal to 10° for rain and 40° for graupel and snow. Density of graupel is equal to the density of solid ice ($\rho_i = 0.92$ g cm^{-3}), whereas density of snow decreases with size as prescribed by Eq. (4.25)

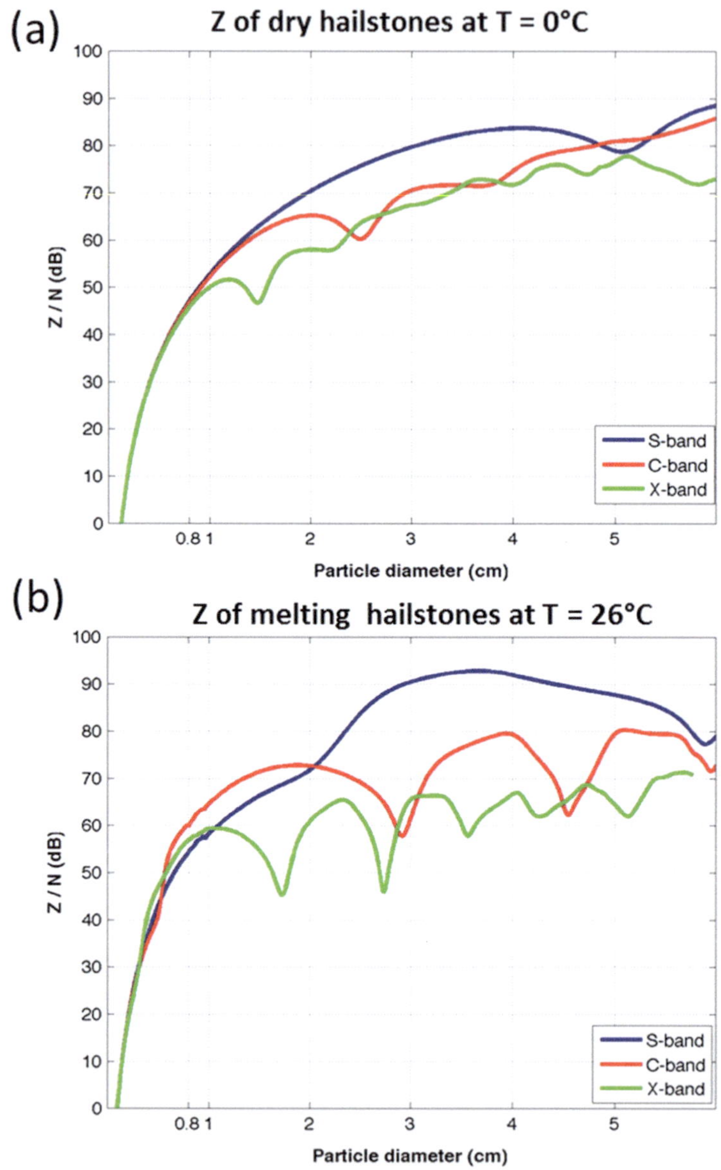

Fig. 5.6 Normalized reflectivity of dry (**a**) and melting (**b**) hail as a function of its size at three different wavelengths: $\lambda = 11.0$ cm (S band), $\lambda = 5.45$ cm (C band), and $\lambda = 3.2$ cm (X band). Mass water fraction f_w of melting hail follows the dependence shown in Fig. 5.8. Hailstones aspect ratio and the width of the canting angle distributions are determined by f_w using (4.33) and (4.35). At $D < 8$ mm, $f_w = 1$, i.e., hailstones are completely melted

difference Z(S)-Z(X) called "hail signal" was utilized for hail detection in early studies of hail with dual-frequency single-polarization radars (Eccles and Atlas 1973).

5.2.4 Reflectivity of Mixed-Phase Hydrometeors

Water-coated graupel particles or hailstones undergoing melting as well as ice-coated supercooled raindrops in the process of freezing can be modeled as two-layer spheroids. Reflectivity of such hydrometeors depends on the mass water fraction. This dependence can be quantified in the Rayleigh approximation using (5.7). Figure 5.7 shows the difference dZ between reflectivity factors for two-layer sphere and liquid sphere of the same diameter as a function of mass water fraction f_w for the case of water-coated ice (solid line) and ice-coated water enclosed in an ice shell (dashed line). As expected, the dZ varies between −7.2 dB (completely frozen sphere) and 0 dB (completely liquid sphere). However, reflectivity factors of water-coated and ice-coated particles are different for a given value of mass water fraction. The dependence for the water-coated ice reflects well-known fact that ice particle with relatively thin film of water reflects microwave radiation very similarly to raindrop of the same size (Battan 1973). Indeed, the reflectivity of water-coated sphere is within 1 dB of the reflectivity of totally liquid sphere if $f_w > 0.3$ or if the thickness of water film exceeds about 10% of the particle radius.

The dependence in Fig. 5.7 is valid only for relatively small particles of the graupel size which backscatter in the Rayleigh regime. Reflectivity of large melting hailstones can be calculated using either Mie solutions for spherical particles (e.g., Battan 1973; Bohren and Huffman 1983) or the T-matrix method for spheroidal hailstones (Aydin and Zhao 1990; Aydin et al. 1991; Vivekanandan et al. 1990, 1993; Depue et al. 2007; Snyder et al. 2010; Ryzhkov et al. 2013). Snyder et al. (2010) show the results for partially melted hail with 10% fractional water content

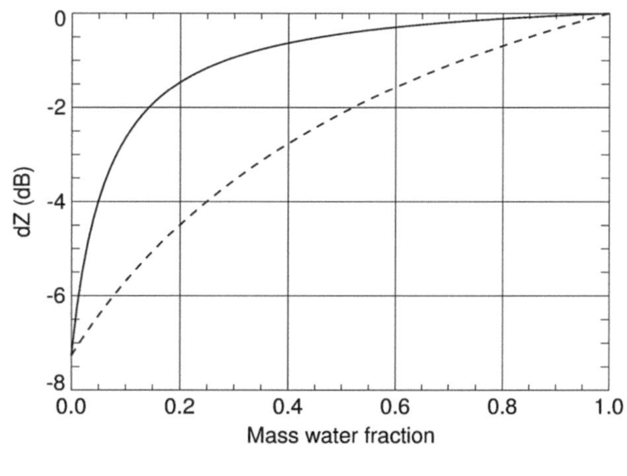

Fig. 5.7 The difference dZ between reflectivity factors of the two-layer sphere consisting of water and ice and liquid sphere with the same diameter. Solid line is for water-coated ice, and dashed line is for ice-coated water in an ice shell. Ice density $\rho_i = 0.92$ g cm^{-3}

at S, C, and X bands which is fixed for all hail sizes. However, in reality the mass water fraction f_w of melting hailstone increases with decreasing size because at the same thickness of water coating the f_w of smaller hailstones is larger and furthermore these stones melt faster. Aydin and Zhao (1990), Aydin et al. (1991), and Ryzhkov et al. (2013) estimate the dependence of f_w on D using the Rasmussen and Heymsfield (1987) thermodynamic model of melting hail. Ryzhkov et al. (2013) show that once the size of maximal raindrop originating from melting hail reaches 8 mm, the $f_w(D)$ dependence of melting hail does not change as further melting progresses. This dependence is presented in Fig. 5.8. Mass water fraction is equal to 1 in the rain part of the size spectrum and rapidly decreases with size in its hail portion (see Chap. 7). As described in Chap. 4, the aspect ratio and the degree of common alignment of melting hailstones linearly depend on f_w; hence the function $f_w(D)$ determines the shape and angular distribution of orientations of hailstones of different sizes.

The dependencies of normalized reflectivity factors at S, C, and X bands on the hydrometeor equivolume diameter for the case of rain/melting hail mixture are shown in Fig. 5.6b. Normalized reflectivities at all three microwave frequencies in the rain part of the spectrum ($D < 8$ mm) are equal to those displayed in Fig. 5.4 for pure rain. It is noteworthy that Z(S) < Z(C), Z(X) for raindrops larger than 6 mm and small melting hailstones with $D < 11$ mm, whereas for dry graupel/hail Z(S) > Z(C), Z(X) in wide range of graupel/hail sizes (Fig. 5.6a).

For larger melting hailstones ($D > 22$ mm), Z(S) may be significantly higher than Z(C) and/or Z(X). Well-pronounced dips in the Z(D) curves are caused by resonance scattering, and the hail sizes corresponding to these minima are larger at longer

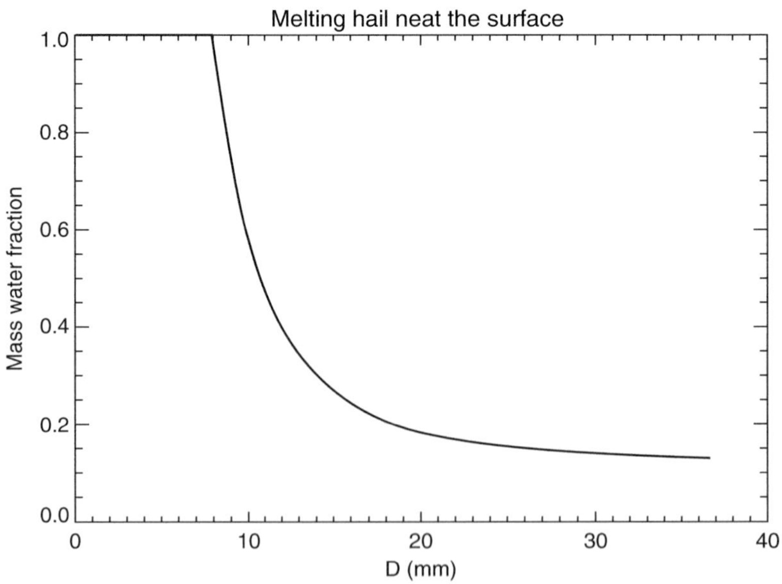

Fig. 5.8 Distribution of mass water fraction across size spectrum of rain/melting hail mixture

wavelength. It is convenient to use a generalized plot in Fig. 5.6b where reflectivities of rain and melting hail are combined to evaluate relative contributions of raindrops and hailstones in rain/hail mixture if size distributions of both hydrometeor types are known.

Another example of large mixed-phase hydrometeors is partially melted snowflakes within the melting layer commonly referred to as “the bright band.” Vertical profiles of reflectivity through the melting layer have been modeled and investigated in numerous studies (Ekpenyong and Srivastava 1970; Yokoyama and Tanaka 1984; Yokoyama et al. 1984; Klaassen 1988; Hardaker et al. 1995; Fabry and Zawadzki 1995; Russchenberg and Ligthart 1996; Fabry and Szyrmer 1999; Zawadzki et al. 2005 among others). The reflectivity factor of melting snowflakes is very sensitive to mass water fraction f_w and snow density both of which vary across the snowflake size spectra. The bright band is usually characterized by a strong maximum of reflectivity. Microphysical aspects of snow melting are discussed in more detail in Chap. 7. Herein, we use a typical f_w-D dependence at the height of the Z maximum within the melting layer which is derived from the model of melting snow assuming that initial density of dry snowflakes above the freezing level is described by (4.25) (see Fig. 5.9). Smaller snowflakes melt faster, and therefore f_w decreases with snowflake size.

Normalized Z as a function of equivolume diameter of wet snowflake is displayed in Fig. 5.10 for S, C, and X bands. It is evident that Z(S) is noticeably higher than Z(C) and Z(X) for melting snowflakes with diameters exceeding 2 mm. This explains the well-known experimental fact that Z enhancement within the bright band is

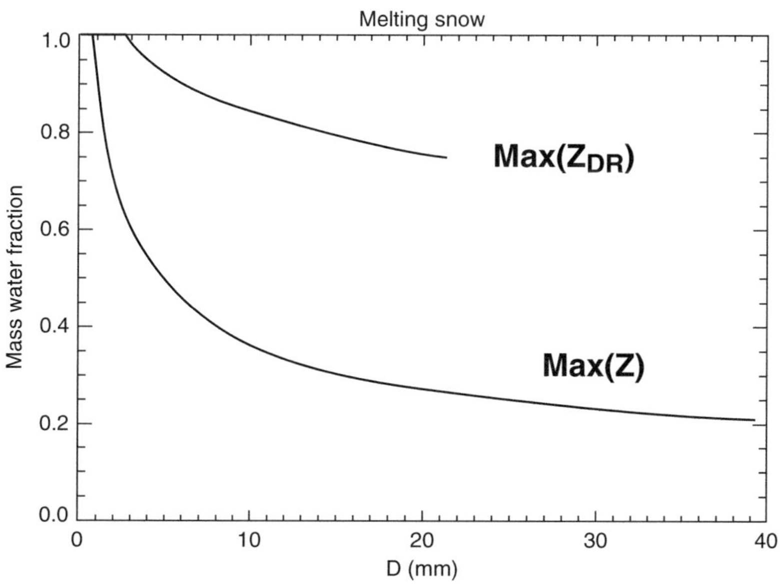

Fig. 5.9 Mass water fraction of melting snowflake as function of its size at the heights of the maximum of Z, Max (Z), and the maximum of Z_{DR}, Max (Z_{DR})

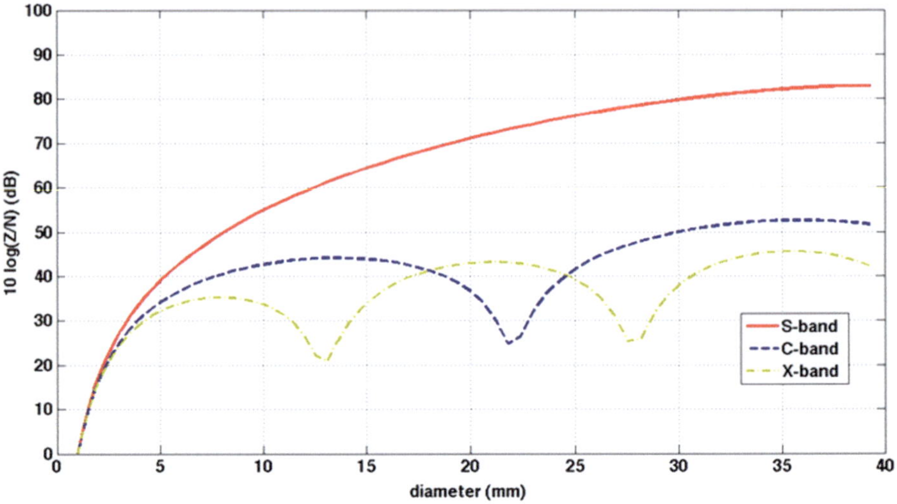

Fig. 5.10 Normalized reflectivity factor of melting snowflakes as a function of their equivolume diameter at S ($\lambda = 11.0$ cm), C ($\lambda = 5.45$ cm), and X ($\lambda = 3.2$ cm) bands at the height of the Z maximum within the melting layer (bright band). The dependence of mass water fraction f_w on diameter is shown in Fig. 5.9 (the Max (Z) curve). The density of dry snowflakes at the top of the melting layer is determined by (4.25)

stronger at longer wavelengths (Yokoyama et al. 1984; Fabry and Szyrmer 1999). The minima in the *Z-D* plots for C and X bands are attributed to the resonance scattering. The curves in Fig. 5.10 are obtained for unrimed snow with initial density described by (4.25). The corresponding *Z*(*D*) dependencies would be different for denser, rimed melting snow as more detailed analysis in Chap. 7 demonstrates.

5.3 Differential Reflectivity

5.3.1 Differential Reflectivity in the Rayleigh Approximation

Differential reflectivity is the most informative polarimetric variable (Zrnic et al. 2001), and it is very important to understand how Z_{DR} is affected by microphysical properties of hydrometeors. According to the definition, differential reflectivity expressed in linear scale is the ratio of reflectivity factors at horizontal and vertical polarizations determined by (5.15):

$$Z_{dr} = \frac{Z_h}{Z_v} = \frac{< |s_b|^2 > -2Re < s_b^*(s_b - s_a) > A_2 + < |s_b - s_a|^2 > A_4}{< |s_b|^2 > -2Re < s_b^*(s_b - s_a) > A_1 + < |s_b - s_a|^2 > A_3}, \tag{5.16}$$

where angular moments A_1–A_4 are defined by (3.45). If hydrometeors have completely random orientations, so that $A_1 = A_2$ and $A_3 = A_4$ (see Chap. 3), then $Z_h = Z_v$, and $Z_{dr} = 1$ (or $Z_{DR} = 10 \log(Z_{dr}) = 0$ dB). Because Z_{dr} is a ratio of Z_h and Z_v, it does not depend on the hydrometeor concentration.

In the case of equioriented hydrometeors, $A_1 = A_3 = 1$ and $A_2 = A_4 = 0$; hence

$$Z_{dr} = \frac{< |s_b|^2 >}{< |s_a|^2 >}. \tag{5.17}$$

For small particles which reflect microwave radiation in the Rayleigh regime, the scattering amplitudes $s_{a,b}$ can be estimated from (5.1) and for the monodispersed size distribution

$$Z_{dr} = \frac{|(\varepsilon - 1)L_a + 1|^2}{|(\varepsilon - 1)L_b + 1|^2}, \tag{5.18}$$

i.e., differential reflectivity depends on dielectric constant and shape parameters $L_{a,b}$. If the particle is spherical, then $L_a = L_b$ and $Z_{dr} = 1$.

For a given shape of the hydrometeor when L_a and L_b are fixed, the magnitude of Z_{dr} strongly depends on dielectric constant ε. Indeed, according to (5.18), Z_{dr} is close to 1 if $|\varepsilon - 1|$ tends to zero as in the case of dry snow so that $|\varepsilon - 1|L_{a,b} << 1$. In the opposite case of very high absolute value of ε, the magnitude of Z_{dr} approaches its upper limit $(L_a/L_b)^2$. Therefore, a raindrop has higher Z_{dr} compared to the snowflake or graupel with the same shape. This fundamental property of Z_{dr} is very beneficial for discrimination between liquid and frozen hydrometeors with dual-polarization radar.

5.3.2 *Differential Reflectivity of Raindrops*

Larger raindrops are more oblate than smaller drops, therefore differential reflectivity of raindrops increases with their size. Size dependencies of Z_{DR} on equivolume diameter of raindrop D for different wavelengths and temperatures are plotted in Fig. 5.11. The values of Z_{DR} are estimated using Eq. (5.16) where the scattering amplitudes $s_{a,b}$ are computed using the T-matrix method and the angular moments A_1–A_4 are calculated assuming that the width of the canting angle distribution σ of raindrops is equal to 10°. Raindrop shape-size relation (4.29) is utilized for Z_{DR} simulations.

Apparent deviations from plain monotonic curves (or "bumps") are visible in the $Z_{DR}(D)$ plots at C and X bands. They are caused by the resonance scattering at certain drop sizes which are different at C and X bands. In the proximity of the resonances, the values of Z_{DR} are noticeably different at the three radar wavelengths and are sensitive to the raindrop temperature. The most remarkable resonance feature

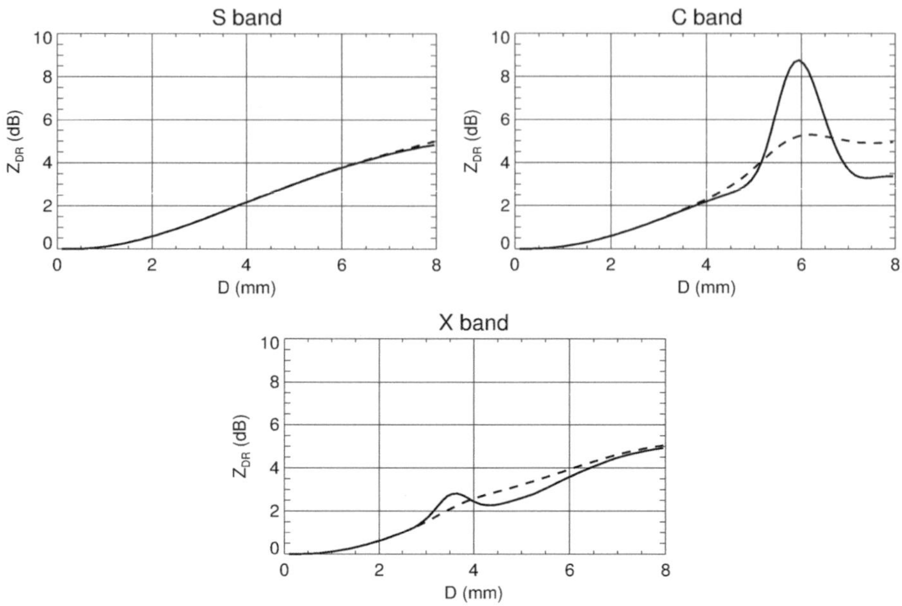

Fig. 5.11 Differential reflectivity as a function of raindrop diameter for different wavelengths and temperatures. Solid lines correspond to $T = 30$ °C and dashed lines are for $T = 0$ °C. Radar wavelengths are 11.0 cm at S band, 5.45 cm at C band, and 3.2 cm at X band

occurs at C band near $D = 6$ mm if the raindrops are warm ($T = 30$ °C). This resonance affects all the polarimetric variables in rain (Keenan et al. 2001). A weaker resonance affecting Z_{DR} takes place at X band at $D = 3.5$ mm. As mentioned in Sect. 5.1 (Fig. 5.3), the resonance effect is stronger for smaller absolute value of imaginary part of the refractive index (or dielectric constant). Because $|\mathrm{Im}(\varepsilon_w)|$ that characterizes absorption decreases with increasing temperature at all three microwave frequencies (Table 4.3), the resonances at higher temperatures are better pronounced. Moreover, since imaginary part of dielectric constant of water at C band is smaller (in the absolute value) than at X band, the resonance effect at C band is stronger.

5.3.3 *Differential Reflectivity of Frozen Particles*

The Z_{DR} of ice particles is significantly lower than that of raindrops with the same shape due to lower dielectric constant of ice. Z_{DR} is further reduced if frozen particles are tumbling and their orientations are more chaotic than raindrops orientations. Combination of these two factors makes differential reflectivity of dry graupel and aggregated snowflakes much lower compared to rain (Fig. 5.5b). Z_{DR} of dry aggregated snow is usually between 0 and 0.3 dB due to very low density of snow aggregates. On the other hand, pristine ice crystals may have extremely high Z_{DR}

due to their very low aspect ratio (for oblates) or very high aspect ratio (for prolates) and relatively high density. Such crystals are usually very small, and simple Rayleigh formulas can be used for quantitative estimation of Z_{DR}.

For ice crystals of oblate shape such as dendrites or plates, Eq. (5.18) can be used for computation of differential reflectivity assuming that oblate crystals have their rotation axis pointed vertically. The dependence of Z_{DR} on the axis ratio *a*/*b* and dielectric constant of solid ice $\varepsilon = 3.17$ for oblate and prolate crystals is shown in Fig. 5.12. The curves indicate that dendrites, hexagonal plates, and elementary needles having axis ratio less than 0.2 (see Fig. 4.20) may have high differential reflectivity within the range between 3 and 10 dB (see also Hogan et al. 2002).

Once pristine ice crystals start aggregating, their density decreases, and the products $L_{a,b}|\varepsilon - 1|$ in (5.18) become much smaller than 1. Therefore, taking into account (4.39), the expression (5.18) for Z_{dr} can be simplified as

$$Z_{dr} \approx 1 + \frac{6\rho_s}{\rho_i}(L_a - L_b)\frac{|\ \varepsilon_i - 1\ |}{|\ \varepsilon_i + 2\ |}, \tag{5.19}$$

and differential reflectivity Z_{DR} expressed in logarithmic scale is linearly proportional to the density of dry snow ρ_s. This is an important property of differential reflectivity which enables using Z_{DR} for evaluating the degree of snow aggregation and distinguishing between rimed and unrimed snow. But high accuracy of Z_{DR} measurements is needed because Z_{DR} of dry snowflakes is usually low due to their small ρ_s and rather chaotic orientation and remains within few tenths of dB.

Dry hailstones have nearly random orientation due to their tumbling falling behavior; hence differential reflectivity of dry hail is usually quite low and has positive sign for oblate hailstones with mean orientation of their rotation axis close to vertical. However, Z_{DR} can change sign at certain resonance sizes of hail even if horizontal dimension of hailstone is larger than vertical (Fig. 5.13a). Therefore, slightly negative Z_{DR} measurements in hail cores above the freezing level usually

Fig. 5.12 Differential reflectivity as a function of the axis ratio of oblate and prolate crystals

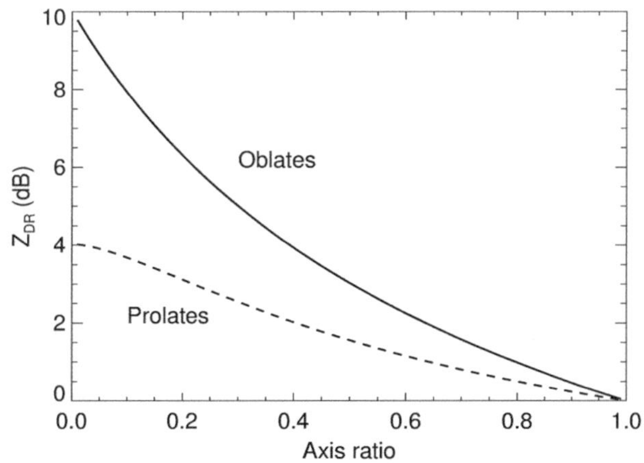

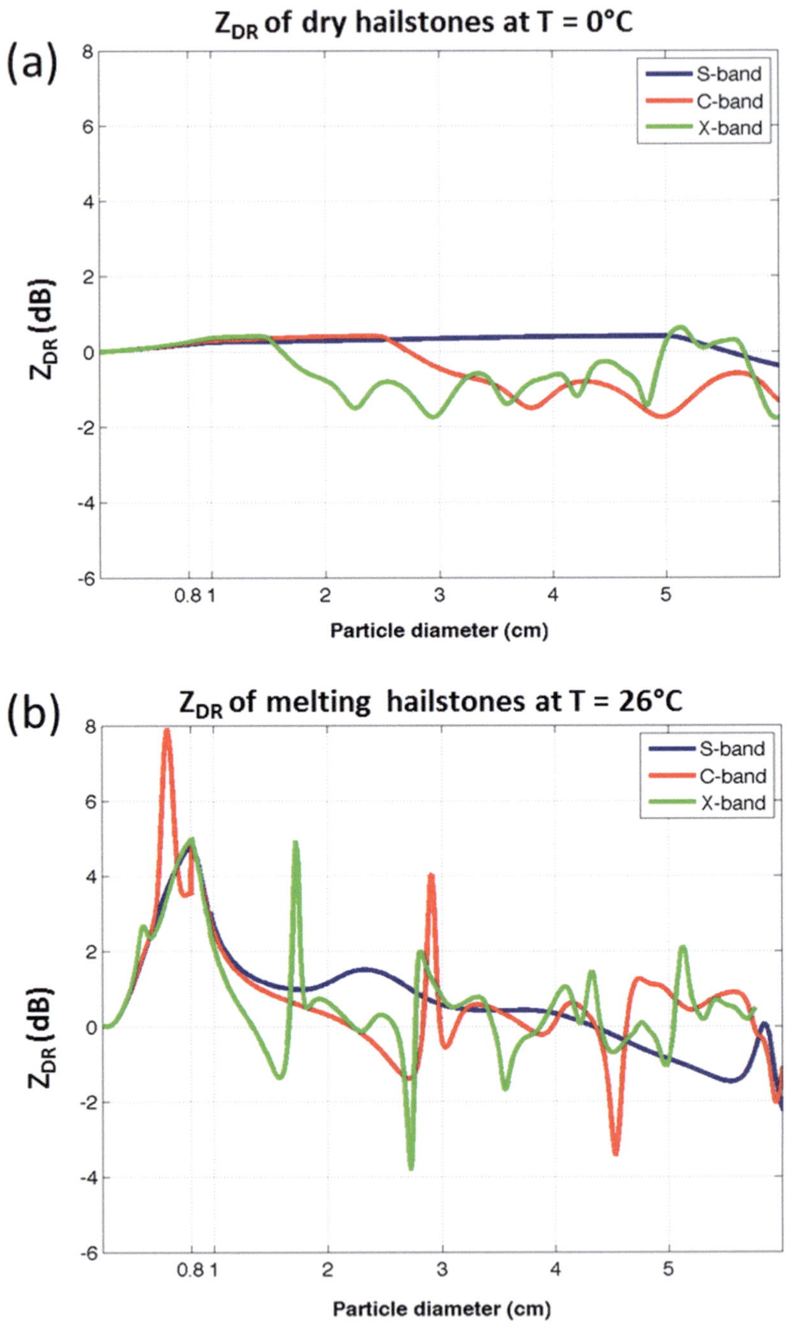

Fig. 5.13 Dependencies of differential reflectivity at S ($\lambda = 11.0$ cm), C ($\lambda = 5.45$ cm), and X ($\lambda = 3.2$ cm) bands on the equivolume diameter of hydrometeor for (**a**) dry hail and (**b**) rain/melting hail mixture. Mass water fraction f_w of melting hail follows the dependence shown in Fig. 5.8 (all

indicate the presence of large hail with resonance sizes. The resonance size of hail at which Z_{DR} changes its sign depends on the wavelength and hail density. As Fig. 5.13a indicates, Z_{DR} becomes negative if $D > 5$ cm at S band, $D > 2.5$ cm at C band, and $D > 1.5$ cm at X band for high density hail ($\rho = 0.92$ g cm^{-3}). Almost linear dependence of the resonance hail size on the radar wavelength is consistent with the interference concept of the resonance effects discussed in Sect. 5.1 and illustrated in Fig. 5.3, according to which the resonance size is proportional to $\lambda/\text{Re}(n)$. Because refractive index (Re (n)) of hail decreases with decreasing density, the resonance sizes tend to increase for softer hail. Computations of Z_{DR} illustrated in Fig. 5.13a are made for the width of the canting angle distributions of hailstones equal to 40°.

More random orientation of hailstones would result in smaller magnitude of Z_{DR} across the whole size spectrum of hail.

5.3.4 Differential Reflectivity of Mixed-Phase Hydrometeors

As in the case of reflectivity, differential reflectivity of water-coated ice spheroidal particle is close to that of liquid spheroidal particle of the same size even for relatively thin water coating (Fig. 5.14, thick solid line). For the same water fraction, Z_{DR} of water-coated spheroid (melting) is larger than Z_{DR} of ice-coated spheroid (freezing) as comparison of thick solid curve and thick dashed curves in Fig. 5.14 shows. Notable is the difference between the values of Z_{DR} computed for water–/ ice-coated hydrometeors and spongy hydrometeors of the same shape and size. Moreover, for a spongy particle, Z_{DR} depends on the choice of the matrix and inclusions in the water-ice uniform mixture due to the difference in dielectric constants computed using Eqs. (4.41) and (4.42).

Water-coated melting hailstones have higher Z_{DR} compared to dry hailstones due to the increase of effective dielectric constant and because the film of water on their surface tends to stabilize their orientation. The difference in Z_{DR} is particularly large for smaller size hail which melts faster and has higher fraction of water as shown in Fig. 5.8. Mass water fraction rapidly decreases with hail size, and thin film of water on the surface of large melting hailstones is not capable to facilitate changing their orientation significantly; thus they are still oriented more chaotically than smaller melting hailstones. As a result, smaller melting hailstones with sizes less than 10–12 mm have relatively high differential reflectivity similar to Z_{DR} of large raindrops, whereas larger melting hailstones are characterized by lower Z_{DR} which is not very different from the one for dry hailstones of similar size (Fig. 5.13b).

Fig. 5.13 (continued) hydrometeors with size less than 8 mm are raindrops). Hailstones aspect ratio and the width of the canting angle distributions are determined by f_w using (4.33) and (4.35). At $D < 8$ mm, $f_w = 1$, i.e., hailstones are completely melted

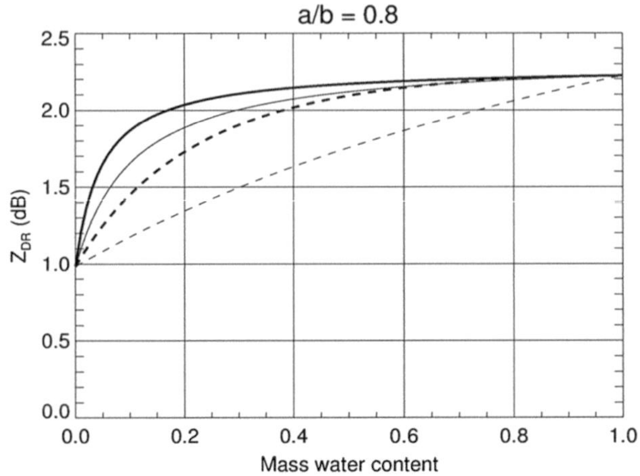

Fig. 5.14 Differential reflectivity as a function of mass water content for water-coated spheroid (thick solid line), ice-coated spheroid (thick dashed line), spongy spheroid with water matrix/ice inclusions (thin solid line), and spongy spheroid with ice matrix/water inclusions (thin dashed line). Computations are made at S band in the Rayleigh approximation using Eq. (5.7)

Additional discussion on differential reflectivity of the mixture of hail and rain can be found in Chaps. 7 and 8.

Differential reflectivity of melting snowflakes can be as high as 3 dB or more. The highest Z_{DR} is usually measured near the bottom of the melting layer in stratiform clouds where large wet snow aggregates are about to melt completely. These aggregates can have significant sizes (exceeding 1 cm) and large aspect ratios which make them reflect radar signals as giant oblate raindrops. Maximal differential reflectivity observed in the melting layer depends on the maximal size and density of snowflakes aloft. The maximal value of Z_{DR} within the melting layer increases with increasing size and decreasing density of snowflakes above the melting layer. Melting of rimed snow results in lower values of Z_{DR} within the melting layer. More details on polarimetric characteristics of the melting layer can be found in Chap. 7.

5.4 Specific Differential Phase

5.4.1 Specific Differential Phase in the Rayleigh Approximation

Specific differential phase K_{DP} plays important role for accurate measurements of precipitation with dual-polarization radar because it is less sensitive than Z to the variability of drop-size distributions and is immune to radar miscalibration, attenuation, and partial beam blockage. For oblate hydrometeors with mean orientation of their rotation axis along the vertical, K_{DP} is given by Eqs. (3.52)–(3.53):

$$K_{DP} = \lambda\,\mathrm{Re}\Big(< s_b^{(0)} > - < s_a^{(0)} >\Big)F_{\mathrm{orient}}, \tag{5.20}$$

where

$$F_{\mathrm{orient}} = \frac{1}{2} r_\sigma (1 + r_\sigma) \tag{5.21}$$

and $r_\sigma = \exp(-2\sigma^2)$.

Equation (5.20) can be modified to yield

$$K_{DP}\left(\mathrm{degkm}^{-1}\right) = \frac{0.18\lambda F_{\mathrm{orient}}}{\pi} \int Re\left[s_b^{(0)}(D) - s_a^{(0)}(D)\right]N(D)dD, \tag{5.22}$$

assuming that the parameter σ characterizing the spread of canting angles does not change with particle size. In (5.22), radar wavelength λ is in mm, scattering amplitudes $s_{a,b}^{(0)}$ are in mm, and $N(D)$ is in $\mathrm{m}^{-3}\ \mathrm{mm}^{-1}$. In the Rayleigh approximation,

$$Re\left[s_b^{(0)}(D) - s_a^{(0)}(D)\right] = \frac{\pi^2 D^3}{6\lambda^2} Re\left[\frac{1}{L_b + \frac{1}{\varepsilon - 1}} - \frac{1}{L_a + \frac{1}{\varepsilon - 1}}\right]. \tag{5.23}$$

If $L_{a,b}\,|\varepsilon - 1| >> 1$ like in the case of raindrops, then

$$\mathrm{Re}\left[\frac{1}{L_b + \frac{1}{\varepsilon - 1}} - \frac{1}{L_a + \frac{1}{\varepsilon - 1}}\right] \approx \frac{L_a - L_b}{L_a L_b} \tag{5.24}$$

and

$$K_{DP} \approx \frac{0.03\pi F_{\mathrm{orient}}}{\lambda} \int \frac{L_a - L_b}{L_a L_b} D^3 N(D)dD. \tag{5.25}$$

In the opposite case when $L_{a,b}\,|\varepsilon - 1| << 1$ (dry aggregated snow),

$$\mathrm{Re}\left[\frac{1}{L_b + \frac{1}{\varepsilon - 1}} - \frac{1}{L_a + \frac{1}{\varepsilon - 1}}\right] \approx \mathrm{Re}[(\varepsilon - 1)]^2 (L_a - L_b) \approx 9\frac{\rho_s^2}{\rho_i^2}\left|\frac{\varepsilon_i - 1}{\varepsilon_i + 2}\right|^2 (L_a - L_b) \tag{5.26}$$

and

$$K_{DP} \approx \frac{0.27\pi}{\lambda \rho_i^2}\left(\frac{\varepsilon_i - 1}{\varepsilon_i + 2}\right)^2 F_{\mathrm{orient}} \int F_{\mathrm{shape}} \rho_s^2(D) D^3 N(D)dD \tag{5.27}$$

where $F_{\mathrm{shape}} = L_a - L_b$.

It is clear from (5.25) and (5.27) that K_{DP} is inversely proportional to the radar wavelength and, similar to reflectivity, directly proportional to the concentration of particles. K_{DP} is equal to zero for spherical or randomly oriented scatterers. This is because $s_a^{(0)} = s_b^{(0)}$ for spherical hydrometeors and the factor F_{orient} in Eq. (5.25) is equal to zero for chaotically oriented scatterers (see Chap. 3). Because of the inverse dependence of ρ_s on the equivolume diameter for aggregated snow, K_{DP} is proportional to a first moment of snow size distribution or its mean diameter.

5.4.2 Specific Differential Phase of Raindrops

In order to separate effects of raindrop concentration and size on K_{DP}, we consider a normalized specific differential phase $K_{\mathrm{DP}}(D)/N$ determined as the ratio of K_{DP} for monodispersed size distribution with equivolume diameter D and total concentration of raindrops N. As in previous simulations for raindrops, it is assumed that drops of a given size have the width of the canting angle distribution equal to 10° and the shape of raindrops is governed by Equation (4.29). In the Rayleigh approximation, normalized K_{DP} monotonically increases with raindrop size as follows from (5.23). This may not be the case for larger drops of resonance sizes at C and X bands (Fig. 5.15). Moreover, K_{DP} can be even negative at C band for raindrops' sizes between 6 and

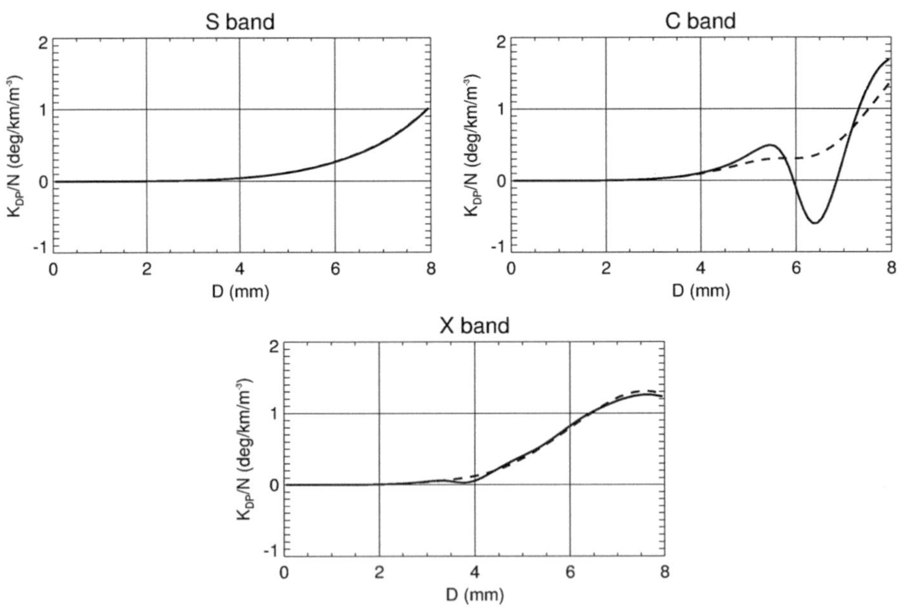

Fig. 5.15 Normalized specific differential phase as a function of equivolume raindrop diameter for different wavelengths and temperatures. Solid curves correspond to $T = 30$ °C; dashed curves are for $T = 0$ °C. Radar wavelengths are 11.0 cm at S band, 5.45 cm at C band, and 3.2 cm at X band

7 mm if their temperature is sufficiently high. This is due to stronger resonances at higher temperatures.

Generally, normalized K_{DP} increases with raindrop size but at a much slower rate than that of the normalized reflectivity factor (Fig. 5.5). Theoretical simulation by Sachidananda and Zrnic (1986) demonstrated that K_{DP} is proportional to $D^{4.24}$. Computations on a large set of DSD from Oklahoma produce $K_{DP}/N \sim D^{\alpha}$ where α varies between 4.3 and 4.9 for $D > 1$ mm, and it is somewhat higher for smaller raindrops, whereas $Z/N \sim D^6$. In other words, K_{DP} is a lower order moment of size distribution $N(D)$ than Z. This has important implications for radar rainfall measurements which is discussed in Chap. 10.

It follows from (5.23) that normalized K_{DP} is roughly proportional to the volume of the scatterer and integral K_{DP} is closely related to liquid water content LWC determined as

$$LWC = \frac{\pi \rho_w}{6} \int D^3 N(D) dD. \tag{5.28}$$

Theoretical simulations of K_{DP} and LWC using large set of disdrometer data in Oklahoma (see more details of this dataset in Chap. 4) indicate that

$$\begin{aligned} K_{DP} &\approx 0.30\, LWC^{1.40} \quad \text{at} \quad \text{S band} \\ K_{DP} &\approx 0.65\, LWC^{1.41} \quad \text{at} \quad \text{C band} \\ K_{DP} &\approx 0.89\, LWC^{1.40} \quad \text{at} \quad \text{X band} \end{aligned} \tag{5.29}$$

where K_{DP} is expressed in deg km^{-1} and LWC is in g m^{-3}. Because K_{DP} is not sensitive to temperature (except for resonance-size raindrops at C band), these formulas are valid at a wide range of temperatures.

5.4.3 *Specific Differential Phase of Frozen Particles*

Normalized K_{DP} of dry graupel/small hail is about two orders of magnitude less than normalized K_{DP} or raindrops with similar size (Fig. 5.5c). One of the reasons for this is lower dielectric constant of ice compared to water which is also responsible for reduction of normalized Z (Fig. 5.5a). Additional factors are more spherical shape and more chaotic orientation of graupel/small hail compared to raindrops. The dependencies of normalized K_{DP} on the equivolume diameter of dry hail in a wide size range (including large) are plotted in Fig. 5.16a for S, C, and X bands. Figure 5.16a shows that K_{DP} of hail becomes increasingly negative for hail sizes exceeding resonance sizes for each radar wavelength. This is somewhat similar to the dependence of Z_{DR} at large hailstone sizes (see Fig. 5.13a).

Specific differential phase of dry aggregated snow is usually very small (Fig. 5.5c) and even smaller than that of the same size graupel due to lower density

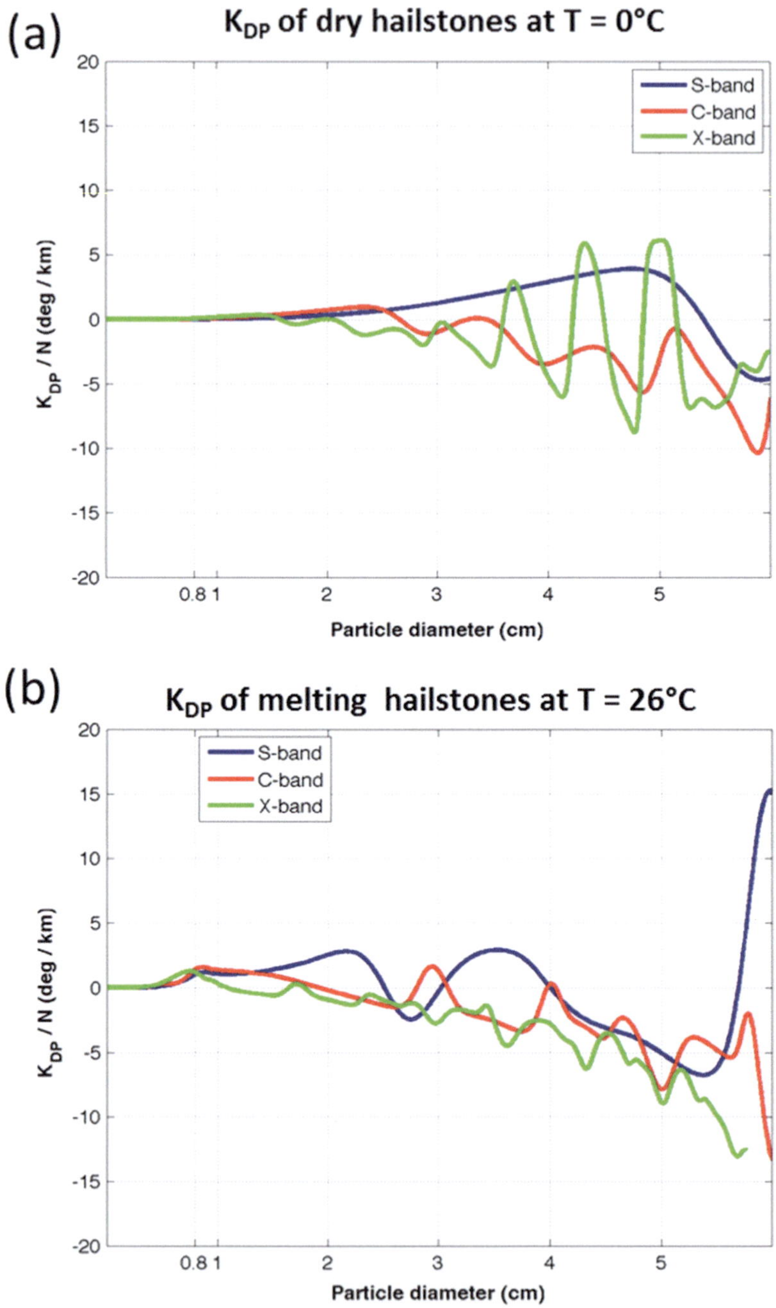

Fig. 5.16 Dependencies of normalized specific differential phase at S (λ = 11.0 cm), C (λ = 5.45 cm), and X (λ = 3.2 cm) bands on the equivolume diameter of hydrometeor for (**a**) dry hail and (**b**) rain/melting hail mixture. Mass water fraction f_w of melting hail follows the dependence

of dry aggregated snow. As follows from (5.27), normalized K_{DP} of snow is proportional to the product $\rho_s^2 D^3$. Because bulk density of snow ρ_s decreases with D as specified by (4.25), this product grows slowly with increasing diameter, and the magnitude of K_{DP} is approximately proportional to the first moment of size distribution of snow particles. Therefore, in the case of aggregated snow, the slope of the size dependency of normalized K_{DP} is much smaller than the slope for raindrops and graupel (Fig. 5.5c). Big difference in the magnitude of K_{DP} between liquid and frozen hydrometeors makes it very useful (along with Z_{DR}) to discriminate between raindrops and such frozen hydrometeors as graupel/small hail and dry aggregated snow with low density.

There is a class of frozen particles for which K_{DP} can be significant and even higher than in rain. These are pristine crystals like plates, columns, needles, and dendrites with high density, large nonsphericity, and high degree of common alignment. The studies of Vivekanandan et al. 1994) and Ryzhkov et al. (1998) at S band indicate that in ice crystals the specific differential phase is almost linearly proportional to ice water content IWC:

$$K_{DP} \approx (0.31 - 0.33) IWC. \tag{5.30}$$

In (5.30), IWC is expressed in g m^{-3}. Because K_{DP} is inversely proportional to the wavelength, and most of pristine crystals are Rayleigh scatterers in the three microwave frequency bands, similar K_{DP}-IWC relations can be obtained at C and X bands provided that the coefficient of proportionality between K_{DP} and IWC is scaled inversely to the wavelength. Relation (5.30) assumes that crystals are oriented with their major dimension in the horizontal plane. In the presence of thunderstorm electric fields capable to orient crystals in the direction of electric field vector, K_{DP} can change the sign if the projection of a crystal onto the vertical is larger than its horizontal projection. In this sense, K_{DP} can provide information about electrical processes in the clouds. Oriented crystals can also cause interesting effects of depolarization in the fields of differential reflectivity; these are described in Chap. 6.

5.4.4 *Specific Differential Phase of Mixed-Phase Hydrometeors*

Melting of hydrometeors affects Z_{DR} and K_{DP} in a similar way. The magnitudes of both variables tend to increase as water coating thickens due to higher effective dielectric constant and more stable and ordered orientation of progressively melting

Fig. 5.16 (continued) shown in Fig. 5.8 (all hydrometeors with size less than 8 mm are raindrops). Hailstones aspect ratio and the width of the canting angle distributions are determined by f_w using (4.33) and (4.35). At $D < 8$ mm, $f_w = 1$, i.e., hailstones are completely melted hydrometeor. Normalized K_{DP} of melting hail is larger than the one of dry hail for hail of relatively smaller size as the comparison of Fig. 5.16a, b indicates. However, resonance diameters of melting hailstones become smaller as melting progresses which results in the prevalence of negative K_{DP} in a wider range of hail sizes.

Melting hail generally coexists with rain which is produced from complete melting of graupel and hailstones of smaller size. It is important to note that the contribution of melting hail with diameters larger than the diameter of biggest stable raindrop (i.e., about 8 mm) to the total K_{DP} of the melting hail/rain mixture is relatively small compared to the contribution of large raindrops. This is not the case for the reflectivity factor which may be dominated by the scattering from larger size hail if it is mixed with rain. Generalized size dependencies of normalized Z and K_{DP} which combine rain (for $D < 8$ mm) and melting hail (for $D > 8$ mm) in Figs. 5.6b and 5.16b explain this difference. Let's, for example, compare Z/N and K_{DP}/N for raindrop of 4 mm equivolume diameter and melting hailstone with 20 mm diameter. The values of normalized Z differ by about 35 dB which is equivalent to the difference of 3162 times in linear scale. On the other hand, the corresponding value of normalized K_{DP} is only 4.5 times larger for the melting hailstone. Such a small increase is easily offset by the difference in the concentration of rain and hail because for typical size distributions of rain and hail particles, the concentration of 20 mm hailstones is usually more than two orders of magnitude smaller than the concentration of 4 mm raindrops (see Chap. 4).

Relative insensitivity of K_{DP} to the presence of hail in the rain/hail mixture makes it a very attractive parameter for discrimination between rain and hail and for rainfall measurements in the situations when rain is mixed with hail (e.g., Balakrishnan and Zrnic 1990; Ryzhkov et al. 2005b) (see Chap. 10).

Specific differential phase of wet snow can be significant due to large size and high dielectric constant of melting snowflakes. Modeling study of the melting layer in stratiform precipitation by Ryzhkov et al. (2008) shows that K_{DP} within the melting layer can be more than two times higher than the corresponding K_{DP} in rain beneath the bright band. These modeling results agree quite well with measurements of K_{DP} in the transition region between rain and snow near the surface (Ryzhkov and Zrnic 1998) and with measured values of K_{DP} within the melting layer (Troemel et al. 2017; Griffin et al. 2018). Wet snowflakes have more spherical shapes and more random orientations than pristine crystals such as plates and columns; hence K_{DP} of wet snow, although higher than the one in stratiform rain, is usually lower than K_{DP} associated with ice crystals observed at subfreezing temperatures, particularly in the dendritic growth layer (see Chap. 7).

5.5 Backscatter Differential Phase

Backscatter differential phase δ is a difference between the phases of the horizontal and vertical components of the backscattered wave upon reflection from the scatterers filling the radar resolution volume so that (see Eq. (3.38))

$$\begin{aligned}\delta &= \arg\left(<S_{hh}^{*}S_{vv}>\right)\\ &= \arg\left(<|s_b|^2>+<|s_b-s_a|^2>A_5-<s_b^{*}(s_b-s_a)>A_1-<s_b(s_b^{*}-s_a^{*})>A_2\right),\end{aligned} \tag{5.31}$$

where definitions of angular moments A_1, A_2, and A_5 are listed in (3.45). For equioriented hydrometeors, $A_1 = 1$, $A_2 = A_5 = 0$, and

$$\delta = \arg\left(< s_b^{*} s_a >\right). \tag{5.32}$$

In the case of fully random orientation of particles, $A_1 = A_2$, and the right-hand side of (5.31) becomes real which means that $\delta = 0$. Hence, backscatter differential phase decreases as hydrometeors become more randomly oriented.

5.5.1 *Backscatter Differential Phase of Raindrops*

The magnitude of backscatter differential phase in rain strongly depends on the radar wavelength and temperature and increases with raindrop size as Fig. 5.17 shows. It is insignificant at S band but may exceed 20–40° for large raindrop sizes at C and X bands. The magnitude of δ is small for drop diameters less than 4.5 mm at C band and 2.5 mm at X band. Notable is a nonmonotonic variation of δ as a function of raindrop size at higher temperatures at C band. This is a direct consequence of resonance scattering by raindrops with sizes larger than 5 mm. Compared to C band, the resonance effects are weaker at X band.

Note that the Rayleigh formulas (5.1) for the scattering amplitudes s_a and s_b can be used for computations of δ from Eq. (5.32) only for very small raindrop sizes which produce backscatter differential phase almost indistinguishable from zero. This stems from the fact that δ crucially depends on the imaginary parts of the scattering amplitudes which are not well approximated by the Rayleigh solution, so that the T-matrix code has to be utilized almost exclusively for computation of the backscatter differential phase.

Because total differential phase measured by the radar is a sum of backscatter differential phase δ and propagation differential phase determined by K_{DP}

$$\Phi_{DP}(r) = \delta + 2\int_0^r K_{DP}(s)ds, \tag{5.33}$$

the contributions from the backscattered and propagation components of Φ_{DP} should be carefully separated before specific differential phase K_{DP} is estimated from the slope of the range dependence of Φ_{DP}. The "bumps" in the radial profiles of Φ_{DP} associated with enhanced δ in the areas containing large drops with higher Z_{DR} are

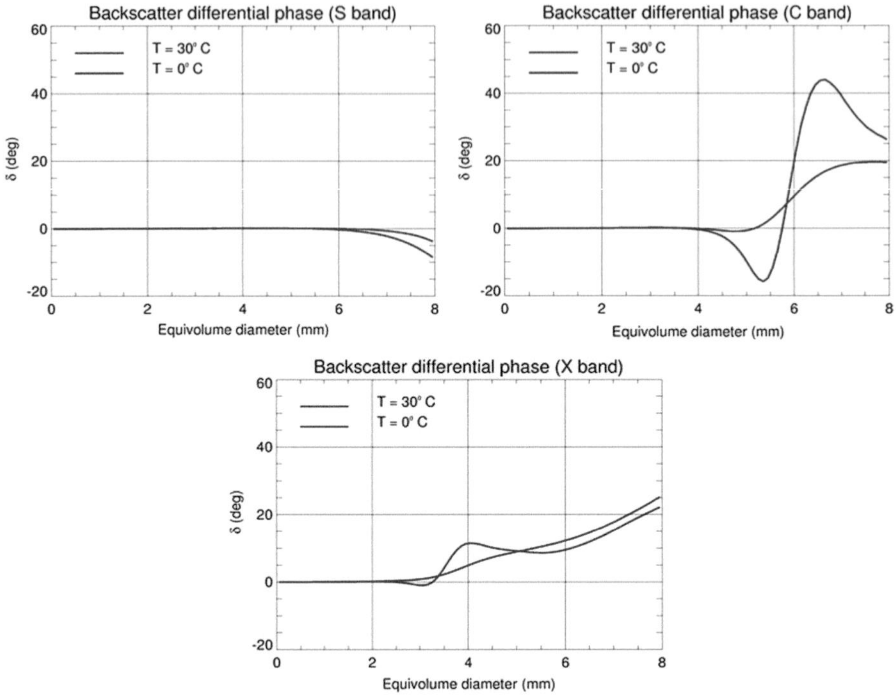

Fig. 5.17 Backscatter differential phase as a function of equivolume raindrop diameter for different radar wavelengths and temperatures. Radar wavelengths are 11.0 cm at S band, 5.45 cm at C band, and 3.2 cm at X band

occasionally observed at shorter radar wavelengths (X and C bands) (e.g., Otto and Russchenberg 2011; Troemel et al. 2013).

5.5.2 *Backscatter Differential Phase of Frozen and Mixed-Phase Hydrometeors*

Backscatter differential phase of dry hydrometeors is usually very low due to small imaginary part of dielectric constant of ice. Thus, δ is negligibly small for most ice hydrometeor types including snow, crystals, and graupel. Nevertheless, large dry hailstones of resonance size may have noticeable backscatter differential phase which is highest at X band as Fig. 5.18a shows. Wet hailstones may have much higher δ that changes its sign at the resonance sizes of melting hail which depend on the radar wavelength (Fig. 5.18b). These sizes are about 5.8 cm for S band, 2.9 cm for C band, and 1.7–1.8 cm for X band. Backscatter differential phase can be significant in wet snow or in the melting layer. Radial profiles of Φ_{DP} are often nonmonotonic within the bright band which is a strong indication of a substantial

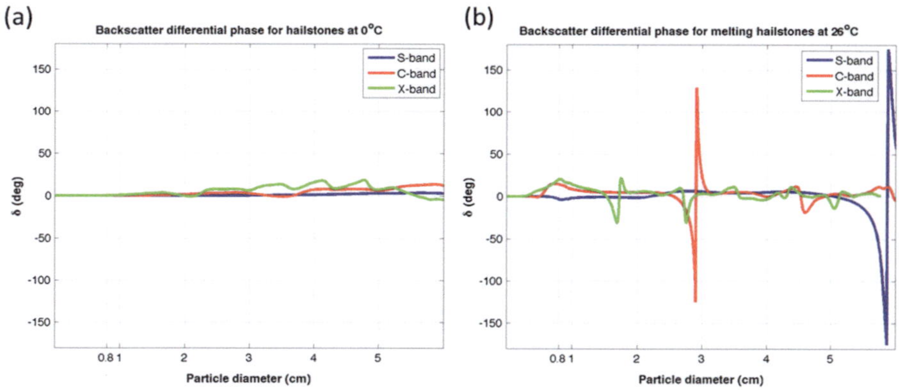

Fig. 5.18 Dependencies of backscatter differential phase at S ($\lambda = 11.0$ cm), C ($\lambda = 5.45$ cm), and X ($\lambda = 5.45$ cm) bands on the equivolume diameter of hydrometeor for (**a**) dry hail and (**b**) rain/melting hail mixture. Mass water fraction f_w of melting hail follows the dependence shown in Fig. 5.8 (all hydrometeors with size less than 8 mm are raindrops). Hailstones aspect ratio and the width of the canting angle distributions are determined by f_w using (4.33) and (4.35). At $D < 8$ mm, $f_w = 1$, i.e., hailstones are completely melted

backscatter differential phase. It is difficult to reliably estimate δ within the melting layer because differential phase is very noisy due to reduced values of the cross-correlation coefficient ρ_{hv} which determines the statistical accuracy of the estimates of all polarimetric radar variables as is shown in Chap. 6.

Nevertheless, the backscatter differential phase within the melting layer can be quantified via azimuthal averaging of the Φ_{DP} data collected at sufficiently high elevation angle following a so-called quasi-vertical profiles (QVP) methodology described in Ryzhkov et al. (2016). Such spatial averaging dramatically reduces the standard deviation of the Φ_{DP} estimate. In the studies of Zrnic et al. (1993), Troemel et al. (2013, 2014, 2017), and Ryzhkov et al. (2016), it was shown that the magnitude of δ within the melting layer (ML) usually varies between 5–10° but occasionally can be significant at S band. The corresponding maximal δ values are somewhat lower at C band and significantly lower at X band. Troemel et al. (2014) hypothesize that the magnitude of δ within the ML is determined by the maximal size of wet snowflakes falling through it. The decrease of δ at shorter radar wavelengths in the ML might be explained by stronger resonance scattering effects for large wet snowflakes similar to the decrease of the maximal reflectivity factor within the bright band observed at shorter wavelengths (see Fig. 5.10).

5.6 Depolarization Ratios

Historically, the first dual-polarization weather radars transmitted waves either with linear or circular polarizations and received both the copolar and cross-polar components of the reflected wave. The ratio of powers of these components is called depolarization ratio. By definition, the linear depolarization ratio (L_{dr}) is the ratio of the power of a cross-polar radar return to the power of a copolar radar return if a wave with horizontal (or vertical) polarization is transmitted (e.g., Bringi and Chandrasekar 2001). Circular depolarization ratio (C_{dr}) is the ratio of the power of the copolar component of the reflected signal (e.g., left-hand circular) to its cross-polar component (e.g., right-hand circular) (McCormick and Hendry 1975). Depolarization ratios are usually represented in a logarithmic scale ($LDR = 10\log(L_{dr})$ and $CDR = 10\log(C_{dr})$).

According to definition, linear depolarization ratio in the absence of propagation effects is given by (3.39)

$$L_{dr} = \frac{<|S_{hv}|^2>}{<|S_{hh}|^2>} = \frac{<|s_b - s_a|^2> A_5}{<|s_b|^2> -2Re<s_b^*(s_b - s_a)> A_2 + <|s_b - s_a|^2>}, \tag{5.34}$$

and it is directly proportional to the angular moment A_5. For equioriented spheroidal hydrometeors ($\sigma_\alpha = 0$), $A_5 = 0$, and $L_{dr} = 0$ if the particle axis of rotation is vertical, and $A_5 \neq 0$ ($L_{dr} \neq 0$) if the orientation of particles is neither vertical nor horizontal. If the orientations of hydrometeors are fully chaotic or chaotic in the horizontal plane with antenna elevation angle $\beta > 0$, then the angular moment A_5 and value of L_{dr} are also different from zero.

For two-dimensional axisymmetric Gaussian distribution of orientation defined by (3.46) in Chap. 3, the expression for A_5 is

$$A_5 = \frac{1}{8}\left(1 - r_\sigma^4\right)\left(\frac{3}{8} + \frac{1}{2}r_\sigma + \frac{1}{8}r_\sigma^4\right), \tag{5.35}$$

which yields $A_5 \approx \sigma_\alpha^2$ for relatively small width of the canting angle distribution σ_α. Hence, L_{dr} is generally higher for more randomly oriented hydrometeors with a given shape and dielectric constant.

It can be shown (Jameson 1987; Ryzhkov 2001; Matrosov et al. 2001) that the circular depolarization ratio C_{dr} depends primarily on the scatterers' shape and phase composition and only slightly on their orientation so that it can be roughly approximated by the formula:

$$C_{dr} \approx \frac{<|s_b - s_a|^2>}{<|s_b + s_a|^2>}. \tag{5.36}$$

Ryzhkov (2001) showed that the difference between *CDR* and *LDR* depends primarily on particle orientations, so joint measurements of the two depolarization ratios can be used to infer hydrometeor fall habits (e.g., Matrosov et al. 2005). If the hydrometeors are oriented randomly, then the difference of *CDR-LDR* is roughly 3 dB. For relatively narrow distributions of the canting angles described by (B.7) (Ryzhkov 2001)

$$\frac{C_{\mathrm{dr}}}{L_{\mathrm{dr}}} \approx \frac{< |s_b|^2 >}{< |s_b + s_a|^2 >} \frac{1}{\sigma_\alpha^2 + < \alpha >^2}, \tag{5.37}$$

where $<\alpha>$ is the mean canting angle. In rain, $<\alpha> \approx 0$, and the factor containing s_a and s_b varies between 0.25 and 0.35; hence the ratio $C_{\mathrm{dr}}/L_{\mathrm{dr}}$ (or the difference of *CDR-LDR* expressed in dB) is a measure of σ_α.

Although circularly polarized meteorological radars have been used in the past, modern operational dual-polarization radars increasingly use simultaneous transmission and reception (SHV) of the waves of horizontal and vertical polarizations with a single transmitter (Doviak et al. 2000) which enables measuring differential reflectivity Z_{DR}, differential phase Φ_{DP}, and copolar correlation coefficient ρ_{hv} but not *LDR*. Some polarimetric weather radars optionally measure *LDR* in a special HSHV mode of operation (Sect. 3.3); then only a horizontally polarized wave is transmitted, and both the horizontally and vertically polarized radar returns are received. This mode of operation, however, requires a polarization switch on transmission, and simultaneous measurements of Z_{DR} and *LDR* are not possible. Most users do not utilize the HSHV mode because switching back and forth between the SHV and HSHV modes slows down the radar data update and causes premature wear of the high-power switch.

Matrosov (2004) and Ryzhkov et al. (2017) suggested a special methodology for estimating the depolarization ratio (*DR*) by dual-polarization radars with simultaneous transmission/reception of orthogonally polarized waves together with traditionally measured Z_{DR}, Φ_{DP}, and ρ_{hv} in a single mode of operation. This depolarization ratio can serve as a proxy for circular depolarization ratio (*CDR*) measured by radars with circular polarization. The suggested methodology implies the use of a high-power phase shifter to control the system differential phase on transmission and a special signal processing to eliminate the detrimental impact of differential phase on the estimate of *DR*.

5.6.1 Depolarization Ratios of Raindrops

Linear depolarization ratio (*LDR*) depends both on the shape and orientation of raindrops, whereas *CDR* is mostly a function of the shape. Because the aspect ratio of raindrops decreases with their size, both *LDR* and *CDR* generally increase with raindrop diameter as shown in Fig. 5.19. For a typical value of $\sigma = 10°$ for rain, *CDR*

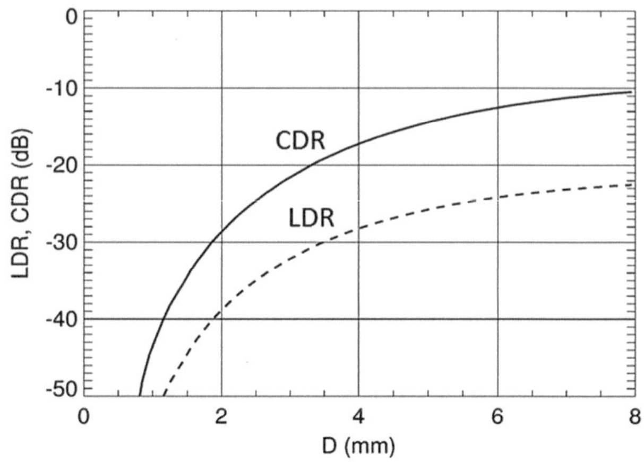

Fig. 5.19 Dependencies of *LDR* and *CDR* on the equivolume raindrop diameter at S band. In the computation of *LDR*, it is assumed that the width of the canting angle distribution σ is equal to 10°

is more than 10 dB higher than *LDR*. It can be also shown that the two depolarization ratios are closely correlated with differential reflectivity Z_{DR} in rain. The *LDR* and *CDR* curves in Fig. 5.19 are presented for monodisperse raindrop size distributions. Integrating over real size spectra generally yields *LDR* in rain lower than −25 dB (Herzegh and Jameson 1992; Kennedy et al. 2001).

5.6.2 Depolarization Ratios of Frozen and Mixed-Phase Hydrometeors

Depolarization ratios of frozen and mixed-phase hydrometeors are very sensitive to their density and water content. In case of dry aggregated snow with low density, simple formulas can be obtained for C_{dr} using Rayleigh approximations for the scattering amplitudes s_a and s_b in Eq. (5.36). If $L_{a,b}\,|\varepsilon - 1| << 1$, then

$$| s_b - s_a | \approx \frac{3}{2}\frac{\pi^2 D^3}{\lambda^2}\frac{\rho_s^2}{\rho_i^2}\left|\frac{\varepsilon_i - 1}{\varepsilon_i + 2}\right|^2 (L_a - L_b) \tag{5.38}$$

(see derivation of Eq. (5.27)) and

$$| s_a + s_b | \approx \frac{\pi^2 D^3}{3\lambda^2}(\varepsilon_s - 1) \approx \frac{\pi^2 D^3}{\lambda^2}\frac{\rho_s}{\rho_i}\left|\frac{\varepsilon_i - 1}{\varepsilon_i + 2}\right|, \tag{5.39}$$

so that

$$\frac{|s_a - s_b|^2}{|s_a + s_b|^2} \sim \rho_s^2(D)(L_a - L_b)^2. \tag{5.40}$$

For unrimed aggregated snowflakes with average aspect ratio of 0.6 (see Chap. 4), the value of $L_a - L_b$ is relatively small (close to 0.2, see Fig. 5.1), while snow density ρ_s is usually lower than 0.1 g cm^{-3} and decreases with the snowflake size (Chap. 4). Therefore, C_{dr} (or *CDR*) is quite low for dry aggregated snow. The linear depolarization ratio *LDR* is even lower than *CDR* (at least by 3 dB) and is usually below −26 dB in dry aggregated snow (Herzegh and Jameson 1992). Rimed snowflakes with higher density usually have a more spherical shape than unrimed snowflakes. There are indications that the net effect determined by the density-dependent and shape-dependent factors in the right side of Eq. (5.40) is the reduction of *CDR* and *LDR* for more rimed snow (Ryzhkov et al. 2017).

Pristine ice crystals such as dendrites, hexagonal plates, and needles have higher density and less spherical shape compared to aggregated snowflakes, and, therefore, their depolarization ratios are much higher. Matrosov et al. (2001) and Matrosov (2015) showed that the dependence of the absolute values of *CDR* (and/or closely related to it slant 45° linear depolarization ratio—*SLDR*) on the antenna elevation angle can be used for discrimination between different ice hydrometeor habits and the estimation of particle shape parameters (e.g., aspect ratios). *CDR* increases with decreasing antenna elevation for planar (oblate) types of snow crystals (hexagonal plates, thick plates, dendrites), whereas the elevation dependency of *CDR* is relatively "flat" for columnar (prolate) type crystals (columns, needles, etc.). At low elevation angles, *CDR* varies over a large range (from the radar system depolarization limit to about −8 dB) depending on the ice habit. According to cloud radar observations by Matrosov et al. (2001, 2012), pristine dendrites and hexagonal plates have highest values of *CDR* up to −8 dB (at low antenna elevations) followed by columns and aggregates of dendrites (−23 to −16 dB) with graupel indicating the lowest *CDR* below −25 dB.

In the early polarimetric radar studies, depolarization ratios have been used primarily for identification of hail and melting layer wherein both *LDR* and *CDR* are much higher than in pure rain and dry aggregated snow (e.g., Minervin and Shupiatsky 1963; Barge 1974; Bringi et al. 1986; Al-Jumily et al. 1991; Holler et al. 1994; Hubbert et al. 1998; Kennedy et al. 2001). Enhanced values of *LDR* and *CDR* in hail and wet snow are attributed to strong resonance scattering effects and the presence of water in melting snow or wet hail. Generally, both depolarization ratios increase with increasing hail size. The dependencies of S-band *CDR* on the diameter of dry hailstones modeled as oblate spheroids with aspect ratios of 0.6, 0.7, and 0.8 are shown in Fig. 5.20 (top panel, Mirkovic 2016). Wet hailstones are usually modeled either as spongy (with water uniformly distributed within the hailstone) or water-coated (with a film of water covering an ice core). For a given mass water fraction, depolarization ratios are higher for water-coated hailstones. Mirkovic (2016) simulated *CDR* for water-coated hailstones for a maximal amount of water that a hailstone can hold before shedding starts. The corresponding mass of water is determined from the equation given by Rasmussen and Heymsfield (1987):

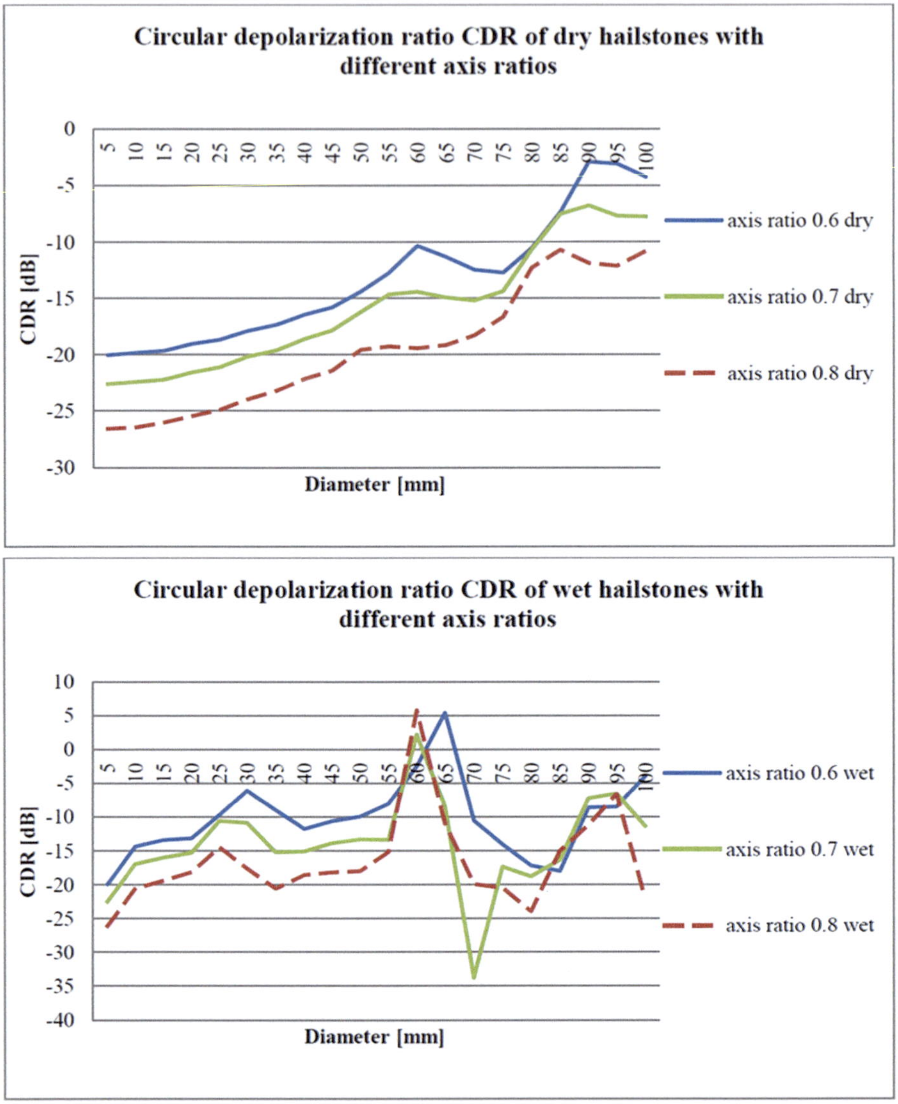

Fig. 5.20 Dependencies of *CDR* on the maximal diameter of dry and wet hailstones for different aspect ratios at S band. Wet hailstones are modeled as water-coated spheroids with the maximal thickness of water film before shedding occurs as determined by Rasmussen and Heymsfield (1987). Adapted from Mirkovic (2016)

$$m_{w\max} = 2.68 \times 10^{-4} + 0.139(m_i + m_{ws}), \tag{5.41}$$

where m_i is the mass of ice and m_{ws} is the mass of retained soaked water, both expressed in kg. The results of the Mirkovic (2016) simulations for three aspect ratios are displayed in the bottom panel of Fig. 5.20. Nonmonotonic dependencies of *CDR* on the diameter of wet hail are attributed to the effects of resonance scattering similar to the corresponding dependencies of Z_{DR} and K_{DP}. The magnitudes of *LDR* in hail are usually 3–5 dB lower than *CDR* depending on the orientations of hailstones.

The melting layer in stratiform precipitation exhibits very pronounced *LDR* and *CDR* signatures characterized by high *LDR* commonly exceeding −18 dB (Herzegh and Jameson 1992; Brandes and Ikeda 2004). The magnitudes of *CDR* are even higher which makes the use of depolarization ratios a very efficient for polarimetric radar detection of the melting layer.

5.7 Specific Attenuation

Specific attenuation of microwave radiation in hydrometeors at horizontal and vertical polarizations is expressed as (1.48)

$$A_{h,v} = -8.686\lambda \mathrm{Im}\left(< S^{(0)}_{\mathrm{hh,vv}} > \right), \tag{5.42}$$

where λ is the radar wavelength and $< S^{(0)}_{\mathrm{hh,vv}} >$is a diagonal element of the forward scattering matrix of the ensemble of hydrometeors in the unit volume (Sect. 1.3). The scattering matrix $\boldsymbol{S}$ depends on size distribution of particles and their orientation. According to (3.44),

$$< S^{(0)}_{\mathrm{hh,vv}} > = < s_b^{(0)} > - < s_b^{(0)} - s_a^{(0)} > A_{2,1}. \tag{5.43}$$

In (5.43), the angular moments A_1 and A_2 are defined in (3.45).

For chaotically oriented hydrometeors, $A_1 = A_2 = 1/3$, and

$$< S^{(0)}_{\mathrm{hh,vv}} > = \frac{2}{3} < s_b^{(0)} > + \frac{1}{3} < s_a^{(0)} > . \tag{5.44}$$

For Gaussian distribution of particle orientations with zero mean canting angle, specific attenuation at horizontal polarization can be expressed as

$$A_h = -8.686\lambda \left\{ \mathrm{Im}\left(< s_b^{(0)} > \right) - \frac{1}{4}\left(1 - r_\sigma^2\right)\left[\mathrm{Im}\left(< s_b^{(0)} - s_a^{(0)} > \right) \right] \right\}, \tag{5.45}$$

where parameter r_σ is defined by (3.54). For equioriented hydrometeors ($r_\sigma = 1$), A_h can be written in the following integral form

$$A_h = -8.686\lambda \int \mathrm{Im}\left[s_b^{(0)}(D)\right] N(D)dD. \tag{5.46}$$

In (5.46), A_h is expressed in dB km^{-1}, λ and $s_b^{(0)}$ are in mm, and $N(D)dD$ is in m^{-3}.

Specific attenuation is directly proportional to the hydrometeor concentration, and it generally increases with the particle size and decreases with increasing wavelength. As in the case of backscatter differential phase, the imaginary part of the forward scattering amplitude $s_b^{(0)}$ is poorly approximated by the Rayleigh formula, and there is no simple closed-form analytical solution for specific attenuation.

5.7.1 Specific Attenuation by Raindrops

The dependencies of normalized specific attenuation $A_h(D)/N$ on raindrop size for three different radar wavelengths and two temperatures, 0 and 30 °C, are displayed in Fig. 5.21. Similar to normalized Z and normalized K_{DP}, specific attenuation increases with raindrop size although with slower rate roughly proportional to $D^{3+\alpha}$ where α is between 0.5 and 0.8. The $A_h(D)$ dependence may become nonmonotonic at resonance sizes which is evident for higher temperatures at C band. Because rain rate is proportional to the 3.67th moment of raindrop size distribution, specific attenuation is approximately a linear function of rain rate (Table 5.2) which makes

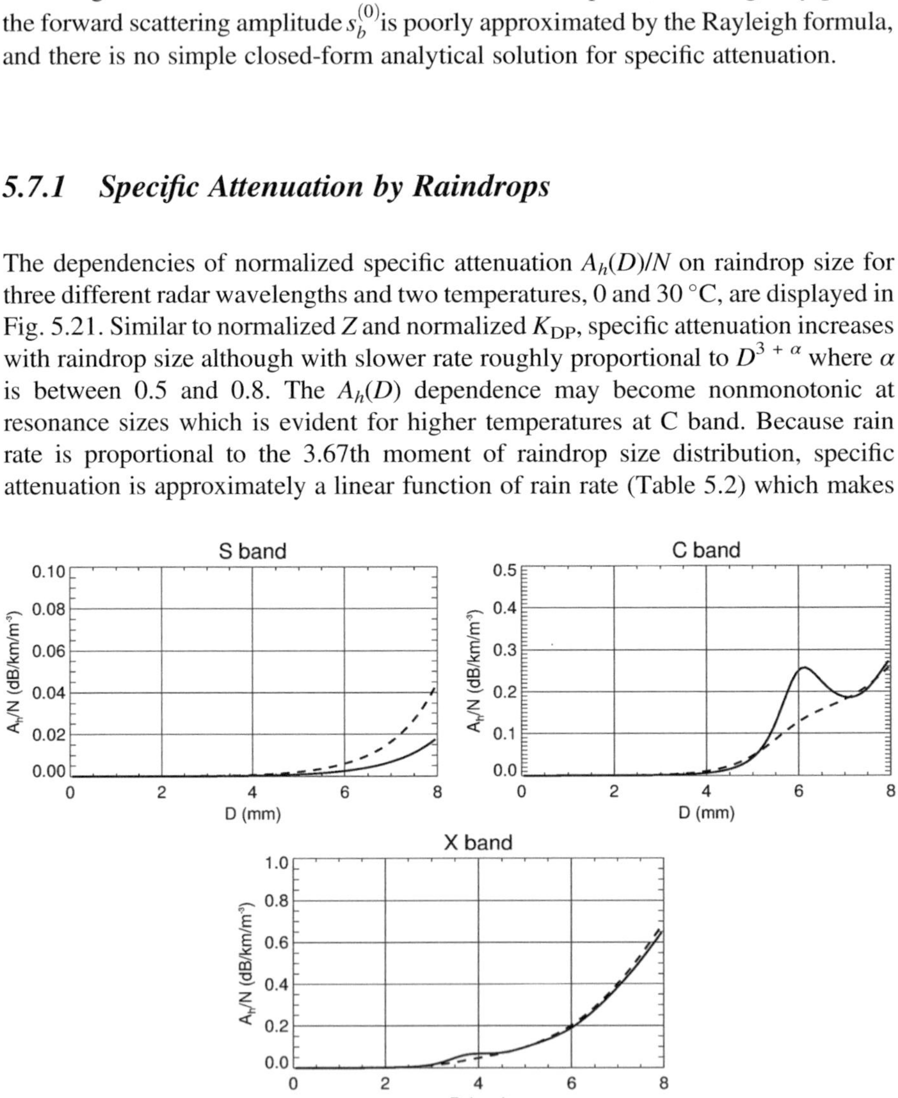

Fig. 5.21 Normalized specific attenuation as a function of equivolume raindrop diameter for different radar wavelengths and temperatures. Solid curves correspond to $T = 30$ °C; dashed curves are for $T = 0$ °C. Radar wavelengths are 11.0 cm at S band, 5.45 cm at C band, and 3.2 cm at X band

Table 5.2 Dependencies of specific attenuation on rain rate at different wavelengths and different temperatures

Temperature	S	C	X
0 °C	$A = 5.37 \times 10^{-4}R^{0.96}$	$A = 2.90 \times 10^{-3}R^{1.06}$	$A = 1.14 \times 10^{-2}R^{1.13}$
10 °C	$A = 4.07 \times 10^{-4}R^{0.96}$	$A = 2.25 \times 10^{-3}R^{1.08}$	$A = 1.00 \times 10^{-2}R^{1.17}$
20 °C	$A = 3.11 \times 10^{-4}R^{0.96}$	$A = 1.75 \times 10^{-3}R^{1.09}$	$A = 8.51 \times 10^{-3}R^{1.21}$
30 °C	$A = 2.46 \times 10^{-4}R^{0.96}$	$A = 1.38 \times 10^{-3}R^{1.09}$	$A = 7.16 \times 10^{-3}R^{1.25}$

A is in dB km^{-1}, R is in mm h^{-1}

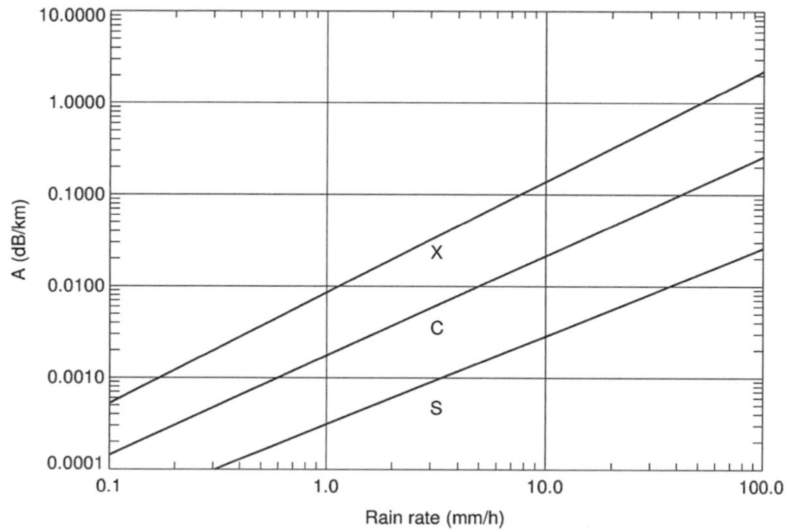

Fig. 5.22 Specific attenuation as a function of rain rate for different radar wavelengths at temperature 20 °C. Radar wavelengths are 11.0 cm at S band, 5.45 cm at C band, and 3.2 cm at X band

it very attractive for estimation of rainfall (see Chap. 10). Specific attenuation is more sensitive to raindrop temperature than Z, Z_{DR}, and K_{DP}. Generally, A_h decreases with increasing temperature for a given rain rate (Table 5.2).

The dependencies of A_h on rain rate R are illustrated in Fig. 5.22 for three radar wavelengths and temperature 20 °C. This plot indicates that specific attenuation in precipitation at $T = 20$ °C exceeds the one in atmospheric gases at rain rates as low as about 1 mm h^{-1} at X band, 4 mm h^{-1} at C band, and 20 mm h^{-1} at S band (see Appendix A).

5.7.2 *Specific Attenuation by Frozen and Mixed-Phase Hydrometeors*

Attenuation (or extinction) of electromagnetic waves in clouds and precipitation is caused by absorption and scattering of microwave radiation by atmospheric particles

(Ishimaru 1991). The absorption is dominant for hydrometeors smaller than wavelength, whereas the "scattering loss" due to the wave energy scattered by hydrometeors in directions different from the direction of wave propagation is prevalent for larger particles. The absorption loss is mainly determined by particles volume, concentration, and water content. Dry atmospheric particles have dielectric constant with small imaginary part (Chap. 4). If dry ice particles are small compared to the wavelength of electromagnetic wave, then the corresponding imaginary part of their forward scattering amplitude $s_{a,b}^{(0)}$ is also small (see Eq. (5.1)). According to Eq. (5.42), this means negligibly low attenuation in the medium filled with dry crystals, snowflakes, and small graupel. This is not the case for larger size graupel and hailstones (dry and wet) as Fig. 5.23 shows. Normalized specific attenuation in hail is much higher than in rain, and it is mainly caused by the "scattering loss." The latter fact explains why dry hailstones may cause higher attenuation than melting hailstones for certain hail sizes (Fig. 5.23).

Attenuation in hail is particularly high at C and X bands. Maximal values of A_h in pure rain rarely exceed 1 dB km^{-1} and 3 dB km^{-1} at C and X bands, respectively. Theoretical simulations by Ryzhkov et al. (2013) and comparative analysis of radar measurements in melting hail at S and C bands in Borowska et al. (2011) show that specific attenuation in melting hail can be 3–4 dB km^{-1} at C band. Similar experimental studies by Tuttle and Rinehart (1983) and Bringi et al. (2009) on the comparison of radar data collected simultaneously at S and X bands reported A_h at X band in wet graupel/hail larger than 4 dB km^{-1}. Significantly higher values may occur, and Snyder et al. (2010) report complete extinction of the X-band radar signal after propagating just 5 km into strong supercell storms containing hail.

Wet snow commonly observed within the melting layer in stratiform precipitation can also produce noticeable attenuation at shorter wavelengths which usually is significantly higher than the corresponding attenuation in pure rain below the bright band. However, attenuation in melting snow is poorly quantified both theoretically and experimentally.

5.8 Specific Differential Attenuation

In anisotropic media, the microwave radiation is attenuated differently at horizontal and vertical polarizations. The difference between specific attenuations of horizontally and vertically polarized signals is called specific differential attenuation A_{DP}. Differential attenuation causes bias in differential reflectivity that should be accounted for in the quantitative use of Z_{DR}.

In Sect. 3.4, it was stated that A_{DP} equals zero for randomly oriented hydrometeors, and

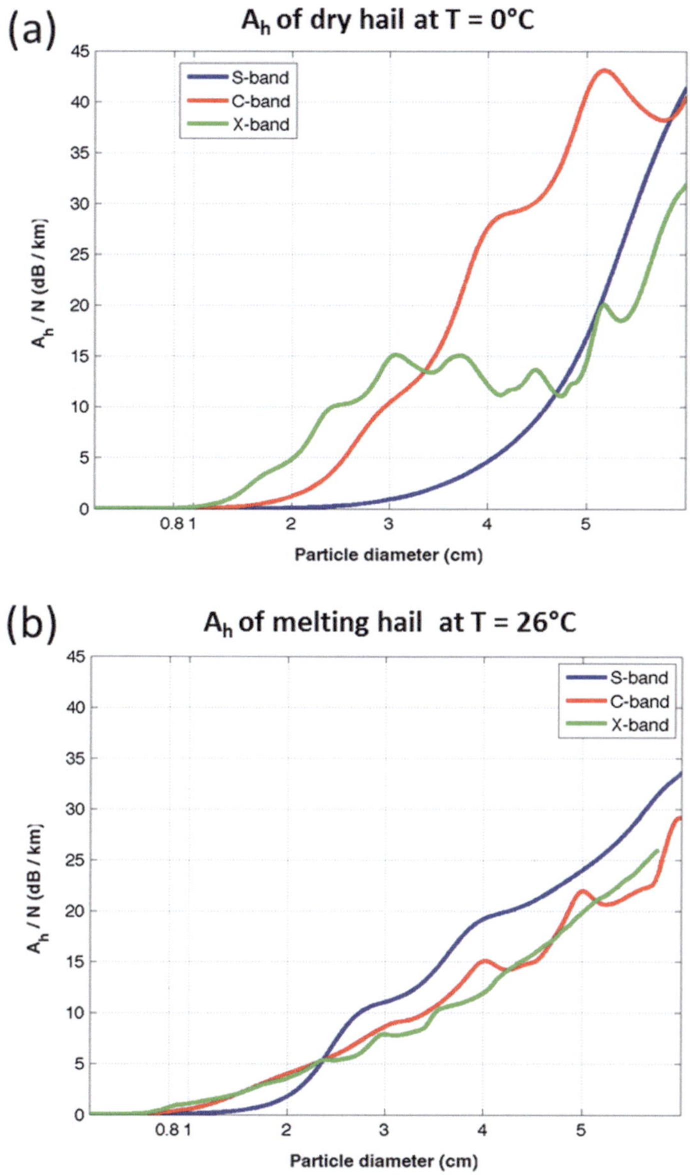

Fig. 5.23 Dependencies of normalized specific attenuation at S ($\lambda = 11.0$ cm), C ($\lambda = 5.45$ cm), and X ($\lambda = 3.2$ cm) bands on the equivolume diameter of hydrometeor for (**a**) dry hail and (**b**) rain/melting hail mixture. Mass water fraction f_w of melting hail follows the dependence in Fig. 5.8 (all hydrometeors with size less than 8 mm are raindrops). Hailstones aspect ratio and the width of the canting angle distributions are determined by f_w using (4.33) and (4.35). At $D < 8$ mm, $f_w = 1$, i.e., hailstones are completely melted

$$A_{DP} = -8.686\lambda \mathrm{Im}\left(< s_b^{(0)} > - < s_a^{(0)} >\right) F_{\mathrm{orient}}$$
$$= -8.686\lambda F_{\mathrm{orient}} \int \mathrm{Im}\left[s_b^{(0)}(D) - s_a^{(0)}(D)\right] N(D) dD, \quad (5.47)$$

in the case of Gaussian distribution of particle orientations, where the orientation factor F_{orient} is defined in (5.21). In (5.47), A_{DP} is expressed in dB km^{-1}, λ and $s_{a,b}^{(0)}$ are in mm, and $N(D)dD$ is in m^{-3}. Similar to specific attenuation A_h, specific differential attenuation is directly proportional to the concentration of hydrometeors.

Figure 5.24 illustrates the dependencies of normalized specific differential attenuation $A_{DP}(D)/N$ on raindrop size for three radar wavelengths and two temperatures, 0 and 30 °C. Notable are negative values of A_{DP} for very large raindrops at C band at higher temperatures which is a result of resonance scattering. Specific differential attenuation A_{DP} is much more sensitive to raindrop shape (and, consequently, to drop size or Z_{DR}) than A_h. Hence, A_{DP} is very low for small raindrops having almost spherical shape but becomes comparable with A_h at large drop sizes as the comparison of Figs. 5.24 and 5.21 demonstrates. Theoretical simulations of A_h and A_{DP} based on the large statistics of raindrop size distributions indicate that the ratio A_{DP}/A_h in rain may vary over a wide range while generally being within the interval 0.1–0.4 for moderate-to-heavy rain with reflectivity exceeding 40–45 dBZ. Maximal

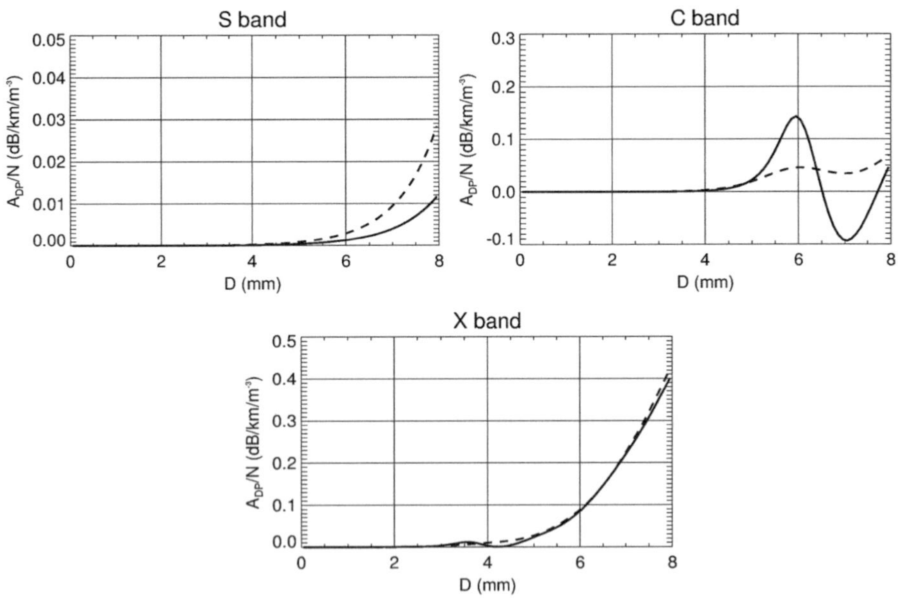

Fig. 5.24 Normalized specific differential attenuation as a function of equivolume raindrop diameter for different radar wavelengths and temperatures. Solid curves correspond to $T = 30$ °C; dashed curves are for $T = 0$ °C. Radar wavelengths are 11.0 cm at S band, 5.45 cm at C band, and 3.2 cm at X band

values of A_{DP} in rain can be as large as 0.04, 0.6, and 1.6 dB km^{-1} at S, C, and X bands, respectively. Hence, it is critically important to correct Z_{DR} for differential attenuation even in relatively light rain at C and X bands. A detailed discussion on the methods for differential attenuation correction is in Chap. 6.

Differential attenuation in dry hydrometeors is much lower than in rain due to smaller imaginary part of their dielectric constant, more random orientation (e.g., graupel and snowflakes), or lower density (dry aggregated snow). Individual hailstones of large size may produce higher A_{DP} than individual raindrops (see Fig. 5.25), but the concentration of hailstones is usually significantly smaller than of raindrops, and relative contribution of hail to differential attenuation is not necessarily high. Ryzhkov et al. (2013) speculate that only melting graupel/hail of relatively small size (up to 20 mm) can produce A_{DP} comparable to the one associated with large raindrops. Smaller size hailstones with high mass water fraction f_w have more orderly orientation than larger hailstones with lower f_w and tend to attenuate the waves as giant oblate raindrops.

Wet snowflakes in the melting layer of stratiform precipitation which is characterized by high differential reflectivity cause tangible differential attenuation at C and X bands exceeding the one in typical rain below the bright band.

5.9 Cross-Correlation Coefficient

Based on statistical analysis, Sachidananda and Zrnic (1985) introduced the cross-correlation coefficient as a polarimetric descriptor of precipitation attributes. They considered the effects of raindrop oscillation and canting angle distribution assuming exponential DSD. The value of the correlation coefficient computed under these conditions exceeded 0.995. It is noteworthy that high-quality polarimetric radars typically measure this value in pure rain.

The cross-correlation coefficient ρ_{hv} is very valuable for radar data quality assessment and classification of radar returns. It is a measure of how synchronously the complex voltages V_{hh} and V_{vv} corresponding to the radar returns at horizontal and vertical polarizations fluctuate in time. For most of weather scatterers, these fluctuate almost in unison, and the magnitude of ρ_{hv} is close to 1. This is not so for radar returns caused by nonmeteorological scatterers (such as ground, insects, birds, chaff, etc.). Herein, we present relatively simple explanation of the physical meaning of the cross-correlation coefficient and its dependence on various physical factors.

According to definition (3.37), the cross-correlation coefficient is

$$|\rho_{\mathrm{hv}}|\, e^{j\bar{\delta}} = \frac{< S_{\mathrm{hh}}^{*} S_{\mathrm{vv}} >}{\left(< |S_{\mathrm{hh}}|^2 >< |S_{\mathrm{vv}}|^2 >\right)^{1/2}}, \tag{5.48}$$

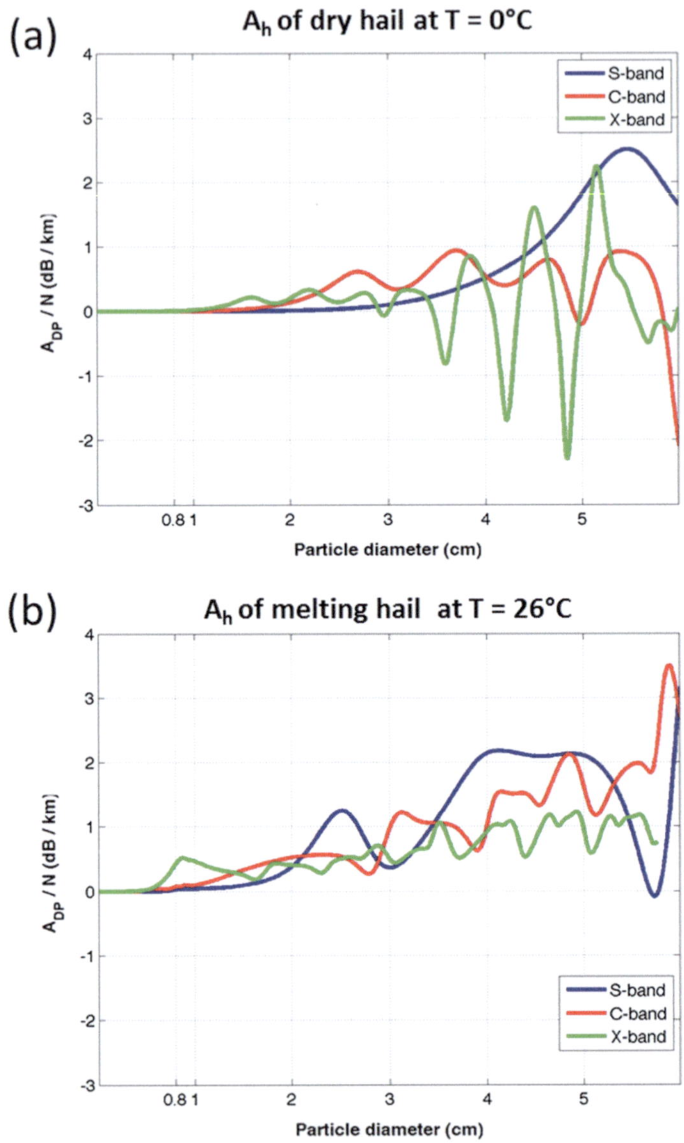

Fig. 5.25 Dependencies of normalized specific differential attenuation at S ($\lambda = 11.0$ cm), C ($\lambda = 5.45$ cm), and X ($\lambda = 3.2$ cm) bands on the equivolume diameter of hydrometeor for (**a**) dry hail and (**b**) rain/melting hail mixture. Mass water fraction f_w of melting hail follows the dependence shown in Fig. 5.8 (all hydrometeors with size less than 8 mm are raindrops). Hailstones aspect ratio and the width of the canting angle distributions are determined by f_w using (4.33) and (4.35). At $D < 8$ mm, $f_w = 1$, i.e., hailstones are completely melted

where angular brackets mean integration over ensemble of scatterers contained in the unit volume. If the radar resolution volume contains only one scatterer or one of the scatterers dominates others, then the brackets in (5.48) can be omitted and $\rho_{hv} = 1$. In the latter case, this can be a stationary man-made target such as building or tower which produces very strong radar return. Moreover, if the scatterers are frozen and stationary in the resolution volume, no matter what their shape or orientation is, the ρ_{hv} is 1. But if these frozen scatterers are advecting through the resolution volume, the ρ_{hv} will be smaller than 1 for the following reason. The scatterers that have entered the resolution volume between the sample time say $(n-1)T_s$ and nT_s create a differential incremental pair of voltages $\Delta V_{hh}(nT_s)$ and $\Delta V_{vv}(nT_s)$ at the time nT_s, whereas the scatterers that have left the resolution volume during the same time period had created the pair $\Delta V_{hh}[(n-1)T_s]$ and $\Delta V_{vv}[(n-1)T_s]$ at the time $(n-1)T_s$. The ratio $\Delta V_{hh}(nT_s)/\Delta V_{hh}[(n-1)T_s]$ differs from $\Delta V_{vv}(nT_s)/\Delta V_{vv}[(n-1)T_s]$, and this reduces the synchronization between the voltages V_{hh} and V_{vv}, causing ρ_{hv} to decrease. Similar reduction occurs if the scatterers change orientation, shape, composition, transformation (such as aggregation, or collision, breakup, etc.). The decrease in correlation is caused by temporal change of the scatterers in the resolution volume between the PRTs. If it weren't for this change, there would be no decrease in ρ_{hv}.

Summarizing, it is fair to say that the cross-correlation coefficient is a measure of variability of hydrometeors' shape/size, orientation, and phase composition within the radar resolution volume. The decrease is enhanced if the size of scatterers is in the resonant range and their shapes are nonspherical causing significant backscatter differential phase. Phase composition and backscatter differential phase may have the largest impact on ρ_{hv}.

In pure rain, the effects of shape and orientation variability are relatively weak, and the theoretical value of ρ_{hv} calculated from the measured drop-size distributions remains above 0.98 at S and X bands (see next section and Fig. 5.28). Theoretical value of the magnitude of ρ_{hv} may drop down to 0.93 at C band due to enhanced diversity of δ if raindrops of resonance sizes (more than 5 mm) are present in the raindrop spectrum. In fact, the measured values of ρ_{hv} can be lower than their theoretical estimates due to statistical measurements errors, effects of beam motion, and effects of nonuniform beam filling which are discussed in Chap. 6. Additionally, raindrops may acquire very complex shapes during coalescence and breakup which are difficult to account for in theoretical simulations. These secondary effects tend to reduce ρ_{hv} as well.

As stated, the mixture of hydrometeors with different phase composition is commonly characterized by reduced values of the cross-correlation coefficient which are significantly lower than minimal values of ρ_{hv} in rain. Typical examples are the mixtures of partially and completely melted snowflakes within the melting layer and raindrops and partially melted hailstones. Typical values of ρ_{hv} within the melting layer of stratiform cloud vary between 0.90 and 0.97, and extremely low values of ρ_{hv} down to 0.3 are occasionally observed in melting hail below the freezing level (Picca and Ryzhkov 2012). Hail in the "cold" part of clouds above the freezing level may also exhibit quite low ρ_{hv} due to the presence of large, resonance-size hailstones which can be dry or water-coated if hail is growing in wet regime (see Chap. 7).

Additional discussion on the cross-correlation coefficient is contained in Appendix D.

5.10 The Differences Between Polarimetric Variables at S, C, and X Bands

In the previous sections, size dependencies of the polarimetric variables have been examined assuming monodisperse size distributions. Herein, we analyze variability of radar observables in rain for real drop-size distributions (DSD) measured by a 2D-video disdrometer. A comprehensive multi-year disdrometer dataset containing 47,114 DSDs collected in central Oklahoma by Schuur et al. (2005) is utilized for simulations of the polarimetric variables assuming the shape-size dependence (4.29) and the width of the canting angle distribution σ_α equal to 10°. Measured DSDs have been truncated at maximal raindrop equivolume diameter of 8 mm. Because weather surveillance radars operate at S, C, and X bands, it is of special interest to compare the radar observables at all three frequency bands for a given DSD dataset.

The scatterplots of Z_{DR}, K_{DP}, ρ_{hv}, δ, and A_h versus Z at horizontal polarization computed for temperature $T = 20$ °C and the radar wavelengths of 11.0, 5.45, and 3.2 cm are displayed in Figs. 5.26, 5.27, 5.28, 5.29, and 5.30. These plots demonstrate the span of polarimetric variables in rain at the three wavelengths. Figure 5.26 shows that Z_{DR} at S and X bands usually does not exceed 4–5 dB in pure rain, whereas significantly higher Z_{DR} can be observed at C band due to resonance effects illustrated in Fig. 5.11. Such enhancement of Z_{DR} is more pronounced at higher temperatures.

Specific differential phase K_{DP} is inversely proportional to λ, and its logarithm well correlates with Z. The variability of K_{DP} at fixed Z increases for heavier rain (Fig. 5.27).

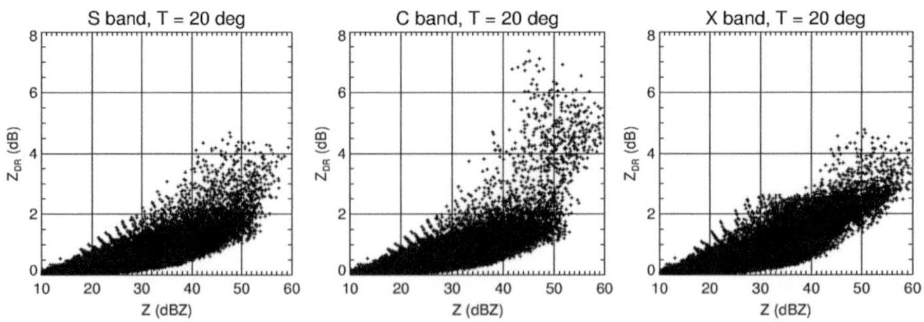

Fig. 5.26 Scatterplots of Z_{DR} versus Z in rain simulated from disdrometer data at S band ($\lambda = 11.0$ cm), C band ($\lambda = 5.45$ cm), and X band ($\lambda = 3.2$ cm) for $T = 20$ °C

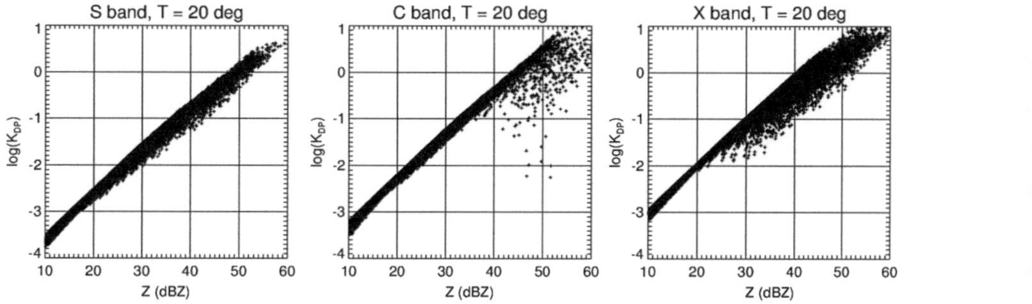

Fig. 5.27 Scatterplots of $\log(K_{DP})$ versus Z in rain simulated from disdrometer data at S band ($\lambda = 11.0$ cm), C band ($\lambda = 5.45$ cm), and X band ($\lambda = 3.2$ cm) for $T = 20$ °C

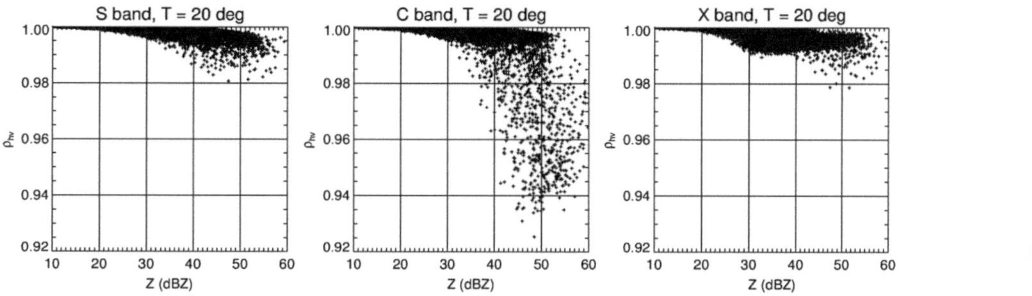

Fig. 5.28 Scatterplots of ρ_{hv} versus Z in rain simulated from disdrometer data at S band ($\lambda = 11.0$ cm), C band ($\lambda = 5.45$ cm), and X band ($\lambda = 3.2$ cm) for T = 20 °C

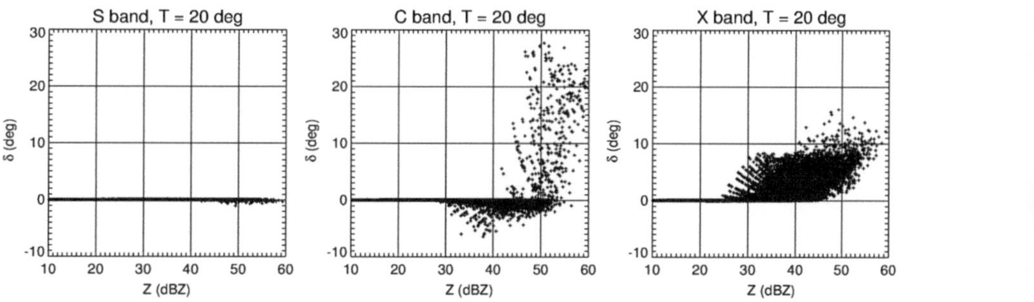

Fig. 5.29 Scatterplots of δ versus Z in rain simulated from disdrometer data at S band ($\lambda = 11.0$ cm), C band ($\lambda = 5.45$ cm), and X band ($\lambda = 3.2$ cm) for $T = 20$ °C

The cross-correlation coefficient for pure rain remains very high for all reflectivities, and there is not much difference between ρ_{hv} at S and X bands. In contrast, ρ_{hv} at C band can drop as low as 0.93 if raindrops with resonance sizes of 5–7 mm are present in the raindrop spectrum (Fig. 5.28).

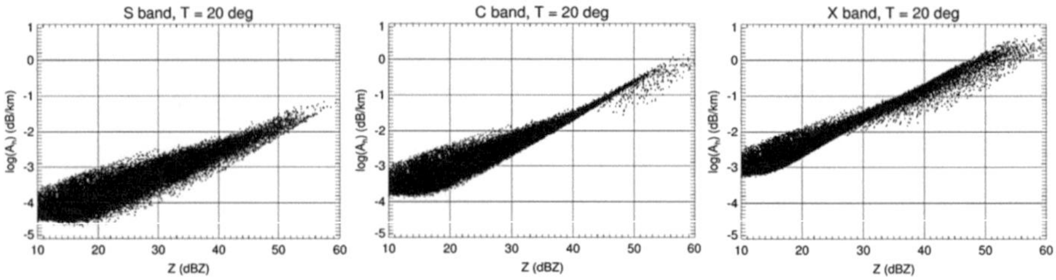

Fig. 5.30 Scatterplots of $\log(A_h)$ versus Z in rain simulated from disdrometer data at S band ($\lambda = 11.0$ cm), C band ($\lambda = 5.45$ cm), and X band ($\lambda = 3.2$ cm) for $T = 20$ °C

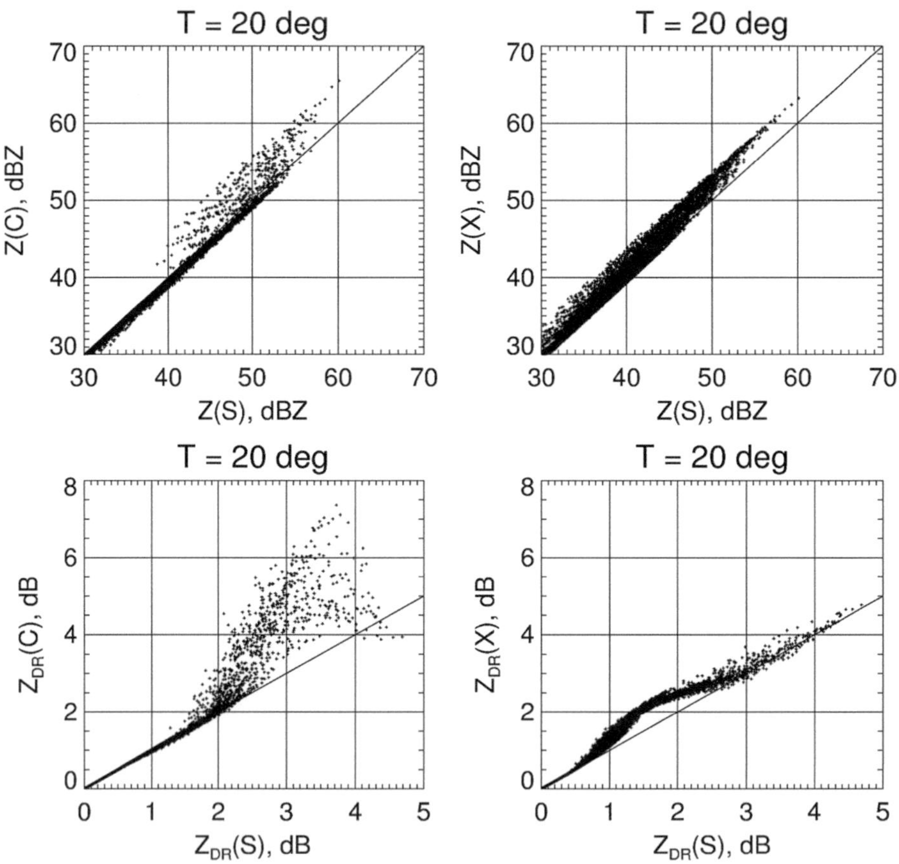

Fig. 5.31 Scatterplots of Z and Z_{DR} at C and X bands versus their values at S band simulated from large disdrometer dataset collected in Oklahoma

The backscatter differential phase δ in rain is vanishingly small at S band for all reflectivities but can be substantial at C and X bands in the presence of large drops in agreement with Fig. 5.17 (see Fig. 5.29).

The specific attenuation A_h steadily decreases with radar wavelength for a given Z but the width of the $Z - \log(A_h)$ scatterplot shows significant variability at different values of Z and radar wavelengths (Fig. 5.30).

The effects of non-Rayleigh scattering from raindrops of large sizes result in noticeable differences between the values of Z at S, C, and X bands for a given DSD or rain rate which are often neglected in the interpretation of the reflectivity data at different radar wavelengths (Fig. 5.31, top panels). Z measured at X band is steadily higher than the one at S band, whereas such difference between $Z(C)$ and $Z(S)$ can be even higher if the raindrops with resonance sizes between 5 and 7 mm substantially contribute to the reflectivity. The contrast between the values of differential reflectivity Z_{DR} at C and S bands is very significant for $Z_{DR}(S) > 2$ dB (bottom left panel in Fig. 5.31).

References

Al-Jumily, K., Charlton, R., & Humphries, R. (1991). Identification of rain and hail with circular polarization radar. *Journal of Applied Meteorology, 30*, 1075–1087.

Aydin, K., & Zhao, Y. (1990). A computational study of polarimetric radar observables in hail. *IEEE Transactions on Geoscience and Remote Sensing, 28*, 412–422.

Aydin, K., Giridhar, V., & Zhao, Y. (1991). Polarimetric C-band radar observables in melting hail: A computational study. In *Preprints. 25th International Radar Conference on Radar Meteorology*, Paris (pp. 733–736).

Balakrishnan, N., & Zrnic, D. (1990). Estimation of rain and hail rates in mixed-phase precipitation. *Journal of the Atmospheric Sciences, 47*, 565–583.

Barber, P., & Yeh, C. (1975). Scattering of electromagnetic waves by arbitrary shaped dielectric bodies. *Applied Optics, 14*, 2684–2872.

Barge, B. (1974). Polarization measurements of precipitation backscatter in Alberta. *Journal de Recherches Atmospheriques, 8*, 163–173.

Battan, L. J. (1973). *Radar observation of the atmosphere*. Chicago, IL: University of Chicago Press. 324 pp.

Bohren, C., & Huffman, D. (1983). *Absorption and scattering of light by small particles*. New York: Wiley. 530 pp.

Borowska, L., Ryzhkov, A., Zrnic, D., Simmer, C., & Palmer, R. (2011). Attenuation and differential attenuation of the 5 cm wavelength radiation in melting hail. *Journal of Applied Meteorology and Climatology, 50*, 59–76.

Brandes, E. A., & Ikeda, K. (2004). Freezing-level estimation with polarimetric radar. *Journal of Applied Meteorology, 43*, 1541–1543.

Bringi, V., & Seliga, T. (1977). Scattering from axisymmetric dielectrics or perfect conductors imbedded in an axisymmetric dielectric. *IEEE Transactions on Antennas and Propagation, 25*, 575–580.

Bringi, V., & Chandrasekar, V. (2001). *Polarimetric Doppler weather radar: Principles and applications*. Cambridge, UK: Cambridge University Press. 636 pp.

Bringi, V., Vivekanandan, J., & Tuttle, J. (1986). Multiparameter radar measurements in Colorado convective storms. Part II: Hail detection studies. *Journal of the Atmospheric Sciences, 43*, 2564–2577.

Bringi, V., Huang, G., May, P., Glasson, K., & Keenan, T. (2009). The estimation of X-band attenuation due to wet ice in the mixed-phase region of convective storms and correction of LDR at X band using the CP-2 radar. In *Proceeding 34th Conference on Radar Meteorology*. Williamsburg, VA: American Meteor Society. 11A.3.

Depue, T., Kennedy, P., & Rutledge, S. (2007). Performance of the hail differential reflectivity (H_{DR}) polarimetric hail indicator. *Journal of Applied Meteorology and Climatology, 46*, 1290–1301.

Doviak, R., & Zrnic, D. (2006). *Doppler Radar and Weather Observations* (2nd ed.). Reprint, Mineola, NY: Dover. 562pp.

Doviak, R., Bringi, V., Ryzhkov, A., Zahrai, A., & Zrnic, D. (2000). Considerations for polarimetric upgrades to operational WSR-88D radars. *Journal of Atmospheric and Oceanic Technology, 17*, 257–278.

Eccles, P., & Atlas, D. (1973). A dual-wavelength radar hail indicator. *Journal of Applied Meteorology, 12*, 847–854.

Ekpenyong, B., & Srivastava, R. (1970). Radar characteristics of the melting layer – a theoretical study. In *14th International Conference on Radar Meteorology,* Tucson, AZ (pp. 161–166).

Fabry, F., & Zawadzki, I. (1995). Long-term radar observations of the melting layer of precipitation and their interpretation. *Journal of the Atmospheric Sciences, 52*, 838–851.

Fabry, F., & Szyrmer, W. (1999). Modeling of the melting layer. Part II: Electromagnetic. *Journal of the Atmospheric Sciences, 56*, 3593–3600.

Griffin, E., Schuur, T., & Ryzhkov, A. (2018). A polarimetric analysis of ice microphysical processes in snow, using quasi-vertical profiles. *Journal of Applied Meteorology and Climatology, 57*, 31–50.

Hardaker, P., Holt, A., & Collier, C. (1995). A melting –layer model and its use in correcting for the bright band in single-polarization radar echoes. *Quarterly Journal of the Royal Meteorological Society, 121*, 495–525.

Herzegh, P., & Jameson, A. (1992). Observing precipitation through dual-polarization radar measurements. *Bulletin of the American Meteorological Society, 73*, 1365–1374.

Hogan, R., Field, P., Illingworth, A., Cotton, R., & Choularton, T. (2002). Properties of embedded convection in warm-frontal mixed-phase cloud from aircraft and polarimetric radar. *Quarterly Journal of the Royal Meteorological. Society, 128*, 451–476.

Holler, H., Bringi, V., Hubbert, J., Hagen, M., & Meischner, P. (1994). Life cycle and precipitation formation in a hybrid-type hailstorm revealed by polarimetric and Doppler radar measurements. *Journal of the Atmospheric Sciences, 51*, 2500–2522.

Huang, G., Bringi, V., & Thurai, M. (2008). Orientation angle distributions of drops after 80 m fall using a 2D-video disdrometer. *Journal of Atmospheric and Oceanic Technology, 25*, 1717–1723.

Hubbert, J., Bringi, V., Carey, L., & Bolen, S. (1998). CSU-CHILL polarimetric radar measurements in a severe hailstorm in eastern Colorado. *Journal of Applied Meteorology, 37*, 748–775.

Ishimaru, A. (1991). *Electromagnetic wave propagation, radiation and scattering* (p. 1991). Englewood Cliffs, NJ: Prentice Hall.

Jameson, A. (1987). Relations among linear and circular polarization parameters measured in canted hydrometeors. *Journal of Atmospheric and Oceanic Technology, 4*, 634–645.

Keenan, T. D., Carey, L. D., Zrnic, D. S., & May, P. (2001). Sensitivity of 5-cm wavelength polarimetric radar variables to raindrop axial ratio and drop size distribution. *Journal of Applied Meteorology, 40*, 526–545.

Kennedy, P., Rutledge, S., Petersen, W., & Bringi, V. (2001). Polarimetric radar observations of hail formation. *Journal of Applied Meteorology, 40*, 1347–1366.

Klaassen, W. (1988). Radar observations and simulations of the melting layer of precipitation. *Journal of the Atmospheric Sciences, 45*, 3741–3753.

Matrosov, S. Y. (2004). Depolarization estimates from linear H and V measurements with weather radars operating in simultaneous transmission – simultaneous receiving mode. *Journal of Atmospheric and Oceanic Technology, 21*, 574–583.

Matrosov, S. Y. (2015). Evaluations of the spheroidal particle model for describing cloud radar depolarization ratios of ice hydrometeors. *Journal of Atmospheric and Oceanic Technology, 32*, 865–879.

Matrosov, S. Y., Reinking, R. F., Kropfli, R. A., Martner, B. E., & Bartram, B. W. (2001). On the use of radar depolarization ratios for estimating shapes of ice hydrometeors in winter clouds. *Journal of Applied Meteorology, 40*, 479–490.

Matrosov, S., Reinking, R., & Djalalova, I. (2005). Inferring fall attitudes of pristine dendritic crystals from polarimetric radar data. *Journal of the Atmospheric Sciences, 62*, 241–250.

Matrosov, S. Y., Mace, G. G., Marchand, R., Shupe, M. D., Hallar, A. G., & Mc Cubbin, I. B. (2012). Observations of ice crystal habits with a scanning polarimetric W-band radar at slant linear depolarization ratios mode. *Journal of Atmospheric and Oceanic Technology, 29*, 989–1008.

McCormick, G., & Hendry, A. (1975). Principles for the radar determination of the polarization properties of precipitation. *Radio Science, 10*, 421–434.

Minervin, V., & Shupiatsky, A. (1963). Radar method for determining the phase state of clouds and precipitation. *Trudy Centralnoj Aerologicheskoj Observatory, 47*, 63–84. in Russian.

Mirkovic, D. (2016). *Computational electromagnetics for polarimetric radar scatterers: An approach to polarimetric variable modeling of hydrometeors and biota*. Saarbrücken, Germany: Lambert Academic Publishing. 272 pp.

Mishchenko, M. (2000). Calculation of the amplitude matrix for a nonspherical particle in a fixed orientation. *Applied Optics, 39*, 1026–1031.

Otto, T., & Russchenberg, H. (2011). Estimation of specific differential phase and differential backscatter phase from polarimetric weather radar measurements of rain. *IEEE Geoscience and Remote Sensing Letters, 8*, 988–992.

Picca, J., & Ryzhkov, A. (2012). A dual-wavelength polarimetric analysis of the 16 May 2010 Oklahoma City extreme hailstorm. *Monthly Weather Review, 140*, 1385–1403.

Rasmussen, R., & Heymsfield, J. (1987). Melting and shedding of graupel and hail. Part I: Model physics. *Journal of the Atmospheric Sciences, 44*, 2754–2763.

Russchenberg, H., & Ligthart, L. (1996). Backscattering and propagation through the melting layer of precipitation: A new polarimetric model. *IEEE Transactions on Geoscience and Remote Sensing, 34*, 3–14.

Ryzhkov, A. (2001). Interpretation of polarimetric radar covariance matrix for meteorological scatterers. Theoretical analysis. *Journal of Atmospheric and Oceanic Technology, 18*, 315–328.

Ryzhkov, A., & Zrnic, D. (1998). Discrimination between rain and snow with a polarimetric radar. *Journal of Applied Meteorology, 37*, 1228–1240.

Ryzhkov, A., Zrnic, D., & Gordon, B. (1998). Polarimetric method for ice water content determination. *Journal of Applied Meteorology, 37*, 125–134.

Ryzhkov, A. V., Giangrande, S. E., Melnikov, V. M., & Schuur, T. J. (2005a). Calibration issues of dual-polarization radar measurements. *Journal of Atmospheric and Oceanic Technology, 22*, 1138–1155.

Ryzhkov, A. V., Giangrande, S. E., & Schuur, T. J. (2005b). Rainfall estimation with a polarimetric prototype of the WSR-88D radar. *Journal of Applied Meteorology, 44*, 502–515.

Ryzhkov, A., Giangrande, S., Khain, A., Pinsky, M., & Pokrovsky, A. (2008). *Exploring model-based polarimetric retrieval of vertical profiles of precipitation*. In Extended Abstracts. 5th European Conference on Radar in Meteorology and Hydrology, Helsinki, Finland, CD-ROM, P6.1.

Ryzhkov, A., Pinsky, M., Pokrovsky, A., & Khain, A. (2011). Polarimetric radar observation operator for a cloud model with spectral microphysics. *Journal of Applied Meteorology and Climatology, 50*, 873–894.

Ryzhkov, A., Kumjian, M., Ganson, S., & Khain, A. (2013). Polarimetric radar characteristics of melting hail. Pt I: Theoretical simulations using spectral microphysical modeling. *Journal of Applied Meteorology and Climatology, 52*, 2849–2870.

Ryzhkov, A., Zhang, P., Reeves, H., Kumjian, M., Tschallener, T., Simmer, C., & Troemel, S. (2016). Quasi-vertical profiles – a new way to look at polarimetric radar data. *Journal of Atmospheric and Oceanic Technology, 33*, 551–562.

Ryzhkov, A., Matrosov, S., Melnikov, V., Zrnic, D., Zhang, P., Cao, Q., Knight, M., Simmer, C., & Troemel, S. (2017). Estimation of depolarization ratio using radars with simultaneous transmission / reception. *Journal of Applied Meteorology and Climatology, 56*, 1797–1816.

Sachidananda, M., & Zrnic, D. S. (1985). Z_{DR} measurement consideration for a fast scan capability radar. *Radio Science, 20*, 907–922.

Sachidananda, M., & Zrnic, D. S. (1986). Differential propagation phase shift and rainfall rate estimation. *Radio Science, 21*, 235–247.

Schuur, T., Ryzhkov, A., & Clabo, D. (2005). *Climatological analysis of DSDs in Oklahoma as revealed by 2D-video disdrometer and polarimetric WSR-88D*. In Preprints, 32nd Conference on Radar Meteorology, CD-ROM, 15R.4.

Snyder, J., Bluestein, H., Zhang, G., & Frasier, S. (2010). Attenuation correction and hydrometeor classification of high-resolution, X-band, dual-polarized mobile radar measurements in severe convective storms. *Journal of Atmospheric and Oceanic Technology, 27*, 1979–2001.

Troemel, S., Kumjian, M., Ryzhkov, A., & Simmer, C. (2013). Backscatter differential phase – estimation and variability. *Journal of Applied Meteorology and Climatology, 52*, 2529–2548.

Troemel, S., Ryzhkov, A., Zhang, P., & Simmer, C. (2014). Investigations of backscatter differential phase in the melting layer. *Journal of Applied Meteorology and Climatology, 53*, 2344–2359.

Troemel, S., Ryzhkov, A., & Simmer, C. 2017. Climatology of vertical profiles of polarimetric radar variables at X band in stratiform clouds. In *38th Conference on Radar Meteorology* (p. 5). Chicago, IL: American Meteor Society. Retrieved from https://ams.confex.com/ams/38RADAR/meetingapp.cgi/Paper/320485

Tuttle, J., & Rinehart, R. (1983). Attenuation correction in dual-wavelength analyses. *Journal of Climate and Applied Meteorology, 22*, 1914–1921.

Waterman, P. (1965). Matrix formulation of electromagnetic scattering. *Proceedings of the IEEE, 53*, 805–812.

Van de Hulst, H. (1981). *Light scattering by small particles*. New York: Dover.

Vivekanandan, J., Bringi, V., & Raghavan, R. (1990). Multiparameter radar modeling and observation of melting ice. *Journal of the Atmospheric Sciences, 47*, 549–563.

Vivekanandan, J., Raghavan, R., & Bringi, V. (1993). Polarimetric radar modeling of mixtures of precipitation particles. *IEEE Transactions on Geoscience and Remote Sensing, 31*, 1017–1030.

Vivekanandan, J., Bringi, V., Hagen, M., & Meischner, P. (1994). Polarimetric radar studies of atmospheric ice particles. *IEEE Transactions on Geoscience and Remote Sensing, 32*, 1–10.

Yokoyama, T., & Tanaka, H. (1984). Microphysical processes of melting snowflakes detected by two-wavelength radar. Part I. Principle of measurement based on model calculation. *Journal of the Meteorological Society of Japan, 62*, 650–667.

Yokoyama, T., Tanaka, H., Nakamura, K., & Awaka, J. (1984). Microphysical processes of melting snowflakes detected by two-wavelength radar. Part II. Application of two-wavelength radar technique. *Journal of the Meteorological Society of Japan, 62*, 668–677.

Zawadzki, I., Szyrmer, W., Bell, C., & Fabry, F. (2005). Modeling of the melting layer. Part III: The density effect. *Journal of the Atmospheric Sciences, 62*, 3705–3723.

Zrnic, D., Balakrishnan, N., Ziegler, C., Bringi, V., Aydin, K., & Matejka, T. (1993). Polarimetric signatures in the stratiform region of a mesoscale convective system. *Journal of Applied Meteorology, 32*, 678–693.

Zrnic, D. S., Ryzhkov, A., Straka, J., Liu, Y., & Vivekanandan, J. (2001). Testing a procedure for automatic classification of hydrometeor types. *Journal of Atmospheric and Oceanic Technology, 18*, 892–913.

Zrnic, D., Doviak, R., Zhang, G., & Ryzhkov, A. (2010). Bias in differential reflectivity due to cross-coupling through the radiation patterns in the simultaneous horizontal / vertical polarization mode. *Journal of Atmospheric and Oceanic Technology, 27*, 1624–1637.

Chapter 6
Data Quality and Measurement Errors

Providing high quality of weather radar data is essential for producing reliable and robust hydrological and meteorological information useful to the scientific and operational communities. The accuracy of quantitative precipitation estimation (QPE) and hydrometeor classification directly depends on the quality of different radar variable estimates. Modern operational Doppler polarimetric radars directly measure reflectivity Z, differential reflectivity Z_{DR}, differential phase Φ_{DP}, cross-correlation coefficient ρ_{hv}, Doppler velocity v, Doppler spectrum width σ_v, and linear depolarization ratio *LDR* (in the *HSHV* mode of operation). Specific differential phase K_{DP} is not directly measured but derived from Φ_{DP}.

The estimates of all these variables are obtained in the radar data processor from the time series of successive radar samples within the dwell time interval and are subject to random fluctuations caused by the statistical nature of the radar signal. The uncertainty of such estimates is characterized by bias (or accuracy) and standard deviation (or precision). The latter one is the measure of the "noisiness" of the estimate or the intensity of its temporal and spatial fluctuations.

Several factors may cause bias in the estimates of different radar variables. These include (1) radar miscalibration, (2) impact of wet antenna radome, (3) attenuation in atmospheric gases and precipitation, (4) partial beam blockage (PBB), (5) ground clutter contamination, (6) low signal-to-noise ratio (SNR), (7) nonuniform beam filling (NBF), (8) depolarization from propagation in oriented ice crystals, and (9) multipath propagation (three-body scattering). These factors affect differently the biases of various radar variables. The sensitivity of these variables to some of the listed factors is summarized in Table 6.1 where "X" sign indicates that the particular radar variable is prone to the particular factor. Some variables (such as v, σ_v, ρ_{hv}, Φ_{DP}, K_{DP}) are immune either to all (like v) or to the most detrimental effects, whereas others (e.g., Z and Z_{DR} in particular) are very vulnerable.

In addition to the adverse effects listed in Table 6.1, possible contamination from nonmeteorological returns caused by insects, birds, chaff, etc. may also degrade the quality of all measured radar variables and cause problems in the interpretation of their meteorological meaning.

A. V. Ryzhkov, D. S. Zrnic, *Radar Polarimetry for Weather Observations*,
Springer Atmospheric Sciences, https://doi.org/10.1007/978-3-030-05093-1_6

Table 6.1 Sensitivity of various radar variables to different adverse effects

Parameter	Radar miscalibration	Wet radome	Attenuation	Beam blockage	Low SNR	NBF	Depolarization
Z	X	X	X	X	X	X	
Z_{DR}	X	X	X	X	X	X	X
$\Phi_{DP}(K_{DP})$						X	X
ρ_{hv}					X	X	
LDR		X	X		X	X	X
v							
σ_v					X		

6.1 Absolute Calibration of *Z*

For most important practical applications of polarimetric weather radar, the reflectivity factor Z should be calibrated with the accuracy of 1 dB and differential reflectivity Z_{DR} with the accuracy of 0.2 dB. These generally enable estimating rainfall within 15% accuracy (Ryzhkov et al. 2005). Better accuracy of the Z_{DR} calibration (0.1 dB) might be needed for measurements of light rain or snow.

Polarimetric diversity provides a new method for absolute calibration of Z which was a long-standing problem for single-polarization radars. This methodology rests on the idea that Z, Z_{DR}, and K_{DP} are interdependent in rain and Z can be estimated from K_{DP} and Z_{DR} which are independent of absolute radar calibration (Gorgucci et al. 1992, 1999; Goddard et al. 1994; Scarchilli et al. 1996; Vivekanandan et al. 2003). The difference between computed and measured values of Z is considered to be the Z bias. The consistency of Z, Z_{DR}, and K_{DP} in rain can be formulated either as linear relation (Ryzhkov et al. 2005)

$$Z = a + b\log(K_{DP}) + cZ_{DR}, \tag{6.1}$$

where Z and Z_{DR} are expressed in logarithmic scale, or as a dependence of the ratio K_{DP}/Z on Z_{DR} as suggested by Goddard et al. (1994) and Illingworth and Blackman (2002).

$$\frac{K_{DP}}{Z} = f(Z_{DR}). \tag{6.2}$$

In (6.2), Z and K_{DP} are in linear scale (i.e., mm^6 m^{-3} and deg km^{-1}, respectively).

The scatterplots of the ratio K_{DP}/Z versus Z_{DR} simulated from large DSD dataset in Oklahoma for three radar wavelengths and two temperatures, 0 °C and 30 °C, are illustrated in Fig. 6.1. Simulations are performed assuming size-shape relation (4.29) for raindrops. It is evident that the dependence in (6.2) on temperature is negligibly small at S band where the effects of resonance scattering are insignificant. However, the temperature becomes an important factor at C band for $Z_{DR} > 2$ dB and should be taken into account for all Z_{DR} at X band. At S or C bands, Z can be estimated from

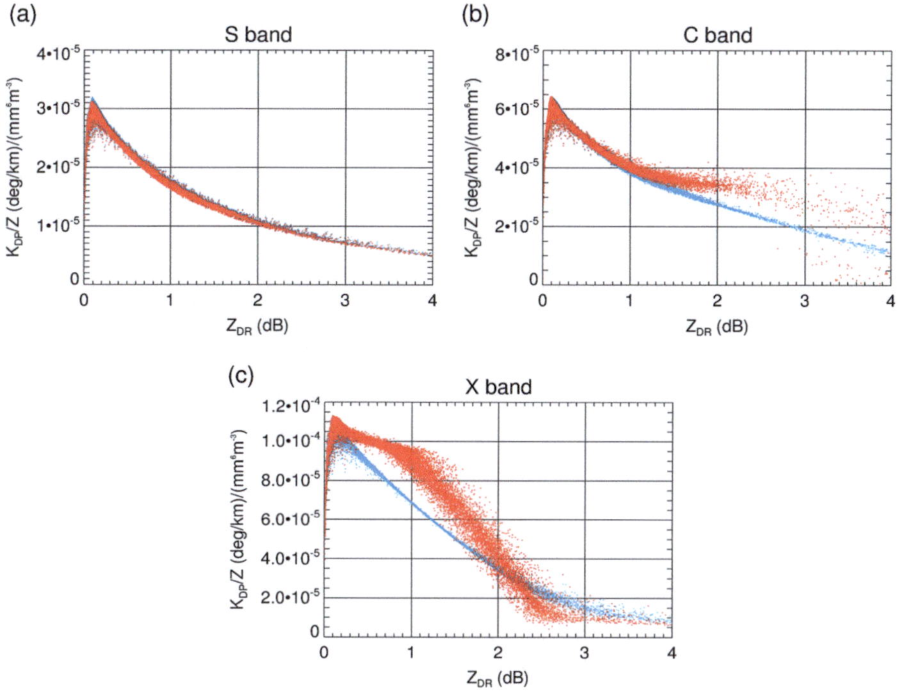

Fig. 6.1 Scatterplots of (**a**) K_{DP}/Z versus Z_{DR} at S band ($\lambda = 11.0$ cm), (**b**) C band ($\lambda = 5.45$ cm), and (**c**) X band ($\lambda = 3.2$ cm) for raindrop temperature 0 °C (blue dots) and 30 °C (red dots) assuming drop diameter-shape relation (4.29)

known K_{DP} and Z_{DR} with the accuracy better than 1 dB if rain does not contain many resonance-size drops. Possible deviation of raindrop shape from the one prescribed by Eq. (4.29) may produce additional small uncertainty of the ratio K_{DP}/Z (e.g., Ryzhkov et al. 2005; Gourley et al. 2009).

The function $f(Z_{DR})$ can be well approximated by a fourth-order polynomial fit in certain range of Z_{DR} so that (6.2) can be presented as (Gourley et al. 2009)

$$\frac{K_{DP}}{Z} = 10^{-5}\left(a_0 + a_1 Z_{DR} + a_2 Z_{DR}^2 + a_3 Z_{DR}^3\right). \tag{6.3}$$

In (6.3), Z_{DR} is in decibels, and the coefficients a_0–a_3 for the S-, C-, and X-band radar wavelengths are listed in Table 6.2. It is important that Eq. (6.3) with coefficients from Table 6.2 is valid in the Z_{DR} range 0.2 dB to 2 or 3 dB and that different consistency relations should be used for different temperatures at X band. If the radar wavelength at a given frequency band (S, C, or X) is not equal to the one for which the values of a_0–a_3 (Table 6.2) are computed, then the wavelength adjustment should be made by dividing the right side of (6.3) with the factor $\lambda/\lambda_{s,c,x}$, where $\lambda_{s,c,x}$ are the radar wavelengths in Table 6.2.

Table 6.2 Coefficients a_0–a_3 in (6.3) for S band ($\lambda = 11.0$ cm), C band ($\lambda = 5.45$ cm), and X band ($\lambda = 3.2$ cm)

Frequency band	Temperature (°C)	Z_{DR} range (dB)	a_0	a_1	a_2	a_3
S	0–30	0.2–3.0	3.19	−2.16	0.795	−0.119
C	0–30	0.2–2.0	6.70	−4.42	2.16	−0.404
X	0	0.2–3.0	11.2	−4.75	0.349	−0.0532
X	10	0.2–3.0	10.9	−2.63	−1.22	0.341
X	20	0.2–3.0	10.4	0.109	−3.01	0.636
X	30	0.2–3.0	9.68	3.07	−4.67	0.869

Because each of the three polarimetric variables in (6.3) has statistical errors and K_{DP} is notoriously noisy in light rain (especially at longer wavelengths), it is instrumental to rewrite (6.2) as

$$K_{DP} = 10^{0.1Z(dBZ)} f(Z_{DR}) \tag{6.4}$$

and integrate both sides of (6.4) over a sufficiently large spatial/temporal domain Ω (Ryzhkov et al. 2005). The integral

$$I_1 = \int K_{DP} d\Omega, \tag{6.5}$$

should be equal to the integral

$$I_2 = \int 10^{0.1Z_m} f(Z_{DR}) d\Omega, \tag{6.6}$$

if measured reflectivity Z_m is perfectly calibrated. The difference between I_1 and I_2 points to Z bias ΔZ which can be estimated as

$$\Delta Z(\mathrm{dB}) = 10\log(I_2/I_1). \tag{6.7}$$

if $Z_m = Z + \Delta Z$. Because approximation (6.3) is valid only in the limited range of Z_{DR} listed in Table 6.2, the integrations (6.5) and (6.6) should be carried out only over the data within the appropriate range of Z_{DR} (e.g., between 0.2 and 2.0 dB at C band). It is also required that data in the domain Ω are not biased by low signal-to-noise ratio or contaminated by scatterers other than raindrops. These requirements are satisfied if SNR > 25 dB and $\rho_{hv} > 0.99$.

The methodology of matching the integrals I_1 and I_2 was first tested at S band on a large polarimetric dataset obtained during the Joint Polarization Experiment in Oklahoma and yielded an accuracy of Z calibration within 1 dB (Ryzhkov et al. 2005). Later on, this technique has been successfully implemented, and similar accuracy of Z calibration was confirmed in the studies by Lang et al. (2009) and Marks et al. (2011) using S-band polarimetric radars in Mexico and Kwajalein.

At shorter wavelengths, areal integration in (6.5) and (6.6) can be replaced by integration along the radials to avoid azimuthal directions with strong attenuation which may bias Z and Z_{DR}. Integrating along the rays is equivalent to matching measured differential phase Φ_{DP} which is proportional to the integral of K_{DP} (i.e., I_1) and Φ_{DP} calculated from the radial profiles of Z and Z_{DR} (i.e., I_2). The idea of matching radial profiles of the measured and calculated differential phase was first proposed by Goddard et al. (1994) and later explored by Gourley et al. (2009) with applications to C-band radar. The mismatch between these two profiles indicates a possible calibration error of Z. This error can be used as an adjustment to Z required to match the two profiles of the differential phase.

To mitigate the impact of attenuation, Z and Z_{DR} should be either corrected for attenuation according to the methods described in Sect. 6.3, or only the data radials with sufficiently small span of Φ_{DP} should be used for calibration. In the Gourley et al. (2009) study, only the rays (or part of the rays) with Φ_{DP} less than 12° were selected for self-consistency check at C band. At X band, the Φ_{DP} constraint is even tougher warranting attenuation correction as a precondition for utilizing the self-consistency method.

More traditional methods of absolute calibration of Z including comparison of the radar-measured Z and its estimate from the raindrop spectrum measured by a disdrometer are discussed by Atlas (2002), Chandrasekar et al. (2015), and Frech et al. (2017).

6.2 Absolute Calibration of Z_{DR}

6.2.1 System Internal Calibration

Relative internal calibration of Z_{DR} can be achieved by measuring the differences between gains and losses in the two orthogonal channels. Because the transmission path and reception path differ, separate relative calibration of each is needed. Thus, the power ratio P_h/P_v downstream of the components that can cause bias in each path needs to be monitored. A change in either ratio would cause a corresponding relative drift in the Z_{DR} bias which is then corrected (Zrnic et al. 2006). An additional step to account for the absolute bias must be made. The procedure is explained next by referring to the diagram in Fig. 6.2.

The relative values of the power ratios (in dB) are measured at two points. One is at the waveguide couplers Tc_h and Tc_v on the transmission side; these extract powers from the corresponding H and V waveguides to establish the relative value in the transmission path. The other point is at the output of the two receivers when the signals are injected into the receiving couplers Rc_h and Rc_v above the low-noise amplifiers.

Let the power ratio of outputs at the couplers Tc_h and Tc_v be

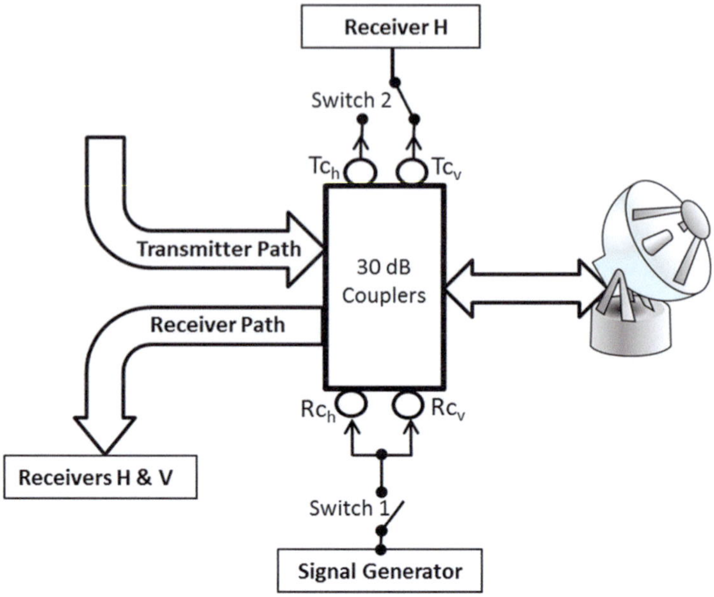

Fig. 6.2 Transmitter and receiver paths to the antenna. The couplers in the transmitter path Tc_h and Tc_v tap the signals from the H and V waveguides close to the antenna; comparison is made sequentially, via Switch 2, in the H receiver. The signal generator's output is split and injected into the receiver couplers Rc_h and Rc_v located above the low-noise amplifiers in the H and V waveguides. During data collection the Switch 1 is open; it closes at the end of volume scans to enable automatic calibration of the receiver path

$$\Delta_T(t_0) = 10\log[P(\mathrm{Tc_h})/P(\mathrm{Tc_v})] \tag{6.8}$$

where t_0 is a reference time stamp; in its proximity few more initial measurements must be made. $\Delta_T(t_0)$ is measured using one receiver (say H) as in Fig. 6.2 by switching between the outputs of Tc_h and Tc_v. That way the receiver's transfer function does not affect the measurement. The $\Delta_T(t_0)$ should be stable over many hours because there are no separate active components in the path up to the couplers.

In the receiver path, a similar procedure is applied (Fig. 6.2). Note that the signal generator power is split (approximately 50:50) and the exact value at the splitter output is immaterial because the measurement is relative. Thus the power ratio is

$$\Delta_R(t_0) = 10\log[P(\mathrm{Rc_h})/P(\mathrm{Rc_v})], \tag{6.9}$$

and it is measured immediately after (6.8) to avoid possible changes between the measurements.

After these two measurements are made, one needs to establish the absolute bias. This is more challenging, and few options have been tried. One is from Bragg scatterers (Sect. 6.2.5) which produce zero Z_{DR}; thus, the overall correct bias is the value of Z_{DR} measured from Bragg scatterers. Let that correct value be $\Delta_C(t_0)$. It needs to be subtracted from the biased estimates denoted with $\hat{Z}_{DR}$ to obtain the corrected differential reflectivity Z_{DR}:

$$Z_{DR} = \hat{Z}_{DR} - \Delta_C(t_0). \tag{6.10}$$

This correction is valid if (6.8) and (6.9) do not change. $\Delta_T(t_0)$ normally does not change, and it suffices to check it at intervals of several hours (eight on the WSR-88D). If a change, denoted with $\Delta_{Tb} = \Delta_T(t) - \Delta_T(t_0)$, does occur, it should be subtracted from (6.10), i.e.:

$$Z_{DR} = \hat{Z}_{DR} - \Delta_C(t_0) - \Delta_T(t) + \Delta_T(t_0). \tag{6.11}$$

The same reasoning applies to the receiving part of the bias which, however, changes more often, and to catch these relatively fast changes, calibration of the receiving path is made at the end of each volume scan. This is automated and produces stable result. Thus, the receiver bias is $\Delta_{Rb}(t_i) = \Delta_R(t_i) - \Delta_R(t_0)$, at the times t_i when volume scans end. The correction requires subtraction of $\Delta_{Rb}(t_i)$ from all the data within the subsequent volume scan and so on.

The sun can be a reference source, and in case the transmitting path is well balanced, the sun flux may be sufficient for absolute calibration. Then the bias Δ_{Sb} revealed from the sun scan can be substituted for $\Delta_C(t_0)$ in (6.10).

6.2.2 "Birdbath" Calibration of Z_{DR}

Because the mean canting angle of raindrops is close to zero (Chap. 4), raindrops appear spherical if viewed at vertical incidence, and the measured Z_{DR} in light rain with vertically pointing antenna should be close to 0 dB. Such calibration technique ("birdbath" calibration) is discussed in Gorgucci et al. (1999), Hubbert et al. (2008), Marks et al. (2011), and Frech et al. (2017). This technique may work well only in light rain and in the absence of contamination from ground clutter via antenna sidelobes. Such contamination can cause azimuthal modulation of Z_{DR} for vertically looking rotating antenna (Gorgucci et al. 1999; Marks et al. 2011). If this is the case, azimuthal averaging is needed for determining the Z_{DR} bias, or spectral filtering of the ground clutter contribution to the radar signal can be applied (Zrnic and Melnikov 2007).

6.2.3 Z-Z_{DR} Consistency in Light Rain

Small raindrops have nearly spherical shape, and it is expected that Z_{DR} in light rain dominated by small-size drops is relatively close to 0 dB. Therefore, light rain may serve as a natural calibrator of Z_{DR}. This, however, is valid only in a general sense because raindrop size distributions associated with intense size sorting within convective updrafts are skewed toward larger drops and high values of Z_{DR} may be measured in the areas of relatively low Z. Figure 6.3 shows Z-Z_{DR} dependencies corresponding to different percentiles of Z_{DR} for a given Z in rain simulated from 47,114 DSDs measured in Oklahoma. The simulations are for S band at $T = 20$ °C. The domain between two dashed curves encompasses Z-Z_{DR} pairs of the whole dataset. Thus, Z_{DR} can be as high as 1 dB for $Z = 20$ dBZ. Nevertheless, in 80% of cases, Z_{DR} at $Z = 20$ dBZ stays below 0.4 dB with average value of 0.23 dB. The results of simulations from the measured DSDs are consistent with real observations of Z and Z_{DR} at S band in central Oklahoma reported by Ryzhkov et al. (2005) (see Fig. 6.4).

Average Z_{DR}-Z relations depend on prevalent rain regime associated with different climate regions (Fig. 6.5). Continental rain typical for Oklahoma (curve 1) is characterized by noticeably higher Z_{DR} than tropical rain in Florida (curve 2) and especially in Taiwan (curve 3) for larger Z. However, the three curves in Fig. 6.5 converge at lower Z; therefore Z_{DR} measurements at $Z < 30$ dBZ can serve for absolute calibration of Z_{DR} regardless of rain type.

The Z-Z_{DR} dependencies (Figs. 6.3, 6.4, and 6.5) are valid at S band. Similar analysis assuming shorter wavelengths produces quite similar results for $Z < 30$ dB (see Table 6.3). The Oklahoma DSD dataset is used for computation of Z_{DR} values in Table 6.3.

The procedure for Z_{DR} calibration based on the radar measurements in rain can be easily automated so that the consistency between measured and expected values of

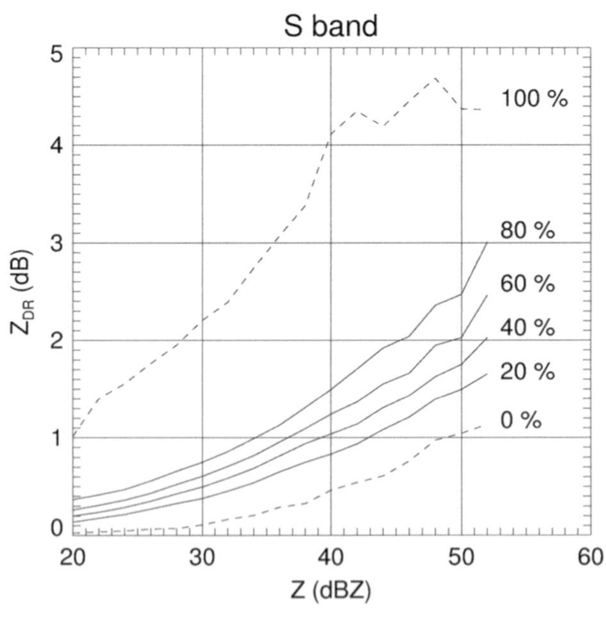

Fig. 6.3 Z-Z_{DR} dependencies corresponding to various percentiles of Z_{DR} for a given Z in rain. Z and Z_{DR} are simulated at S band from 47,144 DSDs measured in Oklahoma

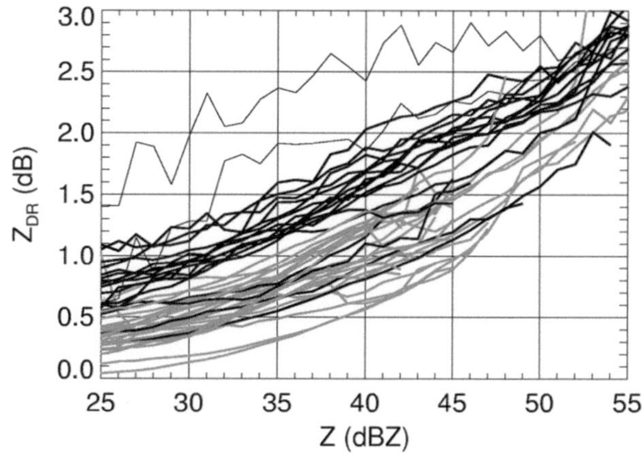

Fig. 6.4 Mean Z-Z_{DR} dependencies retrieved from the KOUN WSR-88D (Norman, OK) observations for each of the 43 h of JPOLE observations (21 rain events). Thick black curves correspond to the cases with rain dominated by large drops, thick gray curves to the rain events with abundance of small drops, and thin curves to the event with large hail. Adapted from Ryzhkov et al. (2005). © American Meteorological Society. Used with permission

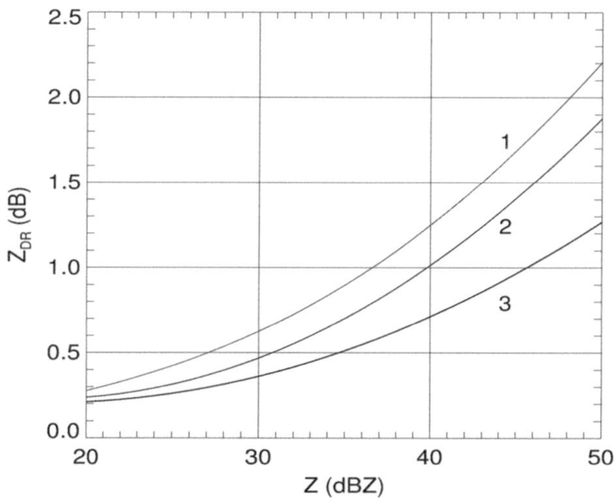

Fig. 6.5 Average dependencies of the median value of Z_{DR} on Z from rain in Oklahoma (1), Florida (2), and Taiwan (3). Curves 1 and 2 are from Cao et al. (2008)

Table 6.3 Median climatological values of Z_{DR} (dB) for different Z (dBZ) at S, C, and X bands in rain ($20 < Z < 30$ dBZ)

Z	20	22	24	26	28	30
Z_{DR}(S)	0.23	0.27	0.32	0.38	0.46	0.55
Z_{DR}(C)	0.23	0.27	0.33	0.40	0.48	0.56
Z_{DR}(X)	0.23	0.28	0.33	0.41	0.49	0.58

Z_{DR} in light rain is checked every radar scan if appropriate data are available. In the automatic calibration routine implemented on the MeteoFrance operational radar network, the measured median Z_{DR} at $Z = 20$–22 dBZ is compared with the reference value 0.2 dB (Figueras i Ventura et al. 2012). It is also possible to estimate the Z_{DR} bias as

$$\Delta Z_{DR} = \frac{1}{6}\sum_{k=1}^{6}\left[Z_{DR}^{(m)}(k) - < Z_{DR}^{(m)}(k) >\right], \tag{6.12}$$

where $<{Z_{DR}}^{(m)}(k)>$ is the median climatological value of Z_{DR} in the kth 2 dB category of Z listed in Table 6.3 and ${Z_{DR}}^{(m)}(k)$ is the value estimated from radar data. Similar to the self-consistency calibration of Z, the data appropriate for calibration of Z_{DR} should be selected where SNR is sufficiently high (SNR $>$ 20–25 dB), differential attenuation is insignificant, and rain scatterers are dominant contributors (i.e., $\rho_{hv} > 0.98$–0.99).

6.2.4 Z_{DR} Calibration Using Dry Aggregated Snow

Dry aggregated snow produces small intrinsic Z_{DR} because of its very low density. The studies by Ryzhkov and Zrnic (1998) and Ryzhkov et al. (2005, 2008) indicate that mean Z_{DR} (i.e., averaged over a sufficiently large spatial/temporal interval) in aggregated snow is usually smaller than 0.2 dB if $Z > 30$ dBZ. Dry aggregated snow near the surface does not occur in warm climatic zones. In addition, such snow should be carefully separated from wet aggregated snow and dry crystallized snow that are characterized by a much higher and more variable Z_{DR}. Nevertheless, dry, low-density, aggregated snowflakes are commonly present above the melting layer in stratiform clouds (provided that $Z > 30$ dBZ). Numerous polarimetric radar measurements show that Z_{DR} drops almost to 0 dB 1–2 km above the 0 °C level where dry aggregated snow is most likely (Brandes and Ikeda 2004).

Quasi-vertical profiles (QVP, Ryzhkov et al. 2016) of Z_{DR} in aggregates above the melting layer are suitable for monitoring deviation from expected low values. Because QVPs made from azimuthal averages over 360° at high elevations, the precision of this measurement is better than 0.1 dB (see Chap. 7).

6.2.5 Z_{DR} Calibration Using Bragg Scatter

Melnikov et al. (2011) suggest using clear-air radar echoes associated with Bragg scattering for absolute calibration of Z_{DR}. Bragg backscatter from refractive index perturbations at 5 cm scales creates sufficiently strong echo in a convective boundary layer to be detected by 10 cm wavelength weather radars. These echoes are characterized by intrinsic Z_{DR} equal to 0 dB and cross-correlation coefficient ρ_{hv} very close to 1 making them easily distinguishable from the echoes caused by biota which have very large Z_{DR} and low ρ_{hv} (see Chap. 9).

An automated algorithm for estimating Z_{DR} bias from the Bragg scatter was developed and extensively tested on the S-band WSR-88D radars (Richardson et al. 2017a, b). The algorithm yields better accuracy of the Z_{DR} bias estimation than the methods based on the Z_{DR} measurements in light rain and dry snow. Strong Bragg scattering usually occurs at the top of the boundary layer because there the gradients of humidity are largest and mixing by turbulence produces strongest returns. This is seen in Fig. 6.6 as a distinct layer of enhanced Z and close to zero Z_{DR}. Application of thresholds ($Z > 10$ dBZ, SNR >15 dB, $\rho_{hv} > 0.98$, and $|v| > 2$ m s^{-1}) and some other criteria identifies data in the layer that are due to the Bragg scatter (Fig. 6.6b, top left); the histogram of Z_{DR} (Fig. 6.6, right panel) is indeed centered on 0 dB.

Comparison of the Z_{DR} bias deduced from the Bragg scatter with the ones from rain and snow indicates consistency between the three methods (Fig. 6.7). Furthermore, Fig. 6.7 illustrates a change of 0.6 dB recorded by the Omaha WSR-88D and a relatively small change of 0.2 dB measured on the Columbia WSR-88D.

Systematic observations in the USA demonstrated that the Bragg scatter methodology is viable at many radar sites throughout the year and provides an important addition to other calibration techniques.

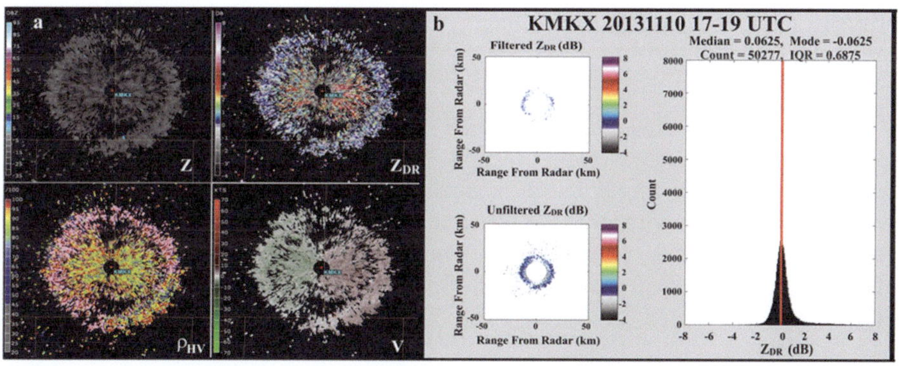

Fig. 6.6 Example of Bragg scattering observed by the KMKX WSR-88D (Milwaukee, WI) on 10 November 2013. (**a**) The fields of Z (upper left), Z_{DR} (upper right), ρ_{hv} (lower left), and Doppler velocity (lower right) are from conical scans at the 3.5° elevation angle (1852 UTC). Maximum range in the image is ~22 km. (**b**) The Z_{DR} histogram (right) is from the data (top left corner) which have passed Bragg detection criteria. Data that have passed the SNR $>$ 2 dB threshold are in the bottom left image. From Richardson et al. (2017a). © American Meteorological Society. Used with permission

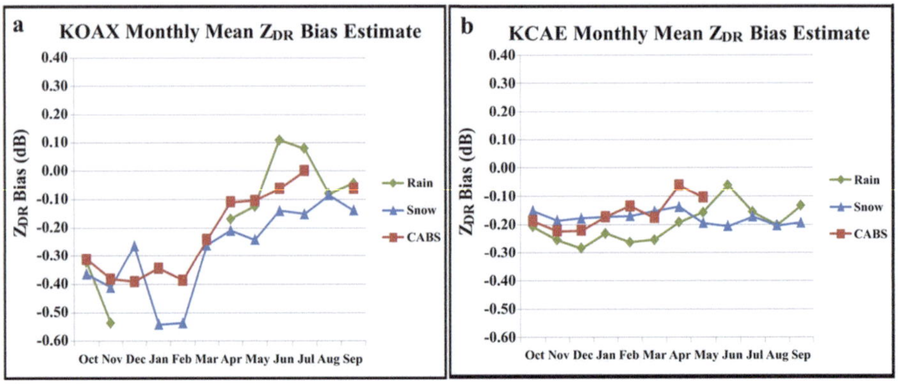

Fig. 6.7 Monthly mean Z_{DR} bias estimates from October 2013 to September 2014 from two WSR-88Ds using rain, snow, and Bragg scatterers as natural calibrators. (**a**) KOAX in Omaha, NE, and (**b**) KCAE in Columbia, SC. Missing data cause discontinuities in the records. The three methods are generally within ±0.1 dB of each other at a given site. Adapted from Richardson et al. (2017b). © American Meteorological Society. Used with permission

6.2.6 Z_{DR} Calibration Using Cross-Polar Measurements

Hubbert et al. (2003) suggest a method for Z_{DR} calibration which utilizes the sun measurements and the ratio of two cross-polar powers of reflected signals from arbitrary weather scatterers. According to the method, true (intrinsic) Z_{DR} is related to the measured $Z_{DR}{}^{(m)}$ as

$$Z_{DR} = Z_{DR}^{(m)} + 2Z_{DR}^{(sun)} + 10\log(P_{hv}/P_{vh}), \tag{6.13}$$

where $Z_{DR}{}^{(sun)}$ is the measured differential reflectivity of sun radiation, P_{hv} is the power of cross-polar component of the radar return if H is transmitted, and P_{vh} is the cross-polar power if V is transmitted. This methodology requires the use of the HSHV/VSVH modes of operation with the capability to transmit both horizontally and vertically polarized waves (see Chap. 2).

6.2.7 Z_{DR} Calibration Using Ground Clutter

Ground clutter returns are ubiquitous and can be a good source for monitoring the state of the radar. These returns often include contributions from moving objects like biological scatterers and vehicles. Furthermore, relative motion of vegetation induced by wind also causes variation of the radar signal. To minimize such effects, the use of strong clutter (>50 dBZ, Borowska and Zrnic 2012) and/or Doppler spectral components within −0.5 to 0.5 m s^{-1} (Melnikov et al. 2017) have been suggested. Borowska and Zrnic (2012) found that the change in the vertical profile of the refractive index due to diurnal variation of air temperature can cause noticeable variation in the mean Z and Z_{DR} from clutter. They recommend using clutter returns originating at ranges closer than 7 km from the radar.

Observing SNR, Z, Z_{DR}, and Φ_{DP} over time is advantageous because changes are easy to spot and can be related to problems with hardware including calibration. Statistical properties of ground clutter are suitable descriptors because they are stable due to the large amount of available clutter data. The example in Fig. 6.8 is obtained

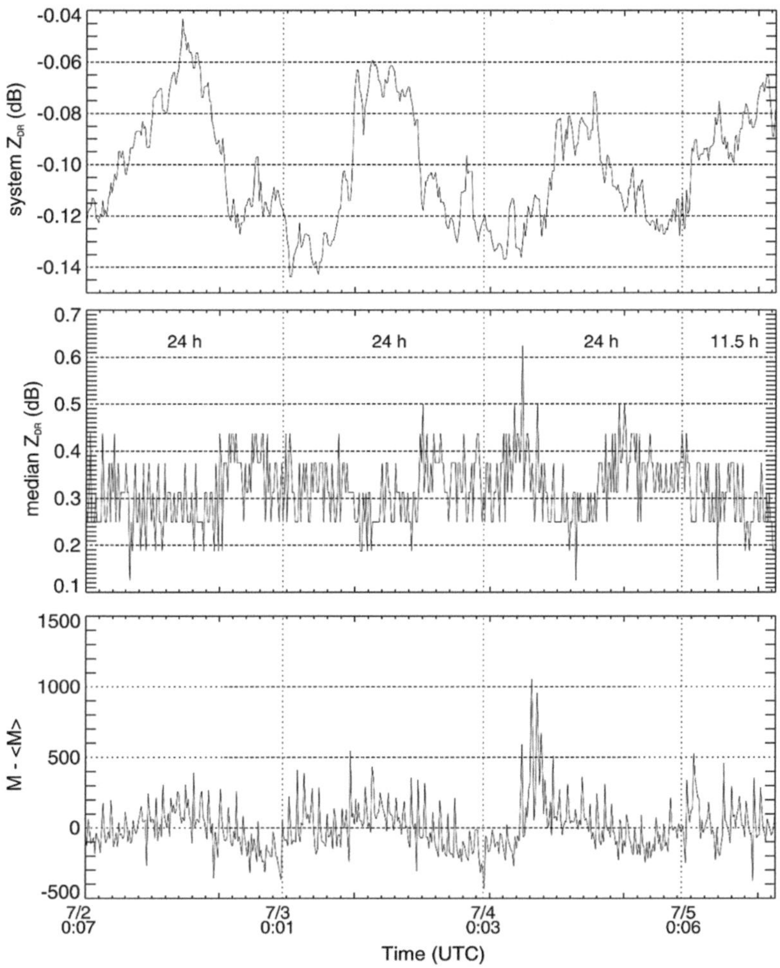

Fig. 6.8 (Top) System Z_{DR} obtained automatically at the end of volume scans using signal generator and a prior initial calibration (Sect. 6.2.1). Dotted vertical lines indicate beginning of a day (UTC time); the dates and minutes of the beginning are indicated along the abscissa. (Middle) Median Z_{DR}. (Bottom) The difference between the number of points M and its average value ($<M> = 2742$). Data obtained with the KOUN WSR-88D radar. The thick marks on the abscissa separate ten consecutive volume scans (accomplished in 97.3 min). From Borowska and Zrnic (2012). © American Meteorological Society. Used with permission

with the KOUN WSR-88D radar on 3 consecutive days in July 2011. The system Z_{DR} shown in the top panel is the bias correction applied to each volume scan after the automatic differential power measurement of the receiver path at the end of each volume scan (see Sect. 6.2.1 and Fig. 6.2). Its diurnal periodicity is attributed to the effects of temperature variations on the receivers. The median values of Z_{DR} from clutter (middle graph in Fig. 6.8) are mainly within 0.2 dB with very few outliers. Noticeable is the increase on the third day. It is caused by the change in the environment. A nearby storm generated a cold pool of air which moved over the radar and enhanced beam bending thus increasing the number of detected clutter returns as seen in the bottom of the figure. This also altered the illumination of the ground via the antenna patterns for horizontal and vertical polarizations which changed the statistics of Z_{DR}. The number of detected clutter data per scan M minus the average number of detected data $\langle M \rangle$ exhibits local periodic maxima early in the morning at the time of minimum temperature. Again, increased beam bending is responsible for this increase. One more peculiarity in this graph is the periodicity of sharp peaks which occur every sixth volume scan (Fig. 6.8, bottom panel). The consecutive volume scans start at azimuths displaced by 60° so that after six scans the beginning azimuths repeat. It was caused by systematic differences in elevation angles which were not strictly constant but exhibited a slight periodic variation with respect to the nominal 0.5°. This problem has been corrected.

Median and modal values of Z_{DR} measured over few months by the KOUN WSR-88D radar exhibit consistency and stability within ±0.1 dB over relatively long periods and different seasons (Melnikov et al. 2017). Few abrupt changes have been recorded and related to hardware failures. These authors also document definite increase in Z_{DR} at the times of rain over the radar. It is due to vertical water streams on the radome that attenuate and diffract more the vertically polarized waves than the horizontally polarized waves (Sect. 6.3).

6.3 The Impact of Wet Radome on the Measurements of Z and Z_{DR}

Radomes are needed for protection of radar antenna and microwave hardware from adverse weather. In the presence of precipitation, a film of water or accumulated snow at the surface of radome causes attenuation of microwave radiation and distortion of the beam that affect differently the horizontally and vertically polarized waves. Therefore, both Z and Z_{DR} can be noticeably biased by precipitation on the radome. The bias in Z can be few dB and depends on the wavelength, radome type, its age and condition (clean or dirty), and precipitation type and intensity. The effect can be significantly reduced if the radome has hydrophobic paint. An overview of the Z biases caused by wet radome is in the paper by Kurri and Huuskonen (2008).

The impact of water/snow accumulation at the radome surface on the Z_{DR} measurements at C band was studied by Hudak et al. (2006) and Frech (2009).

According to Frech (2009), the associated bias in Z_{DR} is positive, gradually increases with rain rates, and reaches a plateau at about 6 mm h^{-1}. The Z_{DR} bias can be as high as 1 dB. Hudak et al. (2006) found even larger Z_{DR} biases (up to 3 dB!) if melting snow accumulates on the radome. Moreover, the Z_{DR} bias may exhibit strong azimuthal modulation which reflects the dependence of the thickness of the snow layer on wind direction. Bechini et al. (2006) also reported azimuthal modulation of radome-induced transmission losses in rain.

The Z_{DR} bias caused by wet snow is illustrated in Fig. 6.9 for winter storm observed with the C-band King radar north of Toronto, Ontario, on 21 January 2006. At 1030 UTC, wet snow with strong northerly winds occurred at the radar site just behind the cold front. The very high values of Z_{DR} are only evident at azimuths corresponding to the inbound wind direction during the time when wet snow was occurring at the site. By 1100 UTC, the period of wet snow had ended and the Z_{DR} anomaly disappeared. The gradual demise of the Z_{DR} anomaly as wet snow was ending is evident in the images taken 10 min apart (Fig. 6.9).

Regardless of the type of precipitation deposit on the radome surface (rain or snow), the related bias in Z_{DR} is generally positive because vertical streams (rivulets) of water caused by rain or melting snow scatter and absorb more the vertically polarized component of the microwave radiation than the horizontally polarized component. An interesting case of the Z_{DR} enhancement caused by intense streams of rainwater at the downwind side of radome is illustrated in Fig. 6.10. The observations have been made with the S-band radar during very heavy rain at the radar site caused by a typhoon in Southern China. A positive Z_{DR} bias of about 1 dB is clearly evident in a narrow azimuthal sector. In this case a positive Z_{DR} bias is observed at the lee side of the radome as opposed to the wet snow case shown in Fig. 6.9. This is because the rainwater is blown off the windward side and accumulates at the lee side, whereas the snow sticks to the windward side.

The biases in Z and Z_{DR} caused by wet radome can be compensated using data-based consistency methodologies. Two examples of the Z_{DR} bias monitoring are presented in Figs. 6.11 and 6.12. In Fig. 6.11, the Z_{DR} bias in the S-band KHTX WSR-88D (Huntsville, AL) is displayed as a function of time (top panel). Time intervals of rain over the radar site are marked with green shading. These correspond to enhancement of average Z in a 5 km radius area centered on the radar (bottom

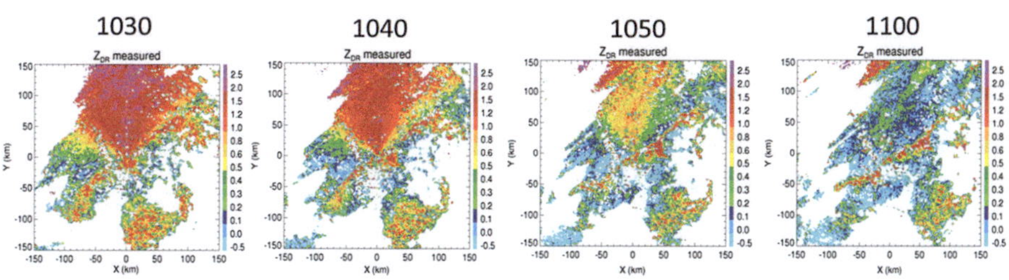

Fig. 6.9 Time series of Z_{DR} images at elevation 0.2° taken every 10 min with the C-band King radar during the wet snow event on 21 January 2006. Notable is remarkable enhancement of Z_{DR} in the northern sector due to accumulation of wet snow on the radome

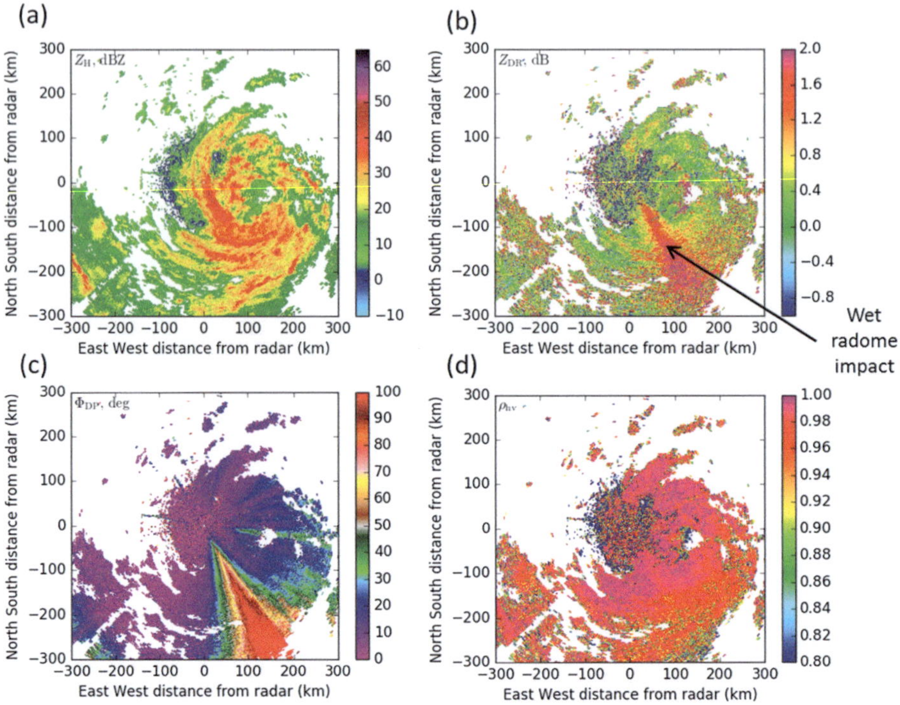

Fig. 6.10 Composite PPI of (**a**) Z, (**b**) Z_{DR}, (**c**) Φ_{DP}, and (**d**) ρ_{hv} measured at El $= 0.48°$ by the S-band MetStar radar in Guangzhou, China, during the typhoon on 1 August 2016. Z_{DR} enhancement in the SE azimuthal sector is caused by rain accumulation on the downwind side of the radome at the time of heaviest rain at the radar site. Courtesy of Guangzhou Meteorological Bureau, China

panel). It is obvious that Z_{DR} bias increases 0.2–0.4 dB when it is raining at the radar site. A similar example on the C-band OU-PRIME radar (Fig. 6.12) exhibits quite large bias (about -1.4 dB). This was on 2 June 2009 shortly after the radar installation. The additional positive bias of 0.6 dB, caused by wet radome during short-lived rainfall over the radar site, is well captured by the technique based on the consistency between Z and Z_{DR} in light rain. The technique demonstrates robust performance and yields quite stable estimate of the Z_{DR} bias for the period of several hours as seen in Figs. 6.11 and 6.12.

6.4 Attenuation Correction

Attenuation of microwave radiation in precipitation may significantly bias the measurements of Z and Z_{DR}, especially at shorter wavelengths. Reliable correction of Z and Z_{DR} is required before utilizing these variables for quantitative rainfall estimation, hydrometeor classification, microphysical retrievals, etc.

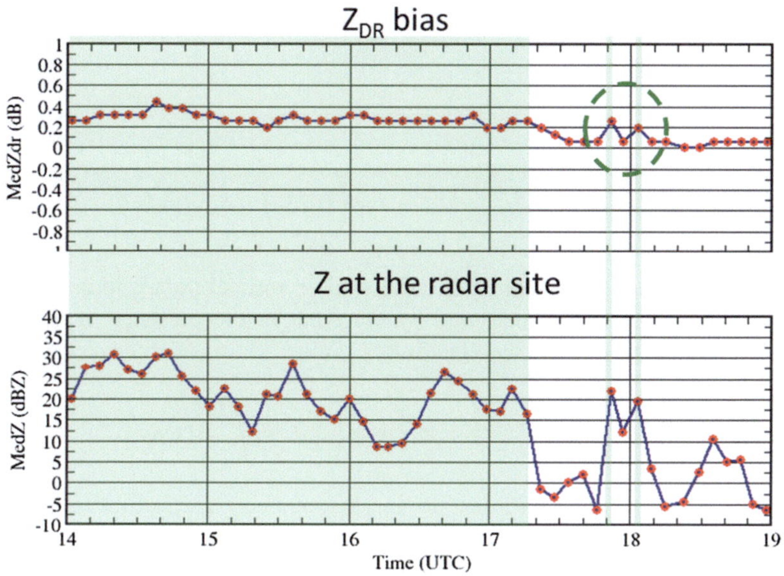

Fig. 6.11 Temporal evolution of Z_{DR} bias and average Z in a 5 km radius area surrounding radar obtained from the measurements by the KHTX WSR-88D (Huntsville, AL) on 1 February 2012 at elevation 0.5°. Green shading indicates periods of rain over the radar

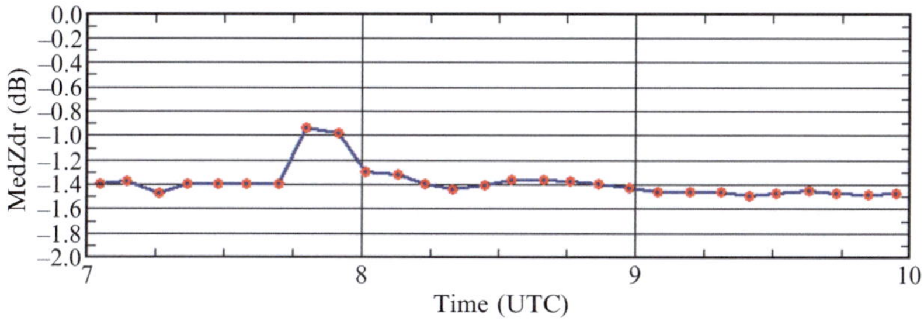

Fig. 6.12 Temporal evolution of the bias in Z_{DR} measured by the C-band OU-PRIME radar (Norman, OK) on 2 June 2009 as revealed from the Z-Z_{DR} consistency check in light rain

Reflectivity factor is related to the received signal power P and the range r according to the radar equation (3.30):

$$P_{\mathrm{hh,vv}}(r) = \frac{C_{\mathrm{r}} Z_{\mathrm{hh,vv}}(r)}{r^2 l_{h,v}^2(r)}, \tag{6.14}$$

where

$$C_r = \frac{\pi^3 P_t g^2 c\tau\theta_1^2 |K_w|^2}{2^{10} \ln 2\lambda^2 l_r}, \tag{6.15}$$

and

$$l_{h,v}(r) = 10^{0.1\int_0^r A_{h,v}(s)ds} = \exp\left(0.23\int_0^r A_{h,v}(s)ds\right), \tag{6.16}$$

and $A_{h,v}$ is the specific attenuation at horizontal or vertical polarization (Sect. 5.7). The reflectivity Z_a directly measured by the radar in the presence of attenuation differs from intrinsic (true) Z by a factor l^2:

$$Z_a(r) = Z(r)l^{-2}(r) = Z(r)\exp\left(-0.46\int_0^r A(s)ds\right), \tag{6.17}$$

where Z_a and Z are expressed in linear scale and the subscripts indicating polarization of the wave are omitted for simplicity. Specific attenuation A in rain can be approximately estimated from Z using a power-law relation

$$A = aZ^b \tag{6.18}$$

in which the factors a and b depend on the size distribution of raindrops and temperature. The factors a and b can be considered constant for certain applications. Combining (6.17) and (6.18) yields

$$Z(r) = Z_a(r)\exp\left(0.46\int_0^r A(s)ds\right) = Z_a(r)\exp\left(0.46\int_0^r a[Z(s)]^b ds\right). \tag{6.19}$$

This is an integral equation with respect to $Z(r)$ which can be solved with different degree of accuracy using various methods. A solution suggested by Hitshchfeld and Bordan (1954) assumes that the factors a and b are constant along the propagation path, therefore

$$Z_{\mathrm{HB}}(r) = Z_a(r)\left[1 - 0.46b\int_0^r a[Z_a(s)]^b ds\right]^{1/b}. \tag{6.20}$$

The Hitshchfeld and Bordan (HB) solution gives a reasonable attenuation correction if attenuation is small (say, less than 10 dB). When attenuation is large, the HB solution usually becomes unstable, that is, slight variation of the parameters a and b or calibration error in Z_a induces a large variation of Z_{HB}. Stabilizing the solution requires introduction of an external constraint such as path-integrated attenuation (*PIA*) which should be estimated from independent measurement. The two-way *PIA* can be obtained as a ratio of reflectivity factors of a reference reflector at distance r_m with known reflectivity $Z^{(\text{ref})}(r_m)$ (i.e., in the absence of precipitation along the propagation path) and attenuated $Z_a(r_m)$:

$$PIA = \frac{Z^{(\text{ref})}(r_m)}{Z_a(r_m)} = \exp\left[0.46 \int_0^{r_m} A(s)ds\right]. \tag{6.21}$$

In logarithmic scale

$$\text{PIA(dB)} = Z^{(\text{ref})}(r_m, dB) - Z_a(r_m, dB) = 2\int_0^{r_m} A(s)ds. \tag{6.22}$$

If the PIA is known, then the radial profile of specific attenuation in the range interval (0, r_m) can be obtained as (Meneghini and Nakamura 1990)

$$A(r) = \frac{a(r)[Z_a(r)]^b C(b, \text{PIA})}{I_a(r_1, r_2) + C(b, \text{PIA}) I_a(r, r_2)}, \tag{6.23}$$

where

$$I_a(r_1, r_2) = 0.46b \int_{r_1}^{r_2} a(s)[Z_a(s)]^b ds, \tag{6.24}$$

$$I_a(r, r_2) = 0.46b \int_{r}^{r_2} a(s)[Z_a(s)]^b ds, \tag{6.25}$$

$$C(b, \text{PIA}) = \exp(0.23b\text{PIA}) - 1. \tag{6.26}$$

The corrected radial profile of Z along the path (0, r_m) is determined from the equation

$$Z(r) = Z_a(r) + 2\int_0^{r} A(s)ds. \tag{6.27}$$

Equation (6.23) is derived assuming that the factor b is constant along the radial, whereas a is variable. If a is constant along the ray, then

$$A(r) = \frac{[Z_a(r)]^b C(b, \text{PIA})}{I(r_1, r_2) + C(b, \text{PIA}) I(r, r_2)}. \tag{6.28}$$

where

$$I(r_1, r_2) = 0.46b \int_{r_1}^{r_2} [Z_a(s)]^b ds, \tag{6.29}$$

$$I(r, r_2) = 0.46b \int_r^{r_2} [Z_a(s)]^b ds, \tag{6.30}$$

and $A(r)$ does not explicitly depend on the value of a.

PIA can be estimated using stable objects such as a mountain, high tower, or ocean surface if near-nadir measurements are made from airborne or satellite platforms. Dual-polarization measurements offer alternate estimate of PIA based on the measurements of differential phase. If A_h and A_{DP} depend linearly on K_{DP} so that the ratios

$$\alpha = A_h/K_{DP} \quad \text{and} \quad \beta = A_{DP}/K_{DP} \tag{6.31}$$

are constant along the radial, then the attenuation-caused biases ΔZ and ΔZ_{DR} can be expressed via the differential phase as

$$\Delta Z_h(r) = 2\int_0^r A_h(s)ds = 2\alpha \int_0^r K_{DP}(s)ds = \alpha \Phi_{DP}(r). \tag{6.32}$$

and

$$\Delta Z_{DR}(r) = 2\int_0^r A_{DP}(s)ds = 2\beta \int_0^r K_{DP}(s)ds = \beta \Phi_{DP}(r), \tag{6.33}$$

and the PIA is proportional to the total span of differential phase

$$\mathrm{PIA} = \alpha[\Phi_{DP}(r_m) - \Phi_{DP}(0)] = \alpha \Delta \Phi_{DP}(0, r_m). \tag{6.34}$$

The linear correction utilizing (6.32) and (6.33) was proposed by Balakrishnan and Zrnic (1989) and Bringi et al. (1990). Testud et al. (2000) suggested using PIA from (6.34) in (6.27); the corresponding attenuation correction procedure is known as ZPHI method and has been examined in detail by Le Bouar et al. (2001). The fact that attenuation biases of Z and Z_{DR} are directly proportional to the differential phase is an advantage of polarimetric radars because it enables accurate quantification of precipitation in the presence of strong attenuation at shorter wavelengths (C and X bands).

The factors α and β in (6.31) are sensitive to the variability of raindrop size distributions and temperature. Typical range of variability at different wavelengths is listed in Table 6.4. Rough approximate correction can be made using "default" or average values in the right column of Table 6.4. It produces substantial improvement in Z and Z_{DR} compared to the absence of correction. The efficiency of default linear correction using (6.32) and (6.33) at C band with $<\alpha> = 0.08$ dB deg^{-1} and $<\beta> = 0.02$ dB deg^{-1} is demonstrated in Figs. 6.13, 6.14, and 6.15 for the case of a tornadic storm in Oklahoma observed with closely located S- and C-band

Table 6.4 Ranges of variability of the factors α and β in rain at S, C, and X bands

S band	
$\alpha = 0.015–0.04$ dB deg^{-1}	$<\alpha> = 0.02$ dB deg^{-1}
$\beta = 0.0025–0.009$ dB deg^{-1}	$<\beta> = 0.004$ dB deg^{-1}
C band	
$\alpha = 0.05–0.18$ dB deg^{-1}	$<\alpha> = 0.08$ dB deg^{-1}
$\beta = 0.008–0.1$ dB deg^{-1}	$<\beta> = 0.02$ dB deg^{-1}
X band	
$\alpha = 0.14–0.35$ dB deg^{-1}	$<\alpha> = 0.28$ dB deg^{-1}
$\beta = 0.03–0.06$ dB deg^{-1}	$<\beta> = 0.05$ dB deg^{-1}

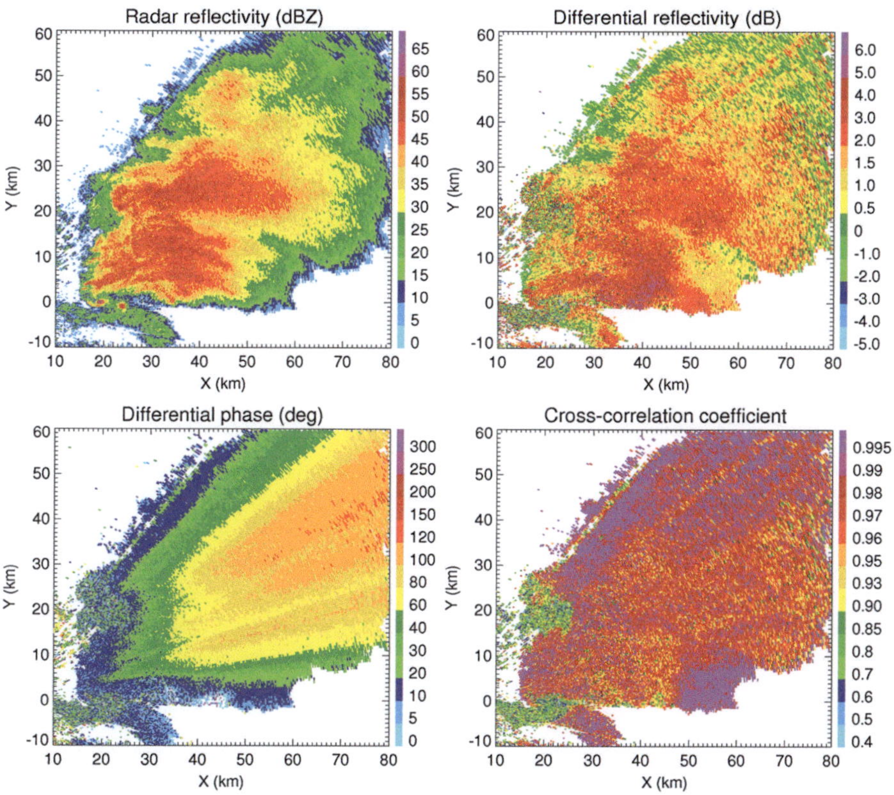

Fig. 6.13 Plot of Z, Z_{DR}, Φ_{DP}, and ρ_{hv} fields measured by the S-band KOUN WSR-88D at elevation 0.5° in the tornadic storm in central Oklahoma on 10 May 2010 at 2045 UTC

polarimetric radars. The fields of Z and Z_{DR} measured with the S-band KOUN WSR-88D do not exhibit the effects of attenuation because it is generally small at S band (Fig. 6.13). Similar fields of Z and Z_{DR} measured by the C-band OU-PRIME radar show large negative biases before attenuation correction is applied (Fig. 6.14). The

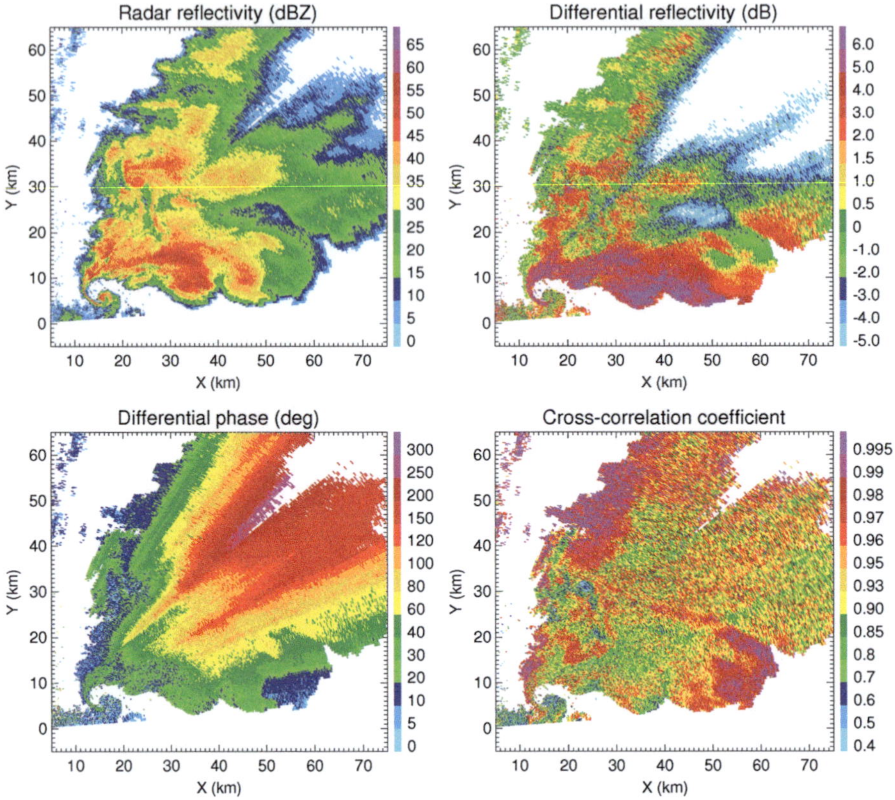

Fig. 6.14 Plot of Z, Z_{DR}, Φ_{DP}, and ρ_{hv} fields measured by the C-band OU-PRIME radar at elevation 0.5° in the tornadic storm in central Oklahoma on 10 May 2010 at 2042 UTC

biases are largest along azimuthal directions where the total differential phase is highest. The corrected fields of Z and Z_{DR} (Fig. 6.15) are consistent with the ones measured with the S-band radar. It is notable that the corrected C-band Z_{DR} values are significantly higher than the corresponding Z_{DR} values at S band in large areas of the storm. This is attributed to the effects of resonance scattering discussed in Chap. 5.

The factors α and β in rain are functions of differential reflectivity and temperature, and these dependencies are illustrated in Fig. 6.16. The scatterplots of α and β versus Z_{DR} are simulated from the Oklahoma disdrometer dataset for temperatures 0 °C and 30 °C. The dependence of α on Z_{DR} is nonmontonic, whereas β generally increases with Z_{DR}. The ratio A_h/K_{DP} is higher in melting hail than in rain because A_h tends to increase for larger melting hail, while K_{DP} decreases and may become even negative (see Figs. 5.16 and 5.23). Borowska et al. (2011) evaluated A_h in melting hail at C band via direct comparisons with simultaneous measurements at S band and reported anomalously high values of α at C band. Some of the results of that study are presented in Fig. 6.17 which shows that average slopes of the A_h-K_{DP} relation for

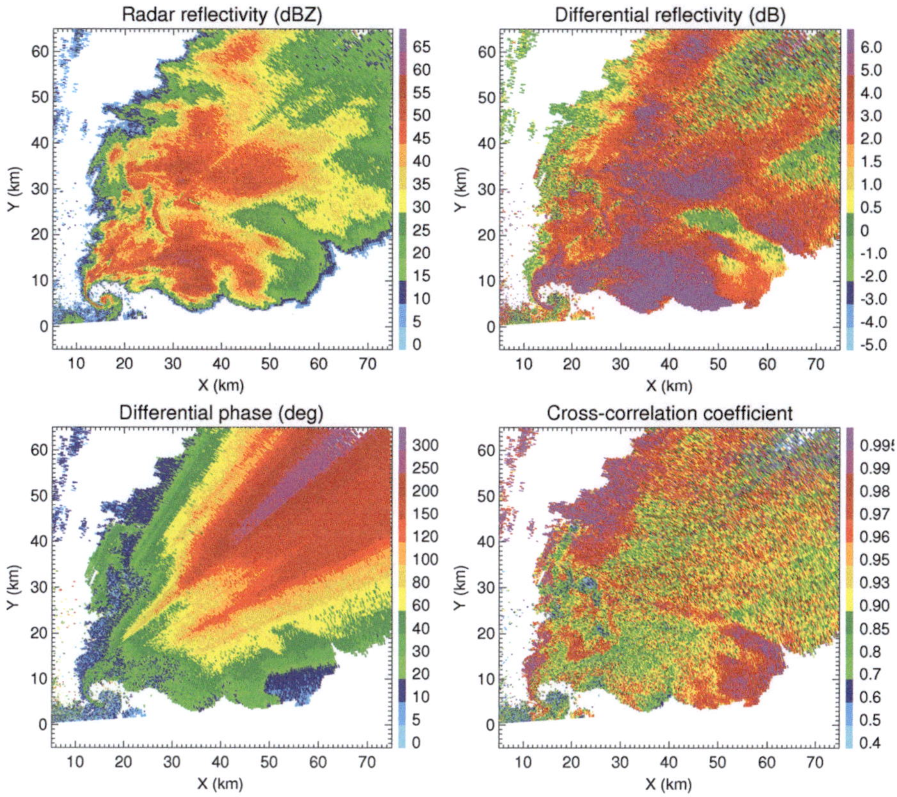

Fig. 6.15 Same as in Fig. 6.14 but after a linear correction of Z and Z_{DR} for attenuation via (6.32) and (6.33)

two cases of melting hail (0.33 dB deg^{-1} and 0.67 dB deg^{-1}) are well above what is expected from rain (Table 6.4).

Finding optimal α and β is a challenge for attenuation correction techniques. First, it is important to determine the average climatological values of α and β for a particular area using local disdrometer measurements. However, this may not be sufficient in the areas of high diversity of precipitation regimes where α and β can change dramatically even in a single storm. There are two approaches to address the problem.

The first one, called a self-consistent method with constraints, aims at estimating optimal α and β for a single ray (Bringi et al. 2001). The method searches for an optimal value α within a predetermined range ($\alpha_{\min}$, $\alpha_{\max}$). For each α value in the predetermined range, a radial profile of specific attenuation $A_h(r,\alpha)$ is estimated using the ZPHI method (6.28), and a model profile of differential phase $\Phi_{DP}{}^{(\text{mod})}$ is calculated as

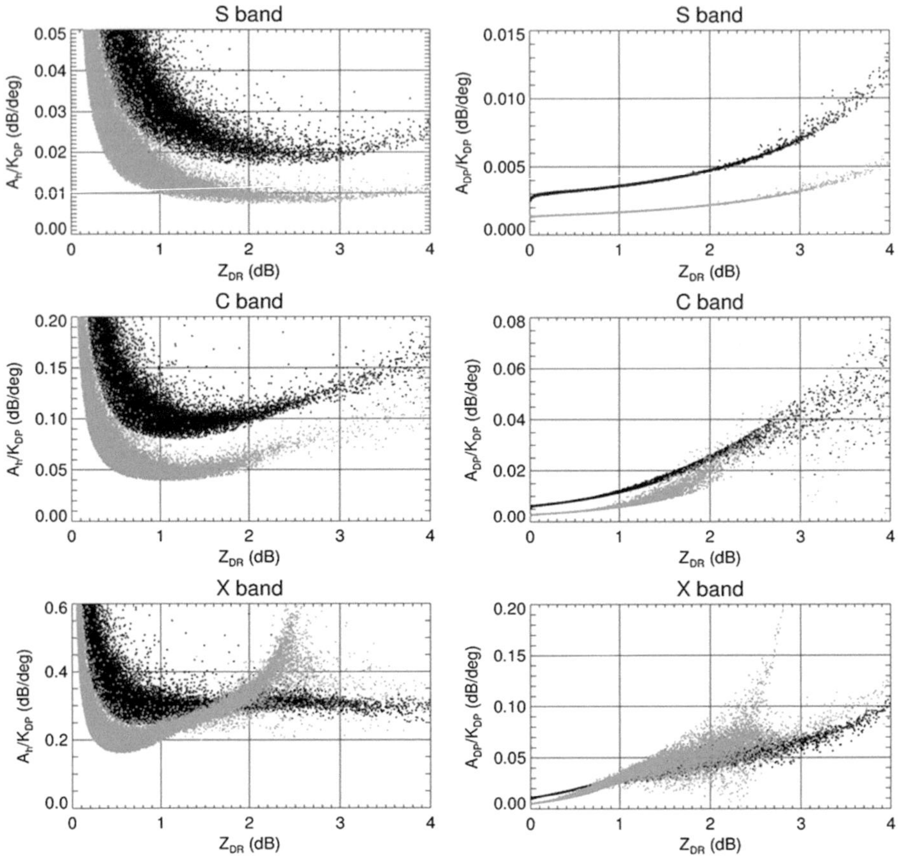

Fig. 6.16 Scatterplots of the ratios A_h/K_{DP} and A_{DP}/K_{DP} versus Z_{DR} in rain at S, C, and X bands for two temperatures: 0 °C (black dots) and 30 °C (gray dots). The polarimetric variables are simulated using a large disdrometer dataset from Oklahoma

$$\Phi_{DP}^{(\mathrm{mod})}(r,\alpha) = 2\int_0^r \frac{A_{\mathrm{h}}(s,\alpha)}{\alpha}ds \tag{6.35}$$

and it is compared with the measured (and filtered) $\Phi_{DP}(r)$ over the entire propagation path in rain. An optimal $\alpha^{(\mathrm{opt})}$ minimizes the difference Δ between the measured and calculated profiles:

$$\Delta = \sum_{i=1}^{N} \mid \Phi_{DP}^{(\mathrm{mod})}(r_{\mathrm{i}},\alpha) - \Phi_{DP}(r_i) \mid , \tag{6.36}$$

where i denotes the range gate index along the propagation path in rain (see conceptual drawing in Fig. 6.18).

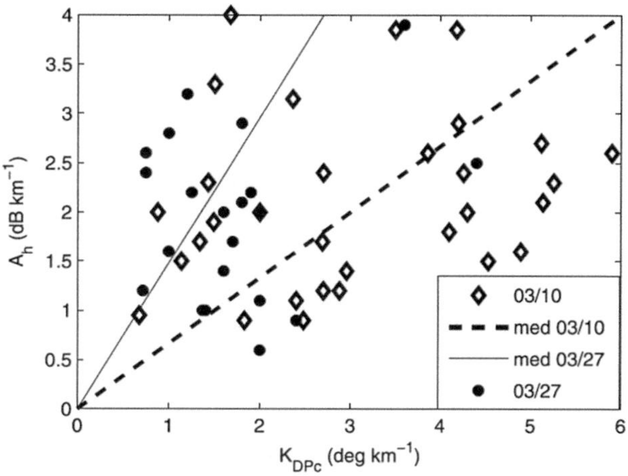

Fig. 6.17 The A_h-K_{DP} scatterplots in melting hail at C band obtained from the data collected on 10 March 2009 (diamonds) and 27 March 2009 (dots) by the OU-PRIME radar (Norman, OK). The dashed and solid median lines are for March 10 and 27, respectively. Adapted from Borowska et al. (2011). © American Meteorological Society. Used with permission

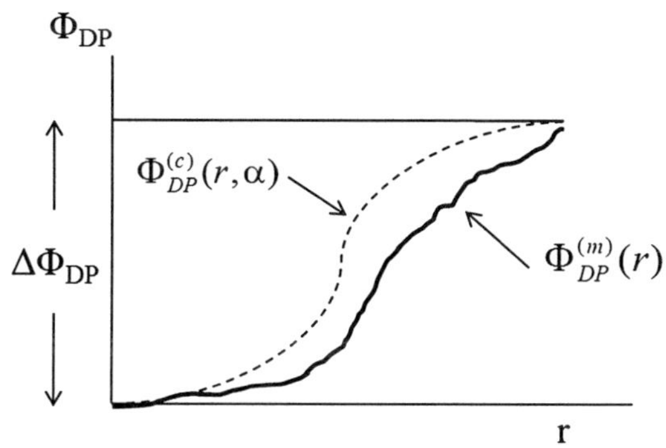

Fig. 6.18 Illustration of the concept for self-consistent method with constraints (Bringi et al. 2001). $\Phi_{DP}^{(m)}$ is the measured profile of differential phase; $\Phi_{DP}^{(c)}$ is the computed profile for a given α

Differential reflectivity is corrected according to the formula:

$$Z_{DR}(r) = Z_{DR}^{(a)}(r) + 2\int_0^r A_{DP}(s)ds \tag{6.37}$$

in which specific differential attenuation $A_{DP}(r)$ is assumed to be proportional to the optimized $A_h[r,\alpha^{(opt)}]$:

$$A_{DP}(r) = \gamma A_h\left[r, \alpha^{(opt)}\right], \tag{6.38}$$

with the coefficient of proportionality γ determined from

$$Z'_{DR}(r_m) = Z_{DR}^{(a)}(r_m) + 2\gamma\int_0^{r_m} A_h\left[s, \alpha^{(opt)}\right]ds. \tag{6.39}$$

In (6.39), $Z_{DR}^{(a)}(r_m)$ and $Z'_{DR}(r_m)$ are measured (biased by differential attenuation) and expected differential reflectivity, respectively, at range r_m (behind attenuating cell) for a non-attenuated reflectivity $Z(r_m)$. That is, the corrected value of Z_{DR} at range r_m has to be equal to what is expected in rain at the reflectivity $Z(r_m)$.

Another attenuation correction scheme introduced by Ryzhkov et al. (2007) and further explored by Gu et al. (2011) separates relative contributions of "hot spots" (i.e., strong convective cells) and the rest of the storm to the path-integrated total and differential attenuation.

The "hot-spot" algorithm implies that

$$\alpha = \alpha_0 + \Delta\alpha \tag{6.40}$$

and

$$\beta = \beta_0 + \Delta\beta, \tag{6.41}$$

in the hot spot (HS) (Fig. 6.19), whereas the "background" values α_0 and β_0 are constant outside hot spots for a given radar scan. These background values can be set equal to average climatological values or can be estimated from the data (Gu et al. 2011). The hot spots are identified using Z and Z_{DR} which are preliminary corrected for attenuation using equations

$$\Delta Z = \alpha_0 \Phi_{DP} \quad \text{and} \quad \Delta Z_{DR} = \beta_0 \Phi_{DP}. \tag{6.42}$$

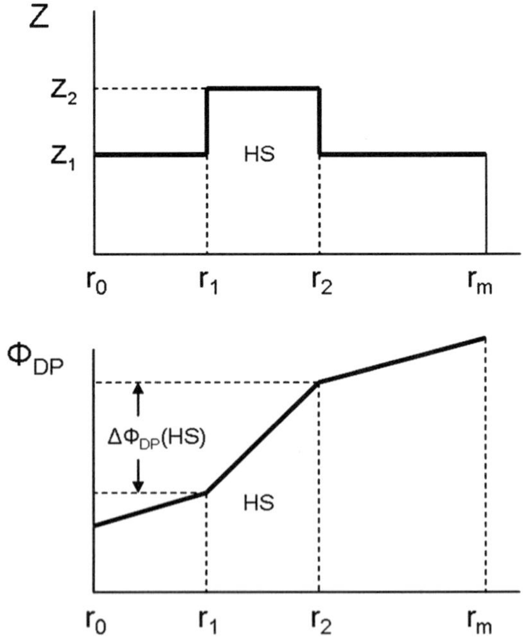

Fig. 6.19 Conceptual plot illustrating the hot spot between ranges r_1 and r_2 in the radial profiles of Z and Φ_{DP}

These preliminary corrected Z and Z_{DR} are utilized to identify hot spots via the following criteria: (1) $Z > Z^{(th)}$ and $\rho_{hv} > 0.7$ everywhere in the hot spot, (2) the maximal value of Z_{DR} within the hot spot exceeds $Z_{DR}{}^{(th)}$, (3) the hot spot has sufficient length (usually 2 km), and (4) the change of Φ_{DP} within the hot spot exceeds $\Delta\Phi_{DP}{}^{(th)}$ (typically 10°). Gu et al. (2011) selected thresholds $Z^{(th)} = 45$ dBZ and $Z_{DR}{}^{(th)} = 3$ dB for C band.

According to the hot-spot method, the radial profile of A_h is estimated from (6.28) but with PIA determined as

$$\mathrm{PIA} = \alpha_0 \Delta\Phi_{DP}(r_0, r_m) + \Delta\alpha\, \Delta\Phi_{DP}(\mathrm{HS}). \tag{6.43}$$

This means that two measured differential phase increments, $\Delta\Phi_{DP}(r_0,r_m)$ and $\Delta\Phi_{DP}(\mathrm{HS})$, are used instead of one. Therefore, the radial profile of A_h becomes dependent on the value of $\Delta\alpha$. The optimal value of $\Delta\alpha$ is determined by the iterative process of incrementing $\Delta\alpha$ until the condition

$$\int_{OHS} A_h(s, \Delta\alpha) ds = \frac{\alpha_0}{2} \Delta\Phi_{DP}(OHS) \tag{6.44}$$

is satisfied. In (6.44), integration is over the ranges outside the hot spots (*OHS*) and

$$\Delta\Phi_{DP}(OHS) = \Delta\Phi_{DP}(r_0, r_m) - \Delta\Phi_{DP}(HS). \tag{6.45}$$

Finally, the corrected reflectivity factor is

$$Z(r) = Z_a(r) + 2\int_{r_0}^{r} A_h\left[s, \Delta\alpha^{(opt)}\right] ds. \tag{6.46}$$

Similar to the Bringi et al. (2001) self-consistent method with constraints, the procedure for the Z_{DR} attenuation correction in the hot-spot technique is based on the comparison of the measured (negatively biased) Z_{DR} behind the attenuating cell with what is expected in light rain in the shadow of this cell (see Fig. 6.20). The attenuation-related bias in Z_{DR} is determined as

$$\Delta Z_{DR}(r) = \begin{cases} \beta_0 \Phi_{DP}(r) & if \quad r < r_1 \\ \beta_0 \Phi_{DP}(r) + \Delta\beta[\Phi_{DP}(r) - \Phi_{DP}(r_1)] & if \quad r_1 < r < r_2 \\ \beta_0 \Phi_{DP}(r) + \Delta\beta \Delta\Phi_{DP}(HS) & if \quad r > r_2 \end{cases} \tag{6.47}$$

The parameter $\Delta\beta$ is estimated from

$$Z_{DR}^{(a)}(r_m) + \beta_0 \Delta\Phi_{DP}(r_0, r_m) + \Delta\beta \Delta\Phi_{DP}(HS) = Z'_{DR}(r_m), \tag{6.48}$$

where the expected value of differential reflectivity in the shadow of an attenuating cell $Z'_{DR}(r_m)$ is determined from the $Z(r_m)$ corrected for attenuation.

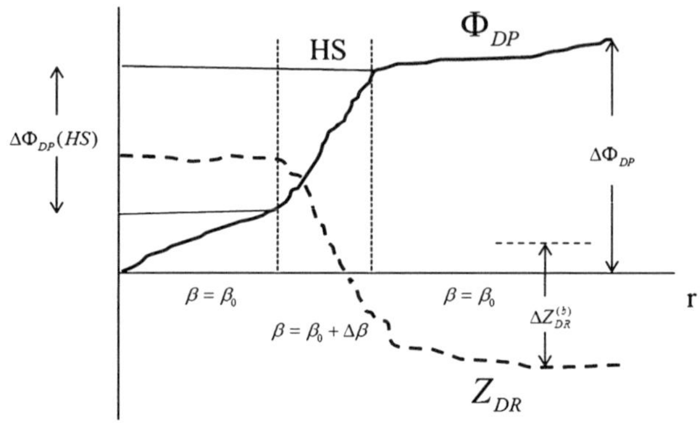

Fig. 6.20 Conceptual plot illustrating correction of Z_{DR} for differential attenuation in the hot-spot method

An example of attenuation correction of Z and Z_{DR} at X band using the self-consistent method with constraints is in Fig. 6.21. The measurements have been made in Japan with MP-X radar during typhoon on 11 September 2001 (Park et al. 2005). Large negative biases in Z and Z_{DR} caused by attenuation in the band of heavy rain (left panels) are efficiently eliminated (right panels). Numerous examples of polarimetric attenuation correction at X band can be also found in the study by Snyder et al. (2010).

The efficiency of attenuation correction using polarimetric measurements at C band is investigated in a number of studies (e.g., Bringi et al. 2001, 2006; Carey et al. 2000; Gu et al. 2011; Le Bouar et al. 2001; Ryzhkov et al. 2007; Tabary et al. 2009; Testud et al. 2000; Vulpiani et al. 2008). Ryzhkov et al. (2007) compare the performance of a simple linear correction scheme with "default" values α and β and the hot-spot correction method for the cases of tropical rain on 17 October 2006 in Alabama and continental rain on 23 April 2007 in Canada. The comparison is illustrated in Fig. 6.22 where black curves denote radial profiles of Z and Z_{DR} before attenuation correction, blue curves indicate these profiles after simple linear correction with default values $<\alpha> = 0.06$ dB deg^{-1} and $<\beta>$ equal to 0.017 dB deg^{-1} in Canada and 0.010 dB deg^{-1} in Alabama, and red curves correspond to attenuation correction results using the hot-spot method. Although maximal values of Φ_{DP} for both storms are quite comparable (more than 400°), differential attenuation in continental rain is almost 10 dB larger due to the presence of large raindrops and melting hail. This is further substantiated by noticeably higher values of corrected Z_{DR} in the Canadian storm. The linear attenuation correction technique yields very similar attenuation bias in Z for both cases (blue curves) exceeding 25 dB at the end of the propagation path. However, the linear scheme underestimates actual attenuation by 11 dB in the continental case and by 5 dB in the tropical case. Because tropical rain contains much smaller number of large drops than the continental one, the difference in attenuation correction results between the linear and hot-spot techniques is significantly smaller in the tropical case.

An example of heavy attenuation at C band in the squall line and its efficient polarimetric correction using the hot-spot method is illustrated in Fig. 6.23. The

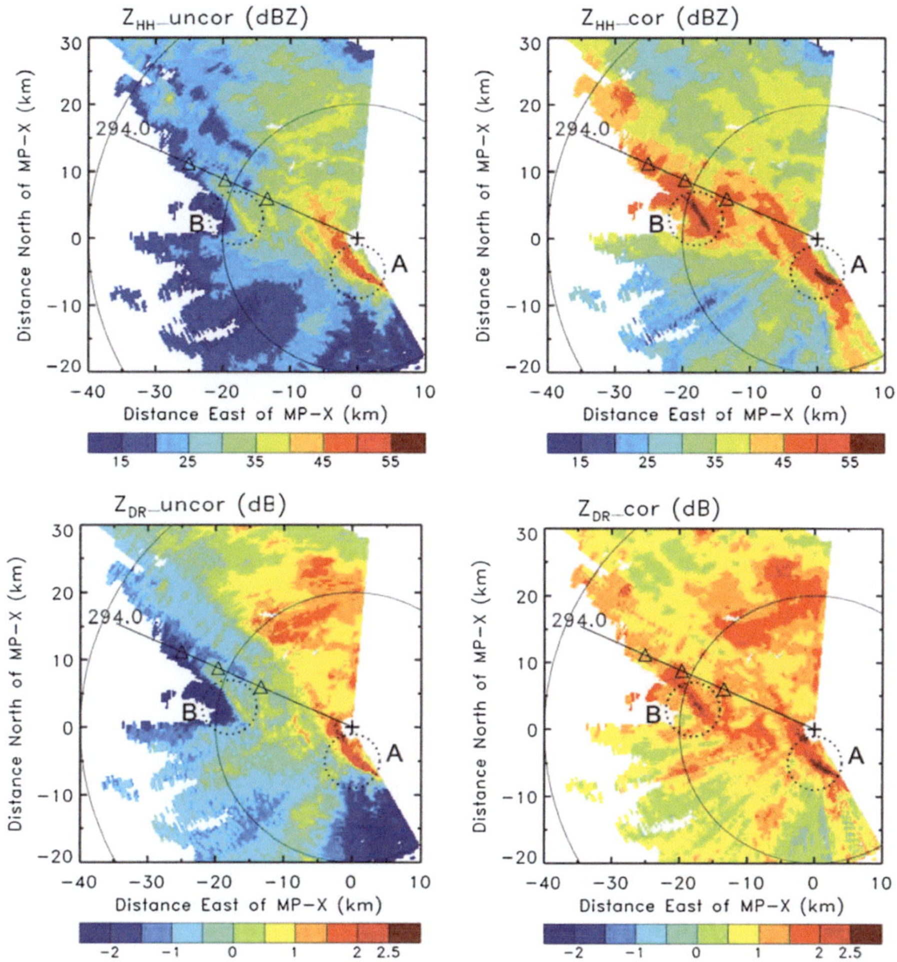

Fig. 6.21 The fields of Z and Z_{DR} at X band before and after correction for attenuation (left and right panels, respectively). The data have been collected in Japan during typhoon on 11 September 2001. Antenna elevation is 2.5°. Adapted from Park et al. (2005). © American Meteorological Society. Used with permission

comparison of Z fields measured by C-band and S-band radars shows extreme attenuation of the C-band radiation along the SE flank of the squall line where differential phase exceeds 400° and Z bias approaches 30–40 dB. The largest recoverable attenuation bias of about 40 dB is estimated along the radial at the azimuth of 257.2° (line in Fig. 6.23) where Φ_{DP} as high as 602° has been measured (Fig. 6.24). Amazingly, the polarimetric algorithm for attenuation correction restores reliably the reflectivity when 99.99% of signal power is lost! Total differential attenuation along the same ray reaches 7 dB (Fig. 6.24). In the northern flank of

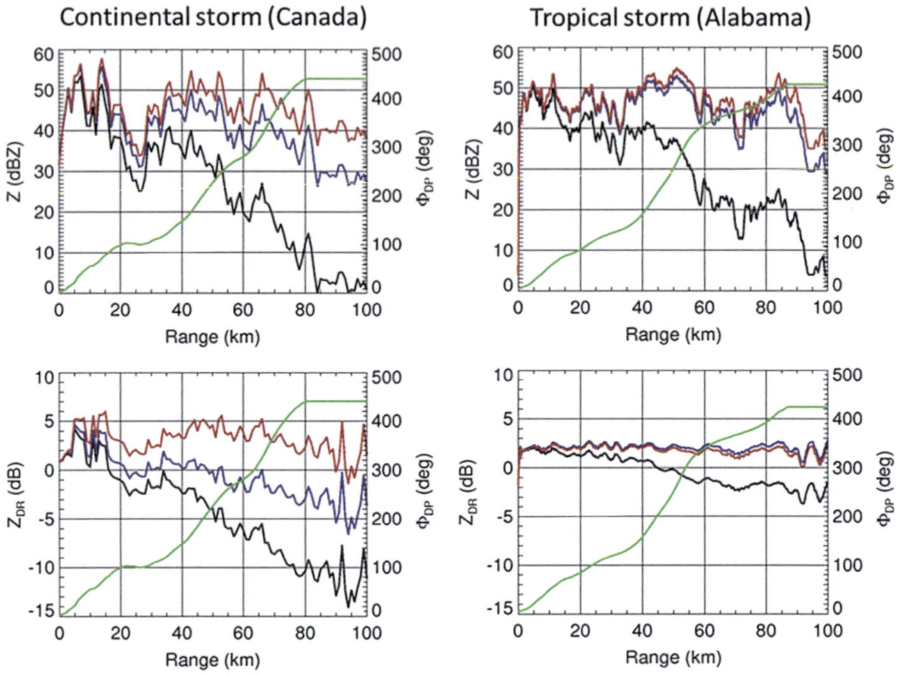

Fig. 6.22 Radial profiles of Z and Z_{DR} before correction for attenuation (black lines), after correction using linear scheme (blue lines), and after correction according to the hot-spot method (red lines). Radial profiles of differential phase are in green. The continental storm occurred in Canada on 23 April 2007; the data from the tropical storm were collected in Alabama on 17 October 2006

the squall line, the C-band signal is completely extinct at about 60 km range; hence attenuation correction is not possible beyond this distance.

Even S-band microwave radiation may experience noticeable attenuation in heavy precipitation if the propagation path through it is sufficiently long (Ryzhkov and Zrnic 1995). These authors report that in extreme cases when the radar beam is aligned with a squall line in Oklahoma, the bias in Z can be over 10 dB, and the corresponding bias in Z_{DR} is 3–4 dB. Higher attenuation-related biases at S band in melting hail are reported by Ryzhkov et al. (2013) who also modified the "hot-spot" method for attenuation correction. Their methodology capitalizes on the estimation of the specific attenuation A in rain behind the hail-containing cell (or hot spot) using the ZPHI formula (6.28). This estimate of A is immune to the Z bias caused by attenuation.

The "true" value of Z in rain behind the hail cell can then be determined from a power-law A-Z relation (6.18). The corresponding attenuation-related bias of Z_{DR} can be determined by comparing the measured and expected (true) Z_{DR} at ranges where the corrected Z is between 20 and 30 dBZ. The expected or reference Z_{DR} in this range of reflectivities is obtained from the corrected Z as

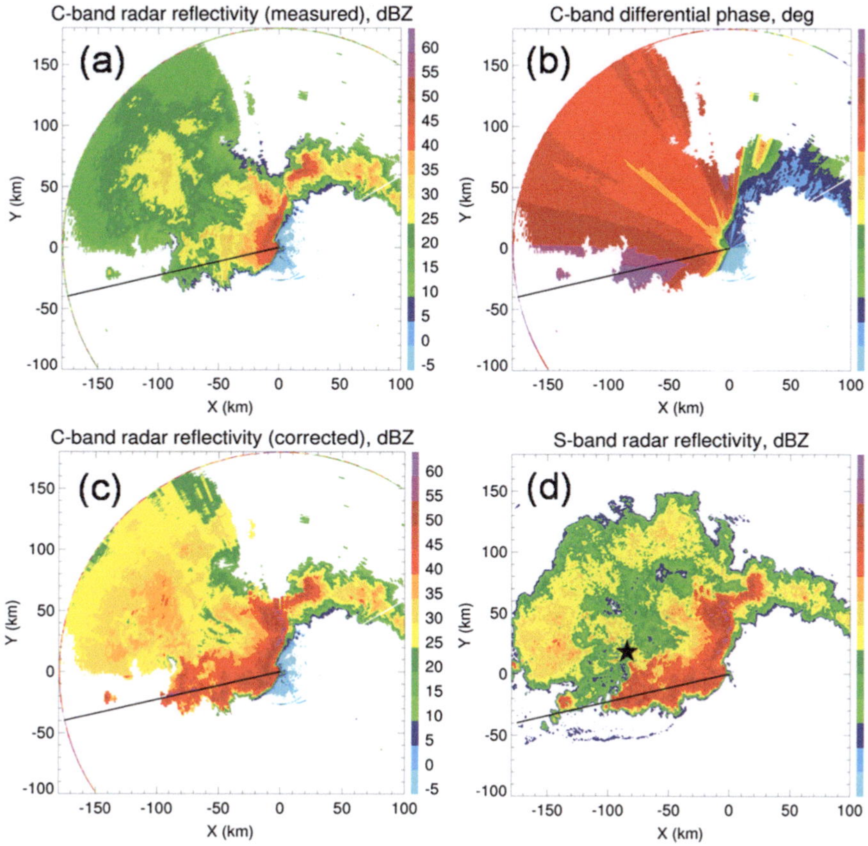

Fig. 6.23 Fields of (**a**) measured and (**c**) corrected Z at C band, (**b**) Φ_{DP} at C band, and (**d**) Z measured by S-band radar (KLOT WSR-88D) in the Chicago area on 5 August 2008 at 0149 UTC. The antenna elevation is 0.5°. The C-band radar is situated at X = 0, Y = 0 km. The star marks the location of the WSR-88D. The straight line indicates azimuth 257.2° for which the radial data are presented in Fig. 6.24. The correction of Z is performed using the hot-spot method. Adapted from Gu et al. (2011). © American Meteorological Society. Used with permission

$$Z_{DR} = 0.585 - 0.0507Z + 0.00165Z^2, \tag{6.49}$$

where Z_{DR} and Z are expressed in dB. The reason for using low reflectivities for such a comparison is that the $Z_{DR}(Z)$ relation is less affected by the DSD variability in lighter rain (see Fig. 6.5). Equation (6.49) has been obtained via simulations using the large DSD dataset collected in Oklahoma.

An example of anomalously large attenuation/differential attenuation experienced by the S-band KICT WSR-88D (Wichita, KS) in a supercell storm is illustrated in Fig. 6.25 where a composite plot of Z, Z_{DR}, Φ_{DP}, and ρ_{hv} at elevation 0.9° is presented. The storm produced large hail, with reported maximal diameters between

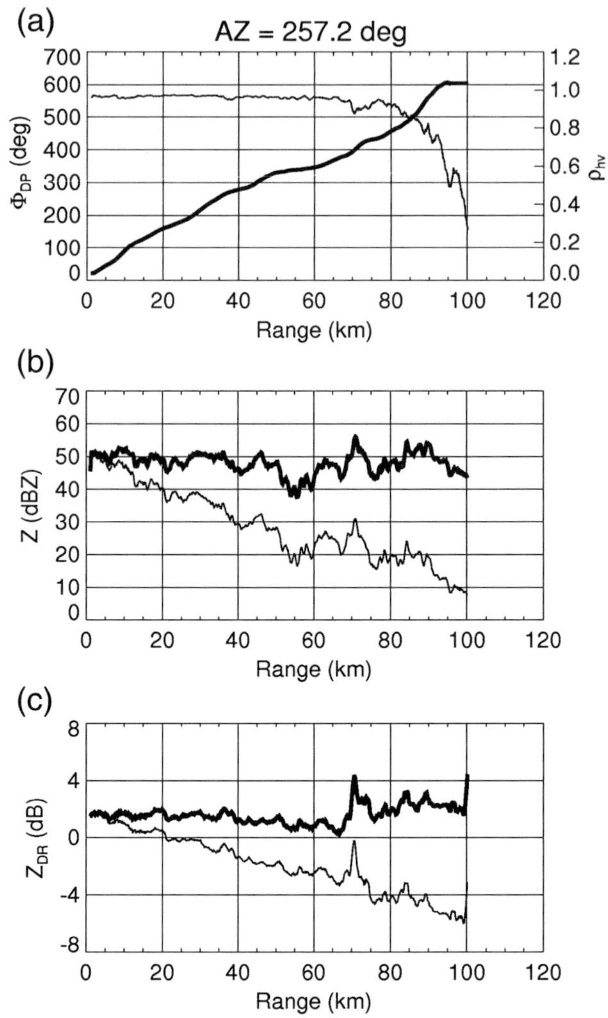

Fig. 6.24 Radial profiles of (**a**) filtered Φ_{DP} (thick line) and ρ_{hv} (thin line), (**b**) measured (thin line) and corrected (thick line) Z, and (**c**) measured (thin line) and corrected (thick line) Z_{DR} at the azimuth marked in Fig. 6.23. Attenuation correction of Z and Z_{DR} is made using the hot-spot method. Adapted from Gu et al. (2011). © American Meteorological Society. Used with permission

2.5 and 4.5 cm. Measured Z and Z_{DR} (before attenuation correction) are displayed in the top panels of Fig. 6.25. Large negative biases of Z and Z_{DR} are clearly visible in the narrow eastern sector where Φ_{DP} is maximal and Z_{DR} drops below -5 dB. The reduction of ρ_{hv} is attributed to nonuniform beam filling (see Sect. 6.8), as ρ_{hv} is not directly affected by attenuation. The fields of Z and Z_{DR} corrected for attenuation are shown in the bottom panels of Fig. 6.25. It is evident that the correction procedure restores efficiently the Z and Z_{DR} fields. Radial profiles of Z and Z_{DR} before and after correction along the radial with maximal attenuation at Az = 88.7° are in Fig. 6.26. It follows from this plot that the magnitudes of maximal attenuation-related biases are 17 and 6 dB for Z and Z_{DR} (at S band!).

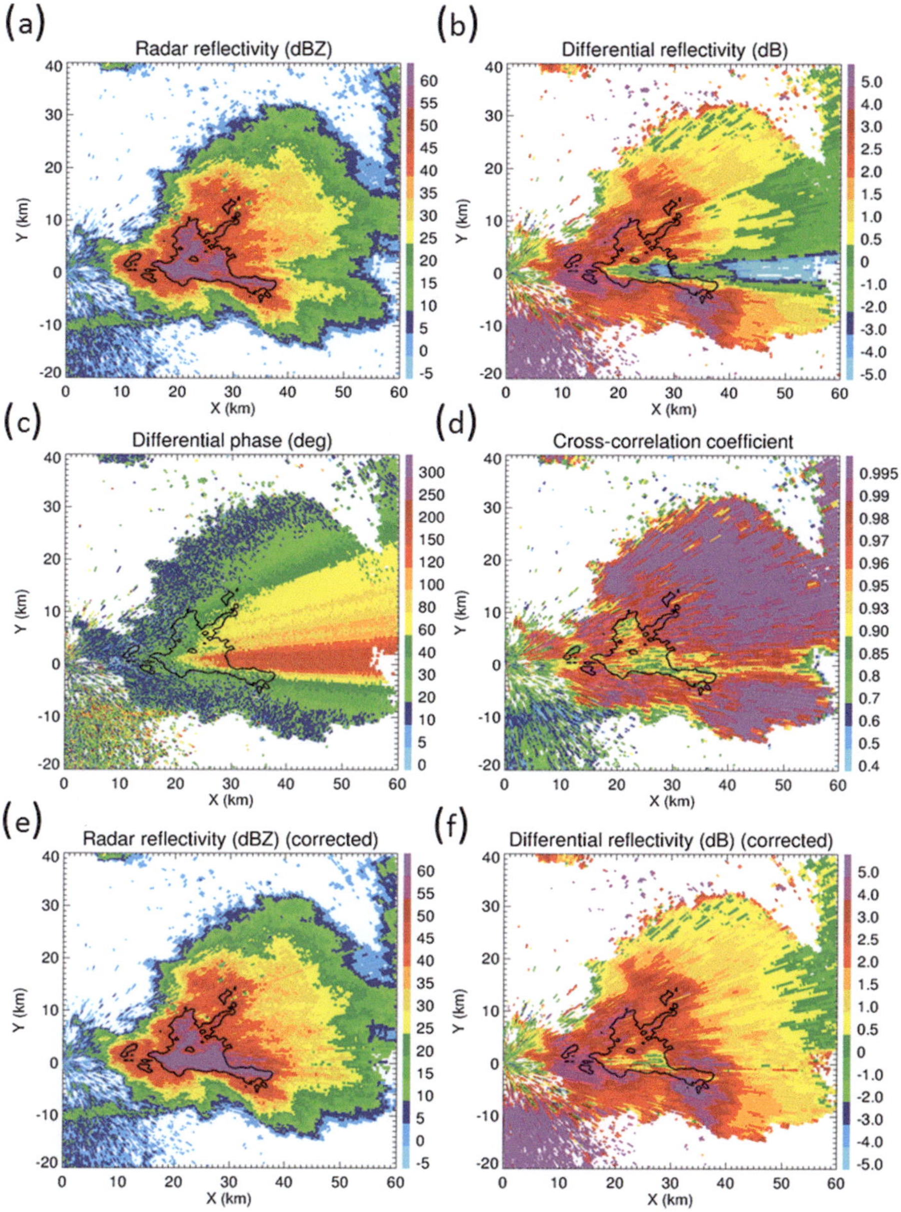

Fig. 6.25 Composite plot of Z and Z_{DR} before attenuation correction (**a**, **b**) and after attenuation correction (**e**, **f**) as well as Φ_{DP} and ρ_{hv} (**c**, **d**) of the hailstorm on 30 May 2012 at 2313 UTC observed by the dual-polarization WSR-88D near Wichita, KS (KICT), at 0.9° elevation. A 55 dBZ contour of reflectivity is overlaid. Adapted from Ryzhkov et al. (2013). © American Meteorological Society. Used with permission

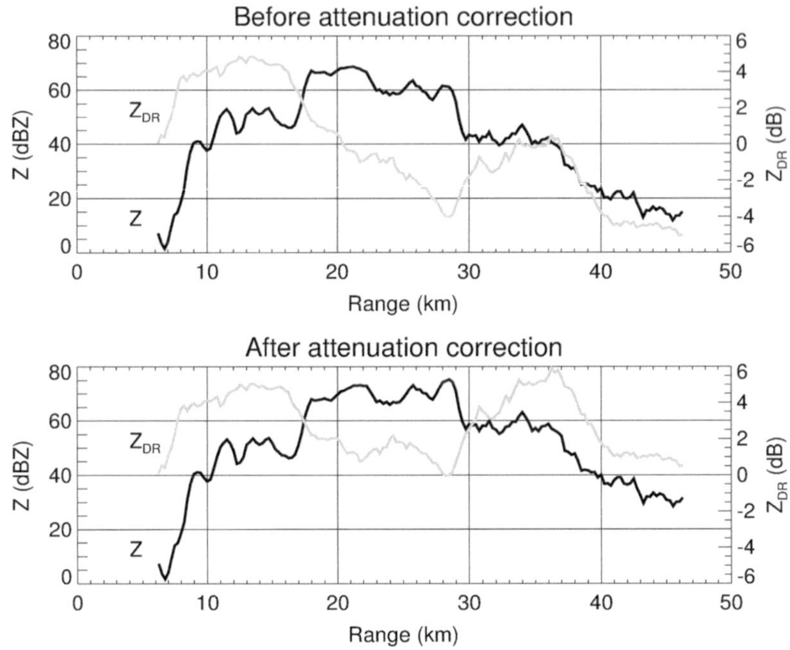

Fig. 6.26 Radial profiles of Z and Z_{DR} in the storm depicted in Fig. 6.25 before and after attenuation correction along the radial at Az = 88.7° where attenuation is maximal. Adapted from Ryzhkov et al. (2013). © American Meteorological Society. Used with permission

6.5 Mitigation of Partial Beam Blockage

Beam blockage caused by terrain and other obstacles such as buildings and trees limits radar coverage and introduces bias in measurements. Therefore, the quality of the weather radar products and quantitative precipitation estimates (QPEs) are compromised. One of the most common methods for mitigation of partial beam blockage (PBB) uses a digital elevation map (DEM) to estimate the degree of beam blockage at particular azimuths and elevations based on the beam's geometry and its occultation (Fig. 6.27). The reflectivity is corrected using the relation

$$Z(\mathrm{dBZ}) = Z_{\mathrm{bl}}(\mathrm{dBZ}) + 10\log(F_{\mathrm{bl}}), \tag{6.50}$$

where the shielding or blocking factor F_{bl} is determined from equations

$$F_{\mathrm{bl}} = 0.5\tanh[0.0277(50 - bd)] + 0.5 \tag{6.51}$$

and

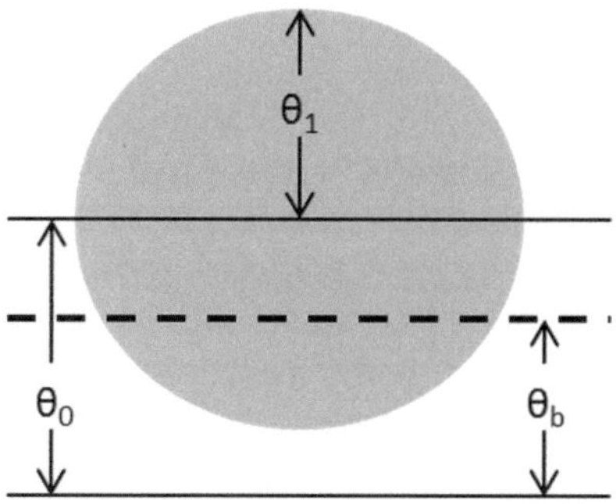

Fig. 6.27 Cross section of radar beam and its occultation

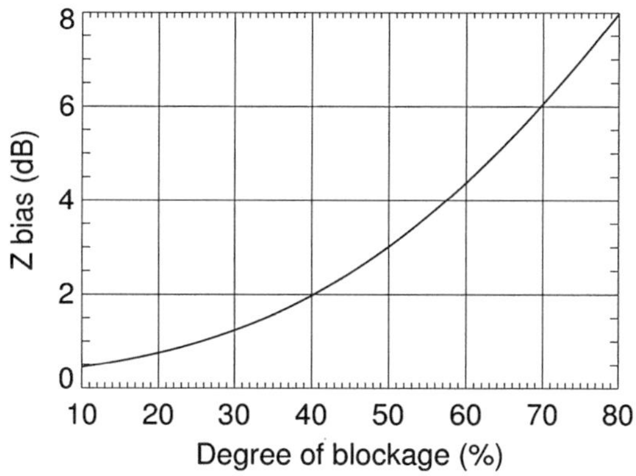

Fig. 6.28 *Z* bias caused by partial beam blockage as a function of degree of blockage

$$bd = 100\frac{\theta_b - \theta_0 + \theta_1/2}{\theta_1}. \tag{6.52}$$

Here bd is the blockage degree expressed in (%), θ_1 is the radar beamwidth (one-way, 3 dB), θ_0 is the elevation of the beam center, and θ_b is the blockage elevation (Fig. 6.27). Equation (6.48) is derived using Gaussian shape of the antenna pattern defined in (2.5). The absolute value of *Z* bias computed from (6.50) and (6.51) as a function of blockage degree *bd* is displayed in Fig. 6.28.

The DEM-based correction may not work well if the degree of blockage exceeds 60%. The degree of beam blockage at grazing elevation angles depends on atmospheric refractive conditions. This may cause errors in *Z* correction, particularly if anomalous propagation occurs (Bech et al. 2003). In addition to larger-scale terrain features, small-scale anthropogenic structures (e.g., towers, buildings) and nearby trees that are not accounted for by DEMs can cause additional occultation of the radar beam.

The problem of the partial beam blockage can be better resolved with a dual-polarization radar than with a single-polarization radar because the former can directly measure differential phase Φ_{DP} and estimate specific attenuation *A* (using Eq. (6.28)) which are not biased by PBB. Zhang et al. (2013), Ryzhkov et al. (2014), and Diederich et al. (2015) explored utilization of differential phase and attenuation to mitigate the effects of PBB. It is evident from (6.28) that the estimate of specific attenuation A from a radial profile of Z and a total span of differential phase $\Delta\Phi_{DP}$ is totally immune to the Z biases caused by attenuation, radar miscalibration, partial beam blockage, and wet radome. Indeed, if attenuated Z (Z_a in (6.28)) expressed in linear scale is multiplied by an arbitrary constant ζ along the propagation path (r_1, r_2), then A remains intact because the numerator and denominator in (6.28) are multiplied by the same factor ζ^b which is canceled in the ratio. This property of the A estimate proves to be very beneficial for quantification of rainfall in the partially blocked areas of radar returns if the A-based algorithm is used for rainfall estimation (see Chap. 10). The reflectivity factor unbiased by PBB can be estimated from A using the Z(A) relation which is an inverted Eq. (6.18).

The performance of this technique is illustrated in Fig. 6.29 where the fields of the measured X-band Z and Z_{DR} (before correction for attenuation and beam blockage)

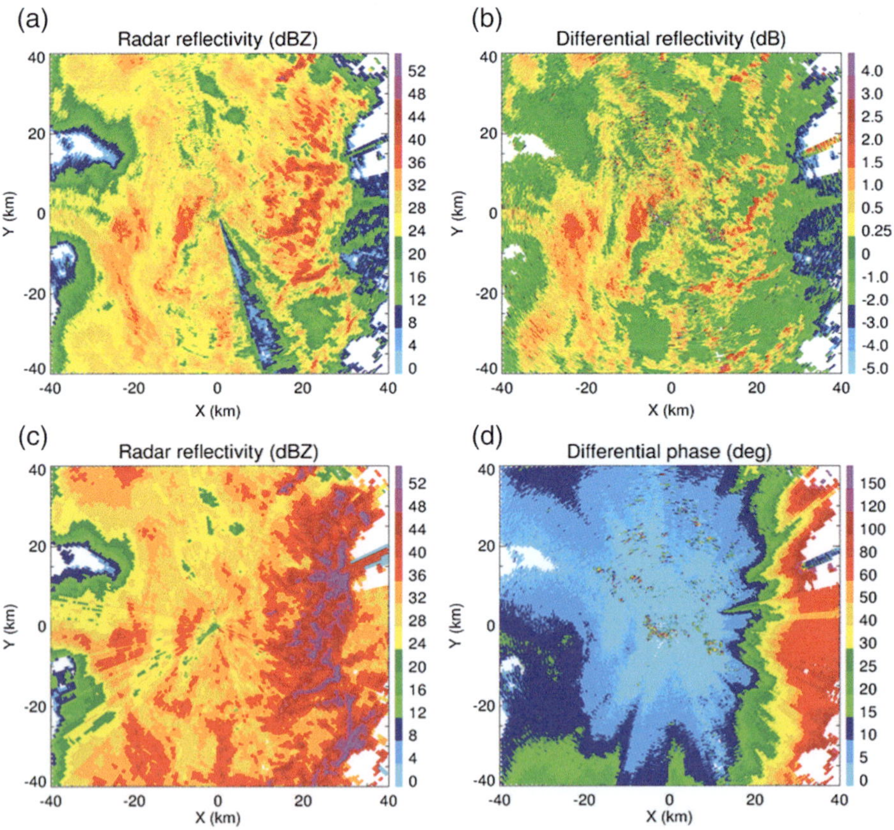

Fig. 6.29 Composite plot of measured Z and Z_{DR} before correction for attenuation and beam blockage (**a**, **b**), Z after correction (**c**), and differential phase (**d**). The measurements were made with the University of Bonn X-band polarimetric radar on 22 June 2011 at 1126 UTC at elevation 1.5°

at antenna elevation 1.5° are displayed along with the fields of Φ_{DP} and reflectivity corrected for attenuation and PBB. It is obvious that the PBB-related Z bias in a narrow SE sector is completely eliminated in the panel (c) of Fig. 6.29. Diederich et al. (2015) made a comprehensive analysis of multi-season polarimetric radar data collected by an X-band radar in Germany and concluded that the overall long-time cumulative accuracy of calibration- and PBB-related biases derived from $Z(A)$ is very good. Using sufficiently long time series of radar data is necessary to reduce the impact of the known variability of the parameters in the $Z(A)$ relations. The Diederich et al. (2015) analysis demonstrates that the accuracy of the Z bias retrieval is comparable to or better than the one obtained from techniques based on the DEM. This holds particularly true if PBB is caused by man-made obstacles or nearby trees which are not accounted for in the DEM.

Differential reflectivity can be also biased by PBB. This, however, may not be as obvious as an impact of PBB on the reflectivity. Indeed, without an in-depth analysis, it is impossible to determine how the measurements of Z_{DR} in Fig. 6.29b are affected. This is generally not the case if PBB is caused by nearby trees as in the situation with the S-band Cimarron polarimetric radar in Oklahoma (Fig. 6.30a). Systematic measurements of the Cimarron Z_{DR} at elevation 0.5° in light rain and snow reveal persistent azimuthal variations associated with PBB (Giangrande and Ryzhkov 2005) (Fig. 6.30b, d). Such azimuthal variability does not exist at the blockage-free antenna elevation of 1.5°, and the difference between the average azimuthal dependencies of Z_{DR} at elevations 1.5° and 0.5° can be used to quantify the Z_{DR} bias at the blocked elevation angle (Fig. 6.30c).

6.6 Ground Clutter Contamination

Land and sea clutter strongly affects Doppler and polarimetric variables if it is mixed with the returns from hydrometeors. Ground clutter is caused by reflections from non-weather surface objects (natural or man-made) mostly located in a relatively close proximity of the radar. However, clutter reflections are often produced at large distances by anomalous propagation during the super-refractive conditions in the lower atmosphere. Moreover, mountains in the line of sight of the radar beam produce strong clutter at any distance. Land clutter is characterized by the Doppler spectrum with zero mean Doppler velocity and narrow spectrum width, whereas sea clutter generally has nonzero mean Doppler velocity and larger spectrum width due to the substantial motion of the scattering sea waves.

Using a static clutter map obtained during clear-air conditions is a traditional way to identify the locations of ground clutter contamination and mitigate its impact on the data by filtering or censoring. Such technique is relatively efficient for a stable local clutter but does not address the problem of sporadic ground clutter caused by anomalous propagation due to super-refractive conditions in the lower atmosphere. Following the need for automatic clutter mitigation and suppression, several techniques based on the analysis of Doppler spectra of the radar signals have been

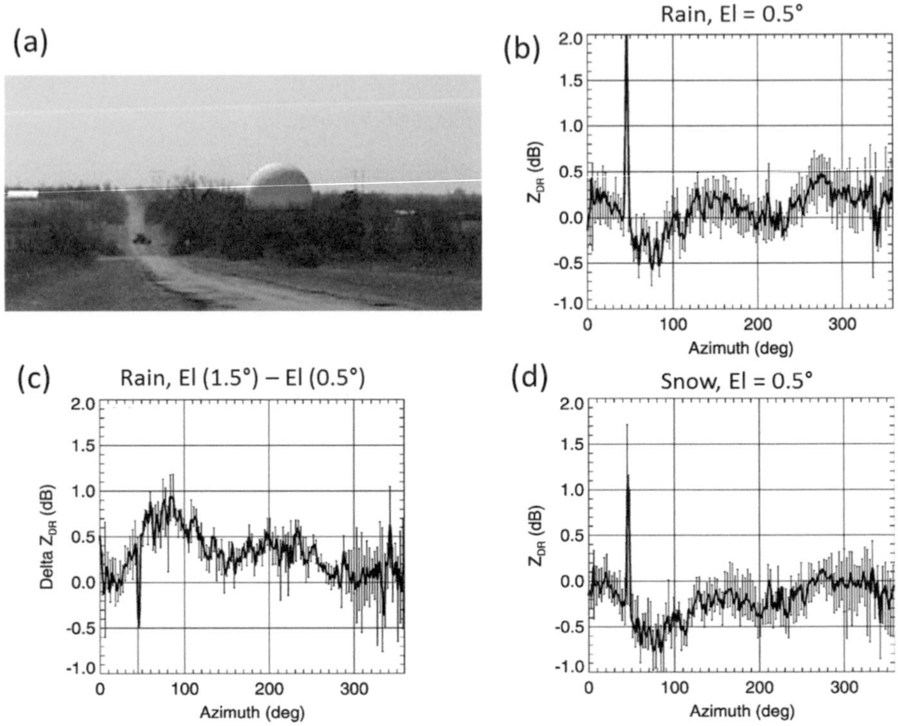

Fig. 6.30 The Cimarron polarimetric radar (OK, panel **a**), mean azimuthal dependencies of Z_{DR} measured at elevation 0.5° by the radar in five light rain events (panel **b**) and seven snow events (panel **d**), and the difference between Z_{DR} at El = 1.5° and 0.5° (panel **c**). Adapted from Giangrande and Ryzhkov (2005). © American Meteorological Society. Used with permission

developed. Of these, the Gaussian model adaptive processing (GMAP) technique (Siggia and Passarelli Jr 2004) has been chosen for the WSR-88Ds. GMAP spectral clutter filter processes signals in the Doppler spectral domain and automatically adapts its suppression depending on the strength of the clutter signal. GMAP assumes that both clutter and hydrometeors are characterized by Gaussian Doppler spectra with different mean Doppler velocity and spectrum width. It also assumes that the clutter spectrum has zero Doppler velocity and approximately known narrow spectrum width σ_c. Similar to the GMAP, a bi-Gaussian model of the mixed Doppler spectrum of the weather and clutter echoes is used in another technique for clutter detection and filtering called spectrum-time estimation and processing (STEP) (Cao et al. 2012).

Hubbert et al. (2009) introduced the clutter mitigation decision (CMD) algorithm which uses parameters of the spatial variability of reflectivity and a measure of the phase variability in the I and Q time series (coherent phase alignment (CPA)) to identify clutter via a fuzzy logic procedure. The phase of clutter signal (i.e., CPA) changes much slower than the one associated with weather echoes. Torres and

Warde (2014) suggested the Clutter Environment Analysis using Adaptive Processing (CLEAN-AP) technique which automates the detection and mitigation of ground clutter contamination using the spectral density of the autocorrelation at lag one instead of the traditional Doppler power spectrum. The book by Zhang (2016) contains a detailed overview of the various methods for clutter detection and mitigation.

The use of polarimetry provides additional opportunities for better identification and mitigation of ground clutter because polarimetric properties of ground clutter and weather scatterers are very different. Differential reflectivity of land clutter has relatively narrow distribution centered either at 0 dB or slightly negative value (Borowska and Zrnic 2012; Melnikov et al. 2017). Z_{DR} of sea clutter varies in a broader range and usually exhibits strong azimuthal dependence related to the directions of sea waves (Chap. 9, Sect. 9.8.1) but is generally negative. Bald hills/mountains or buildings, towers, or masts have intrinsic value of their cross-correlation coefficient close to 1, whereas vegetation, trees, and natural land covers are characterized by ρ_{hv} much lower than most of the weather returns. The methodology for ground clutter identification with polarimetric radars using fuzzy logic is described in Chap. 9.

6.7 Noise Correction

The noise generated by radar receiver, waveguides, and antenna as well as surrounding atmosphere and ground may adversely affect the quality of polarimetric radar measurements if the signal-to-noise ratio (SNR) is low. The estimates of all power-related variables such as Z, Z_{DR}, ρ_{hv}, LDR, CDR, and σ_v can be biased by noise. For example, the noise impact on the estimates of Z_{DR} and ρ_{hv} can be noticeable for SNR values as high as 20 dB. The phase-related measurements of Doppler velocity and differential phase are not biased by noise, but their statistical errors increase at low SNR.

Generally, the noise correction in polarimetric weather radars is implemented in a signal processor by subtracting the noise powers N_h and N_v in the orthogonal receiver channels from the measured powers P_h and P_v so that the key polarimetric variables Z_{DR} and ρ_{hv} are determined as

$$Z_{dr} = \frac{P_h - N_h}{P_v - N_v}, \tag{6.53}$$

where Z_{dr} is the value of differential reflectivity expressed in linear units and

$$\rho_{hv} = \frac{|R_{hv}|}{(P_h - N_h)^{1/2}(P_v - N_v)^{1/2}}. \tag{6.54}$$

In (6.54), R_{hv} is the complex covariance at zero lag determined by (3.14a). Because the noises in the orthogonal receiver channels are uncorrelated, the magnitude of R_{hv} is not biased by noise. Equations (6.53) and (6.54) imply that the noise powers N_h and N_v are known *precisely*. Unfortunately, in practice, this is not always the case, and persistent overestimation/underestimation of Z_{dr} and ρ_{hv} is quite common at low SNR even for well-designed polarimetric radars. It is well known that the atmospheric and ground noises generally vary with azimuth and elevation and it is important to measure noise in real time for each individual radial using the data at range gates free of radar returns (Ivic et al. 2013).

If the noise-related biases persist, then it is advantageous for the user to have an option to take uncorrected (or "measured") values of Z_{dr} and ρ_{hv} and make correction "offline." The measured values of Z_{dr} and ρ_{hv} (${Z_{dr}}^{(m)}$ and ${\rho_{hv}}^{(m)}$, respectively) are

$$Z_{dr}^{(m)} = \frac{P_h}{P_v} = \frac{Z_{dr}(1 + snr^{-1})}{1 + Z_{dr}(\eta\, snr)^{-1}} \tag{6.55}$$

and

$$\rho_{hv}^{(m)} = \frac{|R_{hv}|}{P_h^{1/2} P_v^{1/2}} = \frac{\rho_{hv}}{(1 + snr^{-1})^{1/2}\left[1 + Z_{dr}(\eta\, snr)^{-1}\right]^{1/2}}, \tag{6.56}$$

where Z_{dr} and ρ_{hv} are the intrinsic (true) values of differential reflectivity and cross-correlation coefficient, $snr = 10^{0.1 SNR(dB)}$ is the signal-to-noise ratio at horizontal polarization expressed in linear units, and $\eta = N_h/N_v$ is the ratio of noise powers from the two channels (also in liner units). Solving (6.55) and (6.56) for Z_{dr} and ρ_{hv} yields

$$Z_{dr} = \frac{\eta\, snr\, Z_{dr}^{(m)}}{\eta\, snr + \eta - Z_{dr}^{(m)}} \tag{6.57}$$

and

$$\rho_{hv} = \rho_{hv}^{(m)}\left(1 + \frac{1}{snr}\right)^{1/2}\left(1 + \frac{Z_{dr}}{\eta\, snr}\right)^{1/2}. \tag{6.58}$$

If $Z_{dr} = \eta = 1$ then

$$\rho_{hv} = \rho_{hv}^{(m)}\left(1 + \frac{1}{snr}\right). \tag{6.59}$$

Using Eqs. (6.57)–(6.59) assumes that *snr* (or *SNR*) is determined correctly which implies correct estimation of the noise power at the horizontal channel. A simple check of this assumption can be done by examining the *SNR* dependence of the value

ρ_{hv} computed from (6.59) and obtained from radar data collected in the areas of weather echo where the intrinsic values of the cross-correlation coefficient are expected to be close to 1. If the noise-corrected value of ρ_{hv} does not exhibit noise-related bias at low *SNR* (less than 15–20 dB), then the *SNR* estimate is considered to be correct. If the bias occurs, then the *SNR* should be varied until a "flat" dependence of ρ_{hv} on *SNR* is achieved (see Fig. 6.31). This can be done by varying the radar constant C in the radar equation

$$SNR(\text{dB}) = Z(\text{dBZ}) - 20\log R(\text{km}) + C(\text{dB}). \tag{6.60}$$

Figure 6.31 illustrates different types of the $\rho_{hv}(SNR)$ dependencies for various values of the radar constant. Our experience and Fig. 6.31 demonstrate that such

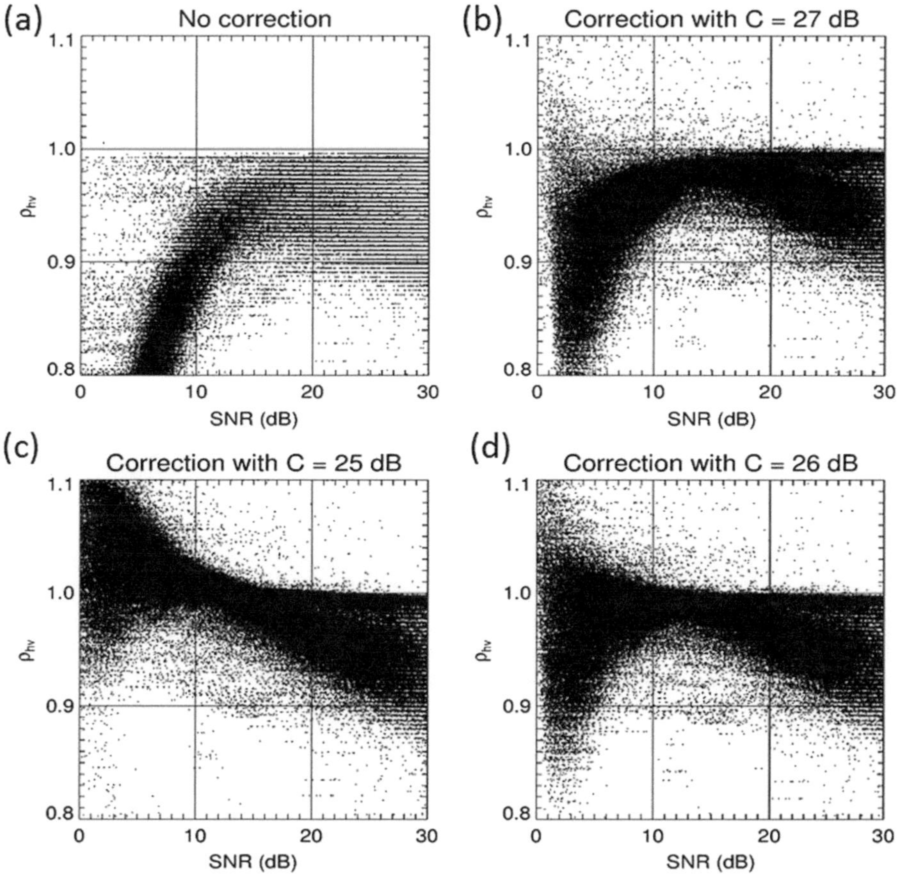

Fig. 6.31 The ρ_{hv}-*SNR* scatterplots without correction for noise (**a**) and with correction with different values of the radar constant C (**b–d**). The correction with C = 26 dB produces the most "flat" ρ_{hv}-*SNR* scatterplot for *SNR* < 20 dB. The data have been collected in stratiform rain with the X-band radar. Lower values of ρ_{hv} for *SNR* > 20 dB indicate melting layer

data-based technique provides estimates of the "true" value of the radar constant with the accuracy better than 1 dB.

An alternative methodology to obtain noise-free estimates of Z_{DR} and ρ_{hv} was suggested by Melnikov and Zrnic (2007). They estimate Z_{DR} and ρ_{hv} using one-lag covariances $R_h(T_s)$, $R_v(T_s)$, $R_{hv1}(T_s)$, and $R_{hv2}(T_s)$ rather than powers P_h and P_v and a zero-lag covariance R_{hv}:

$$Z_{dr} = \frac{| R_h(T_s) |}{| R_v(T_s) |} \tag{6.61}$$

and

$$\rho_{hv} = \frac{| R_{hv1}(T_s) | + | R_{hv2}(T_s) |}{2|R_h(T_s)R_v(T_s)|^{1/2}}, \tag{6.62}$$

where

$$R_{hh}(T_s) = \overline{V^*_{hh}(t)V_{hh}(t+T_s)} = \frac{1}{M-1}\sum_{m=1}^{M-1} V^*_{hh}(m)V_{hh}(m+1), \tag{6.63}$$

$$R_{vv}(T_s) = \overline{V^*_{vv}(t)V_{vv}(t+T_s)} = \frac{1}{M-1}\sum_{m=1}^{M-1} V^*_{vv}(m)V_{vv}(m+1), \tag{6.64}$$

$$R_{hv1}(T_s) = \overline{V_{hh}(t)V^*_{vv}(t+T_s)} = \frac{1}{M-1}\sum_{m=1}^{M-1} V_{hh}(m)V^*_{vv}(m+1), \tag{6.65}$$

$$R_{hv2}(T_s) = \overline{V_{hh}(t+T_s)V^*_{vv}(t)} = \frac{1}{M-1}\sum_{m=1}^{M-1} V_{hh}(m+1)V^*_{vv}(m). \tag{6.66}$$

In (6.63)–(6.66), V_{hh} and V_{vv} are complex voltages in the orthogonal channels defined in Chap. 3. Because noise samples in successive pulses are uncorrelated, all four complex covariances and the estimates (6.61) and (6.62) are not biased by noise. The one-lag algorithm produces high-quality Z_{DR} and ρ_{hv} estimates in the areas of clear-air echoes with *SNR* less than 0 dB (Melnikov et al. 2011).

In addition to the one-lag covariances, the covariances at two lags or higher (i.e., at $2T_s$, $3T_s$, etc.) can be utilized to further improve the accuracy of the estimates of polarimetric variables. The concept of the multilag estimators is explained in detail by Zhang (2017). Note that the use of the one-lag or multilag estimators implies that the correlations between the radar samples spaced at the lag T_s or higher must be significant. If the Doppler spectrum of radar signal is too broad and successive radar samples are poorly correlated, then the multilag estimators fail.

6.8 The Impact of Beam Broadening on the Polarimetric Variables

Progressive beam broadening and stronger impact of nonuniform beam filling (NBF) are the reasons the quality of polarimetric information deteriorates with range. This deterioration is commonly manifested by the gradual decrease of the cross-correlation coefficient ρ_{hv} with range within azimuthal sectors containing strong azimuthal gradients of differential phase Φ_{DP}. An example of the NBF effect is shown in Fig. 6.32 where a composite PPI of Z, Z_{DR}, ρ_{hv}, and Φ_{DP} measured by the KOUN WSR-88D during the mesoscale convective system on 19 May 2010 is presented. The decrease in ρ_{hv} is observed in the elongated area enclosed by the white oval in the azimuthal sector along the leading edge of the squall line where the azimuthal gradient of Φ_{DP} is maximal. Similar NBF-related artifacts are commonly observed in the RHI fields of ρ_{hv} in close proximity to the melting layer (Fig. 6.33). Because the magnitude of ρ_{hv} determines the statistical accuracy of all polarimetric

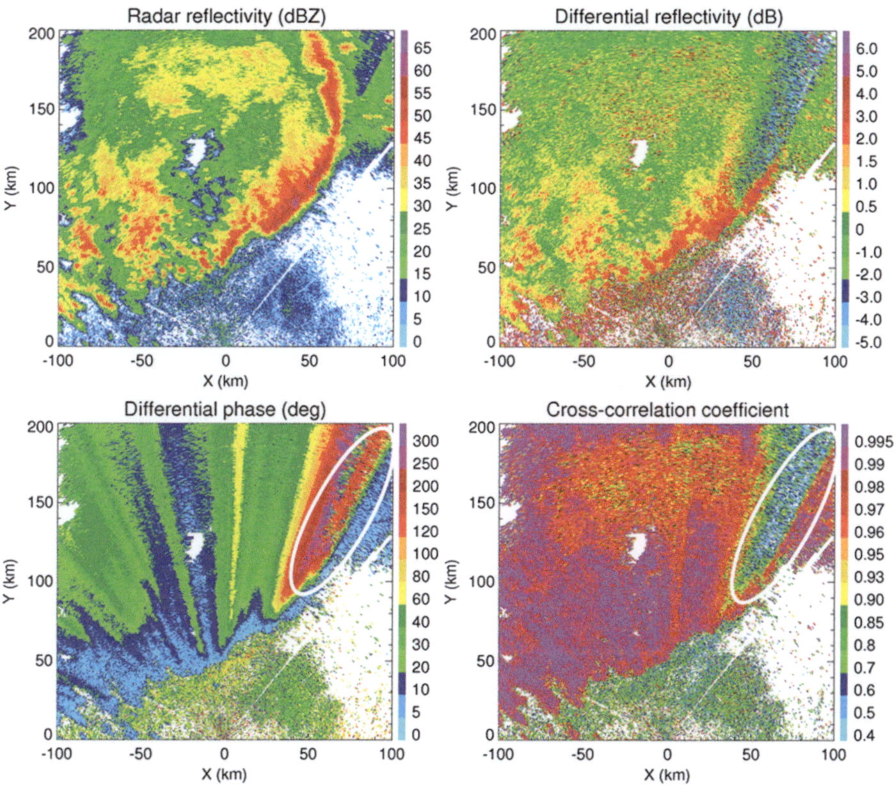

Fig. 6.32 Composite PPI plot of Z, Z_{DR}, ρ_{hv}, and Φ_{DP} measured at antenna elevation 0.5° by the KOUN WSR-88D on 19 May 2010. The area of the nonuniform beam filling (NBF) impact is enclosed in the white ellipse

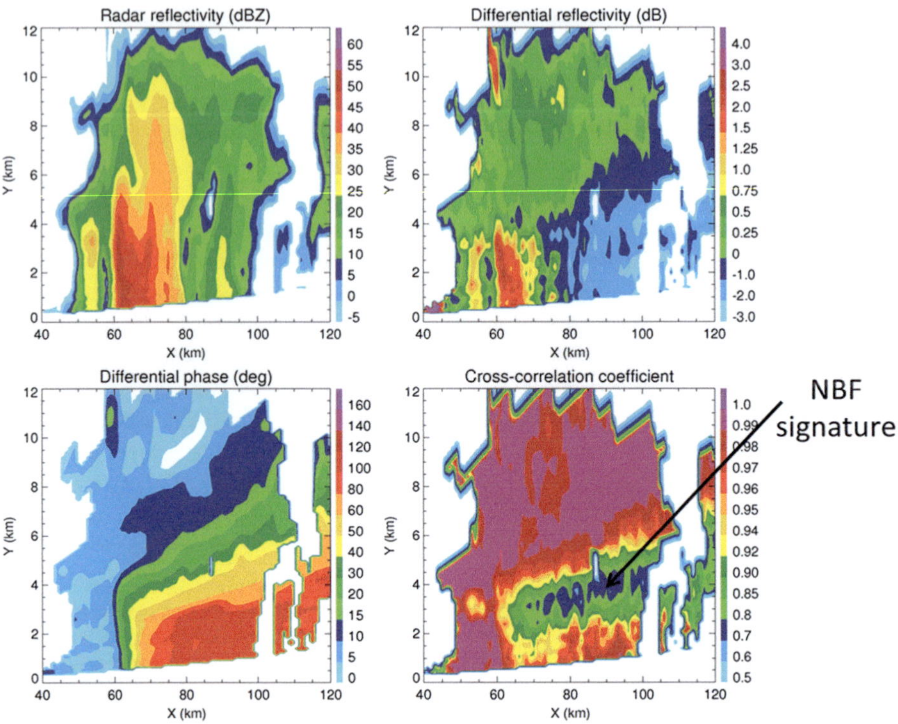

Fig. 6.33 Composite RHI of Z, Z_{DR}, ρ_{hv}, and Φ_{DP} measured by the C-band King City radar in Ontario, Canada

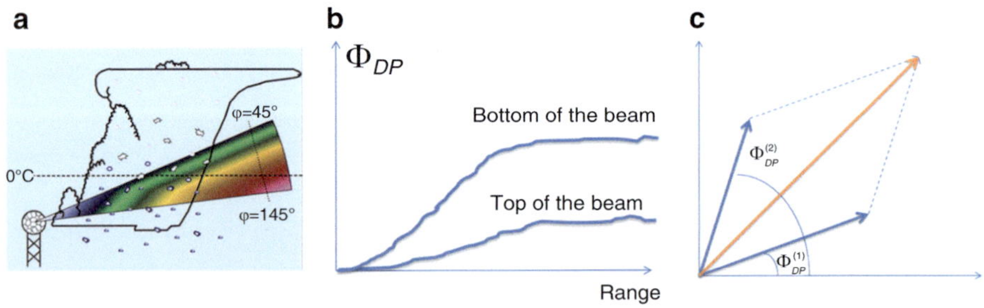

Fig. 6.34 Conceptual plot explaining the reduction of ρ_{hv} due to strong vertical gradient of Φ_{DP} within the radar resolution volume. (**a**) vetrical cross section of the beam traversing the melting layer the top is indicated with the 0°C isoterm. The colors depict differential phase along rays; blue is about 45° and red is 145°. (**b**) The differential phases corresponding to the rays at the bottom and top of the beam. (**c**) vector addition of the signals from the top ray (beam) and bottom ray

variables, the decrease of ρ_{hv} is accompanied by rapid increase of errors of the Z_{DR}, Φ_{DP}, and K_{DP} estimates (Sect. 6.11).

The diagram in Fig. 6.34 explains the nature of the ρ_{hv} reduction in the presence of strong vertical gradient of Φ_{DP} within the radar resolution volume. Such gradients often occur near the melting layer due to significant difference between K_{DP} in rain below the melting layer and ice above it. Therefore, the differential phase from the bottom part of the beam that remains in rain is significantly larger than Φ_{DP} from the top of the beam which contains mostly ice and snow (Fig. 6.34b).

To explain the ρ_{hv} reduction due to the gradient of Φ_{DP} within the beam, we can use a simplistic model of two spheroidal scatterers located within the same radar resolution volume but in the parts of the volume with different values of Φ_{DP}. The H and V voltages of the radar return signals from the two scatterers can be expressed as

$$h = a_{h1}e^{j\varphi_{h1}} + a_{h2}e^{j\varphi_{h2}} \tag{6.67}$$

$$v = a_{v1}e^{j\varphi_{v1}} + a_{v2}e^{j\varphi_{v2}}. \tag{6.68}$$

The cross-correlation coefficient is determined as

$$\begin{aligned} \rho_{hv} &= \frac{\overline{h^* v}}{\left(\overline{|h|^2}\,\overline{|v|^2}\right)^{1/2}} \approx \frac{a_{h1}a_{v1}e^{j(\varphi_{v1}-\varphi_{h2})} + a_{h2}a_{v2}e^{j(\varphi_{v2}-\varphi_{h2})}}{\left[\left(a_{h1}^2 + a_{h2}^2\right)\left(a_{v1}^2 + a_{v2}^2\right)\right]^{1/2}} \\ &= \frac{a_{h1}a_{v1}e^{j\Phi_{DP}^{(1)}} + a_{h2}a_{v2}e^{j\Phi_{DP}^{(2)}}}{\left[\left(a_{h1}^2 + a_{h2}^2\right)\left(a_{v1}^2 + a_{v2}^2\right)\right]^{1/2}} \end{aligned} \tag{6.69}$$

It follows from (6.69) that the absolute value of ρ_{hv} strongly depends on the difference between $\Phi_{DP}^{(1)}$ and $\Phi_{DP}^{(2)}$ with its maximal value achieved if the difference is zero. Indeed, in a special case of two identical spherical scatterers $a_{h1} = a_{v1} = a_{h2} = a_{v2}$ and

$$| \rho_{hv} | = \frac{1}{2} \left| e^{j\Phi_{DP}^{(1)}} + e^{j\Phi_{DP}^{(2)}} \right|. \tag{6.70}$$

i.e., the ρ_{hv} magnitude is the magnitude of the sum of two unit vectors whose directions are the differential phases (Fig. 6.34c). In this case, $|\rho_{hv}| = 1$ if $\Phi_{DP}^{(1)}$ and $\Phi_{DP}^{(2)}$ are equal, and $|\rho_{hv}|$ can be 0 if the vectors have opposite directions. The effect of the ρ_{hv} reduction due to the gradient of Φ_{DP} within the radar resolution volume is very similar to the impact of the gradient of Doppler velocity (or wind shear) on the magnitude of the autocorrelation coefficient $\rho(T)$ which is related to the Doppler spectral width (Doviak and Zrnic 2006). Wind shear causes the decrease of $\rho(T)$ and broadening of the Doppler spectrum.

Beam broadening also causes the biases in the estimates of Z, Z_{DR}, and Φ_{DP} in the presence of spatial gradients of these variables within the radar resolution volume. Ryzhkov (2007) derived relatively simple analytical formulas for the NBF-induced biases of the polarimetric variables assuming linear dependencies of Z, Z_{DR}, and Φ_{DP} in the cross-beam directions:

$$\Delta Z = 0.01\,\theta_1^2\left[\left(\frac{dZ}{d\theta}\right)^2 + \left(\frac{dZ}{d\varphi}\right)^2\right], \tag{6.71}$$

$$\Delta Z_{DR} = 0.02\,\theta_1^2\left(\frac{dZ}{d\theta}\frac{dZ_{DR}}{d\theta} + \frac{dZ}{d\varphi}\frac{dZ_{DR}}{d\varphi}\right), \tag{6.72}$$

$$\Delta\Phi_{DP} = 0.02\,\theta_1^2\left(\frac{d\Phi_{DP}}{d\theta}\frac{dZ}{d\theta} + \frac{d\Phi_{DP}}{d\varphi}\frac{dZ}{d\varphi}\right), \tag{6.73}$$

and

$$\xi = \frac{|\rho_{hv}^{(m)}|}{|\rho_{hv}|} = \exp\left\{-1.37\times 10^{-5}\theta_1^2\left[\left(\frac{d\Phi_{DP}}{d\theta}\right)^2 + \left(\frac{d\Phi_{DP}}{d\varphi}\right)^2\right]\right\}, \tag{6.74}$$

where θ_1 is the one-way 3 dB antenna pattern width, the ratio $|\rho_{hv}^{(m)}|/|\rho_{hv}|$ is the ρ_{hv} reduction factor, and θ and φ are elevation and azimuth. In (6.71)–(6.74), Φ_{DP}, $\Delta\Phi_{DP}$, θ, and φ are expressed in degrees, whereas Z, Z_{DR}, ΔZ, and ΔZ_{DR} are in decibels.

The magnitudes of ΔZ_{DR}, $\Delta\Phi_{DP}$, and ξ depend on the square of the antenna beamwidth. This quadratic dependence favors narrow-beam antennas for polarimetric measurements. A twofold increase of the beamwidth from 1° to 2° causes four times larger biases. The biases of Φ_{DP} and ρ_{hv} are wavelength-dependent because the differential phase and its gradients are inversely proportional to the wavelength λ, whereas the ρ_{hv} bias is approximately proportional to λ^{-2}. Thus, the NBF effects are stronger at shorter wavelengths.

6.9 Depolarization in Ice Crystals and Its Effect on Polarimetric Measurements

Unusually looking radial streaks of large positive and negative differential reflectivity are often observed in the ice regions of clouds probed by dual-polarization radars with simultaneous transmission and reception of orthogonally polarized H and V waves. These streaks are accompanied by differential phase exhibiting nonmonotonic change along radials. An example of this phenomenon is illustrated in Fig. 6.35 where a composite plot of Z, Z_{DR}, ρ_{hv}, and Φ_{DP} at elevation 5.5° is presented for the case of a mesoscale convective system (MCS) observed by the KOUN WSR-88D. Multiple radial streaks of large positive and negative Z_{DR} are evident in the crystal region of the cloud. Notable also is a well-pronounced decrease of Φ_{DP} with range in the northern flank of the squall line. Radially elongated features in Z_{DR} are also clearly seen in the vertical cross section through another MCS observed with the KOUN radar (Fig. 6.36). Most of these streaks in this and similar

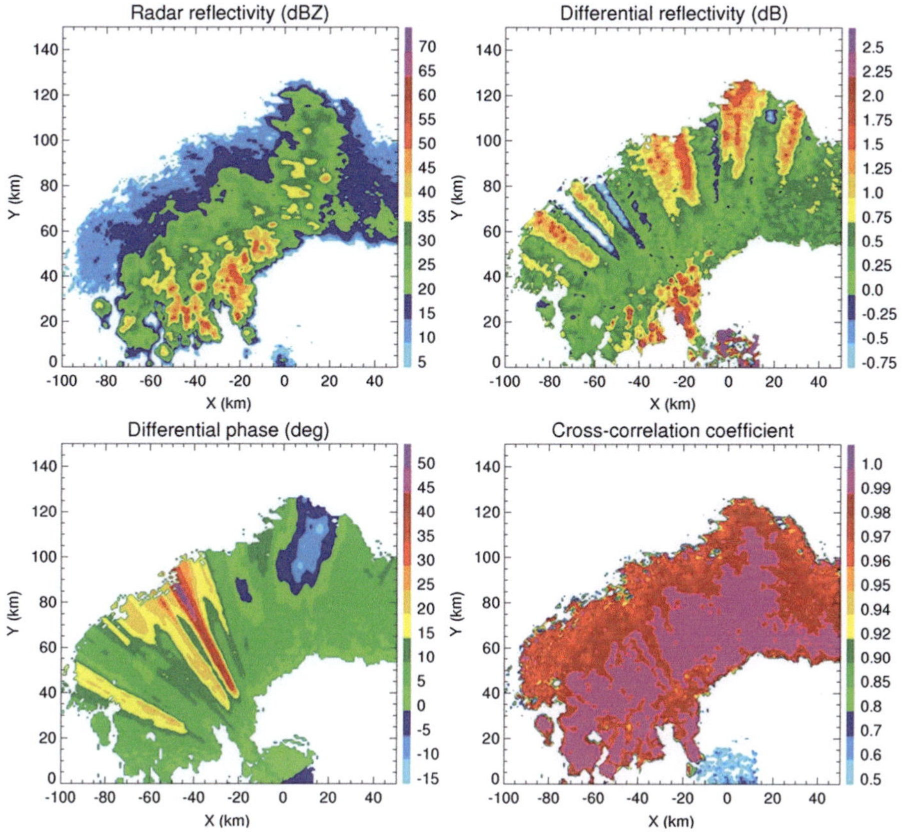

Fig. 6.35 A composite PPI plot of Z, Z_{DR}, Φ_{DP}, and ρ_{hv} at the elevation 5.5°. Data is from the MCS observed with the KOUN WSR-88D in Oklahoma on 21 June 2004. Radial streaks in the Z_{DR} panel indicate depolarization effects by oriented crystals

cases originate at the tops of convective cells. Such a localization and obvious association with the areas of negative K_{DP} point to depolarization by canted crystals as the most likely cause of these signatures.

Ice crystals near the tops of the strong convective updrafts are canted by strong electric fields and produce noticeable cross-coupling between radar signals at horizontal and vertical polarizations if both signals are transmitted and received simultaneously. Although ice graupel growing in the convective updraft is needed for charge separation and production of electric field, it does not cause depolarization and cross-coupling because of its near-spherical shape. However, small pristine crystals with low inertia and pronounced nonsphericity within and in close proximity to the updraft tend to align along the direction of electrostatic field that generally is neither horizontal nor vertical. These crystals act as sources of secondary radiation and gradually change the polarization state of the propagating wave. Note that depolarization occurs in relatively compact areas where ice crystals are canted by

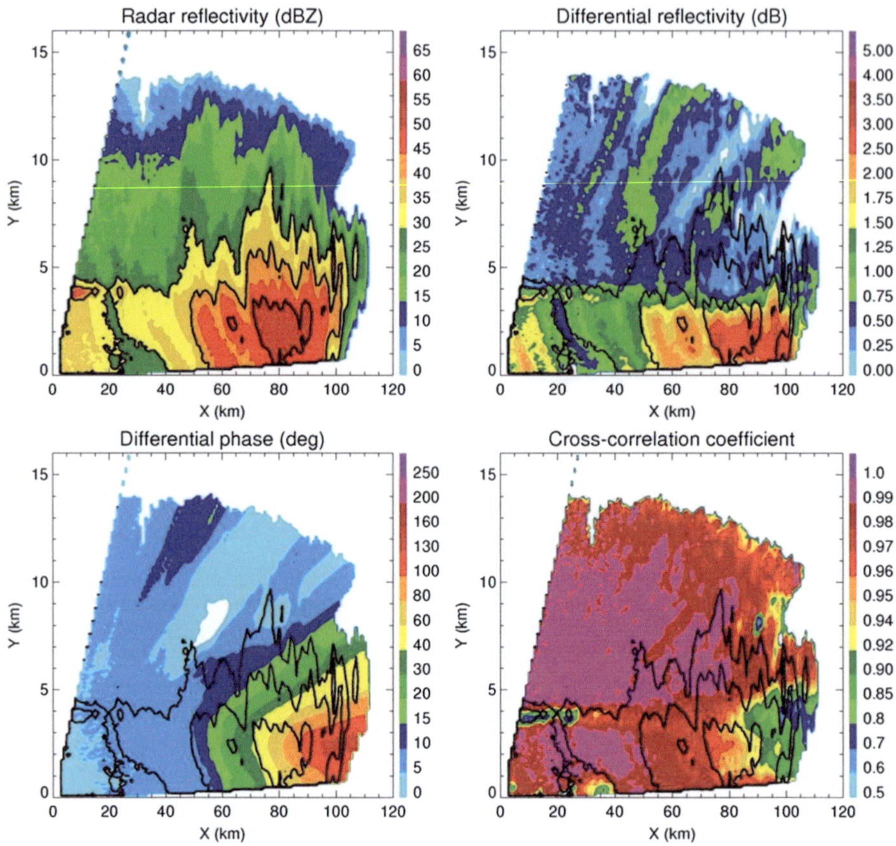

Fig. 6.36 A composite RHI plot of Z, Z_{DR}, Φ_{DP}, and ρ_{hv} from the MCS observed with the KOUN WSR-88D in Oklahoma on 17 June 2004. Contours of $Z = 30$, 40, and 50 dBZ are overlaid

the electric field. However, once the polarization state of the propagating wave has changed, it remains the same along the rest of the ray which explains radial appearances of the depolarization signatures. As demonstrated in the theoretical model of backscattering and propagation through the medium of canted crystals developed by Ryzhkov and Zrnic (2007), such depolarization on propagation is responsible for the observed radial streaks in the fields of Z_{DR} and Φ_{DP}. No depolarization on propagation occurs in rain because the mean canting angle of raindrops is on the average zero.

The depolarization streaks in the Z_{DR} fields are not observed if the orthogonally polarized waves are alternately transmitted and received, because the cross-coupling between H and V waves is significantly reduced. This is clearly demonstrated by Hubbert et al. (2014) using the observations with the NCAR S-Pol radar on which rapid switching between the SHV (simultaneous transmission/reception) and HSHV (alternate transmission/reception) modes of operations is possible (Fig. 6.37).

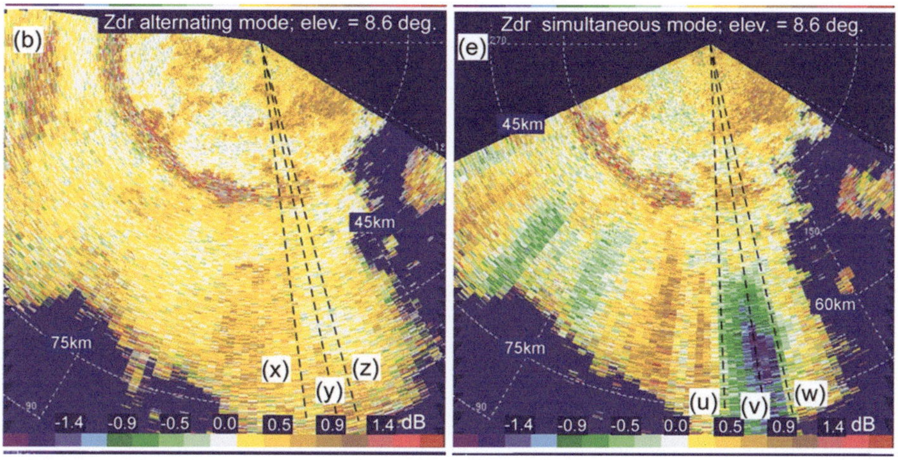

Fig. 6.37 The fields of Z_{DR} measured by the NCAR S-Pol radar in the HSHV mode of operation (left panel) and the SHV mode of operation (right panel). Adapted from Hubbert et al. (2014). © American Meteorological Society. Used with permission

Comparison of the fields of Z_{DR} measured in the thunderstorm at the beam elevation of 8.6° in the HSHV (left panel in Fig. 6.37) and the SHV modes (right panel) separated by 5.5 min shows the presence of the Z_{DR} streaks in the SHV mode and their absence in the HSHV mode.

6.10 Three-Body Scattering

The three-body scatter spike (TBSS) signature is often observed at the rear side of a strong hailstorm and manifests itself as a radial spike extending far beyond the actual storm (Fig. 6.38). TBSS is attributed to the multiple reflection of electromagnetic waves by a hydrometeor to the ground, back to another hydrometeor, and, finally to the radar. Because of this multipath geometry of reflection, an apparent echo is produced well behind the hydrometeor scatterers (e.g., Zrnic 1987; Lemon 1998). The TBSS is also apparent in the Z_{DR} and ρ_{hv} fields (Hubbert and Bringi 2000). The polarimetric TBSS manifests itself as a Z_{DR} enhancement and ρ_{hv} depression just behind the hail core (Figs. 6.38 and 6.39).

The TBSS echo has very distinctive polarimetric characteristics that can be explained by a simple scattering model (see Fig. 6.40). Differential reflectivity of the TBSS is a product of three factors:

$$Z_{dr} \sim \frac{I_h}{I_v}\frac{\sigma_h}{\sigma_v}\frac{l_v}{l_h}, \tag{6.75}$$

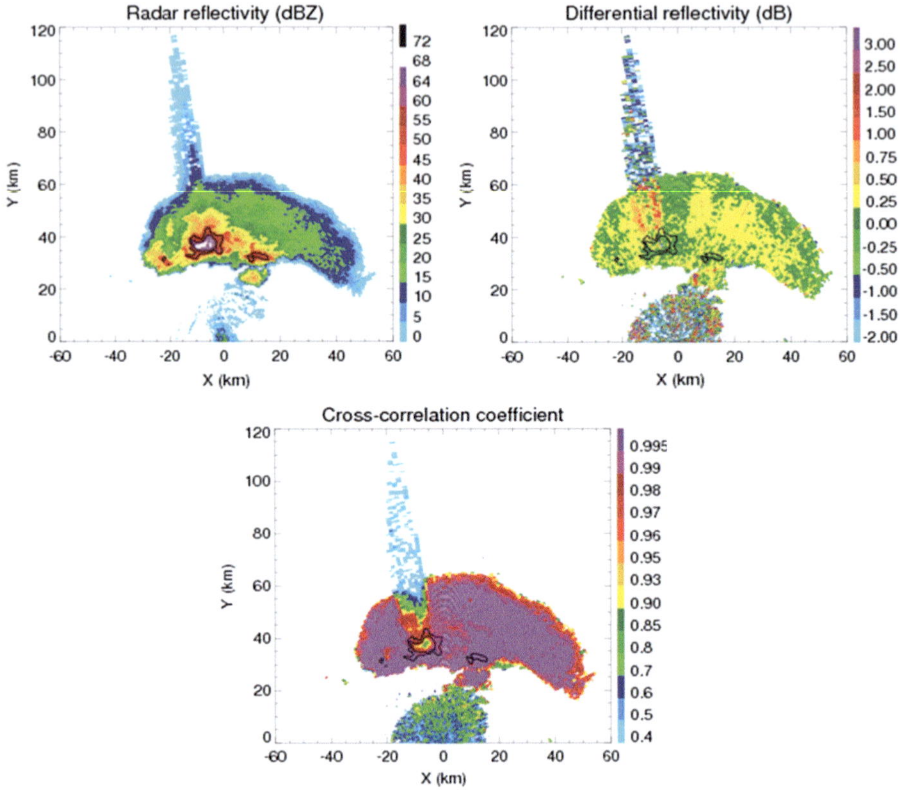

Fig. 6.38 Composite PPI of Z, Z_{DR}, and ρ_{hv} measured at elevation 10.0° by the KOUN WSR-88D on 16 May 2010 during heavy hailstorm in Oklahoma City. Adapted from Picca and Ryzhkov (2012).

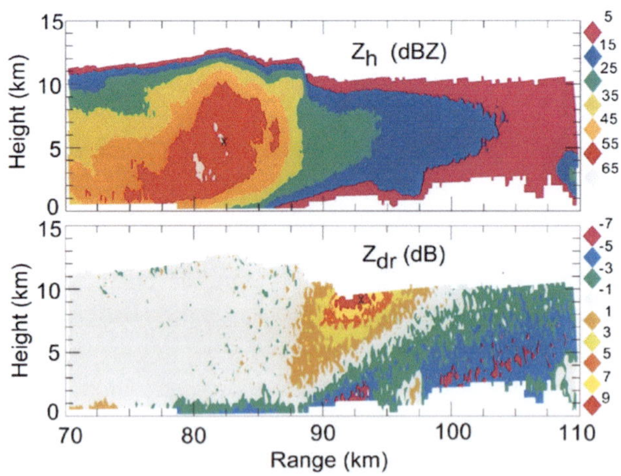

Fig. 6.39 An example of three-body scattering in a hailstorm observed by the DLR's (German Aerospace Agency) C-band radar in Germany. Adapted from Hubbert and Bringi (2000).

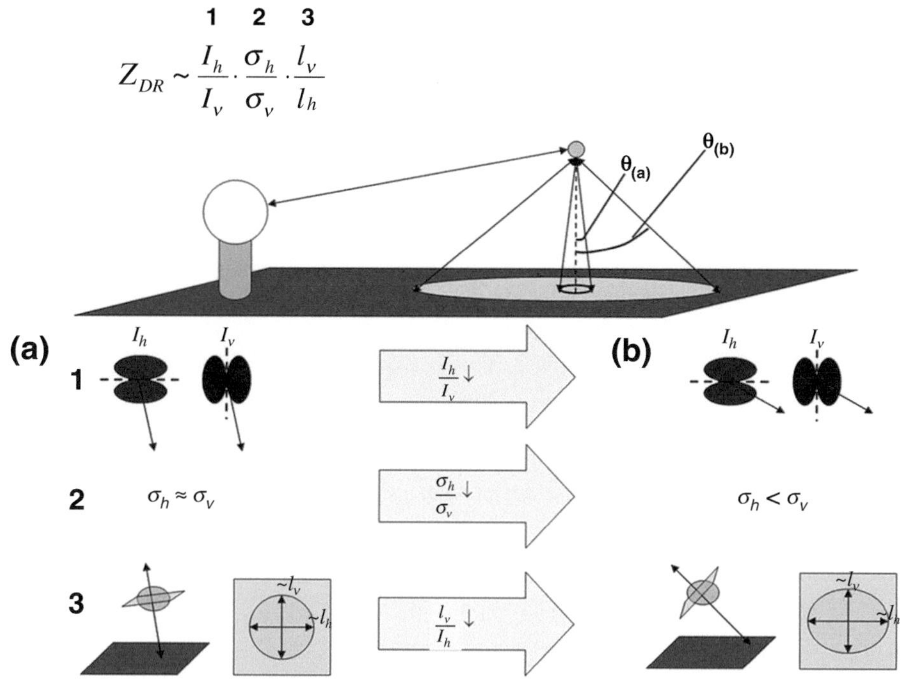

Fig. 6.40 Schematic illustrating the path of electromagnetic radiation that contributes to the TBSS pattern: from the radar to the hailstone, the hailstone to the ground, the ground back to the hailstone, and finally the hailstone back to the radar. Returns come from an oval projection underneath the hailstone, with θ_a representing the angle for a smaller conical region than with θ_b. Also presented are the schematics for the impact of the incidence angle on (1) the returned power ratio resulting from the radiation pattern, (2) the ground backscatter cross sections, and (3) losses from attenuation during scatter to and from the ground. Adapted from Picca and Ryzhkov (2012). © American Meteorological Society. Used with permission

where the ratio I_h/I_v characterizes the difference between radiation patterns of the scatterer at orthogonal polarizations, σ_h/σ_v is the ratio of the cross sections of the ground surface at horizontal and vertical polarizations, and l_v/l_h is the ratio of attenuation factors. All three are functions of the incidence angle θ.

In the Rayleigh approximation, a hailstone illuminated by microwave radiation can be modeled as a composition of horizontally and vertically oriented electric dipoles which are excited by an incident electromagnetic wave and emit secondary radiation. The radiation patterns of the horizontal dipole and vertical dipole excited by the horizontally polarized and vertically polarized components of the incident wave are different (Fig. 6.40). Namely, the vertical dipole does not radiate in the vertical direction down to the ground ($\theta = 0$), but radiates at oblique incidence (i.e., $\theta \neq 0$). Therefore, the reflected signal from the surface just beneath the hailstone ($\theta = 0$) has no vertically polarized component. In reality, however, the TBSS is a result of downward paths over a conical region. Thus, the downward scattered V-polarized component of radiation is nonzero but is still much smaller than the H-polarized component.

In the directions close to nadir ($\theta \approx 0$), the ratios σ_h/σ_v and l_v/l_h in Eq. (6.75) are close to one, and the resulting Z_{DR} is determined primarily by the factor I_h/I_v, which is very large. Hence, the TBSS Z_{DR} is expected to be high at relatively close distances to the hail shaft causing three-body scattering. Extended TBSS spikes indicate that a significant part of the secondary radiation reflects from the ground at a relatively large θ so that the propagation path from the hailstone to the ground and back is longer than in the case of nadir reflection (Fig. 6.40). At larger θ, the ratio I_h/I_v (known as bistatic Z_{DR}; Aydin et al. 1998) decreases. Similarly, the ratio σ_h/σ_v and differential attenuation factor l_v/l_h also decrease with increasing θ (e.g., Ulaby et al. 1982). Therefore, Z_{DR} decreases with increasing θ (or a distance from the radar) and may become negative if the ratio Z_{dr} (i.e., differential reflectivity expressed in linear scale) in (6.75) drops below 1. This is exactly what was observed behind the hail core in Figs. 6.38 and 6.39; Z_{DR} gradually decreases from high positive values down to negative values toward the end of the "flare" echo. Because the radar return associated with "flare" echo involves strong reflection from the surface with low intrinsic ρ_{hv}, the resulting cross-correlation coefficient is also anomalously low (Fig. 6.38).

6.11 Statistical Errors

Radar signals reflected from weather objects are random. This intrinsic randomness is caused by the motions of individual scatterers in the radar resolution volume. Thermal noise generated by the radar itself, surrounding atmosphere, and ground surface also adds to the statistical uncertainty of the estimates of radar variables. To reduce the uncertainty, the estimates of spectral moments and polarimetric variables are calculated from a sequence of M consecutive samples. These samples are correlated, and, therefore, the reduction in the variance of estimates is smaller than what it would be if there was no correlation between the samples. The variance is inversely proportional to the equivalent number of independent pulses M_I which depends on the wavelength λ, Doppler spectrum width σ_v, and pulse repetition period T_s. The statistical accuracy of the polarimetric variables also depends on the correlation between the horizontally and vertically polarized components of the signal which is quantified by the cross-correlation coefficient ρ_{hv}. The statistical errors and biases of the polarimetric variables' estimates are examined in a number of books (e.g., Doviak and Zrnic 2006; Bringi and Chandrasekar 2001; Zhang 2016) and various papers and reports (e.g., Melnikov and Zrnic 2004, 2007). Herein, we provide a brief summary of the statistical errors of the radar variable estimates (standard deviations and biases) and their dependencies on M, λ, σ_v, T_s, ρ_{hv}, and signal-to-noise ratio SNR.

Relatively simple and compact formulas for the standard deviations of the estimates of reflectivity Z, mean Doppler velocity v, spectrum width σ_v, differential reflectivity Z_{DR}, differential phase Φ_{DP}, and cross-correlation coefficient ρ_{hv} can be obtained for high values of signal-to-noise ratio (SNR > 20 dB) if the radar simultaneously transmits and receives H and V waves:

$$SD(Z) = \frac{3.24}{(\sigma_{vn}M)^{1/2}} \quad (\text{dB}), \tag{6.76}$$

$$SD(v) = 0.20\left(\frac{\lambda\sigma_v}{MT_s}\right)^{1/2} \quad (\text{m s}^{-1}), \tag{6.77}$$

$$SD(\sigma_v) = 0.16\left(\frac{\lambda\sigma_v}{MT_s}\right)^{1/2} \quad (\text{m s}^{-1}), \tag{6.78}$$

$$SD(Z_{DR}) = 4.62\left(\frac{1-\rho_{hv}^2}{\sigma_{vn}M}\right)^{1/2} \quad (\text{dB}), \tag{6.79}$$

$$SD(\Phi_{DP}) = 30.3\left(\frac{\rho_{hv}^{-2}-1}{\sigma_{vn}M}\right)^{1/2} \quad (\text{deg}), \tag{6.80}$$

$$SD(\rho_{hv}) = 0.53\frac{1-\rho_{hv}^2}{(\sigma_{vn}M)^{1/2}}, \tag{6.81}$$

where $\sigma_{vn} = 4\sigma_v T_s/\lambda$ is the normalized spectrum width, λ is the wavelength (in m), T_s is the pulse repetition period (in sec), and M is the number of pulses. Equations (6.76)–(6.78) are adapted from Doviak and Zrnic (2006), whereas Eqs. (6.79)–(6.81) are taken from Melnikov and Zrnic (2004, 2007). These formulas are valid if $0.04 < \sigma_{vn} < 0.60$. Specific differential phase K_{DP} is estimated from Φ_{DP} as a slope of linear regression fit to the radial profile of Φ_{DP}. It can be shown that the standard error of the K_{DP} estimate is equal to (Balakrishnan and Zrnic 1990; Ryzhkov and Zrnic 1996)

$$SD(K_{DP}) = \frac{\sqrt{3}SD(\Phi_{DP})}{M_r^{3/2}\Delta r}. \tag{6.82}$$

In (6.82), Δr is the spacing of samples in range, and M_r is the number of range samples in the interval $L = (M_r - 1)\Delta r$ over which the Φ_{DP} slope is estimated.

All standard deviations in (6.76)–(6.82) are directly proportional to $\lambda^{1/2}$ and inversely proportional to $M^{1/2}$. This means that the statistical uncertainty of all radar variables for a given dwell time MT_s and spectrum width σ_v is smaller at shorter wavelength. This is because the consecutive radar samples are less correlated at shorter wavelengths and the corresponding equivalent number of independent samples is larger. It is important that the standard deviations of the three polarimetric variables, Z_{DR}, Φ_{DP}, and ρ_{hv}, depend strongly on the magnitude of the cross-correlation coefficient ρ_{hv} and decrease dramatically if ρ_{hv} increases from, say, 0.95 to 0.99. All standard deviations increase with decreasing SNR, and the corresponding equations can be found in Melnikov and Zrnic (2004).

The dependencies of the standard deviations of the estimates of Z_{DR}, Φ_{DP}, and ρ_{hv} on the Doppler spectrum width for different values of ρ_{hv} and SNR at S band are illustrated in Fig. 6.41. The calculations have been made for a typical surveillance scan of the WSR-88D with pulse repetition frequency PRF = 321 Hz and $M = 17$. The standard deviations of all three variables are quite high for such a short dwell time; therefore additional spatial averaging (typically along a radial) is needed to obtain precise estimates of Z_{DR}, Φ_{DP}, and ρ_{hv}. Additional radial averaging over the range interval of 2 km with range gate spacing of 0.25 km reduces the standard deviations by a factor of 3. Note that the estimates of Z_{DR} and ρ_{hv} have small positive asymptotic biases which are inversely proportional to the number of samples M but can be significant at shorter dwell times (Melnikov and Zrnic 2004). It can be shown that for $M = 17$ the asymptotic biases of Z_{DR} and ρ_{hv} at S band are smaller than 0.1 dB and 0.01, respectively, if $\rho_{hv} > 0.95$, $\sigma_v > 2$ m s^{-1}, and SNR >10 dB.

Melnikov and Zrnic (2007) investigated statistical properties of the one-lag estimator of Z_{DR} and ρ_{hv} (see Sect. 6.7) and found that the standard deviations of the one-lag estimates and conventional (zero-lag) estimates are practically identical at S band if $\sigma_v < 6$ m s^{-1} and SNR > 2 dB.

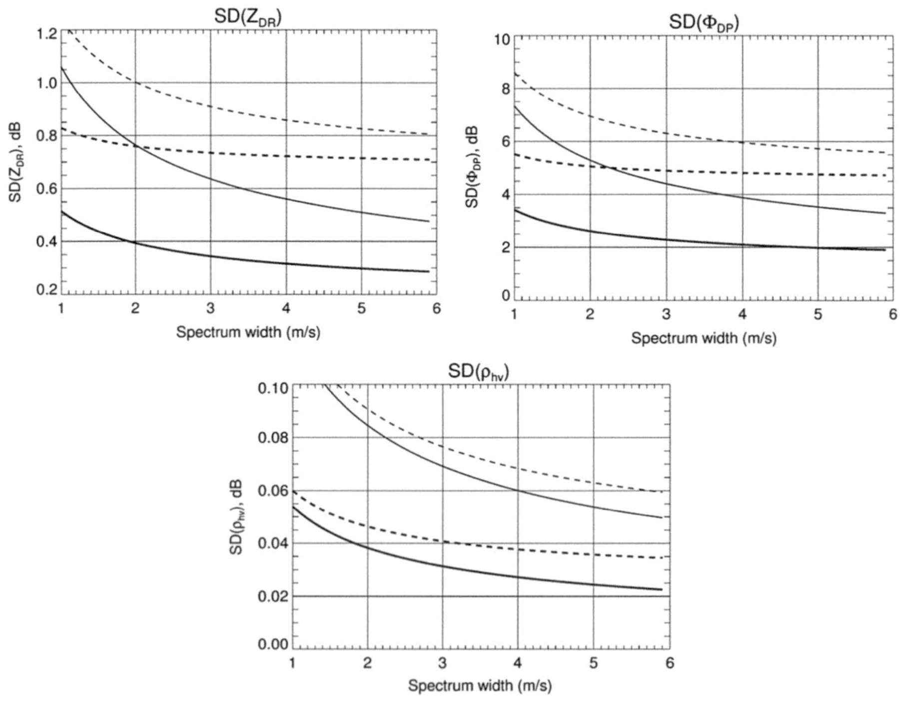

Fig. 6.41 Standard deviations of the estimates of Z_{DR}, Φ_{DP}, and ρ_{hv} as functions of Doppler spectrum width for different values of ρ_{hv} and SNR at S band ($\lambda = 11$ cm) for PRF = 321 Hz and $M = 17$. Solid lines, SNR = 20 dB; dashed lines, SNR = 10 dB; thick lines, $\rho_{hv} = 0.99$; thin lines, $\rho_{hv} = 0.95$

The estimates of Φ_{DP} and K_{DP} are not biased but can be quite noisy in weak weather echoes at longer wavelengths and shorter dwell times. Evaluating statistical errors of K_{DP} is particularly important because this variable is crucial for quantification of precipitation. From (6.82) one can conclude that for SD(Φ_{DP}) equal to 2° (Fig. 6.41), $\rho_{hv} = 0.99$, SNR = 20 dB, $M = 17$, $\Delta r = 0.25$ km, and radial resolution of the K_{DP} estimate $L = 2$ km ($M_r = 9$), the value of SD(K_{DP}) equals 0.5 deg km^{-1}. In the WSR-88D processing chain, K_{DP} with that radial resolution is used if Z > 40 dBZ, whereas the estimate of K_{DP} with three times coarser resolution ($L = 6$ km) is utilized if Z < 40 dBZ (Ryzhkov and Zrnic 1996). In the latter case, SD(K_{DP}) = 0.11 deg km^{-1} is still prohibitively large given that the K_{DP} itself can be much lower than 0.1 deg km^{-1} at S band in light rain and snow. This motivates the use of extra averaging of K_{DP} in a larger spatial/temporal domain as discussed in Chaps. 10 and 11.

References

Atlas, D. (2002). Radar calibration: Some simple approaches. *Bulletin of the American Meteorological Society, 83*, 1313–1316.

Aydin, K., Park, S., & Walsh, T. (1998). Bistatic dual-polarization scattering from rain and hail at S- and C-band frequencies. *Journal of Atmospheric and Oceanic Technology, 15*, 1110–1121.

Balakrishnan, N., & Zrnic, D. (1989). Correction of propagation effects at attenuating wavelengths in polarimetric radars. In *24th Conference on Radar Meteorology* (pp. 287–291). Tallahassee, FL: AMS.

Balakrishnan, N., & Zrnic, D. (1990). Estimation of rain and hail rates in mixed-phase precipitation. *Journal of the Atmospheric Sciences, 47*, 565–583.

Bech, J., Codina, B., Lorente, J., & Bebbington, D. (2003). The sensitivity of single polarization weather radar beam blockage correction to variability in the vertical refractivity gradient. *Journal of Atmospheric and Oceanic Technology, 20*, 845–855.

Bechini, R., Cremonini, R., Gorgucci, E., & Baldini, L. (2006). Dual-pol radar calibration and correction of the bias introduced by nonuniform radome wetting. In *Proceedings of the Fourth European Conference on Radar in Meteorology and Hydrology*. Barcelona, Spain (pp. 593–596).

Borowska, L., Ryzhkov, A., Zrnic, D., Simmer, C., & Palmer, R. (2011). Attenuation and differential attenuation of the 5-cm wavelength radiation in melting hail. *Journal of Applied Meteorology and Climatology, 50*, 59–76.

Borowska, L., & Zrnic, D. (2012). Use of ground clutter to monitor polarimetric radar calibration. *Journal of Atmospheric and Oceanic Technology, 29*, 159–176.

Brandes, E., & Ikeda, K. (2004). Freezing-level estimation with polarimetric radar. *Journal of Applied Meteorology, 43*, 1541–1553.

Bringi, V., & Chandrasekar, V. (2001). *Polarimetric Doppler Weather Radar: Principles and Applications* (p. 636). Cambridge, UK: Cambridge University Press.

Bringi, V. N., Chandrasekar, V., Balakrishnan, N., & Zrnic, D. S. (1990). An examination of propagation effects in rainfall on polarimetric variables at microwave frequencies. *Journal of Atmospheric and Oceanic Technology, 7*, 829–840.

Bringi, V. N., Keenan, T. D., & Chandrasekar, V. (2001). Correcting C-band radar reflectivity and differential reflectivity data for rain attenuation: A self-consistent method with constraints. *IEEE Transactions on Geoscience and Remote Sensing, 39*, 1906–1915.

Bringi, V., Thurai, M., Nakagawa, K., Huang, G., Kobayashi, T., Adachi, A., Hanado, H., & Sekizawa, S. (2006). Rainfall estimation from C-band polarimetric radar in Okinawa, Japan: comparison with 2D-video disdrometer and 400 MHz wind profiler. *Journal of the Meteorological Society of Japan, 84*, 705–724.

Cao, Q., Zhang, G., Brandes, E., Schuur, T., Ryzhkov, A., & Ikeda, K. (2008). Analysis of video disdrometer and polarimetric radar data to characterize rain microphysics in Oklahoma. *Journal of Applied Meteorology and Climatology, 47*, 2238–2255.

Cao, Q., Zhang, G., Palmer, R., Knight, M., May, R., & Stafford, R. (2012). Spectrum – Time Estimation and Processing (STEP) for improving weather radar data quality. *IEEE Transactions on Geoscience and Remote Sensing, 50*, 4670–4683.

Carey, L. D., Rutledge, S. A., Ahijevych, D. A., & Keenan, T. D. (2000). Correcting propagation effects in C-band polarimetric radar observations of tropical convection using differential propagation phase. *Journal of Applied Meteorology, 39*, 1405–1433.

Chandrasekar, V., Baldini, L., Bharadwaj, N., & Smith, P. (2015). Calibration procedures for global precipitation - measurements ground-validation radars. *Radio Science Bulletin, 355*, 45–73.

Diederich, M., Troemel, S., Ryzhkov, A., Zhang, P., & Simmer, C. (2015). Use of specific attenuation for rainfall measurements at X-band radar wavelengths. Part I: Radar calibration and partial beam blockage estimation. *Journal of Hydrometeorology, 16*, 487–502.

Doviak, R., & Zrnic, D. (2006). *Doppler Radar and Weather Observations* (2nd ed.). Reprint, Mineola, NY: Dover Publications, Inc., 562pp.

Figueras i Ventura, J., Boumahmoud, A., Fradon, B., Dupuy, P., & Tabary, P. (2012). Long-term monitoring of French polarimetric radar data quality and evaluation of several polarimetric quantitative precipitation estimators in ideal conditions for operational implementation at C band. *Quarterly Journal of the Royal Meteorological Society, 138*, 2212–2228.

Frech, M. (2009). *The effect of a wet radome on dualpol data quality*. In Proceedings of 34th Conference on Radar Meteorology. Wiliamsburg, VA, P13.15.

Frech, M., Hagen, M., & Mammen, T. (2017). Monitoring the absolute calibration of a polarimetric weather radar. *Journal of Atmospheric and Oceanic Technology, 34*, 599–615.

Giangrande, S., & Ryzhkov, A. (2005). calibration of dual-polarization radar in the presence of partial beam blockage. *Journal of Atmospheric and Oceanic Technology, 22*, 1156–1166.

Goddard, J., Tan, J., & Thurai, M. (1994). Technique for calibration of meteorological radars using differential phase. *Electronics Letters, 30*, 166–167.

Gorgucci, E., Scarchilli, G., & Chandrasekar, V. (1992). Calibration of radars using polarimetric techniques. *IEEE Transactions on Geoscience and Remote Sensing, 30*, 853–858.

Gorgucci, E., Scarchilli, G., & Chandrasekar, V. (1999). A procedure to calibrate multiparameter weather radar using properties of the rain medium. *IEEE Transactions on Geoscience and Remote Sensing, 37*, 269–276.

Gourley, J., Illingworth, A., & Tabary, P. (2009). Absolute calibration of radar reflectivity using redundancy of the polarization observations and implied constraints on drop shapes. *Journal of Atmospheric and Oceanic Technology, 26*, 689–703.

Gu, J.-Y., Ryzhkov, A., Zhang, P., Neilley, P., Knight, M., Wolf, B., et al. (2011). Polarimetric attenuation correction in heavy rain at C band. *Journal of Applied Meteorology, 50*, 39–58.

Hitschfeld, W., & Bordan, J. (1954). Errors inherent in the radar measurement of rainfall at attenuating wavelengths. *Journal of Meteorology, 11*, 58–67.

Hubbert, J., & Bringi, V. (2000). The effects of three-body scattering on differential reflectivity signatures. *Journal of Atmospheric and Oceanic Technology, 17*, 51–61.

Hubbert, J., Bringi, V., & Brunkow, D. (2003). Studies of the polarimetic covariance matrix. Part I: Calibration methodology. *Journal of Atmospheric and Oceanic Technology, 20*, 696–706.

Hubbert, J., Dixon, M., & Ellis, S. (2009). Weather radar ground clutter. Part II: Real-time identification and filtering. *Journal of Atmospheric and Oceanic Technology, 26*, 1181–1197.

Hubbert, J., Ellis, S., Chang, W.-Y., Rutledge, S., & Dixon, M. (2014). Modeling and interpretation of S-band ice crystals depolarization signatures from data obtained by simultaneously transmitting horizontally and vertically polarized fields. *Journal of Applied Meteorology and Climatology, 53*, 1659–1676.

Hubbert, J., Pratte, F., Dixon, M., & Riiling, R. (2008). *The uncertainty of Zdr calibration*. In Preprints. Fifth European Conference on Radar in Meteorology and Hydrology (ERAD 2008). Helsinki, Finland. Available online at http://erad2008.fmi.fi/proceedings/extended/erad2008-0285-extended.pdf

Hudak, D., Rodrigues, P., Lee, G.-W., Ryzhkov, A., Fabry, F., & Donaldson, N. (2006). Winter precipitation studies with a dual polarized C-band radar. *Proceedings of the Fourth European Conference on Radar in Meteorology and Hydrology*. Barcelona, Spain (pp. 9–16).

Illingworth, A., & Blackman, T. (2002). The need to represent raindrop size spectra as normalized Gamma distributions for the interpretation of polarization radar observations. *Journal of Applied Meteorology, 41*, 286–297.

Ivic, I., Curtis, C., & Torres, S. (2013). Radial-based noise power estimation for weather radars. *Journal of Atmospheric and Oceanic Technology, 30*, 2737–2753.

Kurri, M., & Huuskonen, A. (2008). Measurements of the transmission loss of a radome at different rain intensities. *Journal of Atmospheric and Oceanic Technology, 25*, 1590–1599.

Lang, T., Nesbitt, S., & Carey, L. (2009). On the correction of partial beam blockage in polarimetric radar data. *Journal of Atmospheric and Oceanic Technology, 26*, 943–957.

Le Bouar, E., Testud, J., & Keenan, T. (2001). Validation of the rain profiling algorithm "ZPHI" from the C-band polarimetric weather radar in Darwin. *Journal of Atmospheric and Oceanic Technology, 18*, 1819–1837.

Lemon, L. (1998). The radar "three-body scatter spike": An operational large – hail signature. *Weather and Forecasting, 13*, 327–340.

Marks, D., Wolff, D., Carey, L., & Tokay, A. (2011). Quality control and calibration of the dual-polarization radar at Kwajalein, RMI. *Journal of Atmospheric and Oceanic Technology, 28*, 181–196.

Melnikov, V., Doviak, R., Zrnic, D., & Stensrud, D. (2011). Mapping Bragg scatter with a polarimetric WSR-88D. *Journal of Atmospheric and Oceanic Technology, 28*, 1273–1285.

Melnikov, V., & Zrnic, D. (2004). *Simultaneous transmission mode for the polarimetric WSR-88D: Statistical biases and standard deviations of polarimetric variables*. NOAA/NSSL Rep. 84 pp. Available online at https://www.nssl.noaa.gov/publications/wsr88d_reports/SHV_statistics.pdf

Melnikov, V., & Zrnic, D. (2007). Autocorrelation and cross-correlation estimators of polarimetric variables. *Journal of Atmospheric and Oceanic Technology, 24*, 1337–1350.

Melnikov, V., Zrnic, D., Free, A., Ice, R., & Macemon, R. (2017). *Monitoring Radar Calibration using Ground Clutter*. NSSL internal report. Available online at https://www.nssl.noaa.gov/publications/wsr88d_reports/Z_ZDRcal_ROC_MOU_report_2017.pdf

Meneghini, R., & Nakamura, K. (1990). Range profiling of the rain rate by an airborne weather radar. *Remote Sensing of Environment, 31*, 193–209.

Park, S., Maki, M., Iwanami, K., Bringi, V., & Chandrasekar, V. (2005). Correction of radar reflectivity for rain attenuation at X band. Part II: Evaluation and application. *Journal of Atmospheric and Oceanic Technology, 22*, 1621–1632.

Picca, J., & Ryzhkov, A. (2012). A dual-wavelength polarimetric analysis of the 16 May 2010 Oklahoma City extreme hailstorm. *Monthly Weather Review, 140*, 1385–1403.

Richardson, L., Cunningham, J., Zittel, W., Lee, R., Ice, R., Melnikov, V., et al. (2017a). Bragg scatter detection by the WSR-88D. Part I: Algorithm development. *Journal of Atmospheric and Oceanic Technology, 34*, 465–478.

Richardson, L., Zittel, W., Lee, R., Melnikov, V., Ice, R., & Cunningham, J. (2017b). Bragg scatter detection by the WSR-88D. Part II: Assessment of Z_{DR} bias estimation. *Journal of Atmospheric and Oceanic Technology, 34*, 479–493.

Ryzhkov, A., Diederich, M., Zhang, P., & Simmer, C. (2014). Utilization of specific attenuation for rainfall estimation, mitigation of partial beam blockage, and radar networking. *Journal of Atmospheric and Oceanic Technology, 31*, 599–619.

Ryzhkov, A., Kumjian, M., Ganson, S., & Zhang, P. (2013). Polarimetric radar characteristics of melting hail. Pt II: Practical implications. *Journal of Applied Meteorology and Climatology, 52*, 2871–2886.

Ryzhkov, A., Zhang, G., Luchs, S., & Ryzhkova, L. (2008). *Polarimetric characteristics of snow measured by radar and 2D video disdrometer*. In Extended Abstracts, 5th European Conference on Radar in Meteorology and Hydrology, Helsinki, Finland, CD-ROM, P6.5.

Ryzhkov, A., Zhang, P., Hudak, D., Alford, J., Knight, M., & Conway, J. (2007). *Validation of polarimetric methods for attenuation correction at C band*. In 33rd Conference on Radar Meteorology, Cairns, Australia, CD-ROM P11B.12.

Ryzhkov, A., Zhang, P., Reeves, H., Kumjian, M., Tschallener, T., Simmer, C., et al. (2016). Quasi-vertical profiles – a new way to look at polarimetric radar data. *Journal of Atmospheric and Oceanic Technology, 33*, 551–562.

Ryzhkov, A., & Zrnic, D. (1995). Precipitation and attenuation measurements at a 10 cm wavelength. *Journal of Applied Meteorology, 34*, 2121–2134.

Ryzhkov, A., & Zrnic, D. (1996). Assessment of rainfall measurement that uses specific differential phase. *Journal of Applied Meteorology, 35*, 2080–2090.

Ryzhkov, A., & Zrnic, D. (1998). Discrimination between rain and snow with a polarimetric radar. *Journal of Applied Meteorology, 37*, 1228–1240.

Ryzhkov, A. V. (2007). The impact of beam broadening on the quality of radar polarimetric data. *Journal of Atmospheric and Oceanic Technology, 24*, 729–744.

Ryzhkov, A. V., Giangrande, S. E., Melnikov, V. M., & Schuur, T. J. (2005). Calibration issues of dual-polarization radar measurements. *Journal of Atmospheric and Oceanic Technology, 22*, 1138–1155.

Ryzhkov, A. V., & Zrnic, D. S. (2007). Depolarization in ice crystals and its effect on radar polarimetric measurements. *Journal of Atmospheric and Oceanic Technology, 24*, 1256–1267.

Scarchilli, G., Gorgucci, E., Chandrasekar, V., & Dobaie, A. (1996). Self consistency of polarization diversity measurements of rainfall. *IEEE Transactions on Geoscience and Remote Sensing, 34*, 22–26.

Siggia, A., & Passarelli, R. Jr. (2004) Gaussian model adaptive processing (GMAP) for improved ground clutter cancellation and moment calculation. In *Proceedings of ERAD*, Visby, Island of Gotland, Sweden (pp. 63–73).

Snyder, J., Bluestein, H., Zhang, G., & Frasier, S. (2010). Attenuation correction and hydrometeor classification of high-resolution, X-band, dual-polarized mobile radar measurements in severe convective storms. *Journal of Atmospheric and Oceanic Technology, 27*, 1979–2001.

Tabary, P., Vulpiani, G., Gourley, J., Illingworth, A., Thompson, R., & Bousquet, O. (2009). Unusually high differential attenuation at C-band: Results from a two-year analysis of the French Trappes polarimetric radar data. *Journal of Applied Meteorology and Climatology, 48*, 2037–2053.

Testud, J., Le Bouar, E., Obligis, E., & Ali-Mehenni, M. (2000). The rain profiling algorithm applied to polarimetric weather radar. *Journal of Atmospheric and Oceanic Technology, 17*, 332–356.

Torres, S., & Warde, D. (2014). Ground clutter mitigation for weather radars using the autocorrelation spectral density. *Journal of Atmospheric and Oceanic Technology, 31*, 2049–2066.

Ulaby, F., Moore, R., & Fung, A. (1982). Radar Remote Sensing and Surface Scattering and Emission Theory. In *Microwave Remote Sensing: Active and Passive* (Vol. 2). Reading, MA: Addison – Wesley. 1064 pp.

Vivekanandan, J., Zhang, G., Ellis, S., Rajopadhyaya, D., & Avery, S. (2003). Radar reflectivity calibration using differential propagation phase measurement. *Radio Science, 38*(3), MAR 14-1 – MAR 4-14.

Vulpiani, G., Tabary, P., Parent du Chatelet, J., & Marzano, F. (2008). Comparison of advanced radar polarimetric techniques for operational attenuation correction at C band. *Journal of Atmospheric and Oceanic Technology, 25*, 1118–1135.

Zhang, G. (2017). *Weather Radar Polarimetry*. Boca Raton, FL: CRC Press. 304pp.

Zhang, P., Zrnic, D., & Ryzhkov, A. (2013). Partial beam blockage correction using polarimetric radar measurements. *Journal of Atmospheric and Oceanic Technology, 30*, 861–872.

Zrnic, D. (1987). Three-body scattering produces precipitation signature of special diagnostic value. *Radio Science, 22*, 76–86.

Zrnic, D., Melnikov, V., & Carter, J. (2006). Calibrating differential reflectivity on the WSR-88D. *Journal of Atmospheric and Oceanic Technology, 23*, 944–951.

Zrnic, D. S., & Melnikov V. M. (2007). *Ground clutter recognition using polarimetric spectral parameters*. In 33rd Conference on Radar Meteorology, AMS, Cairns, Australia, 2007.

Chapter 7
Polarimetric "Fingerprints" of Different Microphysical Processes in Clouds and Precipitation

Polarimetric variables are sensitive to particle size, shape, orientation, and phase composition all of which change during various microphysical processes. Hence, each process may have its own unique polarimetric "fingerprint" which can be used for identification of a particular process. Herein, we examine microphysical processes involving liquid cloud drops, raindrops, mixed-phase particles, and ice particles. Rain processes include condensation, depositional growth and evaporation, coalescence, breakup, and size sorting. Ice and mixed-phase processes include ice nucleation, depositional growth/evaporation (sublimation), riming/accretion, aggregation, freezing/refreezing, and melting.

7.1 Microphysical Processes Involving Liquid Particles

7.1.1 Condensation/Evaporation

The processes of condensational growth and evaporation of water drops are described by the same equation. The equation for condensational growth/evaporation describing the change of the particle mass m with time can be found in the cloud physics literature (e.g., Rogers and Yau 1989; Pruppacher and Klett 1997; Straka 2009; Lamb and Verlinde 2011; Khain and Pinsky 2018):

$$\frac{dm}{dt} = \frac{4\pi CF(S_w - 1)}{F_K^{(w)} + F_D^{(w)}}, \tag{7.1}$$

where C is the "capacitance" characterizing the size and shape of a particle ($C = D/2$ for spheres with diameter D), S_w is the vapor saturation ratio with respect to water determined as

A. V. Ryzhkov, D. S. Zrnic, *Radar Polarimetry for Weather Observations*, Springer Atmospheric Sciences, https://doi.org/10.1007/978-3-030-05093-1_7

$$S_w = \frac{e}{E_w(T)} = 0.01\mathrm{RH}, \tag{7.2}$$

$$F_K^{(w)} = \frac{L_v^2}{KR_vT^2} \tag{7.3}$$

is the thermodynamic term associated with heat conduction,

$$F_D^{(w)} = \frac{R_vT}{D_vE_w(T)} \tag{7.4}$$

is the thermodynamic term associated with vapor diffusion, e is the water vapor pressure, E_w is the saturated water vapor pressure over flat surface of water, RH is the relative humidity (%), T is the air temperature in °K, L_v is the latent heat of vaporization, R_v is the gas constant for water vapor, K is the thermal conductivity of air, D_v is the diffusivity of water vapor, and F is the ventilation coefficient. $S_w > 1$ in the case of condensation, and $S_w < 1$ in the case of evaporation. In principle, one should distinguish between ventilation coefficients F_h for heat and F_v for vapor given by Pruppacher and Klett (1997)

$$F_h = 0.78 + 0.308N_{\mathrm{sc}}^{1/3}N_{\mathrm{Re}}^{1/2} \tag{7.5}$$

$$F_{\mathrm{v}} = 0.78 + 0.308N_{\mathrm{Pr}}^{1/3}N_{\mathrm{Re}}^{1/2}. \tag{7.6}$$

These ventilation coefficients depend on the Reynolds (N_{Re}), Schmidt (N_{Sc}), and Prandtl (N_{Pr}) numbers. It can be shown that the ventilation coefficients of cloud droplets and raindrops, F_h and F_v, are almost equal; hence, herein, we assume $F_h = F_v = F$. Taking into account that $m = \frac{\pi}{6}\rho_wD^3$, (7.1) can be written with respect to the equivolume diameter of liquid drop D:

$$\frac{dD}{dt} = \frac{8CF(S_w - 1)}{D^2\rho_w\left(F_K^{(w)} + F_D^{(w)}\right)}. \tag{7.7}$$

In case of precipitation evaporation (RH < 100%), it is important to quantify the reduction of the mass and size of raindrops as they fall to the surface in dry air with $S_w < 1$. The change of the mass of an individual raindrop as function of height can be determined taking into consideration that

$$\frac{dm}{dh} = \frac{dm}{dt}/\frac{dh}{dt} = \frac{dm}{dt}\frac{1}{V_{\mathrm{t}}}, \tag{7.8}$$

where V_t is the terminal velocity of a raindrop. Then it follows from (7.1) that

$$\frac{dm}{dh} = \frac{4\pi CF(S_w - 1)}{V_t\left(F_K^{(w)} + F_D^{(w)}\right)}. \tag{7.9}$$

Both F and V_t are functions of the raindrop size (or mass). Ventilation coefficient F changes from 1 to almost 20 if raindrop size increases from 0.1 to 8 mm. The dependence of V_t on raindrop equivolume diameter is determined by Brandes et al. (2002):

$$V_t = \begin{cases} -0.102 + 4.93D - 0.955D^2 + 0.0793D^3 - 0.00236D^4 & D < 6\,\text{mm} \\ 9.17 \quad D > 6\ \text{mm} \end{cases}. \tag{7.10}$$

Terminal velocities of hydrometeors are larger at higher altitudes in the atmosphere where the air density is lower and the dependence is

$$V_t(h) = \left[\frac{\rho_a(0)}{\rho_a(h)}\right]^{1/2} V_t(0). \tag{7.11}$$

Vertical dependencies of the diameters of evaporating raindrops with different initial sizes falling in a dry atmospheric layer are illustrated in Fig. 7.1. If RH = 80% (fixed within the height interval between 0 and $H_0 = 3$ km) and adiabatic lapse rate of temperature is 9.8 deg km^{-1}, then all raindrops smaller than 0.8 mm evaporate completely before reaching the surface. Raindrops with initial size 0.3 mm evaporate completely within the first 0.5 km, whereas a 3 mm raindrop loses only 11% of its mass at the surface which is about five times larger mass loss than the total mass loss of the 0.8 mm raindrop. Larger raindrops lose larger total mass due to evaporation, but its relative mass loss is smaller than the mass loss of smaller particles.

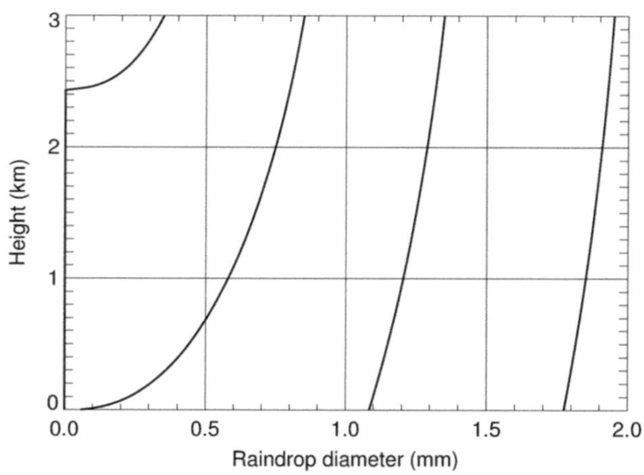

Fig. 7.1 Height dependences of the diameter of evaporating raindrops with different initial sizes. RH = 80%, temperature lapse rate is 9.8 deg km^{-1} (dry adiabatic)

Figure 7.1 shows that the raindrop diameter reduction accelerates at the final stage of evaporation. This is because the derivative *dD/dt* in (7.7) is inversely proportional to the diameter *D*, so that the condensation growth or evaporation is particularly intense for smaller-size drops, i.e., at the beginning of the condensation process and at the end of the evaporation process.

Since raindrops lose their mass and their diameter decreases in the process of evaporation, the corresponding values of Z and K_{DP} decrease. Z_{DR} may increase or decrease depending on the initial size distribution and vertical profile of humidity. In Fig. 7.2 is an example of the vertical profiles of Z, Z_{DR}, K_{DP}, and ρ_{hv} simulated in Kumjian and Ryzhkov (2010) for the case of very dry atmosphere in New Mexico.

The intensity of evaporation process is determined by relative concentration of small drops in the size spectrum. If evaporation is a dominant microphysical process below the melting layer, then the mass flux (or rain rate) decreases toward the surface. For a given air humidity, relative change of rain rate due to evaporation $\Delta R/R$ is higher for rain dominated by smaller raindrops than larger drops. Kumjian and Ryzhkov (2010) demonstrate that the value of $\Delta R/R$ depends on the initial value of Z_{DR} just below the melting layer and is independent of the initial Z aloft. The environmental conditions (including temperature and humidity) and the observed Z_{DR} (which does not change much with height) can be utilized to estimate $\Delta R/R$. As an example (Fig. 7.3), the relative reduction of rain rate is 60% for $Z_{DR} = 0.5$ dB compared to 38% for $Z_{DR} = 1.5$ dB if raindrops fall in a 3-km-deep layer with relative humidity linearly changing from 100% at the top of the layer to 75% at the surface (gray curve in the left panel of Fig. 7.3). Such methodology based on the use

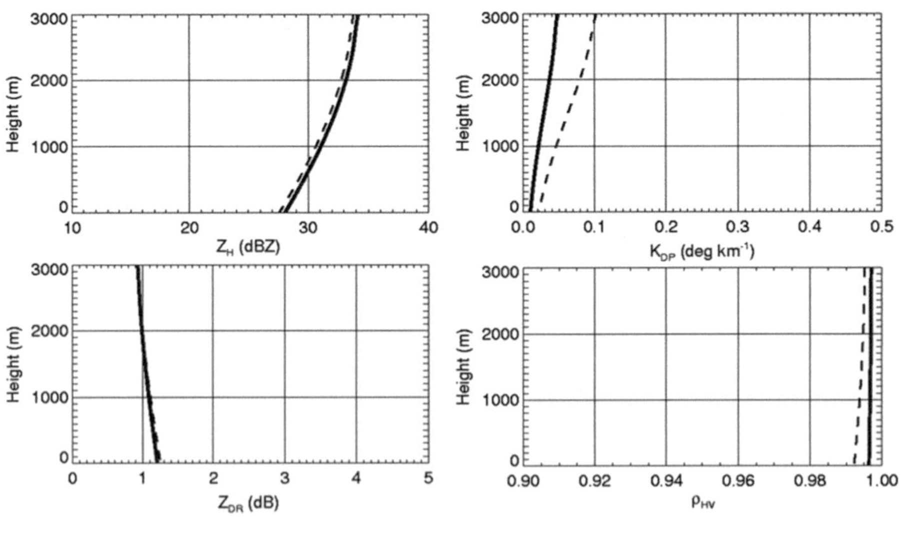

Fig. 7.2 Simulated vertical profiles of polarimetric variables for the case of very dry atmosphere. The solid and dashed lines indicate computations for S and C bands, respectively. The Marshall-Palmer DSD is assumed at the cloud base, with rainfall rate of 5 mm h^{-1} aloft. From Kumjian and Ryzhkov (2010). © American Meteorological Society. Used with permission

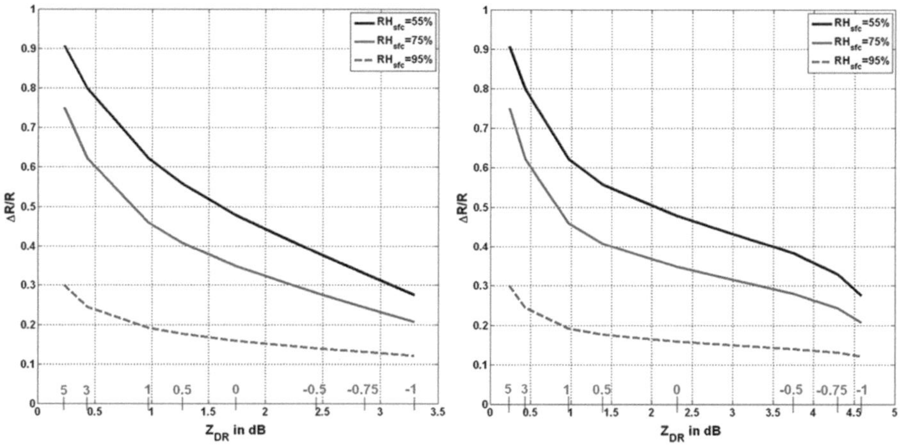

Fig. 7.3 The dependence of the relative change in rainfall rate $\Delta R/R$ on Z_{DR} aloft and surface *RH*. Results are from calculations using a 3-km-deep idealized well-mixed layer with a surface temperature of 30 °C. Each curve corresponds to a different value of the surface *RH*. Only constrained gamma models (with 5 mm h^{-1} rainfall rate) are used in the calculations, and $\Delta R/R$ is independent of *Z* aloft. The gray numbers along the abscissa correspond to the shape parameter μ associated with the given Z_{DR}. Left panel is for S-band values, and right panel is for C-band values. From Kumjian and Ryzhkov (2010). © American Meteorological Society. Used with permission

of Z_{DR} at the top of a dry layer may improve the quality of surface rainfall estimation, especially in regions far away from the radar or in poor low-level radar coverage areas. There the polarimetric radar observations below the cloud base could be used to infer rain rate near the ground.

Typical vertical cross sections of *Z* and Z_{DR} in evaporating rain are displayed in Fig. 7.4. The measurements have been made with the X-band polarimetric radar BoXPol (University of Bonn, Xie et al. 2016). Relative humidity *RH* during the observation period decreased from about 90% at 3 km to about 50% near the surface according to soundings. Following the rain shaft (black solid lines in Fig. 7.4) at a horizontal distance of 1–5 km from the BoXPol, *Z* decreases from ~32 dBZ below the melting layer to ~25 dBZ near the ground. In another rain shaft at 6–9 km away from the radar, *Z* decreases from ~25 to ~15 dBZ. At the same time, Z_{DR} does not change much along the rain shaft trajectory while maintaining relatively low value of about 0.3–0.4 dB in agreement with theoretical predictions from the evaporation model (see Fig. 7.2).

Figure 7.5 illustrates the impact of initial Z_{DR} at 2100 m on the evaporation reduction of the rain rate at 600 m for the same rain event. Rain rates at different altitude levels have been estimated from the measurements by the vertically pointing Micro Rain Radar (MRR) deployed next to BoXPol. Obviously, the initial Z_{DR} aloft at 2100 m mainly determines the rain rate reduction near the surface, which is only marginally affected by *Z* aloft in agreement with theoretical calculations by Kumjian and Ryzhkov (2010). With $Z_{DR} < 0.5$ dB at 2100 m, the rain rate at 600 m is reduced by more than 90%. For the initial $Z_{DR} > 2$ dB, 20–40% of the rainfall may be lost to evaporation.

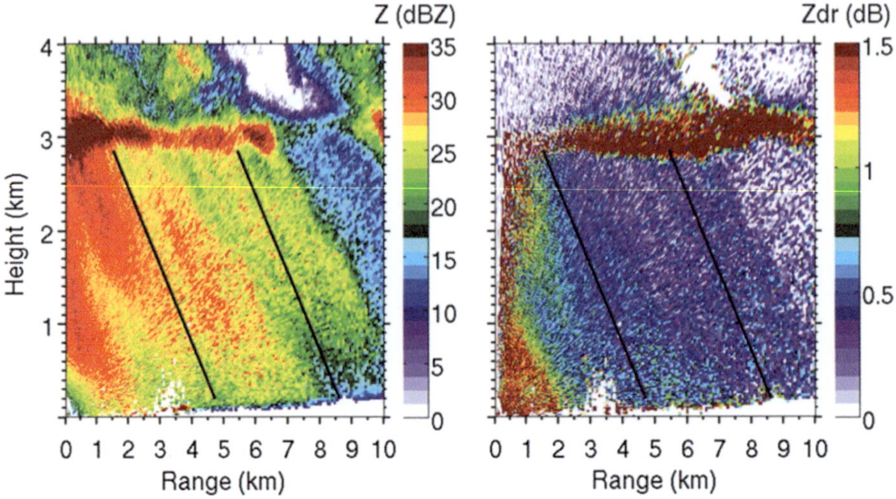

Fig. 7.4 Composite RHI of Z and Z_{DR} observed by the X-band BoXPol radar (University of Bonn) at an azimuth angle of 225° at 1450 UTC 4 July 2014. The solid black lines follow the rain shaft, which is tilted with respect to vertical. From Xie et al. (2016). © American Meteorological Society. Used with permission

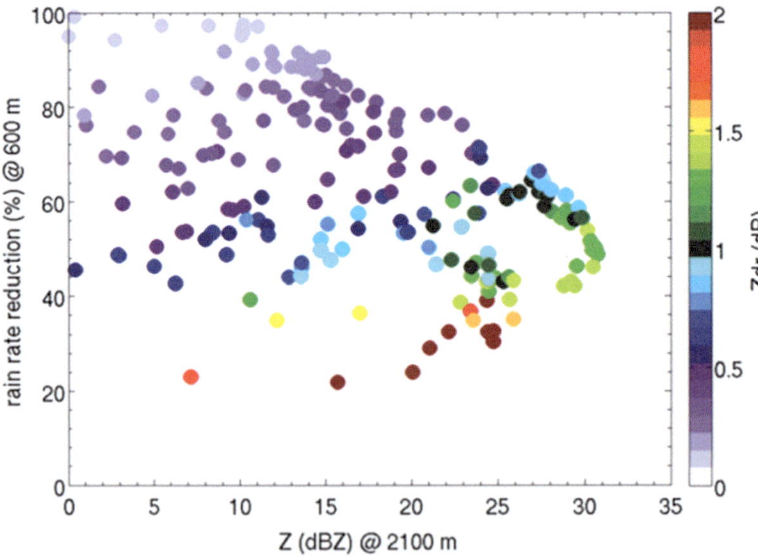

Fig. 7.5 Scatterplot of initial Z at 2100 m and rain rate reduction at 600 m due to evaporation, with the color of the dots indicating the initial Z_{DR} at 2100 m. The values of Z and Z_{DR} have been estimated at 2100 m between 1440 and 1540 UTC 4 July 2014 generated from the MRR-observed DSDs, while the rain rate difference between 2100 and 600 m is derived by the evaporation model. From Xie et al. (2016). © American Meteorological Society. Used with permission

Evaporation of precipitation is considered as one of the most important processes contributing to cooling of the air (Dawson et al. 2010). It contributes to downdraft intensification and thus cold pool generation below precipitation clouds. Cold pools play an important role in storm dynamics, particularly by interacting with the storm inflow as its less dense air is lifted above the cold pools, thus maintaining the updrafts and contributing to the development of new cells. Xie et al. (2016) simulated vertical profiles of cooling rates in evaporating rain and found very strong dependence of the magnitude of the cooling rate and its height dependence on the median volume diameter of raindrop D_0 which can be estimated from differential reflectivity Z_{DR} (see Chap. 11).

7.1.2 Coalescence and Breakup

As precipitation particles fall to the ground, the evolution of the DSD beneath cloud base is governed by four main processes: growth of large drops (and subsequent depletion of small drops) by coalescence, depletion of large drops (and increase in concentration of smaller drops) by collisional and spontaneous breakup, size sorting of drops, and a decrease of drop sizes across the spectrum (and depletion of the smallest drops) by evaporation. In this section, we consider the impact of coalescence and breakup on the polarimetric variables. This can be done in a simplistic thought experiment.

Consider the situation when two raindrops with the same equivolume diameter D_1 collide and coalesce to form one larger drop. The resulting diameter of this large drop is $D_2 = 2^{1/3}D_1$. Assume that the concentration of small drops with diameter D_1 is N_1, so that the mass flux of these equals to $N_1 D_1^3 V_{t1}$. For simplicity, the terminal velocity of the drop is approximated with the power law $V_{t1} = 3.78 D_1^{0.67}$ in which the diameter is in mm (Atlas and Ulbrich 1977). The corresponding reflectivity in the Rayleigh approximation is $N_1 D_1^6$. Upon completion of coalescence, the water mass flux is conserved; therefore, $N_2 D_2^3 V_{t2} = N_1 D_1^3 V_{t1}$. Multiplication of both sides with D_2^3/V_{t2} produces the reflectivity Z_2 in terms of Z_1 as follows:

$$Z_2 = N_2 D_2^6 = N_1 D_1^3 \frac{V_{t1}}{V_{t2}} 2D_1^3 = 2\left(\frac{D_1}{D_2}\right)^{0.67} N_1 D_1^6 \approx 2^{0.78} Z_1, \tag{7.12}$$

i.e., Z_2 is larger than Z_1 by 2.35 dB. Similar considerations can be applied to specific differential phase K_{DP}. Assuming that $K_{DP} \sim D^{\alpha}$ (where α is between 4 and 5), the ratio of K_{DP}s after and before coalescence is equal to $2^{(\alpha/3-1.22)}$ which is larger than 1. Raindrop oblateness increases with size and so does the differential reflectivity Z_{DR}.

It follows from this simple analysis that coalescence increases all three key variables, Z, Z_{DR}, and K_{DP}. Consequently, the opposite effect, raindrop breakup, results in the decrease of Z, Z_{DR}, and K_{DP}. A more detailed analysis of the impact of coalescence and breakup of raindrops on the polarimetric variables is made by Kumjian and Prat (2014) who use a one-dimensional spectral bin microphysical model of a rain shaft.

Coalescence of raindrops is a primary mechanism for generation of tropical (or "warm") rain (Chap. 4). This type of rain can be recognized by strong vertical gradients of Z, Z_{DR}, and K_{DP} near the surface below the cloud base. The vertical structure of precipitation and its temporal evolution can be effectively represented using the "quasi-vertical profiles" (QVP) methodology for processing and visualization of polarimetric radar data (Ryzhkov et al. 2016). It involves azimuthal averaging of the polarimetric variables at high antenna elevation (typically between 10 and 30°) and presenting the QVP profiles in a height-versus-time format. The use of high elevations minimizes the effects of beam broadening and horizontal inhomogeneity. Azimuthal averaging reduces the statistical errors of the estimates without compromising temporal resolution and assigns their average vertical profiles to a conical volume with a vertical axis (Fig. 7.6). The benefits of the QVP technique include ability to monitor temporal evolution of the microphysical processes governing precipitation production with high vertical resolution and to compare polarimetric data with observations made by vertically pointing remote sensors, such as wind profilers, lidars, radiometers, cloud radars, and radars operating on spaceborne and airborne platforms. Tobin and Kumjian (2017) developed an extension of the QVP approach called range-defined QVP methodology by utilizing the data from all available elevation angles.

An example of a composite QVP of Z, Z_{DR}, ρ_{hv}, and K_{DP} measured in the tropical rain associated with the Hurricane Irma that hit Florida on 10 September 2017 is presented in Fig. 7.7. The data at elevation 19.5° have been collected by the KBYX WSR-88D located at Key West, FL. It is evident that the magnitudes of three variables, Z, Z_{DR}, and K_{DP}, increase strongly toward the surface below the melting layer as a result of dominating coalescence process. The enhancement is especially pronounced at about 12 UTC when the reflectivity Z increases from 20 dBZ to 42 dBZ from the altitude of 4 km down to 1 km. Very strong vertical gradients of the

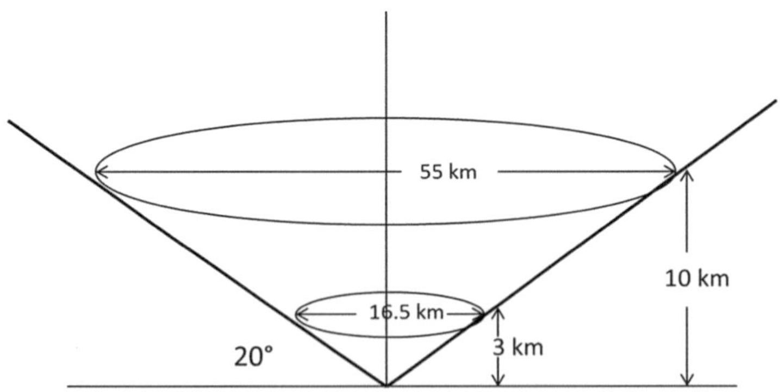

Fig. 7.6 Conical volume representing azimuthally averaged quasi-vertical profiles of radar variables

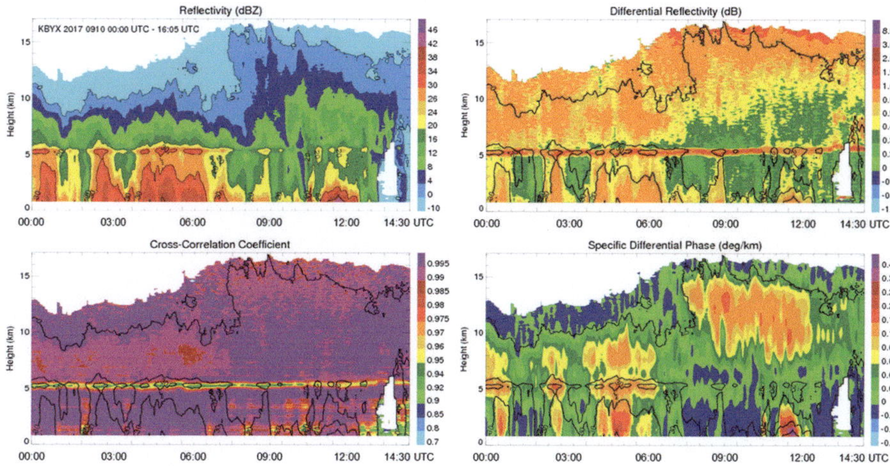

Fig. 7.7 QVPs of Z, Z_{DR}, ρ_{hv}, and K_{DP} measured by the KBYX (Key West, FL) WSR-88D radar during the Hurricane Irma on 10 September 2017. Contours of constant $Z = 0$, 20, 30, and 40 dBZ are overlaid in all panels

polarimetric variables and rain rates in tropical precipitation can cause significant underestimation of rain if the radar beam overshoots a relatively shallow layer of intense surface rain (Wang et al. 2013).

7.1.3 *Size Sorting*

Size sorting of raindrops and hydrometeors is ubiquitous in all precipitation systems. It occurs because hydrometeors exhibit a general increase of terminal fall speed with diameter. Therefore, their trajectories depend on size and the microphysical and dynamical nature of the parent precipitation system (Marshall 1953; Gunn and Marshall 1955; Kumjian and Ryzhkov 2012; Dawson et al. 2015). Most common manifestation of size sorting is the Z_{DR} enhancement combined with the reduction of ρ_{hv} in areas of relatively low Z characterized by low overall concentration of raindrops and drop size spectrum skewed toward larger sizes with relative deficit of smaller drops. Size sorting is generally associated with (a) differential sedimentation, (b) convective updrafts, (c) wind shear, and (d) the difference in the direction of storm motion and wind speed.

The onset of precipitation in clouds starts with the development of particles large enough to sediment relative to cloud droplets and ice crystals. The difference in terminal velocities of large and small particles causes differential sedimentation of precipitating particles. This explains the frequent appearance of big drops beneath developing ordinary convective clouds preceding the onset of heavier precipitation. An example of differential sedimentation signature is presented in Fig. 7.8 where two composite RHIs of Z, Z_{DR}, and ρ_{hv} taken at a time lag of about 2 min are

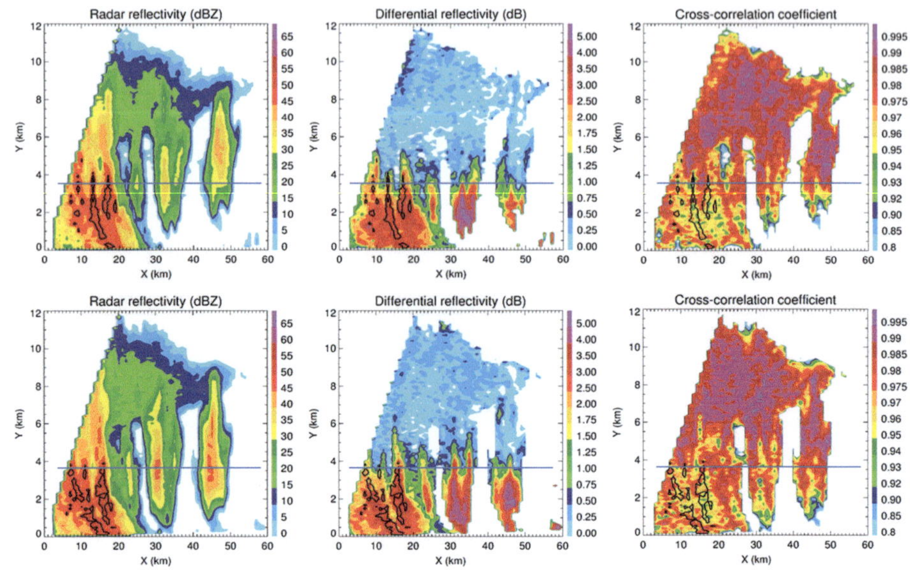

Fig. 7.8 Vertical cross sections of Z, Z_{DR}, and ρ_{hv} in the case of size sorting. S-band KOUN data collected on 13 June 2005 at 0201 UTC (top panels) and 0203 UTC (bottom panels), Az = 191°. Horizontal line marks the freezing level

presented. Two developing convective cells centered at the distance of 33 and 46 km from the radar exhibit classical differential sedimentation signature with extremely high values of Z_{DR} exceeding 5 dB near the bottom of the cells. These anomalously large Z_{DR}s are associated with very low values of Z (less than 20 dBZ) and ρ_{hv} lower than 0.96 and are caused by few largest raindrops which overtake smaller raindrops during their fall.

Another example of differential sedimentation is illustrated in Fig. 7.9 where it is contrasted with the polarimetric signature of coalescence using the data collected by X-band dual-polarization radar. In the left cell in the plot, precipitation still does not reach the ground, and high Z_{DR} combined with low Z is observed at the bottom of this cell. In the neighboring right cell, precipitation shaft is fully formed and displays a typical signature of coalescence characterized by simultaneous increase of both Z and Z_{DR} toward the surface.

The differential sedimentation signature is transient and usually disappears when smaller raindrops start reaching the surface which may take up to 10–15 min after they form in a cloud aloft. The signature has important practical implications and can be used for nowcasting of imminent precipitation enhancement at the surface with 5–15 min lead time by detecting the couplet of high Z_{DR} and low Z at lower antenna scans.

The Z_{DR} column is another size sorting signature that indicates convective updraft. It is identified as a columnar enhancement of Z_{DR} associated with low or moderate Z. If the updraft is strong enough, then smaller-size raindrops cannot fall

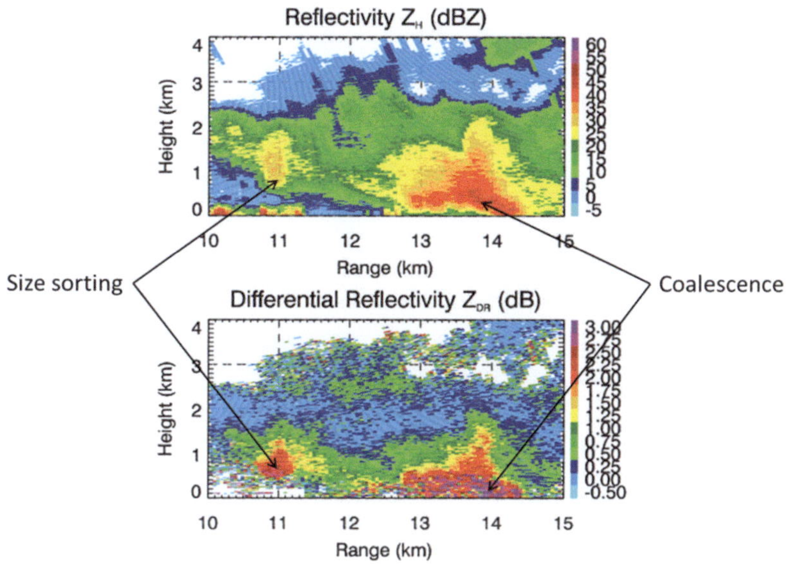

Fig. 7.9 Example of polarimetric signatures of size sorting and coalescence revealed in the X-band BoXPol radar data

through the updraft as do the larger drops which have terminal velocities exceeding the updraft speed. The Z_{DR} columns can be detected at any altitude within the storm where sufficiently strong updrafts exist, but they are most readily detected if they stretch above the environmental freezing level as seen in Fig. 7.10. At a range of about 74 km, along the leading edge of the storm, there exists a tall, columnar region of enhanced Z_{DR}. The 1-dB Z_{DR} contour extends more than 1.5 km above the environmental 0 °C level, which is at ~3.5 km above ground level (AGL). The top of the Z_{DR} column contains supercooled large raindrops which are either grown via coalescence within the updraft or lofted from the base of the updraft up to the levels of subfreezing temperatures where they do not freeze instantaneously (Kumjian et al. 2012, 2014). A more detailed discussion of the Z_{DR} columns can be found in Chap. 8.

Size sorting can be generated and sustained by wind shear (Kumjian and Ryzhkov 2012). The effect of vertical wind shear is to provide nonzero storm-relative flow that advects raindrops away from directly beneath the cloud. Because smaller drops fall slower than larger drops, the smaller drops are exposed to this storm-relative flow for longer periods of time and are thus transported farther downstream than larger drops.

In Kumjian and Ryzhkov (2012), a simple two-dimensional model with spectral bin microphysics is constructed to quantify the changes in polarimetric variables due to the vertical wind shear illustrated in Fig. 7.11. In this model simulation, the storm-relative wind increases from 0 m s^{-1} at cloud base (3 km AGL) to 20 m s^{-1} near the surface. The rain DSD at the top of the domain is assumed to be a Marshall-Palmer

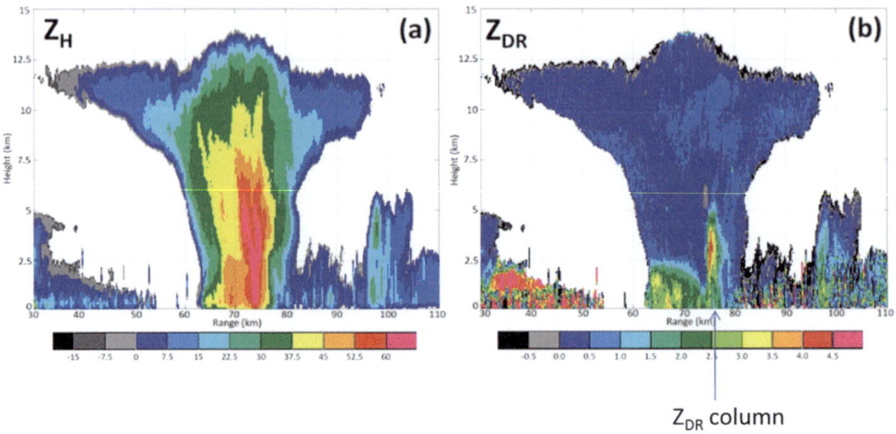

Fig. 7.10 An RHI of (**a**) Z and (**b**) Z_{DR} exhibiting a column. Data were collected by the KOUN WSR-88D on 27 April 2013. The environmental freezing level is at 3.5 km AGL. Adapted from Kumjian et al. (2014). © American Meteorological Society. Used with permission

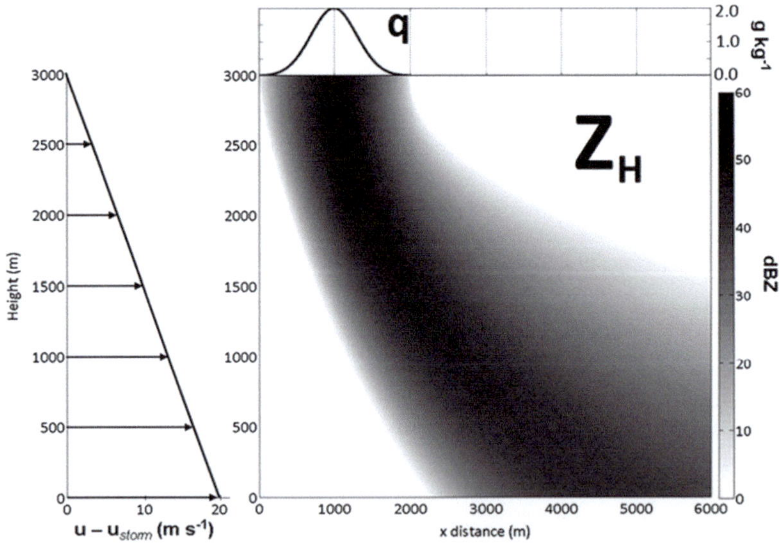

Fig. 7.11 Schematic of the two-dimensional wind shear model configuration: (left) storm-relative wind profile and (right) the domain and resulting Z distribution (shaded; dBZ). The "cloud" rainwater mixing ratio q (g/kg) profile is shown above the domain. Adapted from Kumjian and Ryzhkov (2012). © American Meteorological Society. Used with permission

type with a slope determined by the initial rain mixing ratio q which has a Gaussian spatial distribution (right panel in Fig. 7.11). Simulated fields of Z, Z_{DR}, K_{DP}, and ρ_{hv} in a vertical cross section are shown in Fig. 7.12. Left-to-right advection of the precipitation is evident in all fields. In the Z field, the largest values are confined to

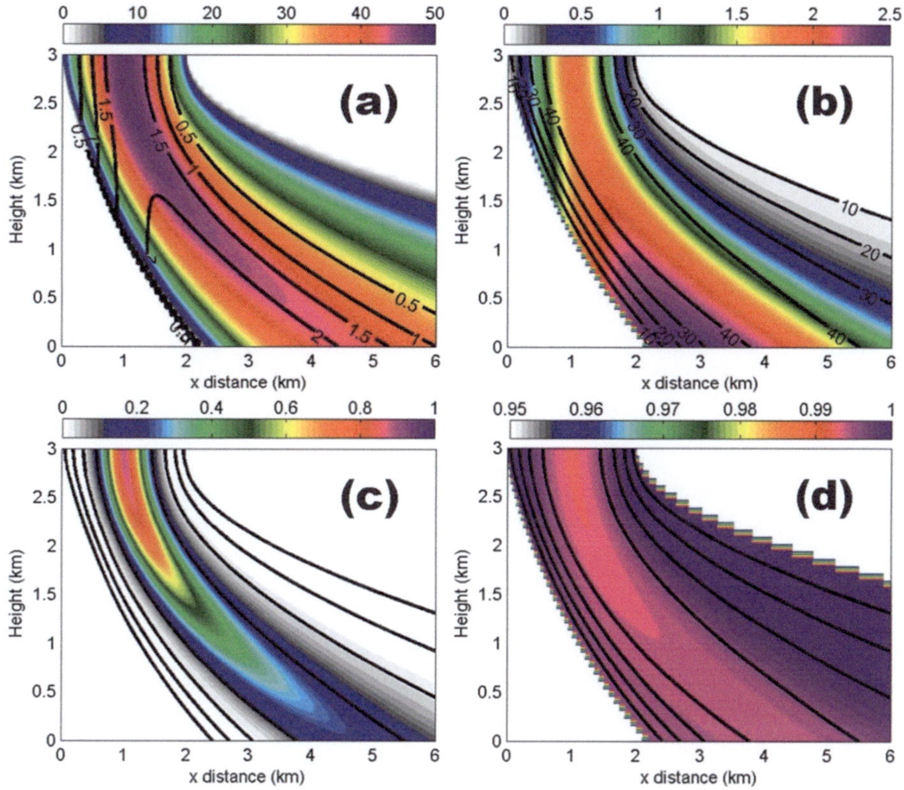

Fig. 7.12 Results from the two-dimensional wind shear model using the bin formulation: panels show the 2D fields of (**a**) Z, (**b**) Z_{DR}, (**c**) K_{DP}, and (**d**) ρ_{hv}. Overlaid on (**a**) are the Z_{DR} contours (0.5–2.5 dB in 0.5 dB increments); (**b**)–(**d**) have Z contours (10–40 dBZ, in 10 dBZ increments) overlaid. Adapted from Kumjian and Ryzhkov (2012). © American Meteorological Society. Used with permission

the center of the rain shaft. The Z_{DR} field reveals where the impact of the size sorting is most readily apparent: the highest values are toward the ground and at the leading edge of the echo along the gradient of Z. K_{DP} closely follows the Z pattern, as expected. The ρ_{hv} field exhibits slightly decreased values toward the leading edge.

The results of model simulations are supported by the observations with the Bonn X-band polarimetric radar (Fig. 7.13). The RHI scan captures an isolated precipitation cell producing light to moderate rainfall into an environment with vertical wind shear. The observed fields of Z and Z_{DR} beneath the melting layer are qualitatively similar to the modeled rain shaft in Fig. 7.11. Namely, the highest Z_{DR} is located at the leading edge of the shaft along a gradient in Z (located at about 29 km in range), whereas the higher Z is offset and coincident with lower Z_{DR} (30–31 km in range).

A case of linear wind shear was examined in the previous example. If the wind direction changes with height which is very common in supercell storms, then size sorting is manifested by the "Z_{DR} arcs" commonly observed near the right (left) edge of the forward flank of right (left)-moving supercells (Kumjian and Ryzhkov 2009).

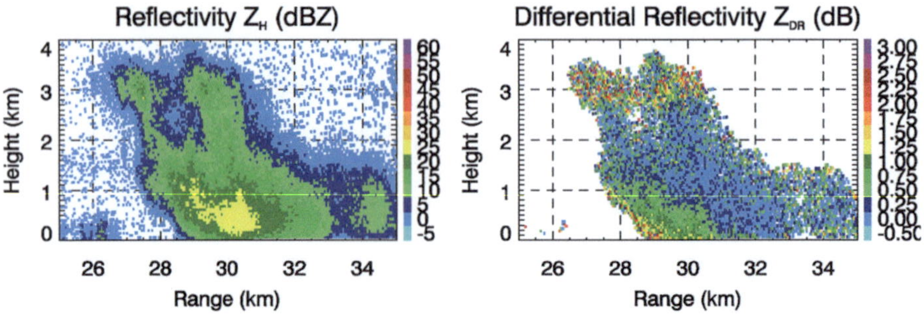

Fig. 7.13 RHI scans of (left) Z and (right) Z_{DR} taken at 0404 UTC 22 June 2011 from the Bonn X-band polarimetric radar. Adapted from Kumjian and Ryzhkov (2012). © American Meteorological Society. Used with permission

A positive correlation has been established between the magnitude of Z_{DR} in the Z_{DR} arc and the magnitude of the storm-relative environmental helicity (SREH) which has important prognostic value regarding the tendency of the supercell development. The Z_{DR} arc is a very important polarimetric attribute of supercell storms which are examined in Chap. 8.

Strictly speaking, wind shear is not required for size sorting (and associated Z_{DR} enhancement) to occur. Dawson et al. (2015) show that the fundamental source of size sorting is the storm-relative wind field itself and, in particular, its mean taken over the depth of the sorting layer.

In a common real-life situation when a driver of a car enters a thunderstorm and experiences splashes of first huge raindrops on the windshield, multiple size sorting-producing factors work simultaneously: differential sedimentation, convective updraft driven by the thunderstorm inflow, and strong wind shear at the leading edge of the storm. All of these contribute to the appearance of anomalously high Z_{DR} coupled with low Z at lower height levels along the leading edge.

It is important to note that single-polarization weather radars can't detect the effects of size sorting, whereas dual-polarization radars are very efficient in identifying and quantifying size sorting and its implications on the microphysical parameterization of the storm-scale numerical weather prediction models.

7.2 Microphysical Processes Involving Ice and Mixed-Phase Particles

Ice and mixed-phase particles are characterized by a huge variety of shapes, orientations, and phase composition which are strongly affected by various microphysical processes. The ice habit diagram in Fig. 7.14 gives an idea about the diversity of ice crystals in the atmosphere. Two key factors determining ice habit are ambient temperature and supersaturation with respect to ice. For example, pristine dendrite crystals grow in a temperature range between -10 and -20 °C at sufficiently high

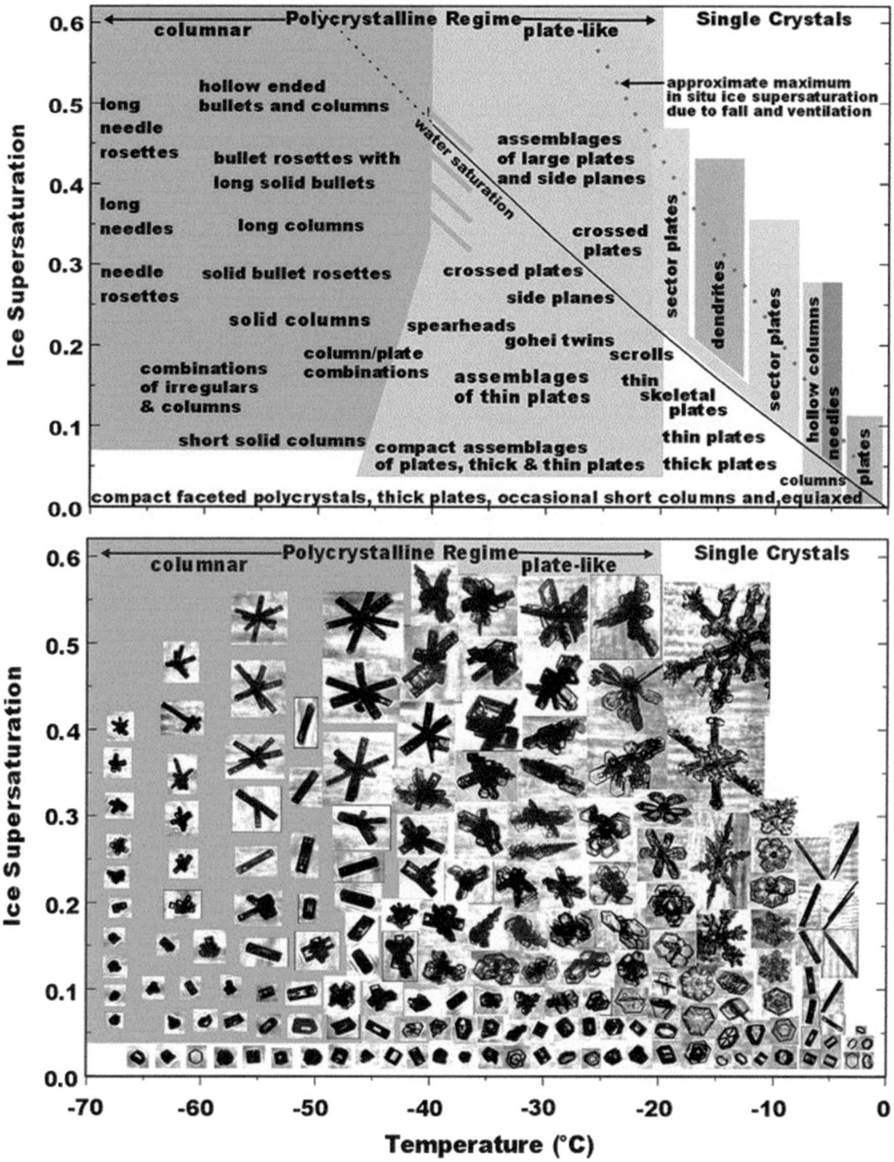

Fig. 7.14 Habit diagram in text and pictorial format for atmospheric ice crystals. Adapted from Bailey and Hallett (2009). © American Meteorological Society. Used with permission

ice supersaturation, whereas hexagonal plates are typically found in the same temperature range but for lower supersaturation with respect to ice. An important water saturation boundary corresponding to relative humidity of 100% with respect to water (top panel in Fig. 7.14) separates different crystal habits across the whole temperature range.

Ice can grow directly from water vapor or from liquid drops at subfreezing temperatures. The first mechanism requires an ice nucleus, which is an aerosol particle with a special crystallographic structure. *Deposition* nucleation of ice occurs when water vapor deposits directly on an ice nucleus. Freezing of supercooled liquid drops of various sizes represents another mechanism of ice initiation or nucleation. Liquid water droplets can freeze spontaneously at sufficiently low temperatures below −40 °C in the absence of impurities or ice nuclei. This process is called *homogeneous freezing nucleation.* For a liquid drop to start freezing at warmer temperatures, it has to contain an ice nucleus which becomes active at certain temperature below 0°C (*immersion freezing*). *Contact freezing* occurs when a supercooled liquid drop (without ice nucleus inside) hits an external ice nucleus which triggers drop freezing.

7.2.1 Depositional Growth/Sublimation

Small-size ice crystals primarily grow by vapor deposition before they become sufficiently large and other processes of ice growth like accretion (riming) or aggregation take over. The equation for depositional growth/sublimation has the form very similar to (7.1) for diffusional growth of liquid drops; it is

$$\frac{dm}{dt} = \frac{4\pi CF(S_i - 1)}{F_K^{(i)} + F_D^{(i)}}, \tag{7.13}$$

where

$$S_i = \frac{e}{E_i(T)}, \tag{7.14}$$

$$F_K^{(i)} = \frac{L_s^2}{KR_vT^2}, \tag{7.15}$$

$$F_D^{(i)} = \frac{R_vT}{D_vE_i(T)}, \tag{7.16}$$

E_i is the saturated water vapor pressure over flat surface of ice, and L_s is the latent heat of sublimation.

Many ice particles have very nonspherical shape with the aspect ratio as low as 10^{-2} which has significant impact on their capacitance C in (7.13) as opposed to liquid drops for which it is approximately equal to $D/2$. If ice particles are modeled by spheroids, their capacitance can be estimated as

$$C_{\text{ob}} = \frac{D}{2} r^{-1/3} \frac{\varepsilon_{\text{ob}}}{\arcsin(\varepsilon_{\text{ob}})} \tag{7.17a}$$

for oblate spheroids and

$$C_{\text{pr}} = \frac{D}{2} r^{2/3} \frac{\varepsilon_{\text{pr}}}{\ln\left[r\left(1 + \varepsilon_{\text{pr}}\right)\right]} \tag{7.17b}$$

for prolate spheroids. In (7.17a) and (7.17b),

$$\varepsilon_{\text{ob}} = \left(1 - r^2\right)^{1/2} \quad \varepsilon_{\text{pr}} = \left(1 - r^{-2}\right)^{1/2}, \tag{7.18}$$

where D is the equivolume diameter and $r = a/b$ is the aspect ratio of the spheroid (a is a rotation axis) so that $r < 1$ for oblates and $r > 1$ for prolates. The dependence of the capacitance of ice particle with $D = 200$ μm on its aspect ratio is presented in Fig. 7.15. It is obvious that the capacitance of the anisotropic ice crystal can be about three times higher than the one of a spherical particle with the same mass and density. The product of the capacitance C and ventilation coefficient F in (7.13) can also change by about a factor of 3–4 depending on the ice habit (Westbrook and Heymsfield 2011).

Another important factor affecting the growth rate of ice crystals is $S_i - 1$ in (7.13). If the air is saturated with respect to water (i.e., RH = 100%), then $e = E_w$ and $S_i - 1$ is proportional to the difference between water and ice saturation pressures $E_w - E_i$ which depends on the temperature (Fig. 7.16). This difference is maximal between −16 °C and −8 °C where the most anisotropic ice crystals

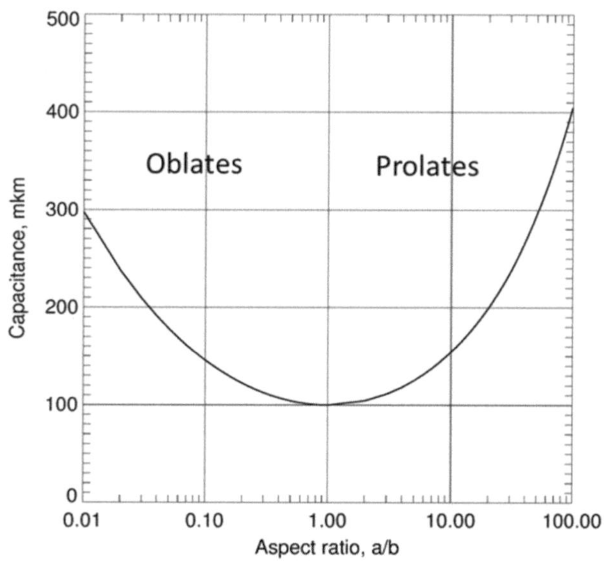

Fig. 7.15 Dependence of ice crystal capacitance on the aspect ratio

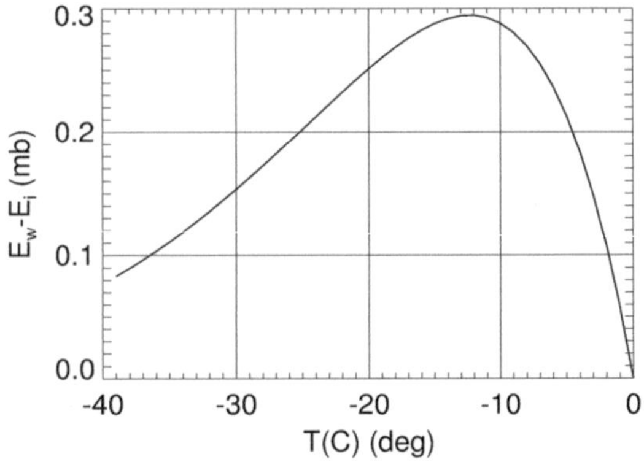

Fig. 7.16 The difference between saturation water vapor pressures over water and ice

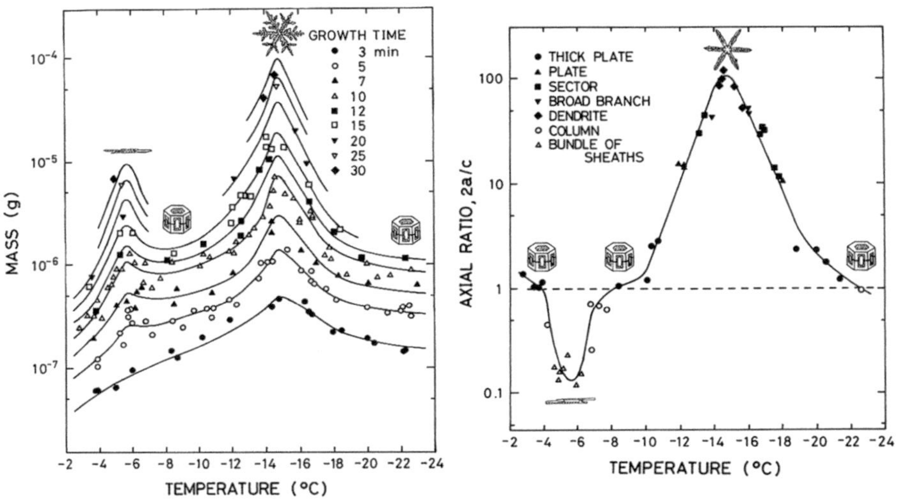

Fig. 7.17 Variation of crystal mass with temperature at different growth times (left panel). Variation of crystal axis ratio with temperature after 10 min of growth (right panel). Adapted from Fukuta and Takahashi (1999). © American Meteorological Society. Used with permission

(approximately oblate) tend to grow according to the Bailey-Hallett habit diagram (Fig. 7.14). The product $CF\,(S_i - 1)$ in (7.13) is highest in this temperature interval, and there also the ice crystals' growth rate is particularly intense.

This fact is supported by the chamber measurements of Takahashi et al. (1991) and Fukuta and Takahashi (1999) illustrated in Fig. 7.17. These measurements show that primary growth maxima occur near -15 °C and -6 °C temperatures that correspond to the dendrite/hexagonal plate and needle regimes. Indeed, ice crystals with the same initial mass acquire an order of magnitude larger mass at $T = -15$ °C than, say, at $T = -24$ °C after 10–20 min growth. Moreover, the fastest growing crystals have the most nonspherical shape (Fig. 7.18, bottom panel) which has direct implications on their polarimetric characteristics. Dendrites and hexagonal plates growing at −15 °C have their aspect ratio as low as 0.01 after 10 min of growth, while needles growing at −6 °C have aspect ratio of 10 after similar depositional growth time.

The Takahashi et al. (1991) and Fukuta and Takahashi (1999) observational results were reproduced in the modeling studies of Chen and Lamb (1994) and Westbrook and Heymsfield (2011) (Fig. 7.18). The Chen and Lamb's calculations of

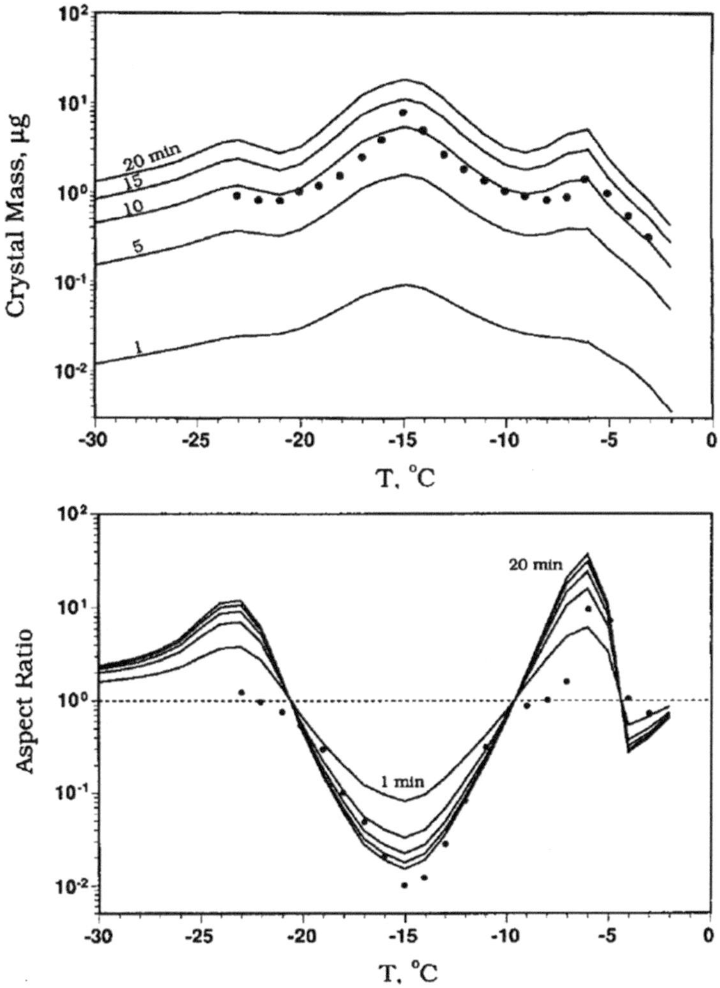

Fig. 7.18 Theoretical simulations of the temperature dependencies of crystal mass (top panel) and aspect ratio (bottom panel) after 1, 5, 10, 15, and 20 min growth. Observational results of Takahashi et al. (1991) after 10 min growth are shown by filled circles. Adapted from Chen and Lamb (1994). © American Meteorological Society. Used with permission

the ice crystal mass dependence on temperature after 10 min of growth time are in a very good agreement with the Takahashi et al. (1991) observations (filled circles in the upper panel in Fig. 7.18). The same authors made a very important conclusion that ice crystals become more anisotropic as they grow (bottom panel in Fig. 7.18) because the most intense growth occurs near their edges (as opposed to basal surfaces) where the gradients of the local field of water vapor pressure are highest. Hence, polarimetric signatures of ice crystals which grow by deposition of water vapor become more pronounced as they grow.

Intense diffusional growth of ice crystals inevitably depletes the excess of available water vapor surrounding crystals so that its pressure can eventually drop below ice saturation pressure E_i and further crystal growth stops. However, this may not be the case in the presence of supercooled liquid droplets coexisting with ice. Once water vapor pressure drops below E_w, liquid droplets start to evaporate releasing additional vapor which keeps water vapor pressure above E_i. The net result is that the ice crystals grow at the expense of the evaporating liquid droplets, until the droplets disappear. This mechanism is called the Wegener-Bergeron-Findeisen (WBF) process. The WBF process is particularly efficient at the temperatures where the difference $E_w - E_i$ is highest, i.e., in the interval between -8 °C and -16 °C.

In summary, physical factors affecting diffusional growth of ice and the results of observational and modeling studies indicate that most vigorous growth of ice is expected in a relatively narrow temperature interval between -10 °C and -15 °C where dendrites grow if supersaturation with respect to ice is sufficiently high. This is confirmed in the majority of dual-polarization radar measurements by significant enhancement of Z_{DR} and K_{DP}, noticeable drop of ρ_{hv}, and strong vertical gradient of Z within the dendritic growth layer (DGL) (Kennedy and Rutledge 2011; Andrić et al. 2013; Bechini et al. 2013; Schrom et al. 2015; Williams et al. 2015; Moisseev et al. 2016; Griffin et al. 2018). Strong DGL polarimetric signatures are documented at S, C, and X bands.

The DGL signature is best exposed in the height vs time format of the quasi-vertical profiles (QVP) of polarimetric variables (Ryzhkov et al. 2016; Griffin et al. 2018). The composite QVP of Z, Z_{DR}, ρ_{hv}, and K_{DP} of a cold-season frontal stratiform cloud observed by the S-band WSR-88D is presented in Fig. 7.19. The melting layer is clearly marked by enhanced Z, and Z_{DR} as well as reduced ρ_{hv} at the height of about 2 km. Notable is clear delineation of the melting layer in Z_{DR} and ρ_{hv}, during the interval between 0 and 4 UTC in which the traditional "bright band" in Z is absent. This underlines the benefit of using polarimetric measurements for identification of the melting layer.

Another distinctive feature, a dendritic growth layer, is centered at the height of about 5 km at the temperatures between $-10°$ and $-15°$ and is manifested by high Z_{DR}, low ρ_{hv}, and strong vertical gradient of Z. The highest values of Z_{DR} are observed during the "shallow" phase of the storm with warmer cloud tops, whereas the highest K_{DP} occurs during the period of much taller cloud with colder cloud tops. Essentially, Z_{DR} and K_{DP} in the DGL are anticorrelated and depend on the temperature at the top of the cloud.

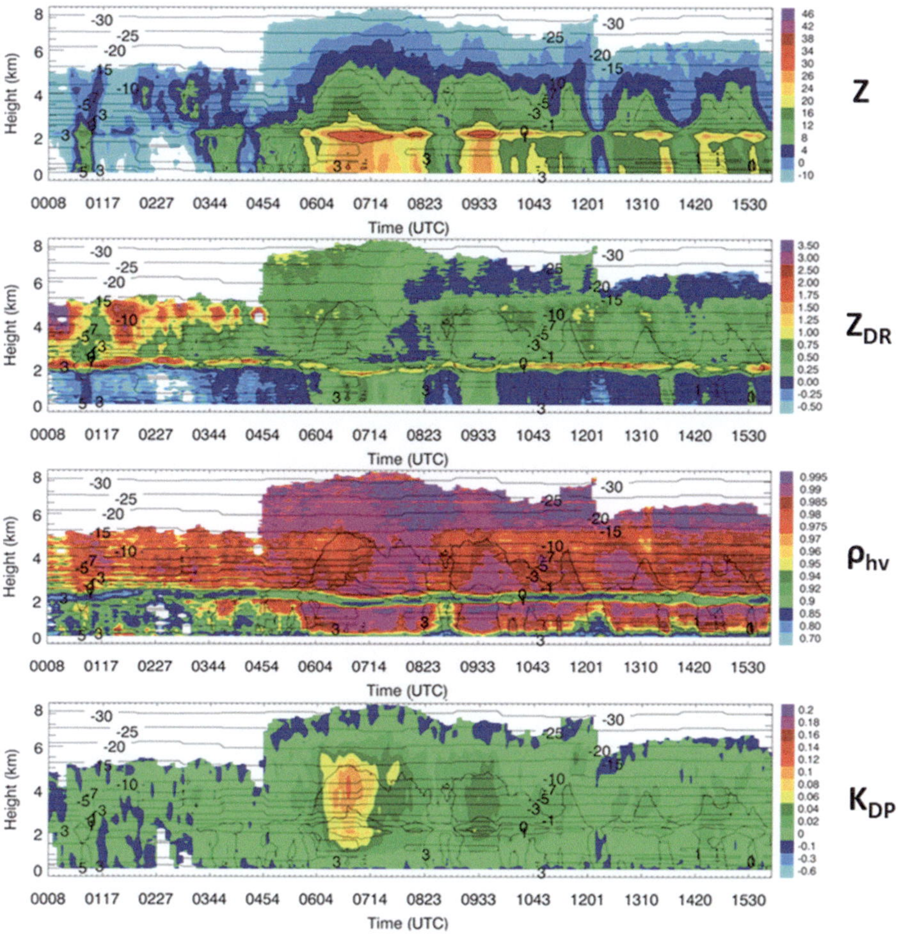

Fig. 7.19 Quasi-vertical profiles of Z, Z_{DR}, ρ_{hv}, and K_{DP} retrieved from the KFFC (Atlanta GA) WSR-88D data collected at antenna elevation 9.9° on 2 February 2014. Contours of HRRR model wet-bulb temperature (°C) are overlaid in each plot. Also, Z is contoured at 10, 20, 30, and 40 dBZ. Courtesy of Drs. P. Zhang and E. Griffin

To quantify the QVP observations within DGL, Griffin et al. (2018) computed 90th percentile maxima of Z_{DR}, K_{DP}, and Z in the DGL in five cold-season storms (Fig. 7.20). The data demonstrate the distinct correlations of the cloud top temperature with Z_{DR} and K_{DP} signatures in DGLs. The statistics also show that larger Z_{DR} is associated with smaller Z in the DGL, while larger K_{DP} is usually combined with larger Z_H in the DGL.

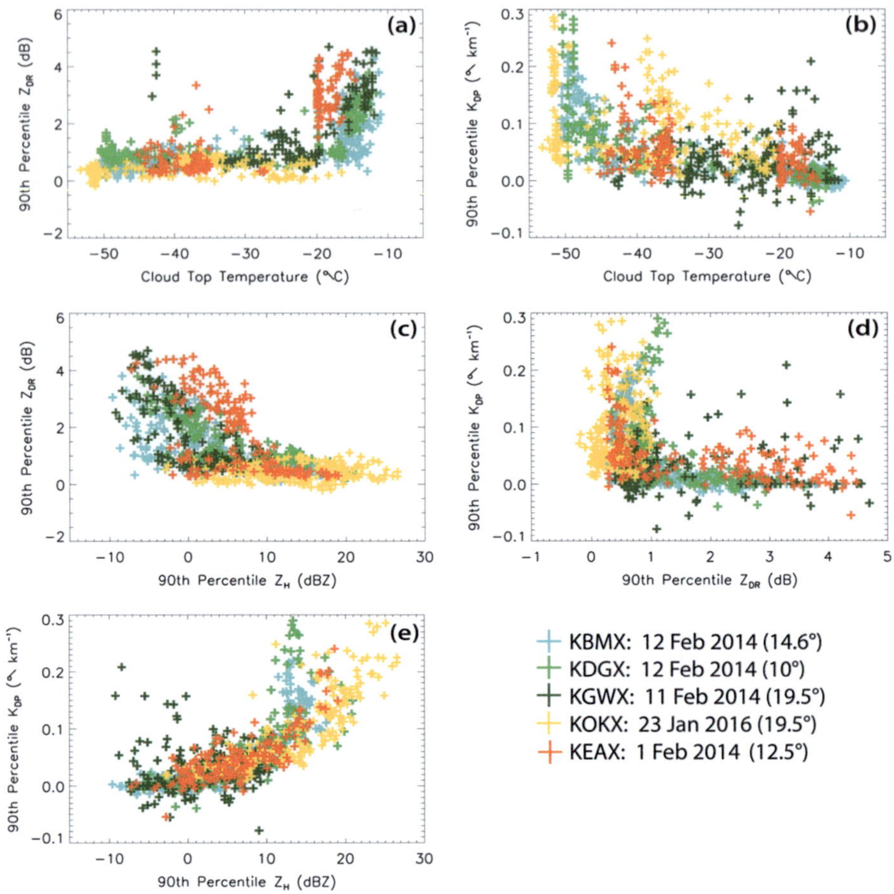

Fig. 7.20 Composite scatterplots of (**a**) cloud top temperature (°C) vs 90th percentile maximum Z_{DR} in the DGL, (**b**) cloud top temperature (°C) vs 90th percentile maximum K_{DP} in the DGL, (**c**) 90th percentile maximum Z_H in the DGL vs 90th percentile maximum Z_{DR} in the DGL, (**d**) 90th percentile maximum Z_{DR} in the DGL vs 90th percentile maximum K_{DP} in the DGL, and (**e**) 90th percentile maximum Z_H in the DGL vs 90th percentile maximum K_{DP} in the DGL. Data from the events observed with the KBMX (Birmingham AL), KDGX (Jackson MS), KGWX (Columbus OH), KOKX (Upton NY), and KEAX (Kansas City MO) WSR-88Ds are indicated by the light blue, light green, dark green, yellow, and orange symbols in each plot, respectively. Adapted from Griffin et al. (2018). © American Meteorological Society. Used with permission

The QVP results can be interpreted taking into account the distinct polarimetric characteristics of quasi-spherical or isometric (I-type) and dendritic (D-type) ice particles. I-type particles include a broad category of snow aggregates and ice crystals with irregular or nearly spherical shapes and can produce moderate Z_{DR} and significant K_{DP} (if the concentration of isometric ice is sufficiently high).

The D-type crystals are comprised of highly oblate (dendrites or hexagonal plates) or prolate (needles) hydrometeors that have very anisotropic shape and higher density than I-type ice particles. These ice crystals can exhibit extremely large Z_{DR} and tangible K_{DP} (if their concentration is sufficiently high). As opposed to the I-type crystals, which can be generated at any level in a cloud, the D-type crystals are generated only in certain temperature ranges, with dendrites and hexagonal plates between -20 and -10 °C and needles between -3 and -8 °C. We submit that in the regions of high Z_{DR}, D-type ice is dominant, whereas in the regions of high K_{DP}, the I-type ice prevails which does not exclude the presence of D-type crystals as well.

7.2.2 Aggregation of Snowflakes

Once ice particles collide, they may stick together and make a bigger particle—their aggregate. Aggregation occurs when some of ice particles grow large enough and acquire larger terminal velocity compared to other ice particles which increases the likelihood of collision. Large aggregates of pristine ice crystals are commonly called snowflakes. The intensity of aggregation depends on the crystal habits and temperature. Intricate crystals, such as dendrites, aggregate or adhere very efficiently, whereas solid plates tend to rebound. The aggregation efficiency of two colliding crystals increases with increasing temperature, with adhesion being particularly likely above -5°C at which temperatures ice surfaces become "stickier."

Aggregation of ice crystals is similar to coalescence of raindrops, and we can use a thought experiment similar to the one described in Sect. 7.1.2 to evaluate the tendencies in the radar polarimetric variables caused by aggregation of snowflakes. When two snowflakes with the same equivolume diameter, density, and shape aggregate, their resulting equivolume diameter increases by a factor $2^{1/2}$ instead of $2^{1/3}$ for raindrops because the density of snowflakes decreases inversely proportional to their diameter (i.e., $\rho \sim D^{-1}$). This is also why the reflectivity of snow is proportional to D^4 compared to D^6 for raindrops (Chap. 5). Due to these physical properties of snow, the mass flux conservation relation between snow with sizes D_1 and sizes D_2 is $N_2 D_2^2 V_{t2} = N_1 D_1^2 V_{t1}$. After multiplying both sides with D_2^2/V_{t2} similarly to deriving (7.12), the relation between the reflectivities corresponding to these two sizes becomes

$$Z_2 = N_2 D_2^4 = N_1 \frac{V_{t1}}{V_{t2}} D_1^2 2 D_1^2 = 2\left(\frac{D_1}{D_2}\right)^{0.18} N_1 D_1^4 = 2^{0.91} N_1 D_1^4 = 2^{0.91} Z_1. \quad (7.19)$$

In (7.19), the terminal velocity of snowflake V_t is roughly proportional to $D^{0.18}$ (Zawadzki et al. 2010). It follows from (7.19) that upon aggregation, the reflectivity increases by 2.74 dBZ.

Because the aggregated snowflakes have lower density, their differential reflectivity decreases due to the decrease in the magnitude of the refractive index. This is opposite to the case of raindrops for which coalescence increases Z_{DR}. It is significant that the specific differential phase of snow also decreases when aggregation occurs. Indeed, because $K_{DP} \sim D$ (see Eq. (5.27)) for snowflakes, a derivation similar to the one that produced (7.19) can be applied to K_{DP}. It yields

$$K_{DP2} = N_2 D_2 = N_1 \frac{V_{t1}}{V_{t2}} \frac{D_1}{2^{1/2}} = 2^{-0.59} N_1 D_1 = 2^{-0.59} K_{DP1}. \qquad (7.20)$$

Thus, K_{DP} after aggregation is about 34% smaller than before aggregation. This tendency is opposite to the one caused by coalescence of raindrops.

Large snowflakes growing via aggregation have larger Reynolds numbers which affect their falling behavior and orientation which becomes more chaotic. This is an additional factor which reduces Z_{DR} and K_{DP} in the process of aggregation. Nevertheless, it is a tremendous drop in the bulk density of particles during aggregation (almost two orders of magnitude down to 0.01 g cm^{-3}) that is a dominant factor driving Z_{DR} and K_{DP} to almost zero values.

The tendencies of increasing Z and decreasing Z_{DR} and K_{DP} due to aggregation of ice crystals can be well illustrated in the QVP representation of polarimetric data from pure snow (Fig. 7.21). The dendritic growth layer is detectable at the heights between 5 and 6 km via the Z_{DR} and K_{DP} enhancement and ρ_{hv} reduction. As in a previous QVP example (Fig. 7.19), such increase of Z_{DR} in the DGL is less prominent during the period when higher K_{DP} indicates the dominance of the I-type ice initially formed near the cloud top. Steady increase of Z and ρ_{hv} accompanied by gradual reduction of Z_{DR} and K_{DP} toward the surface is a clear-cut polarimetric "fingerprint" of snow aggregation. Notable are the appearances of the elevated melting layer at about 2020 UTC and 2300 UTC marked by the local increase of Z_{DR} and the ρ_{hv} depression coincident with the pockets of warm air at the height of 2.5 km retrieved from the HRRR model. It is not possible to detect such temperature inversions in the Z fields.

Similar tendencies are also visible in the hemispherical RHIs of Z and Z_{DR} measured by an X-band dual-polarization radar in polar clouds at the North Slope of Alaska (Oue et al. 2016) (see Fig. 7.22). In polar regions, the overall concentration of ice crystals is usually much lower than in midlatitudes due to very low concentration of ice nuclei there. This explains why pristine crystals do not aggregate much and the depositional growth from water vapor is often a dominant process of ice formation in the polar regions. Consequently, pristine ice crystals in polar clouds may be characterized by very high values of Z_{DR} (up to 6–7 dB) within the whole depth of the cloud (left panel in Fig. 7.22). This is in striking contrast with a typical midlatitude pattern illustrated in Figs. 7.19 and 7.21. Once aggregation starts, then Z increases, and Z_{DR} decreases toward the surface in a manner similar to the tendencies observed in midlatitudes (right panel in Fig. 7.22).

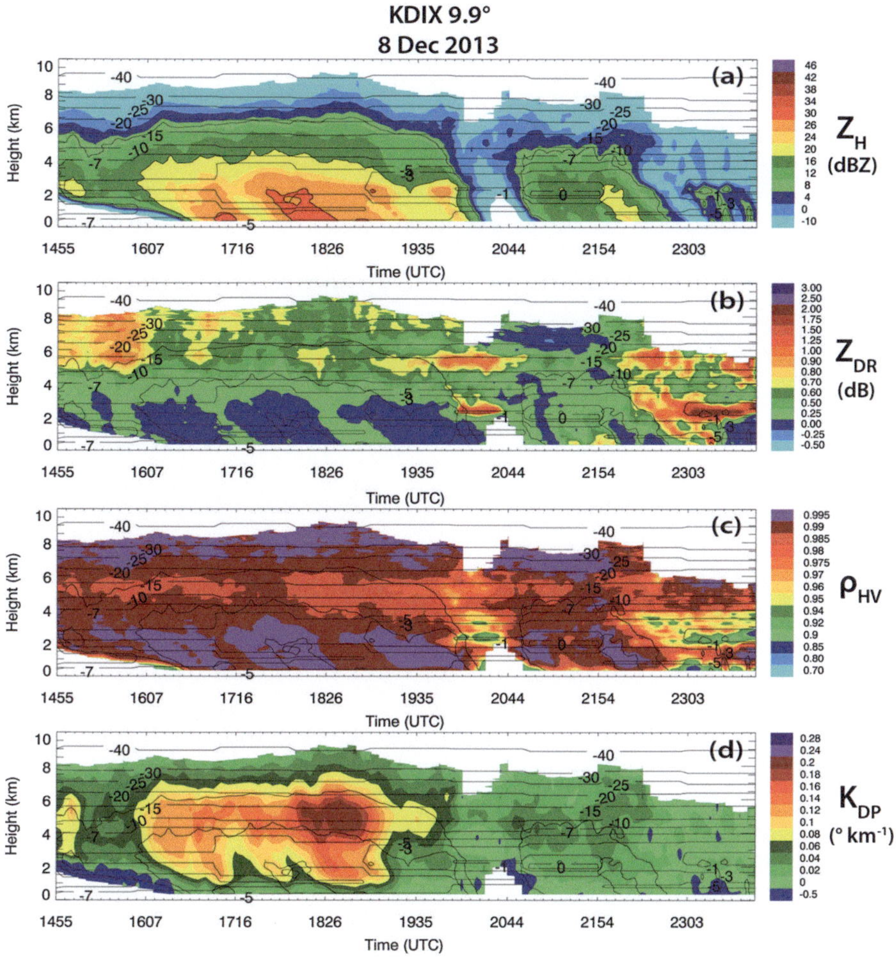

Fig. 7.21 Quasi-vertical profiles of (**a**) Z, (**b**) Z_{DR}, (**c**) ρ_{hv}, and (**d**) K_{DP} retrieved from the KDIX (Fort Dix NJ) WSR-88D data collected at antenna elevation 9.9° on 8 December 2013 in pure snow. Contours of HRRR model wet-bulb temperature (°C) are overlaid in each panel. Courtesy of Drs. P. Zhang and E. Griffin

7.2.3 Riming/Accretion

Ice crystals can grow by capturing small supercooled cloud droplets. This process is called riming or accretion. Riming plays important role in snow formation. Moisseev et al. (2017) found that riming is responsible for 5–40% of snowfall mass after examining 22 snow events in Finland. Other authors (e.g., Grazioli et al. 2015) cite even larger percentages. Mosimann et al. (1994) quantify riming process in terms of 5 degrees of riming f_{rim} as illustrated in Fig. 7.23.

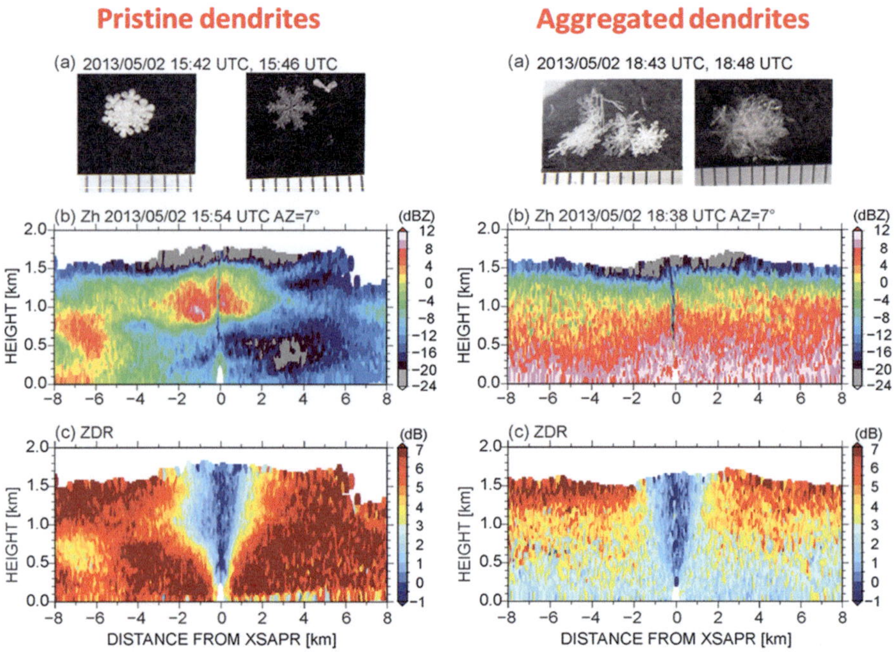

Fig. 7.22 Hemispheric RHIs of (**b**) Z (top panels) and (**c**) Z_{DR} (bottom panels) measured by the X-band radar in Alaska on 2 May 2013 during precipitation periods dominated by (**a**) pristine dendrites (left panels) and (**a**) aggregated dendrites (right panels). Adapted from Oue et al. (2016). © American Meteorological Society. Used with permission

Unrimed crystals are characterized by $f_{\text{rim}} = 1$, whereas for extreme case of riming (i.e., graupel), $f_{\text{rim}} = 5$. Riming tends to increase the density and aspect ratio of ice particles, i.e., makes them denser and more spherical. Although the impact of riming on density and aspect ratio is very difficult to quantify, some simplistic models can estimate roughly the effect of riming on the polarimetric variables. Zawadzki et al. (2005) assume that the density of rimed crystal is equal to the product of the density of original unrimed crystal and f_{rim} up to the density accepted for graupel (0.4–0.5 g cm^{-3}) (see Eq. (4.26)).

Identifying polarimetric fingerprints of snow riming is difficult. On the one hand, riming increases the particle density and therefore the values of Z_{DR} and K_{DP}. In Chap. 5, it is shown that Z_{DR} of low-density snow is approximately linearly proportional to the snow density (Eq. (5.19)) for a given snowflake shape (or aspect ratio). On the other hand, riming makes ice particles more spherical (Matrosov et al. 2001, 2012) which promotes the opposite tendency, i.e., decrease of Z_{DR} and K_{DP}. The net effect is mostly a decrease of Z_{DR} as polarimetric measurements testify although the opposite tendency is occasionally observed. Vogel et al. (2015), Ryzhkov et al. (2016), Kumjian et al. (2016), and Giangrande et al. (2016) claim that Z_{DR} of rimed snow is slightly lower than Z_{DR} of aggregated snow, but this difference is quite small (typically 0.2–0.4 dB). Snow riming usually

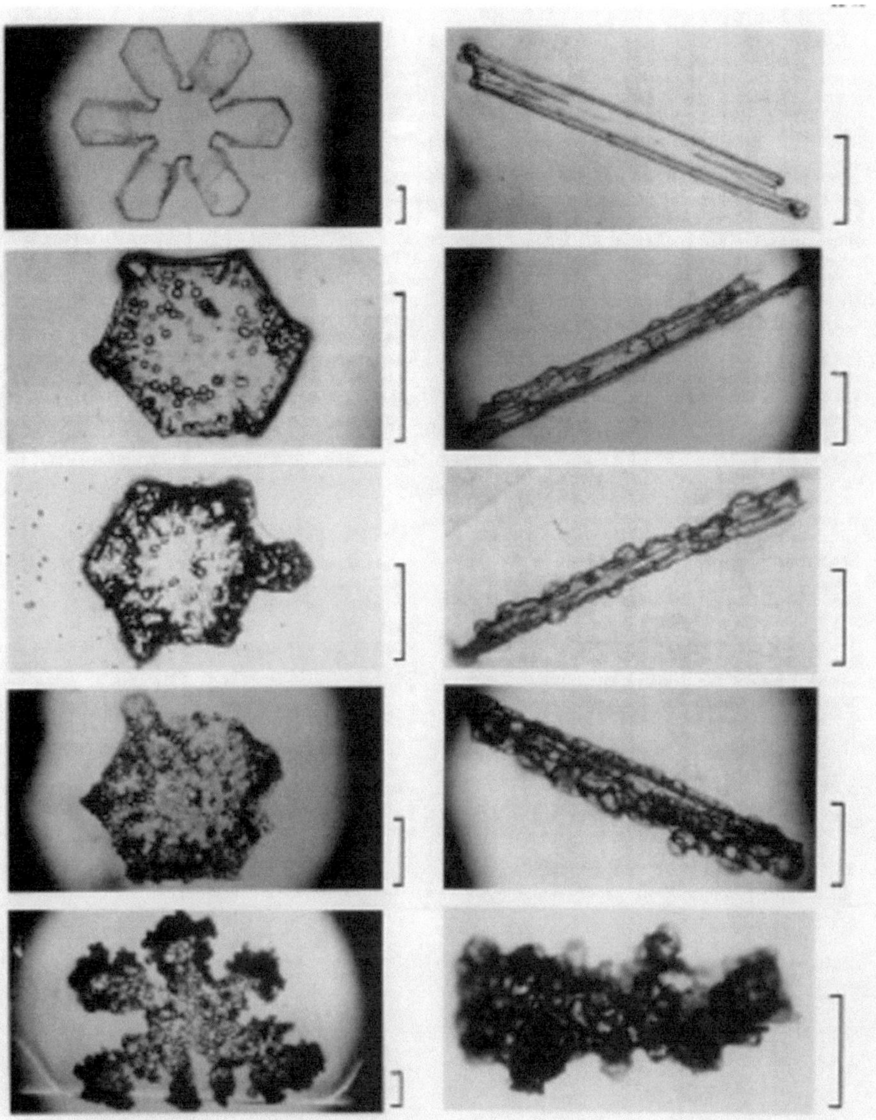

Fig. 7.23 Photographs of snow crystals representing the first five riming categories from top to bottom. The top panels show unrimed crystals ($f_{\mathrm{rim}} = 1$), whereas the bottom panels show heavily rimed crystals ($f_{\mathrm{rim}} = 5$). Left column, planar crystals; right column, long column and needle crystals. The scale to the right of each photo is 300 μm. Adapted from Mosimann et al. (1994). © American Meteorological Society. Used with permission

occurs in two stages. During the first stage, rimed water fills the cavities of the porous snow particle without much change of its size and shape (Erfani and Mitchell 2017). Therefore the density effect may be dominant, and Z_{DR} may increase during the first stage of riming. Once all internal vaults are filled up, the particle start growing from the outside, and the shape effect may dominate which causes the decrease of Z_{DR}.

Z_{DR} usually exhibits lowest value very close to 0 dB in rimed snow just above the melting layer that is apparently "sagging" during intense riming periods (Kumjian et al. 2016). These "sagging" episodes associated with the reduction of Z_{DR} above the freezing level are clearly visible in the examples of the QVP profiles presented in Fig. 7.24 (between 21 and 23 UTC) and Fig. 7.19. Saggy melting layer signatures are produced when dense, fast-falling, and more isometric rimed snowflakes descend into the melting layer and completely melt at lower altitudes compared to unrimed snow. Because the correlation coefficient ρ_{hv} commonly increases with decreasing Z_{DR}, the Z_{DR} depression caused by riming usually coincides with increased ρ_{hv} above the freezing level as can be seen in Figs. 7.19 and 7.24. Kumjian et al. (2016) and Ryzhkov et al. (2016) attribute sagging of the melting layer to riming as one of its possible causes. However, recent theoretical study of Carlin (2018) demonstrates that such sagging can be due to heavy aggregation of snow aloft in the absence of riming.

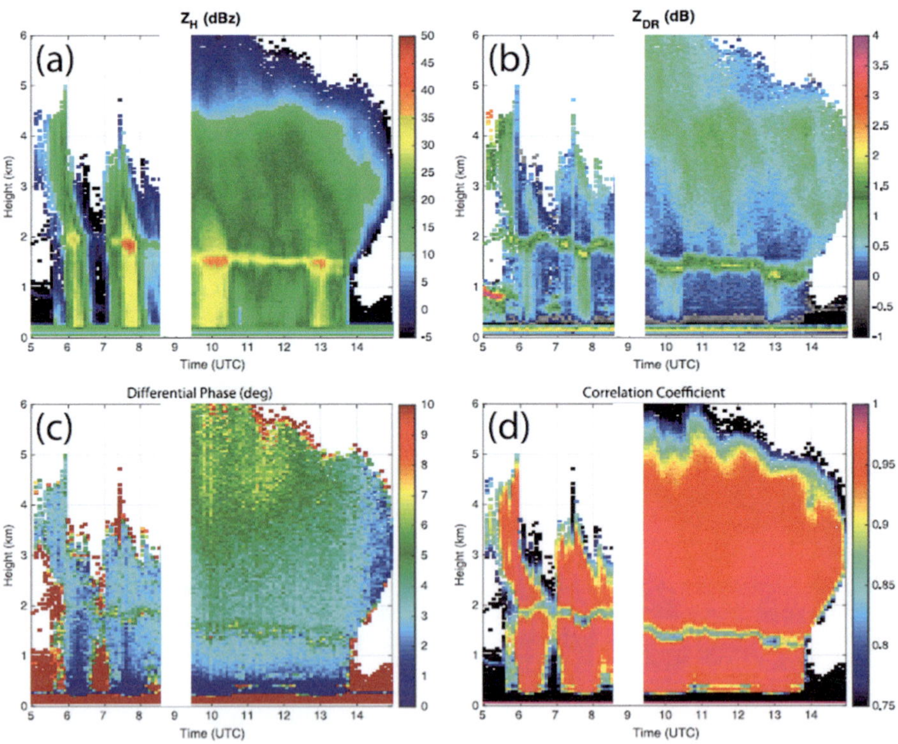

Fig. 7.24 Time series of quasi-vertical profiles of (**a**) Z, (**b**) Z_{DR}, (**c**) Φ_{DP}, and (**d**) ρ_{hv} illustrating "sagging" bright band associated with riming aloft. The data have been collected by a C-band radar in northern Oklahoma. Adapted from Kumjian et al. (2016). Copyright 2016; with permission from Elsevier

Although riming is generally accompanied by the decrease of Z_{DR} and increase of ρ_{hv}, the tendency can be opposite if riming causes generation of secondary ice. One of the important mechanisms for secondary ice production is the Hallett-Mossop process (Hallett and Mossop 1974). Once an ice particle is large enough to collide with supercooled cloud droplets, it accretes rime ice, some of which may shed small ice splinters which can further grow by deposition as very nonspherical ice particles (usually needles). Hallett and Mossop showed that approximately one splinter is released from a rimed particle for every 250 droplets if the particle is at least 0.5 mm in diameter and accreted droplets are larger than 25 μm. The process is active only in a narrow interval of temperatures between −3 °C and −8 °C where the resulting splinters grow as needles (see Fig. 7.14) which are characterized by high Z_{DR} and K_{DP}. Initial splinters are much smaller than a primary ice from which they are released, but their concentration can be quite high which may result in significant K_{DP} and enhanced Z_{DR}. Differential reflectivity Z_{DR} of the mixture of small ice needles and more isometric and large primary ice depends on their relative masses, whereas K_{DP} is more immune to the presence of quasi-spherical ice particles. Hence the Hallett-Mossop secondary ice production caused by riming can be marked by noticeable K_{DP} and increased Z_{DR} if riming is not strong and the contribution of secondary ice to Z_{DR} is comparable to the one from the primary rimed ice particles of larger size. The latter condition is usually satisfied for relatively weak riming. This was confirmed in the recent studies of Vogel et al. (2015), Kumjian et al. (2016), Giangrande et al. (2016), and Sinclair et al. (2016).

If a convective updraft is relatively strong and riming is heavy, Z_{DR} of snow (or graupel) remains very low and close to 0 dB. Generation of large graupel or hail is caused by extreme riming in very intense convective updrafts. If hail is dry, then its Z_{DR} and K_{DP} are close to zero (Chap. 5). The cross-correlation coefficient ρ_{hv} of dry graupel/hail is usually high (>0.95) (Kaltenboeck and Ryzhkov 2013; Ortega et al. 2016) but is generally lower than in pure rain. The situation changes dramatically if graupel or hail contains some water. This happens at positive temperatures when hail is melting or at negative temperatures if hail is growing in the wet regime. In both cases, water first appears at the surface of hailstone and then soaks air cavities inside it if its initial density is less than the density of solid ice (0.92 g cm^{-3}) or forms a liquid shell at its surface if hail density is equal to 0.92 g cm^{-3}. As discussed in Chap. 5, the magnitudes of Z_{DR} and K_{DP} in wet hail can be very large, and both may become negative if hailstones are really big. The corresponding values of ρ_{hv} are usually very low.

Wet growth of hail occurs if its surface temperature is either zero or positive although the temperature of ambient air is well below 0 °C. Surface warming is caused by the release of latent heat of fusion due to the freezing of supercooled cloud droplets collected by hailstone. Hence wet growth of hail is caused by extreme riming in strong convective updrafts. Analysis of surface energy balance indicates that only sufficiently large hailstones can grow in a wet growth regime. Heating rate due to accretion on spherical hailstones is given by the following equation (e.g., Lamb and Verlinde 2011):

$$\frac{dQ_L}{dt} = \frac{\pi}{4} D^2 E_c \ LWC \ V_t(D)\left[L_f - c_w(T_h - T_a)\right]. \tag{7.21}$$

In (7.21), D is the hailstone diameter, E_c is the collection efficiency, LWC is the cloud water content, L_f is the latent heat of fusion, c_w is the specific heat of water, $V_t(D)$ is the hailstone terminal velocity, and T_h and T_a are the surface temperature of a hailstone and ambient air. The term containing c_w accounts for warming of accreted water. This added heat warms the surface which also causes some heat to leave the surface by conduction and sublimation. The heat rates of conduction and sublimation are

$$\frac{dQ_v}{dt} = 2\pi DKF_h(T_h - T_a) \tag{7.22}$$

and

$$\frac{dQ_s}{dt} = 2\pi DL_v D_v F_v(\rho_{\mathrm{vh}} - \rho_{\mathrm{va}}). \tag{7.23}$$

In (7.22) and (7.23), K is the thermal conductivity of air, D_v is the diffusivity of water vapor in air, L_v is the latent heat of vaporization, F_h and F_v are the thermal and vapor ventilation coefficients, and ρ_{vh} and ρ_{va} are the vapor densities near hailstone surface and in the ambient air.

The surface of hailstone is warming if the heat released from accretion exceeds the heat loss via conduction and sublimation:

$$\frac{dQ_L}{dt} > \frac{dQ_v}{dt} + \frac{dQ_s}{dt}. \tag{7.24}$$

Once the surface temperature of hailstone reaches 0 °C, wet growth of hail starts. The necessary condition for this requires that left and right sides of (7.24) are equal at $T_h = 0$ °C. This condition is known as the Schumann-Ludlam threshold for wet growth. It follows from (7.21) that heating rate due to accretion is proportional to $D^{2.8}$ because $V_t(D) \sim D^{0.8}$ for hail (Lamb and Verlinde 2011). The dependencies of the cooling rates due to conduction and sublimation on the hailstone diameter are determined by the products of D and $F_{h,v}(D)$. It can be shown that for typical hail sizes, $F_{h,v} \sim (N_{\mathrm{Re}})^{1/2} \sim [DV_t(D)]^{1/2} \sim D^{1.9}$, i.e., the cooling rates increase more slowly than the warming rate dQ_L/dt with the hailstone diameter. This means that the condition (7.24) holds only for sufficiently large hail. The Schumann-Ludlam condition for wet growth is illustrated in Fig. 7.25 where the minimal graupel/hail diameter for wet growth is depicted as a function of the environmental temperature T_a and supercooled water content LWC. For example, at $T_a = -20$ °C and $LWC = 4$ g m^{-3}, only hailstones larger than 17 mm can grow in the wet regime.

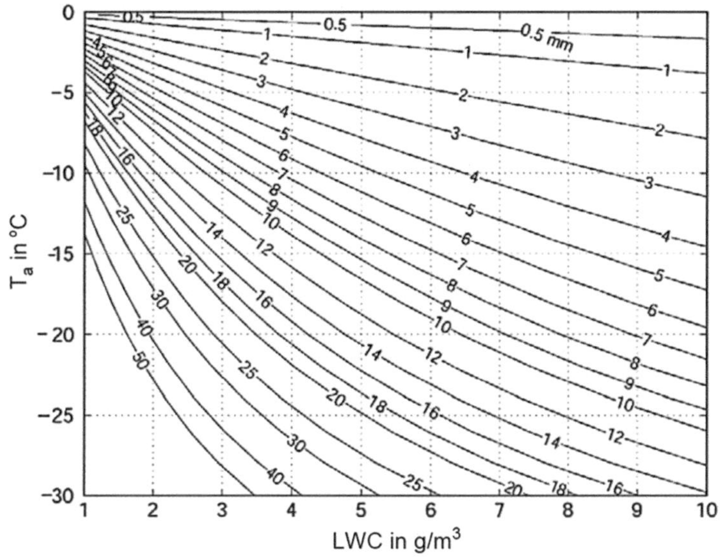

Fig. 7.25 Threshold graupel/hail diameter for wet growth as a function of the environmental temperature T_a and supercooled water content *LWC*. Adapted from Khain et al. (2011). © American Meteorological Society. Used with permission

Wet growth of hail can be identified by large absolute values of Z_{DR} (positive or negative) combined with very low values of ρ_{hv} above the freezing level. Because hail growth via accretion occurs in convective updrafts, typical polarimetric signatures of wet hail are usually observed near the tops or in the close proximity of the Z_{DR} columns. Because wet growth of hail is possible only for large hailstones, the corresponding polarimetric signature can be used not only for hail detection but also for determination of its size and for locating particularly strong convective updrafts. Polarimetric signatures associated with wet hail growth are illustrated in Chap. 8.

7.2.4 *Melting of Snowflakes and Graupel/Hail*

Melting of hydrometeors produces very pronounced polarimetric signatures due to strong changes of dielectric constant, size, density, and shape. Polarimetric characteristics of the melting layer contain very important information about microphysical properties of ice aloft and rain below. Appropriate interpretation of the vertical profiles of polarimetric variables through the melting layer provides invaluable insight into microphysical processes governing precipitation formation.

In case of stratiform rain formed from a low-density ice aloft, complete melting occurs in a relatively thin layer of the atmosphere and is associated with a maximum of reflectivity known as the "bright band." In the situation of convective rain, ice aloft is dominated by higher density graupel or hail falling with large terminal velocities, and their melting takes place in a much deeper layer often stretching down to the surface. In this section, we consider the stratiform and convective cases separately.

7.2.4.1 Melting of Snowflakes: Bright Band

The bright band signature in the vertical cross sections of reflectivity in stratiform clouds has been recognized in the very early weather radar observations and was investigated in a large number of studies. Fabry and Zawadzki (1995) provide one of the most comprehensive analyses and interpretation of long-term observations of the bright band. Three major physical factors determine the depth and intensity of the bright band. One is the change in dielectric constant of melting hydrometeors as a function of a mass or volume water fraction in a mixed-phase particle. The absolute value of the dielectric constant ε rapidly increases with increasing mass water fraction f_w and is accompanied by reflectivity enhancement of a mixed-phase particle with preserved mass (Chap. 5). The second factor is the reduction of the particle size due to the density change. This reduces the reflectivity. The third factor is the change in the particle concentration. Total concentration of raindrops is much smaller than the concentration of snowflakes from which raindrops originate. This is because the terminal velocity of a raindrop is much larger than the one of a snowflake with the same mass and a conservation of particle concentration flux in the process of melting can be assumed:

$$N(D_s)V_s(D_s)dD_s = N(D_w)V_w(D_w)dD_w, \tag{7.25}$$

where the subscripts s and w stand for snow and water. Although the conservation condition (7.25) may not necessarily hold exactly, snow rate just above the melting layer and rain rate immediately below the melting layer are usually very similar (Heymsfield et al. 2015).

The terminal velocity of a melting snowflake V_{ms} depends on its size, density, and mass water fraction. Szyrmer and Zawadzki (1999) and Zawadzki et al. (2005) recommend expressing the terminal velocity of melting snowflakes V_{ms} via the terminal velocity of the corresponding raindrops (to which snowflakes melt) V_w and the mass melted water fraction f_w as

$$V_{\text{ms}} = \frac{V_w(D_w)}{a - bf_w(1 + f_w)}, \tag{7.26}$$

where

$$a = 1.26\rho_s^{-1/3}, b = (a - 1)/2. \tag{7.27}$$

In (7.26) and (7.27), D_w is the melted diameter of the snowflake, and ρ_s is the density of the dry snowflake (g cm^{-3}) before melting starts. In our analysis, we use approximation of Brandes et al. (2002) for the terminal velocity of raindrops (Eqs. (7.10) and (7.11)).

The density of dry snowflake depends on its equivolume diameter D_s and degree of riming f_{rim}. Herein, we use the following formula for ρ_s (Brandes et al. 2007) (see also (4.26)):

$$\rho_s(D_s) = \alpha D_s^{-\beta}, \tag{7.28}$$

where α is directly proportional to the degree of riming f_{rim}, D_s is expressed in mm, and ρ_s is in g cm^{-3}. If the size distribution of snowflakes above the freezing level $N(D_s)$ is known, then the corresponding size distribution of raindrops below the melting layer $N(D_w)$ can be estimated from (7.25) as

$$N(D_w) = N(D_s)\frac{V_s(D_s)}{V_w(D_w)}\frac{dD_s}{dD_w}. \tag{7.29}$$

Considering that

$$D_w = D_s[\rho_s(D_s)/\rho_w]^{1/3}, \tag{7.30}$$

where ρ_w is the water density and

$$\frac{V_s(D_s)}{V_w(D_w)} = \frac{[\rho_s(D_s)]^{1/3}}{1.26}, \tag{7.31}$$

one can derive the size distribution of raindrops resulting from melting snowflakes using (7.26). For example, if the size distribution of dry snowflakes above the freezing level is exponential

$$N_s(D_s) = N_{0s}\exp(-\Lambda_s D_s) \tag{7.32}$$

and the snow density ρ_s is determined by (7.28), then

$$N(D_w) = \frac{N_{0s}}{1.26(1-\beta/3)}\exp\left(-\alpha^{-\frac{1}{3-\beta}}\Lambda_s D_W^{\frac{3}{3-\beta}}\right), \tag{7.33}$$

i.e., the resulting raindrop size distribution is not exponential and strongly depends on the degree of snow riming f_{rim}. For example,

$$N(D_w) = 1.19N_{0s}\exp\left(-\alpha^{-0.5}\Lambda_s D_W^{1.5}\right) \tag{7.34}$$

for $\beta = 1.0$. Inversely, the size distribution of snow above the freezing level can be roughly estimated from the measured raindrop size distributions below the melting layer (see also Zhang et al. (2011)). Figure 7.26 shows average size distributions of raindrops for $Z = 25$ dBZ and 35 dBZ estimated from the analysis of large dataset in Oklahoma (Schuur et al. 2005 and Chap. 10) (solid curves) and the corresponding size distributions of unrimed ($f_{rim} = 1$) and heavily rimed ($f_{rim} = 5$) snowflakes (dashed and dash-dotted curves, respectively) assuming conservation of mass flux through the melting layer. The snow size distributions are significantly wider, especially for unrimed snow.

Vertical profiles of the polarimetric variables through the melting layer depend on the profile of the mass water fraction f_w which can be estimated from the heat balance equation for a melting snowflake as follows:

$$L_f m \frac{df_w}{dt} = 4\pi C[F_h K(T - T_0) + F_v D_v L_v(\rho - \rho_0)], \tag{7.35}$$

where m is the mass of the melting snowflake, $f_w = m_w/m$ is the mass fraction of melted water, C is the capacitance of the melting snowflake, K is the thermodynamic conductivity of air, D_v is the diffusion coefficient of water vapor in air, L_f is the latent heat of fusion (melting), L_v is the latent heat of vaporization (condensation), F_h and F_v are the thermal and vapor ventilation coefficients, and $T - T_0$ and $\rho - \rho_0$ are the temperature and vapor density differences between the ambient air and the surface of the melting particle. The vapor density difference is expressed as

$$\rho - \rho_0 = \frac{e(T)}{R_v T} - \frac{E_w(T_0)}{R_v T_0} = \frac{RH}{100}\frac{E_w(T)}{R_v T} - \frac{E_w(T_0)}{R_v T_0}, \tag{7.36}$$

where e is the water vapor pressure, E_w is the saturation water vapor pressure, RH is the relative humidity (in %), and R_v is the water vapor gas constant. Because

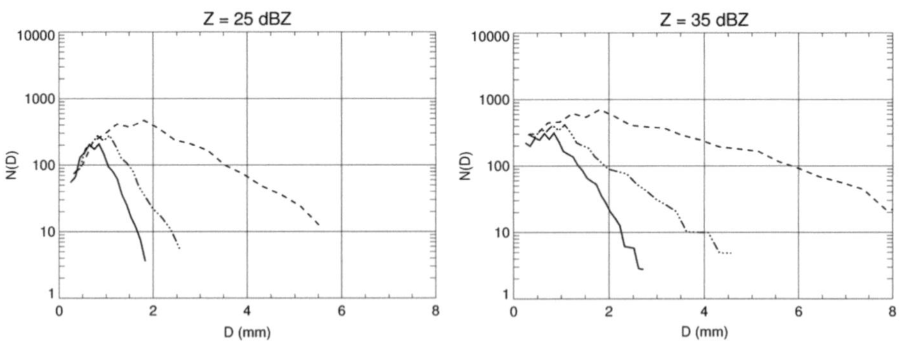

Fig. 7.26 Raindrop size distributions (solid curves) and the corresponding size distributions of snowflakes for unrimed ($f_{rim} = 1$) (dashed curves) and heavily rimed ($f_{rim} = 5$) (dash-dotted curves) snow aloft for the reflectivities in rain at the surface equal to 25 and 35 dBZ

$$\frac{df_w}{dt} = \frac{df_w}{dh}\frac{dh}{dt} = \frac{df_w}{dh}V_{\rm ms}, \tag{7.37}$$

where $V_{\rm ms}$ is the terminal velocity of a melting snowflake, and assuming that $F_h = F_v = F$, we can rewrite Eq. (7.36) as an equation for the vertical gradient of the mass water fraction:

$$\frac{df_w}{dh} = \frac{4\pi CF}{L_f m V_{\rm ms}}\left\{K(T - T_0) + \frac{D_v L_v}{R_v}\left[\frac{RH}{100}\frac{E_w(T)}{T} - \frac{E_w(T_0)}{T_0}\right]\right\}. \tag{7.38}$$

From (7.38), it follows that melting of snow starts immediately at $T = 273$ °K (0 °C) only if $RH = 100\%$, i.e., air is saturated. If the air is subsaturated ($RH < 100\%$), then melting begins at slightly positive temperature T_m for which right-hand side of (7.38) is equal to zero or wet-bulb temperature $T_{\rm wb}$ is equal to zero. The values of T_m for different RH are listed in Table 7.1. In most clouds, relative humidity within the melting layer does not fall below 90%, but in the presence of strong entrainment of dry ambient air (like in the stratiform parts of the mesoscale convective systems), it can drop as low as 60–70%. If $RH < 100\%$, the sublimation of melting snowflakes causes diabatic cooling of surrounding air and formation of the "isothermal layer" with the temperature of about 0 °C which further delays the melting of snowflakes.

The falling distance of a snowflake before it completely melts depends on its original size, density, temperature lapse rate, and humidity. Smaller snowflakes melt faster into drops which quickly acquire large terminal velocity. The dependence of the maximal size of liquid drops on the fall distance from the top of the melting layer is illustrated in Fig. 7.27 for the case of unrimed snow ($f_{\rm rim} = 1$), temperature lapse rate of 6.5 °C km^{-1}, and relative humidity of 100%. The terminal velocities of snow are much smaller than those of raindrops. Figure 7.28 demonstrates that at 400 m below the top of the melting layer, where the maximal size of raindrops is about 2 mm, the difference between terminal velocities of raindrops $V_{\rm w}$ and wet snowflakes $V_{\rm ms}$ can be as large as 5 m s^{-1}. This favors collisions between small raindrops and partially melted snowflakes and accretion of the latter (Troemel et al. 2014).

At any given height, raindrops originating from fully melted snowflakes and partially melted snowflakes have very distinct size distributions (Fig. 7.29). The size distributions in Fig. 7.29 are calculated assuming that size distribution of melted diameters of snowflakes is of Marshall-Palmer type corresponding to rain rate of 5 mm h^{-1}. It is clear from Fig. 7.29 that the size distribution of hydrometeors is close

Table 7.1 Temperature at which melting starts for different relative humidities

RH (%)	T_m (°C)
100	0
90	0.65
80	1.37
70	2.17
60	3.05

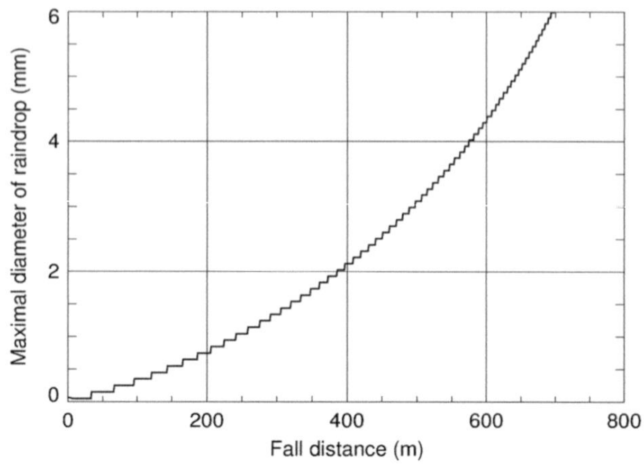

Fig. 7.27 Maximal diameter of raindrop originated from melted snowflake as a function of fall distance from the top of the melting layer for unrimed snow, with a temperature lapse rate of 6.5 °C km^{-1} and relative humidity of 100%

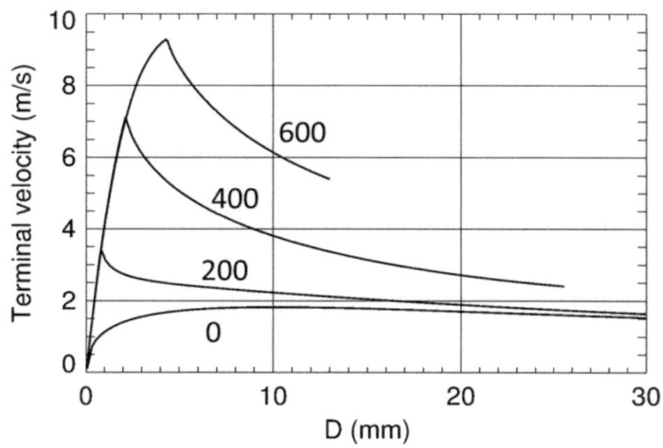

Fig. 7.28 Terminal velocities of raindrops, V_w depicted with the ascending curve, and partially melted snowflakes, V_{ms} depicted with the descending curves, at the top of the melting layer (bottom curve) and at 200, 400, and 600 m below the top. Assumed are unrimed snow aloft, temperature lapse rate of 6.5 °C km^{-1}, and relative humidity of 100%

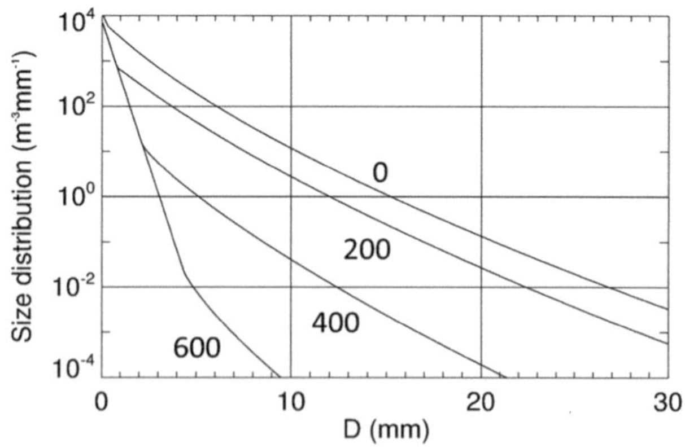

Fig. 7.29 Size distributions of raindrops and partially melted snowflakes at the top of the melting layer (upper curve) and at 200, 400, and 600 m below the top of the melting layer. The Marshall-Palmer distribution of melted diameters with equivalent rain rate 5 mm h^{-1} is assumed

to exponential only at the top and bottom of the melting layer. Inside the melting layer, it is close to bi-exponential, which agrees with the observations of Barthazy et al. (1998) and Goeke and Waldvogel (1998).

Modeling of the melting layer/bright band signatures was a subject of a large number of studies (e.g., Yokoyama and Tanaka 1984; Klaasen 1988; Russchenberg and Ligthart 1996; Fabry and Szyrmer 1999; Olson et al. 2001; Battaglia et al. 2003; Zawadzki et al. 2005). In most of these, polarimetric variables were not simulated with the exception of Russchenberg and Ligthart (1996) and D'Amico et al. (1998) who modeled vertical profiles of Z_{DR} and linear depolarization ratio *LDR* assuming a monodisperse size distribution.

The model of melting snowflakes was used by Ryzhkov et al. (2008) and Carlin (2018) to model vertical profiles of Z, Z_{DR}, K_{DP}, and ρ_{hv} in the melting layer imposing a size distribution of snow just above the ML and a degree of riming f_{rim} as follows. The size distribution of snowflakes at the top of the ML (i.e., at the freezing level) was selected using (7.29) so that the distribution of raindrops beneath the ML is the Marshall-Palmer (Chap. 4). Initial snow density was determined as in (7.28). A melting snowflake represents a mixture of air, solid ice, and water. It was assumed that the density of dry snow (i.e., combination of air and solid ice) mixed with water changes in the process of melting depending on the diameter of ice core according to (7.28), i.e., the density of ice is larger at the end of the melting process. Mass water fraction for each bin size was computed by solving (7.38).

Polarimetric variables are very sensitive to the dielectric constant ε of melting snowflakes which depends on how the mixture of air, ice, and water is incorporated into the electromagnetic model of ε. Eqs. (4.41)–(4.42) have been used by Ryzhkov et al. (2008) for computations of ε of wet snow. Polarimetric variables also depend on the shapes and orientations of melting particles which cannot be obtained from the thermodynamic model of melting. Snowflakes were modeled as oblate spheroids with the aspect ratio 0.8 for dry snowflakes at the top of the melting layer. It was assumed that the aspect ratio of the melting snowflake changes linearly with f_w until it reaches the shape of a raindrop with the same mass. Similar assumptions were made with respect to the width of the canting angle distribution σ. The parameter σ changes gradually from 50° in dry snow to 10° in rain (Ryzhkov et al. 2011).

Simulated profiles of Z, Z_{DR}, K_{DP}, and ρ_{hv} at S band through the melting layer for unrimed ($f_{rim} = 1$) and heavily rimed ($f_{rim} = 5$) snow are displayed in Fig. 7.30 under the assumption that the rain rate below the melting layer is 5 mm h^{-1}. It is evident that the profiles of all these variables are very sensitive to the degree of riming. Larger Z and K_{DP} enhancement, stronger Z_{DR} maximum, and deeper ρ_{hv} minimum are observed if the snow density is low, and the depth of the melting layer is smaller if $f_{rim} = 1$. The increase of the temperature lapse rate decreases the depth of the melting layer, whereas the decrease in the relative humidity *RH* causes subsidence of the melting layer.

The modeling results (Fig. 7.30) are consistent with radar observations. The range-defined composite QVPs within the melting layer in a stratiform part of the mesoscale convective system observed by the KVNX (Enid OK) WSR-88D on 20 May 2011 in northern Oklahoma are generated using the methodology of

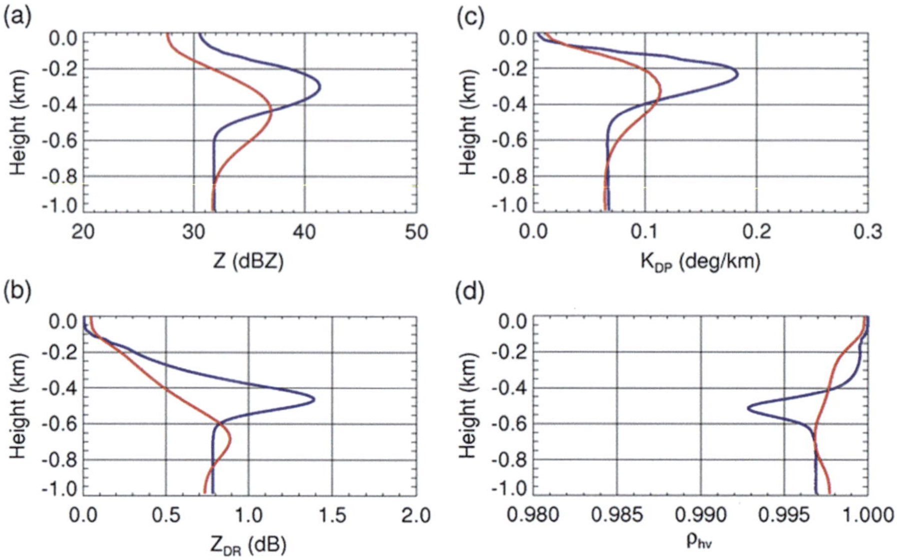

Fig. 7.30 Model vertical profiles of (**a**) Z, (**b**) Z_{DR}, (**c**) K_{DP}, and (**d**) ρ_{hv} at S band within the melting layer for unrimed snow ($f_{rim} = 1$, blue curves) and heavily rimed snow ($f_{rim} = 5$, red curves). Freezing level is at 0.0 km height, temperature lapse rate is 6.5°km^{-1}, and rain rate below the ML is 5 mm h^{-1}

Tobin and Kumjian (2017) (Fig. 7.31). The signature of the melting layer is well pronounced at the height of about 3 km, particularly at the beginning of a sampling period at 1304 UTC. The resulting range-defined QVPs of Z, Z_{DR}, and K_{DP} measured at 1304 UTC (black curves in Fig. 7.32) compare well with the simulated vertical profiles (red curves). The one-dimensional melting snow model was initialized with the mean particle size distribution in the 0– -2 °C layer measured onboard the research aircraft during the period between 1332 UTC and 1335 UTC. The maxima of Z and K_{DP} occur near the top of the ML, whereas the maximum Z_{DR} is at the very bottom of it (more than 300 m below the Z maximum). The minimum of the observed ρ_{hv} and the maximum Φ_{DP} (backscatter differential phase δ) are in between in the middle of the ML (not shown in Fig. 7.32). The maximal values of all three radar variables in the observed and simulated profiles match very well, and the difference between the heights of the K_{DP} and Z_{DR} maxima might be attributed to the impact of beam broadening.

In the stratiform parts of MCSs, the polarimetric signatures of the melting layer are particularly strong due to vigorous aggregation of snow which is favored by a very high concentration of primary nucleated ice. A large part of this primary ice is seeded by the leading convective squall lines. The signatures are somewhat weaker in typical frontal mixed-phase stratiform clouds or orographically forced stratiform precipitation. An example of the latter is in Fig. 7.33 where the vertical profile of linear depolarization ratio *LDR* is plotted in addition to the other four variables. *LDR*

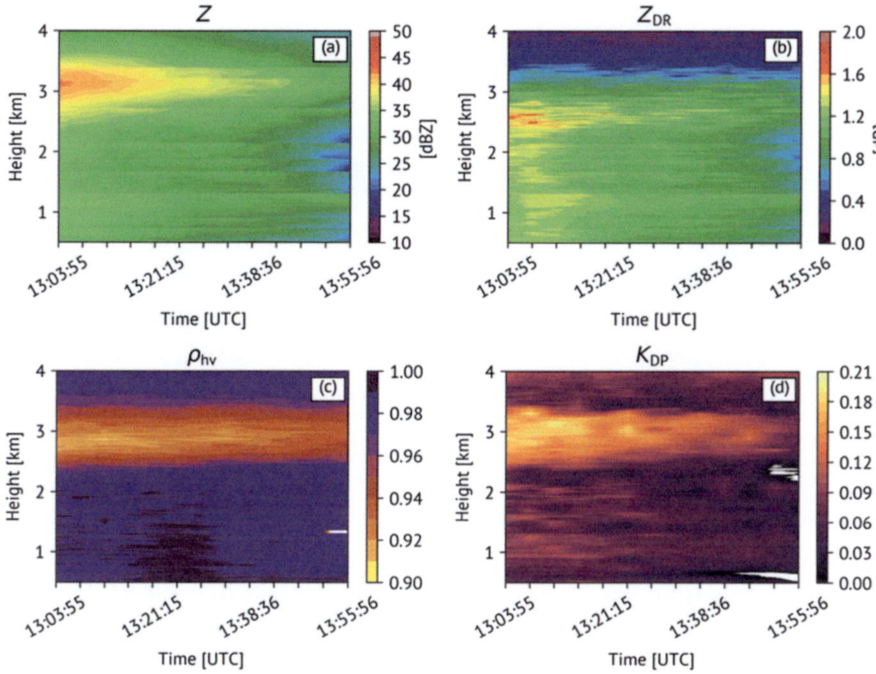

Fig. 7.31 Range-defined QVP time series of (**a**) Z, (**b**) Z_{DR}, (**c**) ρ_{hv}, and (**d**) K_{DP} from KVNX (Enid OK radar) on 20 May 2011 between 13:03:55 and 13:55:56 UTC from the 90° to 180° sector and averaged between 30 and 70 km from the radar site

in the melting layer is usually anticorrelated with ρ_{hv} so that the *LDR* maximum coincides with the ρ_{hv} minimum.

Simulations illustrated in Figs. 7.30 and 7.32 assume that one snowflake produces one raindrop and there are no interactions between melting particles. The model reproduces the peaks in Z and Z_{DR} as well as the dip in ρ_{hv} at the right places within the melting layer (e.g., the Z_{DR} peak below the Z peak). Vertical profiles of Z also look realistic and agree with the ones reported in the literature. This model, however, commonly does not reproduce the very low values of ρ_{hv} and high Z_{DR} often observed at the bottom of the melting layer. The scatterplot of maximal Z_{DR} versus minimal ρ_{hv} in the melting layer from several stratiform rain events in Oklahoma (Fig. 7.34, crosses) reveals strong anticorrelation between ${Z_{DR}}^{(max)}$ and ${\rho_{hv}}^{(min)}$. Occasionally, ${\rho_{hv}}^{(min)}$ is 0.82, and the corresponding ${Z_{DR}}^{(max)}$ is 2.5 dB, whereas the simulated ρ_{hv}s in Fig. 7.30d are well above 0.99 and the Z_{DR}s are smaller than 1.5 dB. To explain the extreme values of ρ_{hv} and Z_{DR}, aggregation of melting snowflakes should be taken into account. Aggregation also explains routinely observed increase in the size of the melting snowflakes in the middle of the melting layer, whereas the model without aggregation implies monotonic decrease of the maximal particle size as melting progresses. The more sophisticated models of Giangrande (2007) and Troemel et al. (2014) include riming and aggregation of

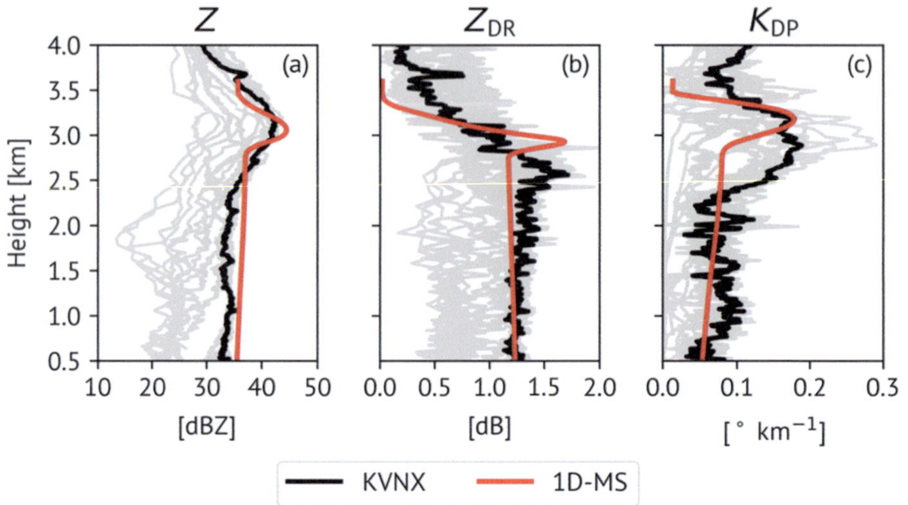

Fig. 7.32 Comparison of (**a**) Z, (**b**) Z_{DR}, and (**c**) K_{DP} from a one-dimensional melting snow model simulation (red) and the range-defined QVP at 13:03:55 UTC (black) (see Fig. 7.31). The gray lines show the constituent QVPs from each elevation angle. The one-dimensional melting snow model was initialized with the mean observed particle size distribution in the 0 °C and −2 °C layer between 13:32:24 and 13:34:48 UTC fit to a gamma distribution and the measured environmental temperature and humidity profile from the aircraft

wet snowflakes. Collisions between large wet snowflakes and riming via collection of small raindrops originating from smaller completely melted snowflakes may increase the snowflakes mass water content and additional aggregation within the melting layer. Once the ratio $D|\varepsilon|^{1/2}/\lambda$ approaches unity due to the increase of both D and ε, the Z_{DR} and ρ_{hv} quickly reach extreme values because of resonance scattering. The asterisks in Fig. 7.34 indicate the results of modeling by Giangrande (2007) for different intensities of aggregation. Substantially higher Z_{DR} and, correspondingly, much lower ρ_{hv} are simulated compared to the values generated by the model without aggregation. However, ρ_{hv} produced by a more sophisticated model is still higher than the observed ones. This might be attributed to the extreme complexity of the actual shape of the melting snowflakes which is not accounted for in a simplified spheroidal model.

7.2.4.2 Melting of Graupel and Hail

In contrast to melting snow, melting graupel and hail do not produce the bright band in reflectivity, the well-pronounced Z_{DR} maximum, and ρ_{hv} minimum in the vertical profiles. This is because the density of graupel and hail is much higher than the density of snow and their terminal velocities are also significantly higher. Therefore, hailstones completely melt in a much deeper layer of the atmosphere which may extend to the surface. In the latter situation, hailstones are observed at the ground.

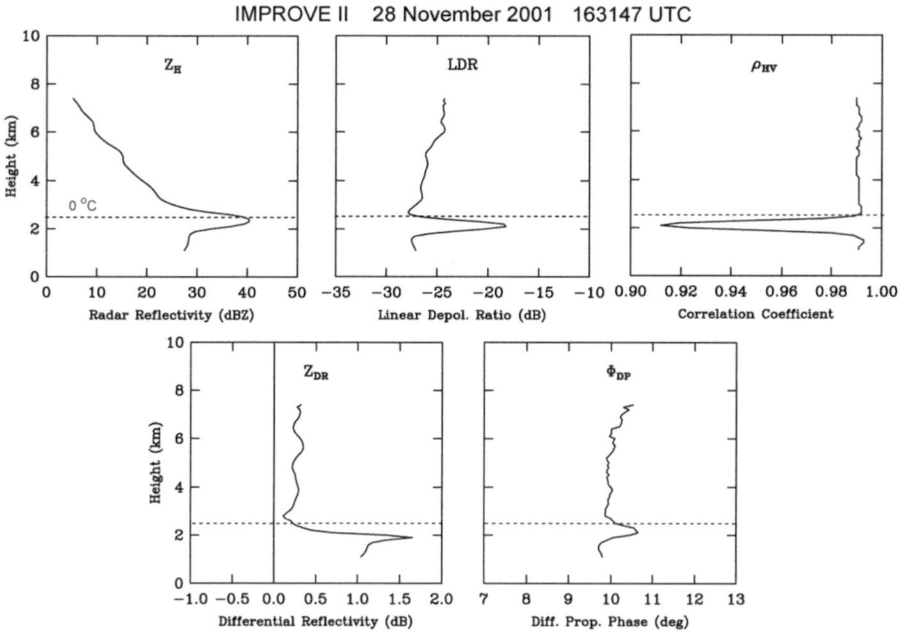

Fig. 7.33 Vertical profiles of Z, Z_{DR}, LDR, ρ_{hv}, and Φ_{DP} within the melting layer measured by the NCAR S-Pol radar on 28 November 2001 during the IMPROVE II field campaign. Adapted from Brandes and Ikeda (2004). © American Meteorological Society. Used with permission

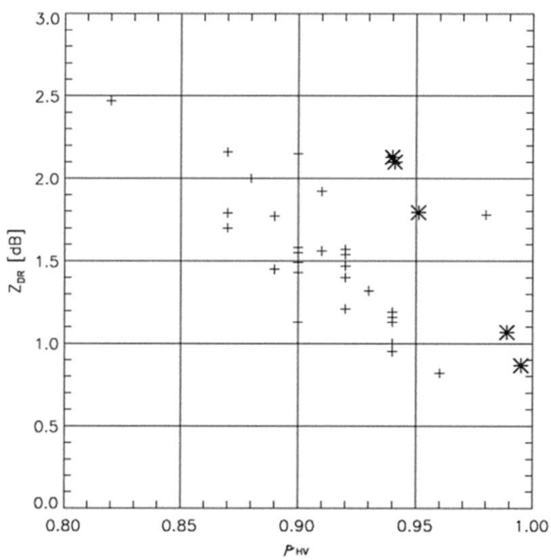

Fig. 7.34 Scatterplots of maximal Z_{DR} vs minimal ρ_{hv} in the melting layer from S-band measurements (crosses). Asterisks indicate the results of modeling which includes different intensities of aggregation (Giangrande 2007)

The tendency for extending the melting downward and dissolving the extrema of Z, Z_{DR}, and ρ_{hv} by increased riming of the ice particles is visible in Fig. 7.30.

Melting of individual graupel or hail particles occurs in several steps. If the density of hail is equal to the density of solid ice (0.92 g cm^{-3}), the melted water accumulates on the surface of the hailstone and sheds after its mass exceeds the "shedding threshold." If the initial density of graupel/hail is lower than 0.92 g cm^{-3} (which is typical for graupel), then melted water first soaks into the interior of the particle and starts accumulating on the surface only after all air cavities are filled with melted water. Again, excessive water sheds after reaching the critical mass m_{wmax} (Rasmussen and Heymsfield 1987; Phillips et al. 2007):

$$m_{w\max} = 2.68 \times 10^{-4} + 0.139\ (m_i + m_{\mathrm{ws}}) \tag{7.39}$$

where m_i is the mass of ice and m_{ws} is the mass of retained soaked water, both expressed in kg.

The thermodynamic model of melting hailstones was first developed by Rasmussen and Heymsfield (1987). This model was later used for simulating vertical profiles of polarimetric variables in melting hail by Aydin and Zhao (1990), Aydin et al. (1991), Vivekanandan et al. (1990), and Ryzhkov et al. (2013a).

The dependencies of diameters of melting hailstones and their ice cores on height for high-density hail retrieved from the Rasmussen and Heymsfield model are illustrated in Fig. 7.35a (Ryzhkov et al. 2013a). The computations assume that the freezing level is at 4 km, temperature lapse rate is 6.5 °C km^{-1}, and relative humidity is 100%. If the hailstone is spongy (i.e., originally has lower density), then the effective diameter of the ice core is defined as an equivolume diameter of a spheroid with the mass of unmelted ice and density of solid ice. For a given size of

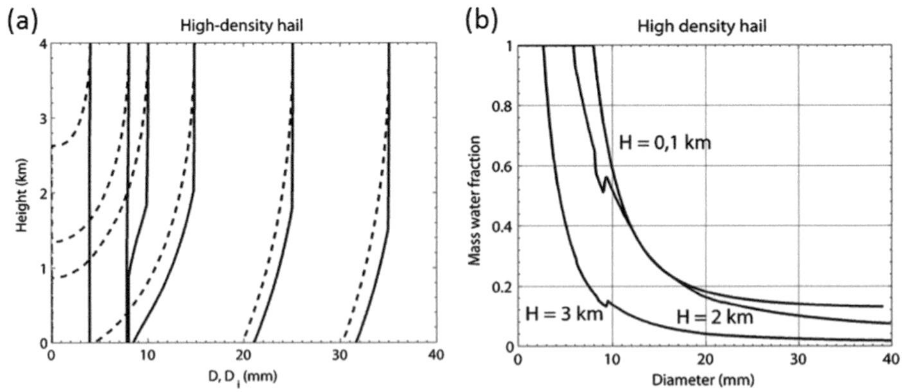

Fig. 7.35 (**a**) Dependence of diameters of melting hailstones (solid curves) and effective diameters of their ice cores (dashed curves) on height. (**b**) Distribution of mass water fraction across the size spectrum at three different heights. Adapted from Ryzhkov et al. (2013a). © American Meteorological Society. Used with permission

dry hailstone aloft, lower-density hail melts faster, and its total diameter and effective diameter of ice core decrease more rapidly toward the ground.

Figure 7.35a shows that hailstones with initial diameters (i.e., diameters aloft) less than 14 mm (for high-density hail) melt completely before reaching the ground, whereas ice cores of much bigger hailstones remain large. Water shedding off larger melting hailstones causes reduction of mass and diameter. At each height below the freezing level, mass water fraction f_w is equal to 1 for totally melted particles and gradually decreases with size of partially melted graupel and hail (Fig. 7.35b). Figure 7.35 also shows that hailstones with initial sizes between 8 and 14 mm end as 8 mm raindrops enhancing their total concentration near the ground. Nonetheless, these and larger raindrops are unstable and tend to break up spontaneously.

Vertical profiles of polarimetric variables in melting hail depend on the initial size distribution of dry graupel/hail at the top of the melting layer (i.e., freezing level). In situ measurements of size distributions of ice particles aloft in hailstorms often reveal a bi-exponential type of particle spectra with different slopes for graupel and hail (Smith et al. 1976; Spahn and Smith 1976). In the simulations by Ryzhkov et al. (2013a), the bi-exponential size distribution of graupel/hail at the freezing level is

$$N(D) = N_g \exp(-\Lambda_g D) + N_h \exp(-\Lambda_h D) \tag{7.40}$$

where subscripts "g" and "h" stand for graupel and hail. The choice of parameters N_h, Λ_h, and D_{max} (maximal hail size at which size distribution is truncated) is dictated by the need to match resulting size distributions of ice cores close to the surface with the observed hail size distributions reported in literature (Ulbrich and Atlas 1982; Cheng and English 1983; Cheng et al. 1985). In the study of Ryzhkov et al. (2013a), the radar characteristics of melting graupel/hail are computed for four different size distributions of ice at the freezing level $H = 4$ km (shown in Fig. 4.16) with the following parameters:

1. No hail aloft and at the surface ($N_h = 0$).
2. "Small" hail. Hail is present aloft (at $H = 4$ km) but is totally melted at the surface ($D_{max} = 14$ mm, $\Lambda_h = 0.99$ mm^{-1}, $N_h = 200\Lambda_h^{4.11}$ m^{-3} mm^{-1}).
3. "Moderate" hail. Larger hail aloft with $D_{max} = 24$ mm so that maximal hail size at the surface is about 19 mm ($\Lambda_h = 0.42$ mm^{-1}, $N_h = 400\,\Lambda_h^{4.11}$ m^{-3} mm^{-1}).
4. "Large" hail. $D_{max} = 35$ mm so that maximal size of hail at the surface is about 30 mm ($\Lambda_h = 0.27$ mm^{-1}, $N_h = 800\,\Lambda_h^{4.11}$ m^{-3} mm^{-1}).
5. For the size distribution of graupel, $N_g = 8000$ m^{-3} mm^{-1} and $\Lambda_g = 1.6$ mm^{-1}.

The D_{max} and Λ_h at the freezing level are selected so that the product $D_{max}\Lambda_h$ corresponding to the size distributions of ice cores at the surface is equal to 7.9—its most likely value (Ulbrich and Atlas 1982) (see Fig. 4.17). The initial bi-exponential size distribution of hydrometeors at the freezing level is modified in the process of melting because the sizes of melting particles and their terminal velocities change and new liquid particles are generated by shedding and breakup. All raindrops in the combined rain/hail size spectrum are formed from three sources: (1) direct melting of graupel/smaller-size hail, (2) water shedding off larger hailstones, and (3) spontaneous breakup of 8 mm raindrops. The size distribution of raindrops from source (1) is determined by the assumed size distribution of graupel/small hail aloft, whereas size distributions of raindrops coming from sources (2) and (3) are prescribed using the total masses of shed drops and fragments of drop breakup.

An example of vertical profiles of total mass associated with ice, melted water, shed water, and breakup water is presented in Fig. 7.36 for the case of "large" high-density hail. Shed water and breakup water start accumulating at heights 2 km and 3 km below the freezing level, respectively. According to several observational studies (e.g., Carras and Macklin 1973; Joe et al. 1976; Lesins et al. 1980; Rasmussen et al. 1984), shed drops are typically between 0.5 and 2.0 mm in diameter, and the modal diameter is about 1 mm.

The forward-scattering operator described by Ryzhkov et al. (2011) is used to compute the polarimetric variables from the output of one-dimensional (1D) cloud model of melting hail (Ryzhkov et al. 2013a). Melting hailstones and raindrops are modeled as spongy, water-coated, or liquid oblate spheroids with aspect ratios and orientations determined by the mass water fraction as in the case of melting snow described in Sect. 7.2.4.1. It is assumed that melting hailstones are characterized by a Gaussian distribution of orientations with the mean orientation of rotation axis along the vertical. The Maxwell Garnett mixing formula (4.41) is used for the dielectric constant, where water is treated as the matrix and ice/air as inclusions (Chap. 4). The melting particle is modeled as a water-coated spheroid once all air pockets are filled and melted water starts accumulating at the surface. The inner core is treated as a spheroid with uniformly filled water matrix and ice inclusions. The T-matrix code for two-layer spheroids is used for scattering computations.

Vertical profiles of Z, Z_{DR}, K_{DP}, A_h, and A_{DP} computed from the output of the 1D model for the cases of "large" and "small" high-density melting hail are displayed in Figs. 7.37 and 7.38. The initial size distributions of the "large" and "small" hail are plotted in Fig. 4.16. Vertical dependencies of Z at S and C bands for large hail are

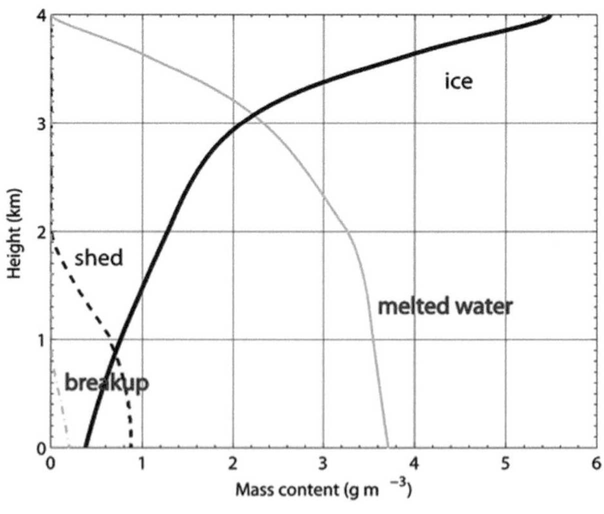

Fig. 7.36 Vertical profiles of mass content associated with ice (thick solid black curve), melted water (solid gray curve), shed water (dashed black curve), and water resulting from breakup of large raindrops (dash-dotted gray curve in lower-left corner) in the case of large high-density hail. From Ryzhkov et al. (2013a). © American Meteorological Society. Used with permission

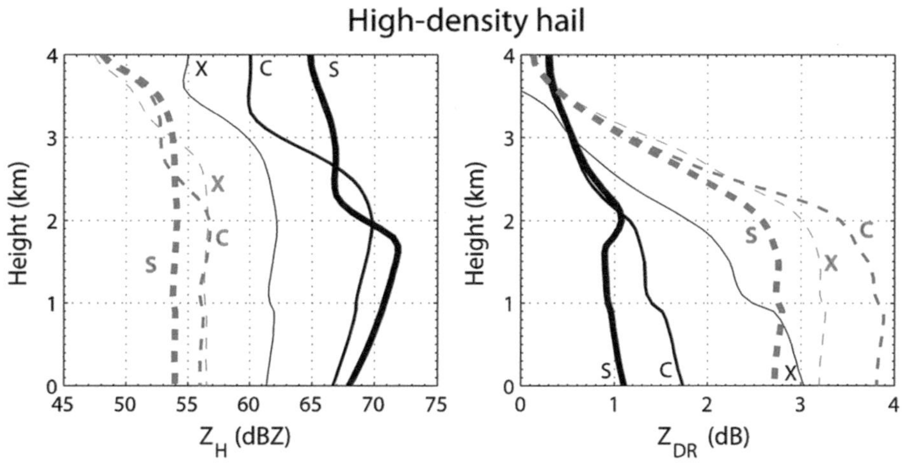

Fig. 7.37 Simulated vertical profiles of Z and Z_{DR} at S, C, and X bands of melting large hail (solid curves) and small hail (dashed curves). The thickest curves are for S band, and the thinnest are for X band. Adapted from Ryzhkov et al. (2013a). © American Meteorological Society. Used with permission

characterized by a maximum at about 2 km below the freezing level, where shedding of water from larger hailstones starts (Fig. 7.37). As melting progresses, the Z first gradually increases due to the increase of the dielectric constant, and then it decreases once shedding starts and the size of larger hailstones diminishes. In the simulations for "small" hail, Z at S and C bands is 10–15 dB lower, and the Z maximum mostly disappears because of a smaller impact of shedding. The value of Z at X band is 5–10 dBZ lower than at S and C bands in the case of "large" hail; this is attributed to stronger resonance effects at X band. Softer hail with lower density produces smaller Z at all heights due to the lower dielectric constant of spongy hail (Ryzhkov et al. 2013a).

Z_{DR} of melting hail increases with decreasing height within the first 2 km below the freezing level and then stabilizes without much change toward the ground (Fig. 7.37). As expected, Z_{DR} near the surface is significantly lower for larger maximal hail size ("large" hail) than for smaller maximal hail size ("small" hail). For "small" hail, Z_{DR} near the surface at C band approaches 4 dB, which is higher by more than 1.0 dB than the corresponding Z_{DR} at S band. Relatively high Z_{DR} of melting hail near the surface is attributed to the contribution of large raindrops originated from melting hailstones. The magnitude of the corresponding Z_{DR} is determined by the strength of the wavelength-dependent resonance scattering effects in rain.

Vertical profiles of K_{DP}, A_h, and A_{DP} for "large" and "small" hail at C band are plotted in Fig. 7.38. K_{DP} rapidly increases with decreasing height within the first 1.5–2 km below the freezing level as smaller-size graupel/hail transforms into raindrops. The increase diminishes below 2 km. Both A_h and A_{DP} reach their maxima at the height about 2 km below the freezing level. This is consistent with

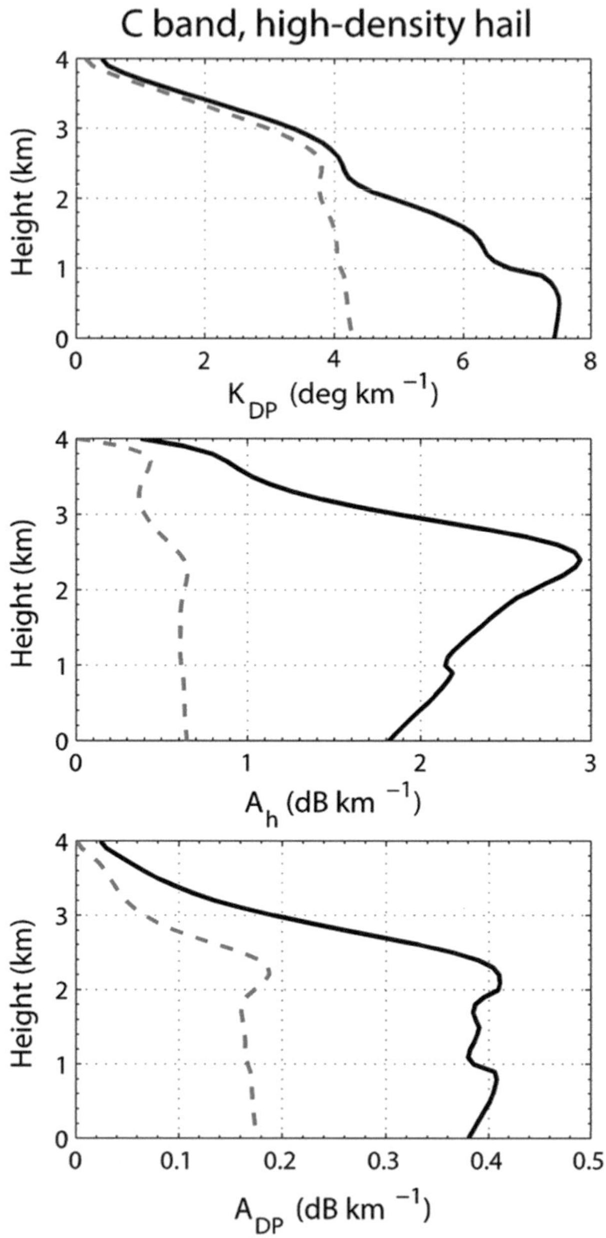

Fig. 7.38 Simulated vertical profiles of K_{DP}, A_h, and A_{DP} for large (solid curves) and small (dashed curves) high-density hail at C band. From Ryzhkov et al. (2013a). © American Meteorological Society. Used with permission

observational evidence that the largest attenuation/differential attenuation at C band occurs at some height above the ground (e.g., Borowska et al. 2011; Kaltenboeck and Ryzhkov 2013). The difference between vertical profiles of attenuation variables for "large" and "small" hail is quite dramatic (especially for A_h) at all three wavelengths. This is consistent with the physical expectation that larger hail size contributes significantly to the overall extinction of electromagnetic waves.

Vertical cross sections of Z, Z_{DR}, and ρ_{hv} measured simultaneously by S-band and C-band polarimetric radars in hail-bearing storm in Oklahoma are in Fig. 7.39. The storm occurred on 27 March 2009. Hailstones with diameters between 1.9 and 2.2 cm were reported at the ground. Maximal values of Z exceeding 60 dBZ are indicative of hail aloft and near the surface. Classical Z_{DR} "depressions" exhibiting low Z_{DR} below the environmental freezing level are visible at both S band and C band in the middle of hail core at ranges 75 and 71 km, respectively. Differential reflectivity at C band increases rapidly toward the ground and reaches values larger than 5 dB near the surface where the resonance-size raindrops are mixed with remaining partially melted hailstones. The corresponding Z_{DR} at S band is much lower in agreement with the results of theoretical simulations (Fig. 7.37). Cross-correlation coefficient ρ_{hv} is much lower in hail at C band (<0.8) than at S band (>0.96).

Very strong attenuation and differential attenuation in hail at C band are also evident in the RHI plots of Z and Z_{DR}. Although attenuation/differential attenuation at S band can be significant in hail (Chap. 6; Ryzhkov et al. 2013b), it is much stronger at C band as comparison of left and right panels in Fig. 7.39 demonstrates. The reflectivity at C band is negatively biased behind the hailstorm by 10–20 dB, and Z_{DR} below -5 dB is observed behind the storm at C band. A detailed comparison of S-band and C-band polarimetric variables in hailstorms is in Picca and Ryzhkov (2012) and Kaltenboeck and Ryzhkov (2013).

7.2.5 *Freezing/Refreezing of Raindrops and Snowflakes*

The difference between processes of melting and freezing is that melting of ice particles starts immediately after wet-bulb temperature of the air T_w becomes positive, whereas freezing of liquid particle does not necessarily occur once T_w drops below zero. This explains abundance of supercooled liquid water at very low negative temperatures. Drop freezing can be homogeneous or heterogeneous (Pruppacher and Klett 1997). The latter one requires the presence of foreign body such as ice nucleus (IN) submersed in the liquid drop (immersion freezing) or ice particle/ice nucleus with which the drop collides (contact freezing). Clean liquid droplet may undergo homogeneous freezing which occurs only at very low temperatures (usually lower than -37 °C). Refreezing takes place when ice particle melts completely or partially and then refreezes after being brought into the areas of subfreezing temperatures. Herein, we consider two important microphysical processes. One is freezing of raindrops lofted in convective updrafts to the levels of negative environmental temperatures. The other is refreezing of snowflakes which first melt (completely or partially) in an elevated warm layer and then refreeze in the subfreezing atmospheric layer underneath. Both microphysical processes produce well-pronounced polarimetric signatures.

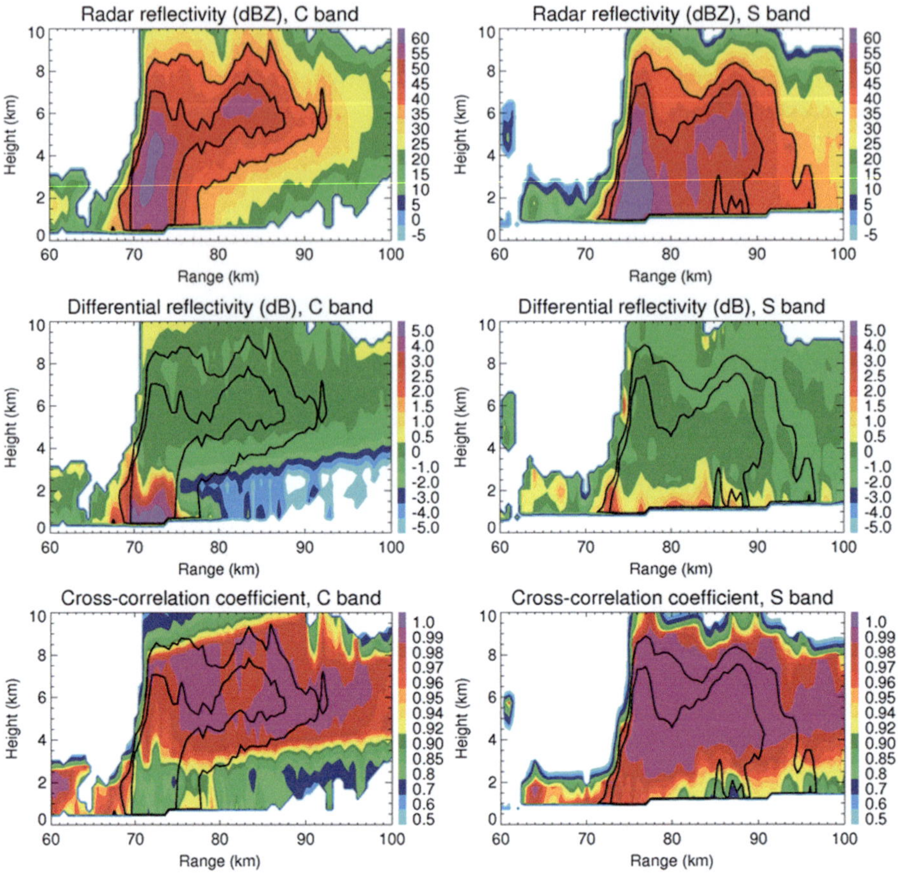

Fig. 7.39 The RHI vertical cross sections of Z, Z_{DR}, and ρ_{hv} measured by the C-band OU PRIME radar and S-band KOUN WSR-88D across the hailstorm on 27 March 2009. Overlaid contours correspond to reflectivities of 40 and 50 dBZ. From Borowska et al. (2011). © American Meteorological Society. Used with permission

7.2.5.1 Freezing of Raindrops in Convective Updrafts

Raindrop lofted by a convective updraft above the freezing level starts freezing at the moment when an ice nucleus inside the drop becomes active or when raindrop collides with ice nucleus or ice particle in the surrounding air. Ice nucleation is a stochastic process, and not all drops contain ice nuclei (IN) needed to initiate nucleation. Large raindrops have higher probability of containing IN and usually nucleate much faster than the small droplets which can be transported to the cloud top without glaciation.

According to the Bigg's theory of stochastic immersion nucleation, the temperature at which 50% of raindrops of a certain size nucleate can be obtained from (Bigg 1953a, b; Pruppacher and Klett 1997; Kumjian et al. 2012)

$$T_{\text{wi}} = -\frac{1}{\alpha}\ln\left(\frac{\alpha \Gamma_w w' \ln 2}{V_d B}\right), \tag{7.41}$$

where $\alpha = 0.65°\text{C}^{-1}$, $B = 2.0 \times 10^{-4}\ \text{cm}^{-3}\ \text{s}^{-1}$, Γ_w is the temperature lapse rate, $w' = w - V_t$ is the updraft-modified vertical velocity of the raindrop, and V_d is the drop volume. If $\Gamma = 6.5°\text{km}^{-1}$ and updraft velocity $w = 12\ \text{m s}^{-1}$, then $T_{\text{wi}} = -11.6°$ for raindrop with diameter of 3 mm which is lofted above the freezing level located at 4 km.

The freezing process occurs in two stages. The first stage is assumed to happen instantaneously upon nucleation, wherein ice crystal growth occurs in the core of the liquid drop, converting only a small portion of the liquid water into ice. The subsequent freezing occurs from the outside in, as an ice shell forms around the droplet and grows inward (Fig. 7.40). The latter stage can take up to several minutes for the largest drop sizes (Fig. 7.41). Heat balance equations governing freezing of raindrops are in Kumjian et al. (2012) and Phillips et al. (2015).

According to Phillips et al. (2015) who also took into account the impact of accretion of supercooled cloud water droplets on the particle growth, the freezing time is approximately proportional to the initial size, raised to the power of about 1.4, and increases with increasing of the liquid water content (*LWC*) of the ambient air. During the freezing time, large raindrops can be lofted at appreciable heights by

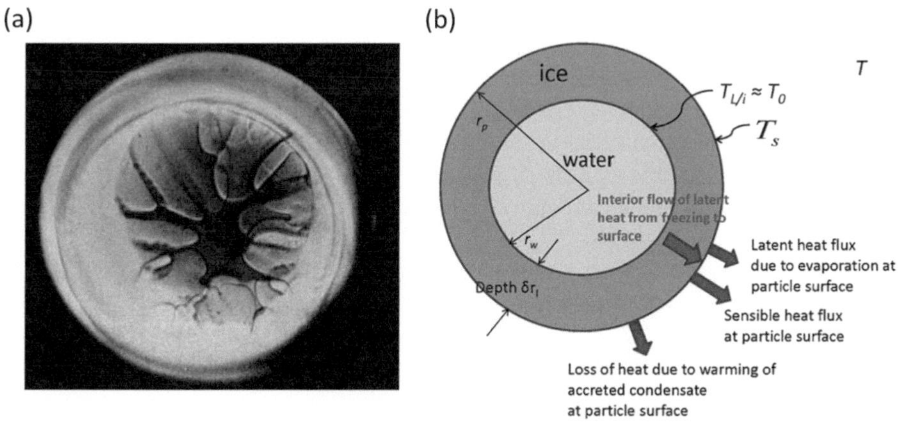

Fig. 7.40 (**a**) Photograph of a 1 mm drop undergoing freezing, from Johnson and Hallett (1968). The outer ice shell has just begun to form, and dendritic growth is evident in the drop core. The photograph was taken 10 s after nucleation, and complete freezing took another 40 s. (**b**) Schematic geometry of a freezing raindrop and heat fluxes forming the heat budget. From Phillips et al. (2015). © American Meteorological Society. Used with permission

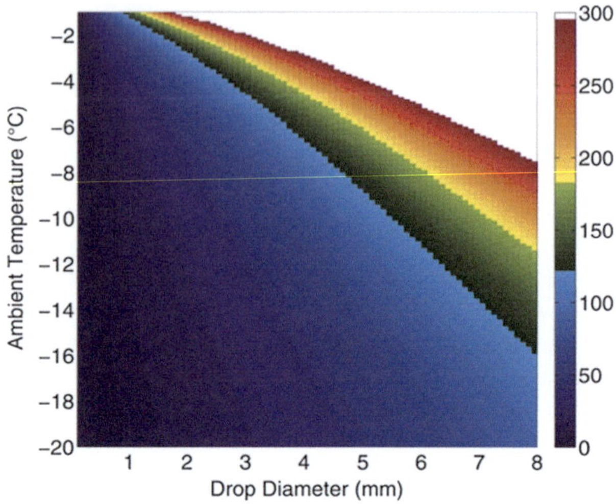

Fig. 7.41 Time for raindrops to freeze in s, following the assumptions and calculations of Pruppacher and Klett (1997). The drops are assumed to be nucleated and remain at a given ambient temperature. Calculations are based on the Pruppacher and Klett (1997), Eqs. (16)–(36). The color bar indicates time in seconds, and white shading is for times over 300 s. Adapted from Kumjian et al. (2012). © American Meteorological Society. Used with permission

strong updrafts while maintaining oblate shape and high dielectric constant which result in significant positive differential reflectivity.

For scattering computations, freezing raindrops can be modeled as two-layer spheroids (Fig. 7.40), with an outer ice shell (the thickness of which is determined explicitly from the heat balance equations) surrounding an inner core consisting of either pure water or water mixed with ice (to represent an embedded ice germ). In the model simulations by Kumjian et al. (2012), it is assumed that the shape of the freezing raindrop remains intact but its orientation depends on the fraction of the unfrozen water similarly to the model of melting hailstones. That is, raindrops become more randomly oriented in the process of freezing. These simulations were made for various initial size distributions of raindrops and demonstrate that Z_{DR} of raindrops lofted in a convective updraft remains high before ice nucleation occurs and then gradually decreases with height reaching low values typical of dry graupel or hail when raindrops are completely frozen. The height level of complete freezing can be 2–3 km above the environmental zero-degree isotherm and strongly depends on the updraft speed so that the updraft intensity can be potentially determined from the height of the Z_{DR} column.

A simple model of raindrop freezing developed in Kumjian et al. (2012) can realistically reproduce vertical profiles of Z_{DR} in convective updrafts. For example, observations and model simulations have been compared for the storm observed by the Bonn X-band polarimetric radar, BoXPol (Fig. 7.42). Two relatively modest convective cells are displayed in the RHI of Z (left panel). The left one, closer to the radar, is associated with clearly defined signature of the Z_{DR} column which stretches above the environmental freezing level at 1.8 km AGL. Vertical profiles of Z_{DR} extracted from the multiple RHI scans compare favorably with the modeled Z_{DR} profile if a gamma raindrop size distribution with $\Lambda = 1.4\ \mathrm{mm}^{-1}$ and $\mu = -0.495$ and the updraft speed of $12\ \mathrm{m\ s}^{-1}$ are assumed in the model (Fig. 7.43).

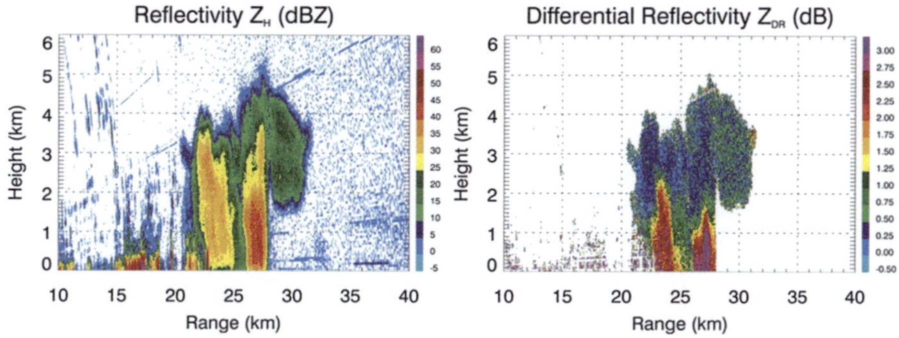

Fig. 7.42 Vertical cross sections of Z_H (left) and Z_{DR} (right) taken along the 309.5° azimuth on 24 June 2011, at 1034 UTC by the X-band BoXPol radar at the University of Bonn. The Z_{DR} column of interest is centered at about 23 km range. Adapted from Kumjian et al. (2012). © American Meteorological Society. Used with permission

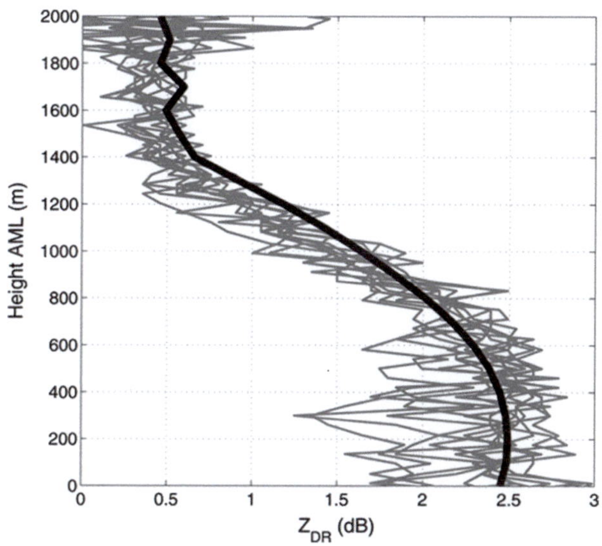

Fig. 7.43 Fifteen observed vertical profiles of Z_{DR} extracted from the RHI in Fig. 7.42 within the Z_{DR} column centered at the range of about 23 km (gray curves), compared to the model Z_{DR} profile (thick black curve). Adapted from Kumjian et al. (2012). © American Meteorological Society. Used with permission

The Z_{DR} model of Kumjian et al. (2012) describes only the freezing process and does not consider interaction of freezing raindrops with supercooled cloud droplets or graupel and hail which may form in a wet growth near the top of the Z_{DR} column in deeper convection. Water-coated graupel or hail may exhibit Z_{DR} similar to the one of freezing raindrops and affect the Z_{DR} signature from deep convective storms.

A more sophisticated microphysical model of convective updraft integrated into the HUCM spectral bin cloud model is described in Phillips et al. (2015); see also Chap. 8.

7.2.5.2 Refreezing of Snowflakes

If air temperature decreases monotonically with height, the snowflakes falling through the melting layer always convert to raindrops which fall to the surface as rain (Sect. 7.2.4.1).

However, vertical profile of temperature is often nonmonotonic during transitional winter events when the temperature inversion with positive temperature aloft is associated with the cold subfreezing layer next to the surface. Then dry snowflakes melt either completely or partially in the elevated melting layer and may refreeze in the cold surface layer. Partially melted snowflakes start refreezing instantaneously while falling into the subfreezing layer. Completely melted snowflakes (rain or drizzle particles) may not refreeze in the cold layer because they have to nucleate first which requires even colder temperature to activate the ice nuclei inside the drops. Detecting such transformation of supercooled raindrops (which constitute freezing rain) to ice pellets is very important because freezing rain poses considerable hazards and destructive potential, while ice pellets are relatively harmless.

Frequently, refreezing of supercooled raindrops enhances Z_{DR} as seen in Fig. 7.44. This is similar to enhancement produced by the melting layer. Two examples of inner ring of high Z_{DR} are in Fig. 7.44 from two winter storms in Oklahoma both characterized by nonmonotonic vertical profiles of temperature. During the storm on 24 December 2009 (observed with the C-band polarimetric radar), the standard Z_{DR} signature of the melting layer is indicated as an outer ring at the 4.3° scan and corresponds to the height exceeding 1 km in full agreement with the sounding profile of temperature in the left top panel in Fig. 7.44. The second circular ring of enhanced Z_{DR} is closer to the radar, and its altitude is about 500 m where the minimal temperature in the subfreezing surface layer is indicated by the sounding. A similar refreezing signature, although less pronounced, is detected in the storm on 27 January 2009 (observed with the S-band KOUN WSR-88D). Quasi-vertical profiles of Z and Z_{DR} taken at three different times and displayed in Fig. 7.45 confirm a sharp maximum of Z_{DR} at the height of 500 m accompanied by the corresponding drop in Z of 5–7 dB. The latter is easily explained by the change of dielectric constant of water to the one of solid ice during refreezing (see Chap. 5).

The increase of Z_{DR} during refreezing of raindrops happens because smaller supercooled raindrops freeze faster than larger ones and their contribution to reflectivities Z_H and Z_V drops by about 7 dB due to the change of dielectric constant. Hence the total Z_{DR} approaches the value determined by largest partially frozen raindrops with high intrinsic Z_{DR}. This is similar to the Z_{DR} enhancement by size sorting whereby small-size raindrops are removed from the spectrum (Sect. 7.1.3). Once all supercooled raindrops refreeze, Z_{DR} quickly drops because the intrinsic Z_{DR} of ice pellets or graupel is close to zero.

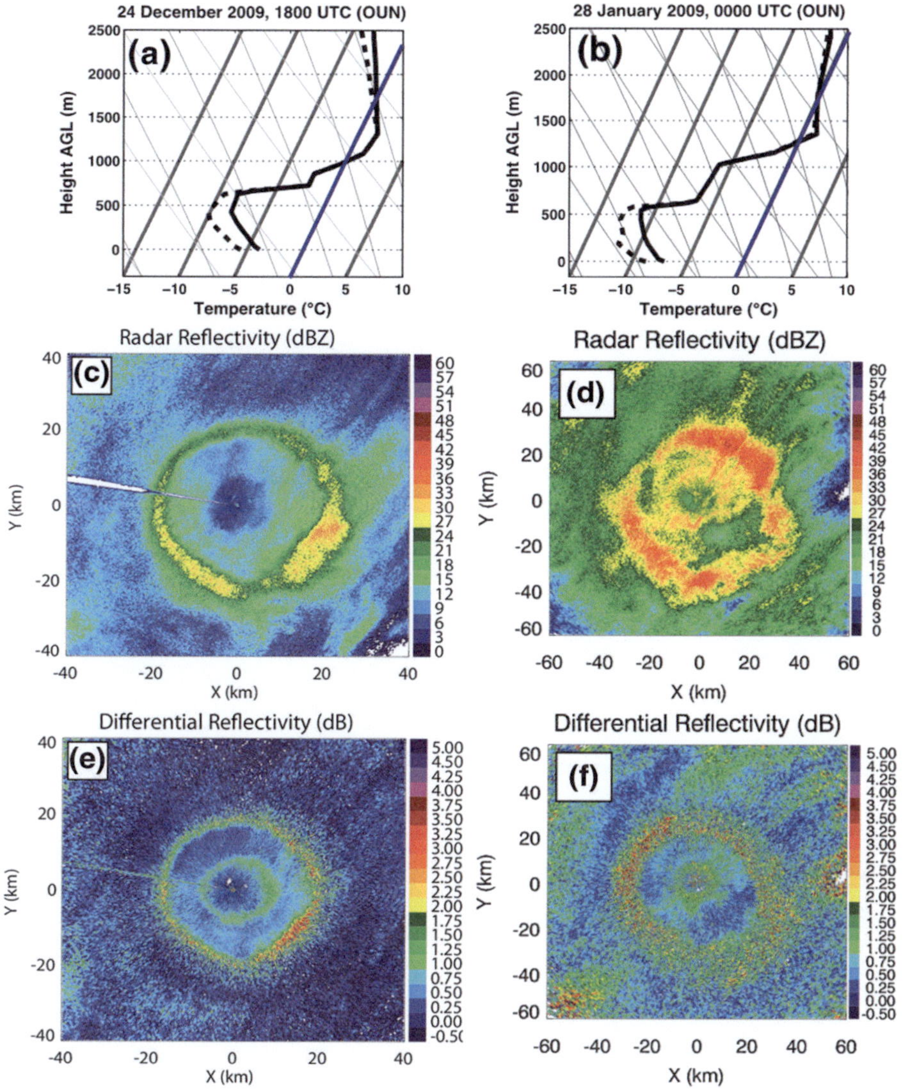

Fig. 7.44 Observed sounding temperatures (solid curves) and dew point temperatures (dashed curves) and PPIs of Z and Z_{DR} from two winter events: (**a**) on 24 December 2009 (left column) and (**b**) 27 January 2009 (right column). C-band PPIs of (**c**) Z and (**e**) Z_{DR} at elevation 4.3° at 1700 UTC are displayed in the left column. S-band PPIs of (**d**) Z and (**f**) Z_{DR} at elevation 3.4° at 2005 UTC are displayed in the right column. Adapted from Kumjian et al. (2013). © American Meteorological Society. Used with permission

Quasi-vertical profiles of Z_{DR} represented in a height vs time format are particularly advantageous for identification of the refreezing signature. An example of the refreezing layer at about 500 m above the surface observed by the KTLX WSR-88D radar in Oklahoma City and retrieved at elevation 6.4° is shown in the QVP height-

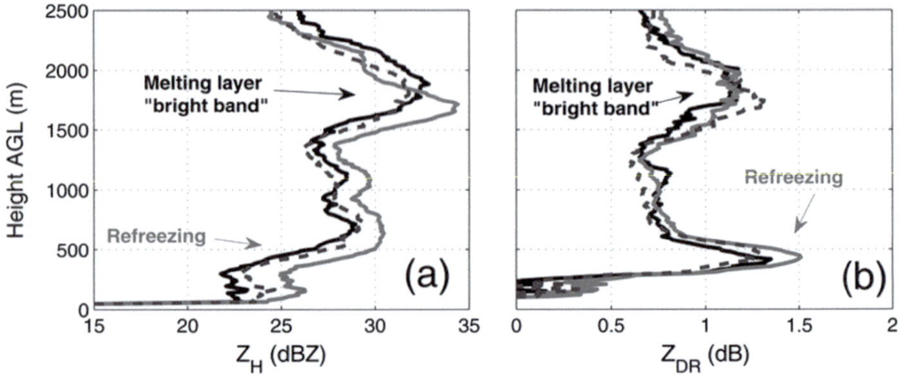

Fig. 7.45 Quasi-vertical profiles of (**a**) Z and (**b**) Z_{DR} taken from the S-band KOUN 3.3° elevation PPI scans on 27 January 2009 at three different times (1939 UTC in black, 1955 UTC in gray, and 2010 UTC in dashed gray). The melting layer bright band and low-level refreezing signature are annotated. Adapted from Kumjian et al. (2013). © American Meteorological Society. Used with permission

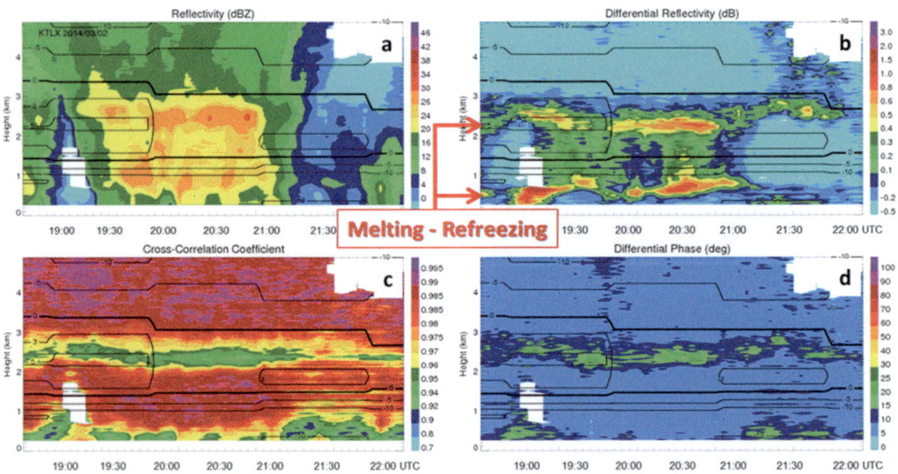

Fig. 7.46 The height vs time representation of quasi-vertical profiles of (**a**) Z, (**b**) Z_{DR}, (**c**) ρ_{hv}, and (**d**) Φ_{DP} retrieved from the KOUN WSR-88D collected at elevation 6.4° during the freezing rain event in central Oklahoma on 2 March 2014. The enhanced Z_{DR} indicates the melting and refreezing layers shown by the arrows in (**b**). Adapted from Ryzhkov et al. (2016). © American Meteorological Society. Used with permission

time plot in Fig. 7.46. The melting layer associated with an elevated temperature inversion between 2 and 3 km is identified by the typical increase of Z, Z_{DR}, and backscatter differential phase δ as well as the drop in ρ_{hv}. The series of vertical profiles of wet-bulb temperature retrieved from the HRRR model (Fig. 7.47) for this event indicates a classical temperature stratification favorable for a freezing rain—ice pellets transition. A deep, warm layer lies aloft, and a very cold surface layer with minimal temperature down to almost −15 °C remains during the period

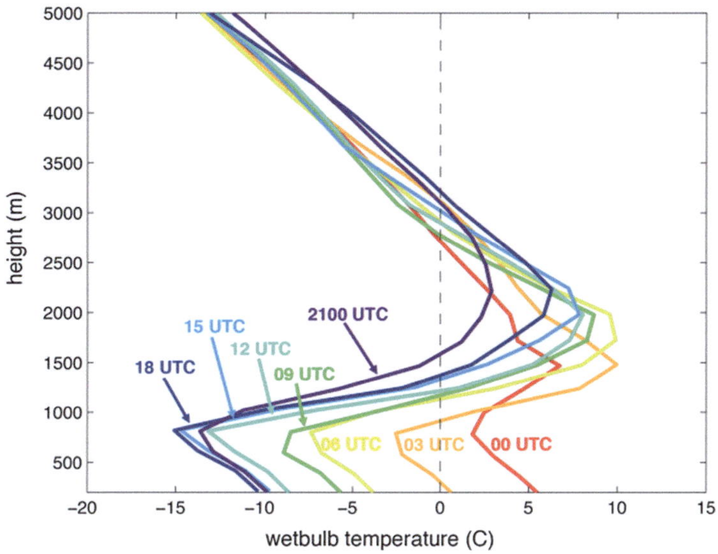

Fig. 7.47 Vertical profiles of the HRRR-analyzed wet-bulb temperature at KTLX from 0000 to 2100 UTC on 2 March 2014. Adapted from Ryzhkov et al. (2016). © American Meteorological Society. Used with permission

from 1900 to 2100 UTC when the Z_{DR} refreezing signature is observed in the QVP plot (Fig. 7.46). Note that the refreezing signature is often transient and poorly visible at the lowest beam elevations. Hence, the QVP provides a unique opportunity to observe its onset and demise, which has important practical implications for classification of winter precipitation types.

A realistically looking refreezing signature in Z_{DR} was simulated using a simple one-dimensional model with spectral (bin) microphysics which explicitly treats the processes of melting/refreezing. The model sets the initial size distribution above the melting layer, considers the density of dry snowflakes, and describes the evolution of mass water fraction and density of hydrometeors as they fall to the ground separately in 80 size intervals (Ryzhkov et al. 2014; Reeves et al. 2016). The model quantifies mixing ratios and precipitation rates associated with liquid and solid parts of precipitation. Hence the model can identify the dominant type of precipitation at the surface and estimate relative proportions of liquid and solid precipitation in the mixture (Reeves et al. 2016). The model assumes no interactions between particles such as riming or aggregation. That is, one snowflake aloft produces a single (generally mixed-phase) particle at the surface. Other required model inputs include vertical profiles of temperature, relative humidity, pressure, and assumption about the degree of riming of snowflakes. It is assumed that snow density decreases with size following Brandes et al. (2007) and increases with the degree of riming.

Vertical profiles of mass water fraction have been used by Ryzhkov et al. (2014) to compute vertical profiles of reflectivity Z and differential reflectivity Z_{DR}. The particles shape, orientation, density, and phase composition were explicitly

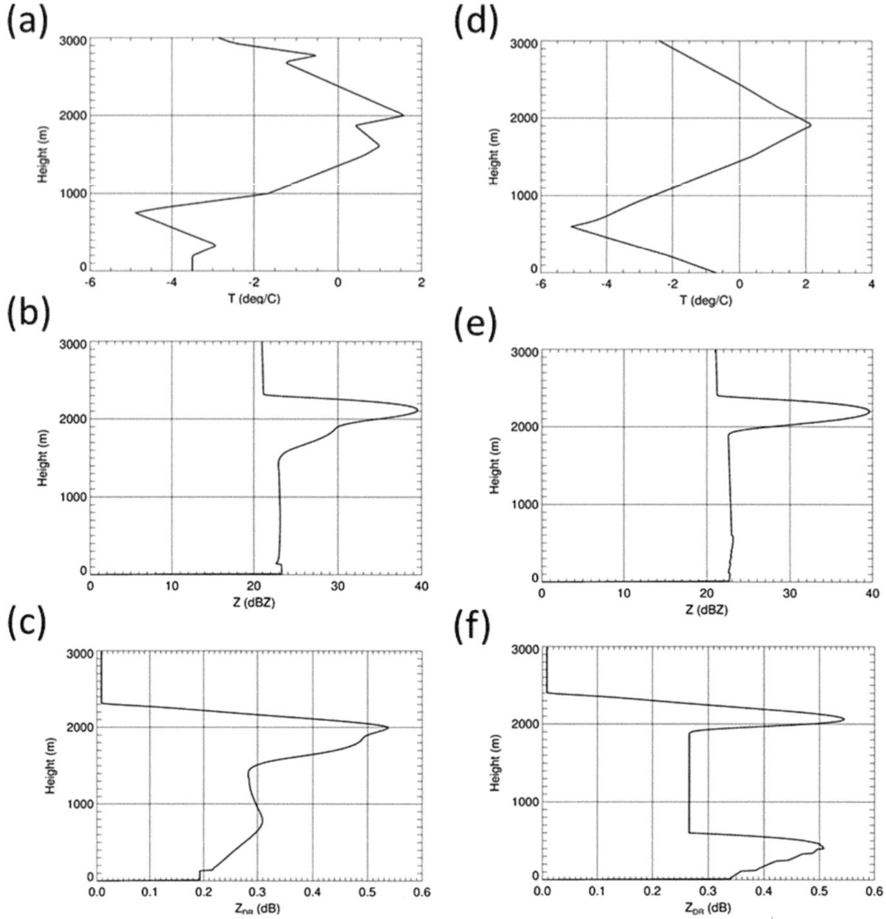

Fig. 7.48 Vertical profiles of temperature, *Z*, and Z_{DR} simulated using a 1D spectral bin model of melting and refreezing for the case of pure freezing rain (left panels: (**a**), (**b**), and (**c**)) and freezing rain and ice pellets underneath (right panels: (**d**), (**e**), and (**f**))

determined as functions of mass water fraction. The forward radar observation operator from Ryzhkov et al. (2011) yielded the vertical dependencies of *Z* and Z_{DR} at S band assuming typical temperature profiles conducive to the formation of freezing rain and ice pellets (Fig. 7.48a, d). Furthermore, the ice nucleation temperature T_{in} was set to −5 °C, and all raindrops originated from completely melted snowflakes and started refreezing regardless of size at temperatures below −5 °C. The model reproduces Z_{DR} enhancement in the vertical profile in the right column in Fig. 7.48f for the temperature profile crossing −5 °C.

In fact, the assumption $T_{in} = -5$ °C is questionable, and the exact range of temperatures when ice nucleation of freezing raindrops occurs is not known. However, the Reeves et al. (2016) analysis of the melting—refreezing model performance for a large number of winter events indicates that the best statistical agreement between the model classification output and ASOS measurements at the surface is achieved for a slightly lower value of $T_{in} = -6$ °C. This was also later confirmed in the study of Troemel et al. (2017).

7.3 Summary

The impact of different microphysical processes on the polarimetric variables Z, Z_{DR}, K_{DP}, and ρ_{hv} can be summarized in a pictorial diagram (Fig. 7.49) (Kumjian 2012). Each of the processes in this diagram has a unique "fingerprint" which characterizes the tendencies and magnitudes of the radar variables affected by the particular process. Such simplified representation can help meteorologists recognize a dominant microphysical process affecting precipitation formation in a particular area of the storm and interpret observed polarimetric variables and their temporal evolution. Cloud modelers can use this information for improving microphysical parameterization of various microphysical processes in storm-scale numerical weather prediction models and validate their performance after model output is converted into the fields of polarimetric variables using forward radar operators (e.g., Jung et al. 2008, 2010; Ryzhkov et al. 2011).

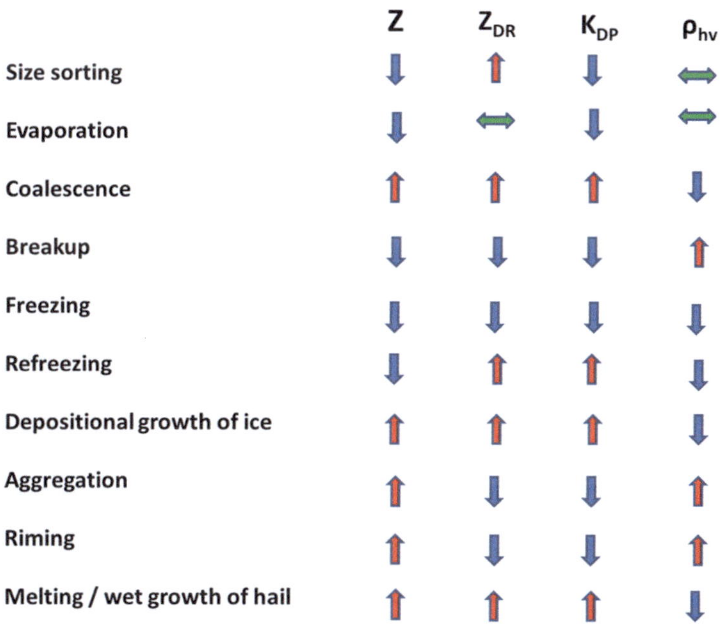

Fig. 7.49 Diagram of the qualitative polarimetric fingerprints of different microphysical processes. The direction of the arrows indicates the direction of the change in the variable caused by the given process. Sideways arrows indicate changes can be in either direction. From Kumjian (2012), with changes

References

Andrić, J., Kumjian, M., Zrnić, D., Straka, J., & Melnikov, V. (2013). Polarimetric signatures above the melting layer in winter storms: An observational and modeling study. *Journal of Applied Meteorology and Climatology, 52*, 682–700.

Atlas, D., & Ulbrich, C. (1977). Path- and area-integrated rainfall measurement by microwave attenuation in the 1 – 3 cm band. *Journal of Applied Meteorology, 16*, 1322–1331.

Aydin, K., Giridhar, V., & Zhao, Y. (1991). Polarimetric C-band radar observables in melting hail: A computational study. In Preprints. *25th International Conference on Radar Meteorology*. Paris (pp. 733–736).

Aydin, K., & Zhao, Y. (1990). A computational study of polarimetric radar observables in hail. *IEEE Transactions on Geoscience and Remote Sensing, 28*, 412–422.

Bailey, M., & Hallett, J. (2009). A comprehensive habit diagram for atmospheric ice crystals: Confirmation from the laboratory, AIRS II, and other field studies. *Journal of the Atmospheric Sciences, 27*, 919–926.

Barthazy, E., Henrich, W., & Waldvogel, A. (1998). Size distribution of hydrometeors through the melting layer. *Atmospheric Research, 47–48*, 193–208.

Battaglia, A., Kummerow, C., Shin, D., & Williams, C. (2003). Constraining microwave brightness temperatures by radar brightband observations. *Journal of Atmospheric and Oceanic Technology, 20*, 856–871.

Bechini, R., Baldini, L., & Chandrasekar, V. (2013). Polarimetric radar observations in the ice region of precipitating clouds at C-band and X-band radar frequencies. *Journal of Applied Meteorology and Climatology, 52*, 1147–1169.

Bigg, E. (1953a). The supercooling of water. *Proceedings of the Physical Society. Section B, 66*, 88–694.

Bigg, E. (1953b). The formation of atmospheric ice crystals by the freezing of droplets. *Quarterly Journal of the Royal Meteorological Society, 79*, 510–519.

Borowska, L., Ryzhkov, A., Zrnić, D., Simmer, C., & Palmer, R. (2011). Attenuation and differential attenuation of the 5-cm-wavelength radiation in melting hail. *Journal of Applied Meteorology and Climatology, 50*, 59–76.

Brandes, E., Ikeda, K., Zhang, G., Schoenhuber, M., & Rasmussen, R. (2007). A statistical and physical description of hydrometeor distributions in Colorado snowstorms using a video-disdrometer. *Journal of Applied Meteorology, 46*, 634–650.

Brandes, E. A., & Ikeda, K. (2004). Freezing-level estimation with polarimetric radar. *Journal of Applied Meteorology, 43*, 1541–1543.

Brandes, E., Zhang, G., & Vivekanandan, J. (2002). Experiments in rainfall estimation with a polarimetric radar in a subtropical environment. *Journal of Applied Meteorology, 41*, 674–685.

Carlin, J. (2018). *The use of polarimetric radar data for informing numerical weather prediction models*. PhD thesis. University of Oklahoma. 252 p.

Carras, J. N., & Macklin, W. C. (1973). The shedding of accreted water during hailstone growth. *Quarterly Journal of the Royal Meteorological Society, 99*, 639–648.

Chen, J.-P., & Lamb, D. (1994). The theoretical basis for the parameterization of ice crystal habits: Growth by water deposition. *Journal of the Atmospheric Sciences, 51*, 1206–1221.

Cheng, L., & English, M. (1983). A relationship between hailstone concentration and size. *Journal of the Atmospheric Sciences, 40*, 204–213.

Cheng, L., English, M., & Wong, R. (1985). Hailstone size distributions and their relationship to storm thermodynamics. *Journal of Climate and Applied Meteorology, 24*, 1059–1067.

D'Amico, M., Holt, A., & Capsoni, C. (1998). An anisotropic model of the melting layer. *Radio Science, 33*, 535–552.

Dawson, D., Mansell, E., & Kumjian, M. (2015). Does wind shear cause hydrometeor size sorting? *Journal of the Atmospheric Sciences, 72*, 340–348.

Dawson, D., Xue, M., Milbrandt, J., & Yau, M. (2010). Comparison of evaporation and cold pool development between single-moment and multimoment bulk microphysical schemes in idealized simulations of tornadic thunderstorms. *Monthly Weather Review, 138*, 1152–1171.

Erfani, E., & Mitchell, D. (2017). Growth of ice particle mass and projected area during riming. *Atmospheric Chemistry and Physics, 17*, 1241–1257.

Fabry, F., & Szyrmer, W. (1999). Modeling the melting layer. Part II: Electromagnetic. *Journal of the Atmospheric Sciences, 56*, 3593–3600.

Fabry, F., & Zawadzki, I. (1995). Long-term radar observations of the melting layer of precipitation and their interpretation. *Journal of the Atmospheric Sciences, 52*, 838–851.

Fukuta, N., & Takahashi, T. (1999). The growth of atmospheric ice crystals: A summary of findings in vertical supercooled cloud tunnel studies. *Journal of the Atmospheric Sciences, 56*, 1963–1979.

Giangrande, S. (2007). *Investigation of polarimetric measurements of rainfall at close and distant ranges*. PhD thesis, University of Oklahoma, 236 p.

Giangrande, S., Toto, T., Bansemer, A., Kumjian, M., Mishra, S., & Ryzhkov, A. (2016). Insights into riming and aggregation processes as revealed by aircraft, radar, and disdrometer observations for a 27 April 2011 widespread precipitation event. *Journal of Geophysical Research, 121*, 5846–5863.

Goeke, S., & Waldvogel, A. (1998). Studies of snowflake aggregation efficiencies within the melting layer. In Preprints, *Conference on Cloud Physics* (pp. 458–461), Everett, WA, American Meteor Society.

Grazioli, J., Lloyd, G., Panziera, L., Hoyle, C., Connoly, P., Henneberger, J., et al. (2015). Polarimetric radar and in situ observations of riming and snowfall microphysics during CLACE 2014. *Atmospheric Chemistry and Physics, 15*, 13787–13802.

Griffin, E., Schuur, T., & Ryzhkov, A. (2018). A polarimetric analysis of ice microphysical processes in snow, using quasi-vertical profiles. *Journal of Applied Meteorology and Climatology, 57*, 31–50.

Gunn, K., & Marshall, S. (1955). The effect of wind shear on falling precipitation. *Journal of Meteorology, 12*, 339–449.

Hallett, J., & Mossop, S. (1974). Production of secondary ice particles during the riming process. *Nature, 249*, 26–28.

Heymsfield, A., Bansemer, A., Poellot, M., & Wood, N. (2015). Observations of ice microphysics through the melting layer. *Journal of the Atmospheric Sciences, 72*, 2902–2928.

Joe, P., List, R., Kry, P., de Quervain, M., Lui, P., Stagg, P., et al. (1976). Loss of accreted water from a growing hailstone. In *Proceedings of the International Cloud Physics Conference* (pp. 264–269), Boulder, CO.

Johnson, D., & Hallett, J. (1968). Freezing and shattering of supercooled water drops. *Quarterly Journal of the Royal Meteorological Society, 94*, 468–482.

Jung, Y., Xue, M., & Zhang, G. (2010). Simulations of polarimetric radar signatures of a supercell storm using a two-moment bulk microphysical scheme. *Journal of Applied Meteorology and Climatology, 49*, 146–163.

Jung, Y., Zhang, G., & Xue, M. (2008). Assimilation of simulated polarimetric radar data for a convective storm using the Ensemble Kalman Filter. Part I: Observation operators for reflectivity and polarimetric variables. *Monthly Weather Review, 136*, 2228–2245.

Kaltenboeck, R., & Ryzhkov, A. (2013). Comparison of polarimetric signatures of hail at S and C bands for different hail sizes. *Atmospheric Research, 123*, 323–336.

Kennedy, P. C., & Rutledge, S. A. (2011). S-band dual-polarization radar observations of winter storms. *Journal of Applied Meteorology and Climatology, 50*, 844–858.

Khain, A., & Pinsky, M. (2018). *Physical processes in clouds and cloud modeling* (Vol. 686). Cambridge, UK: Cambridge University Press.

Khain, A., Pokrovsky, A., Rosenfeld, D., Blahak, U., & Ryzhkov, A. (2011). The role of CCN in precipitation and hail in a mid-latitude storm as seen in simulations using a spectral (bin) microphysics model in a 2D dynamic frame. *Atmospheric Research, 99*, 129–146.

Klaasen, W. (1988). Radar observation and simulation of the melting layer of precipitation. *Journal of the Atmospheric Sciences, 45*, 3741–3753.
Kumjian, M. (2012). *The impact of precipitation physical processes on the polarimetric radar variables*. PhD Thesis, University of Oklahoma.
Kumjian, M., Ganson, S., & Ryzhkov, A. (2012). Freezing of raindrops in deep convective storms: A microphysical and polarimetric model. *Journal of the Atmospheric Sciences, 69*, 3471–3490.
Kumjian, M., Khain, A., Benmoshe, N., Ilotoviz, E., Ryzhkov, A., & Phillips, V. (2014). The anatomy and physics of Z_{DR} columns: Investigating a polarimetric radar signature with a spectral bin microphysical model. *Journal of Applied Meteorology and Climatology, 53*, 1820–1843.
Kumjian, M., Mishra, S., Giangrande, S., Toto, T., Ryzhkov, A., & Bansemer, A. (2016). Polarimetric radar and aircraft observations of saggy bright band during MC3E. *Journal of Geophysical Research-Atmospheres, 121*, 3584–3607.
Kumjian, M., & Prat, O. (2014). The impact of raindrop collisional processes on the polarimetric radar variables. *Journal of the Atmospheric Sciences, 71*, 3052–3067.
Kumjian, M., & Ryzhkov, A. (2009). Storm-relative helicity revealed from polarimetric radar measurements. *Journal of the Atmospheric Sciences, 66*, 667–685.
Kumjian, M., & Ryzhkov, A. (2010). The impact of evaporation on polarimetric characteristics of rain. Theoretical model and practical implications. *Journal of Applied Meteorology and Climatology, 49*, 1247–1267.
Kumjian, M., & Ryzhkov, A. (2012). The impact of size sorting on the polarimetric radar variables. *Journal of the Atmospheric Sciences, 69*, 2042–2060.
Kumjian, M., Ryzhkov, A., Reeves, H., & Schuur, T. (2013). Dual-polarization radar observations of hydrometeor refreezing in winter storms. *Journal of Applied Meteorology and Climatology, 52*, 2549–2566.
Lamb, D., & Verlinde, J. (2011). *Physics and chemistry of clouds* (p. 584). Cambridge, UK: Cambridge University Press.
Lesins, G. B., List, R., & Joe, P. I. (1980). Ice accretions. Part I: Testing of new atmospheric icing concepts. *Journal de Recherches Atmospheriques, 14*, 347–356.
Marshall, J. (1953). Precipitation trajectories and patterns. *Journal of Meteorology, 10*, 25–29.
Matrosov, S. Y., Mace, G. G., Marchand, R., Shupe, M. D., Hallar, A. G., & Mc Cubbin, I. B. (2012). Observations of ice crystal habits with a scanning polarimetric W-band radar at slant linear depolarization ratios mode. *Journal of Atmospheric and Oceanic Technology, 29*, 989–1008.
Matrosov, S. Y., Reinking, R. F., Kropfli, R. A., Martner, B. E., & Bartram, B. W. (2001). On the use of radar depolarization ratios for estimating shapes of ice hydrometeors in winter clouds. *Journal of Applied Meteorology, 40*, 479–490.
Moisseev, D., Lautaportti, S., Tyynela, J., & Lim, S. (2016). Dual-polarization radar signatures in snowstorms: Role of snowflake aggregation. *Journal of Geophysical Research-Atmospheres, 121*, 12644–12655.
Moisseev, D., von Lerber, A., & Tiira, J. (2017). Quantifying the effect of riming on snowfall using ground-based observations. *Journal of Geophysical Research – Atmospheres, 122*, 4019–4037.
Mosimann, L., Weingartner, E., & Waldvogel, A. (1994). An analysis of accreted drop sizes and mass on rimed snow crystals. *Journal of the Atmospheric Sciences, 51*, 1548–1558.
Olson, W., Bauer, P., Viltard, N., Johnson, D., Tao, W., Meneghini, R., et al. (2001). A melting-layer model for passive-active microwave remote sensing applications. Part I: Model formulation and comparison with observations. *Journal of Applied Meteorology, 40*, 1145–1163.
Ortega, K., Krause, J., & Ryzhkov, A. (2016). Polarimetric radar characteristics of melting hail. Part III: Validation of the algorithm for hail size discrimination. *Journal of Applied Meteorology and Climatology, 55*, 829–848.
Oue, M., Galletti, M., Verlinde, J., Ryzhkov, A., Lu, Y., & Bharadwaj, N. (2016). Use of X-band differential reflectivity measurements to study shallow Arctic mixed-phase clouds. *Journal of Applied Meteorology and Climatology, 55*, 403–424.

Phillips, V., Khain, A., Benmoshe, N., Ilotoviz, E., & Ryzhkov, A. (2015). Theory of time-dependent freezing. Part II. Scheme for freezing raindrops and simulations by a cloud model with spectral bin microphysics. *Journal of the Atmospheric Sciences, 72*, 262–286.

Phillips, V., Khain, A., & Pokrovsky, A. (2007). The influence of melting on the dynamics and precipitation production in maritime and continental storm clouds. *Journal of the Atmospheric Sciences, 64*, 338–359.

Picca, J., & Ryzhkov, A. (2012). A dual-wavelength polarimetric analysis of the 16 May 2010 Oklahoma City extreme hailstorm. *Monthly Weather Review, 140*, 1385–1403.

Pruppacher, H., & Klett, J. (1997). Microphysics of clouds and precipitation (2nd ed., p. 976). Dordrecht, The Netherlands: Kluwer Academic.

Rasmussen, R., & Heymsfield, A. (1987). Melting and shedding of graupel and hail. Part I: Model physics. *Journal of the Atmospheric Sciences, 44*, 2754–2763.

Rasmussen, R., Levizzani, V., & Pruppacher, H. (1984). A wind tunnel and theoretical study of the melting behavior of atmospheric ice particles. Part III: Experiment and theory for spherical ice particles of radius > 500 μm. *Journal of the Atmospheric Sciences, 41*, 381–388.

Reeves, H., Ryzhkov, A., & Krause, J. (2016). Discrimination between winter precipitation types based on spectral-bin microphysical modeling. *Journal of Applied Meteorology and Climatology, 55*, 1747–1761.

Rogers, R., & Yau, M. (1989). *A short course in cloud physics*. New York: Pergamon Press.

Russchenberg, H., & Ligthart, L. (1996). Backscattering by and propagation through the melting layer of precipitation: A new polarimetric model. *IEEE Transactions on Geoscience and Remote Sensing, 34*, 3–14.

Ryzhkov, A., Giangrande, S., Khain, A., Pinsky, M., & Pokrovsky, A. (2008). *Exploring model-based polarimetric retrieval of vertical profiles of precipitation*. In Extended Abstracts, 5th European Conference on Radar in Meteorology and Hydrology, Helsinki, Finland, CD-ROM, P6.1.

Ryzhkov, A., Kumjian, M., Ganson, S., & Khain, A. (2013a). Polarimetric radar characteristics of melting hail. Pt I: Theoretical simulations using spectral microphysical modeling. *Journal of Applied Meteorology and Climatology, 52*, 2849–2870.

Ryzhkov, A., Kumjian, M., Ganson, S., & Zhang, P. (2013b). Polarimetric radar characteristics of melting hail. Pt II: Practical implications. *Journal of Applied Meteorology and Climatology, 52*, 2871–2886.

Ryzhkov, A., Pinsky, M., Pokrovsky, A., & Khain, A. (2011). Polarimetric radar observation operator for a cloud model with spectral microphysics. *Journal of Applied Meteorology and Climatology, 50*, 873–894.

Ryzhkov, A., Reeves, H., Krause, J., & Burcham, H. (2014). *Discrimination between winter precipitation types based on explicit microphysical modeling of melting and refreezing in the polarimetric hydrometeor classification algorithm*. In Extended abstracts. 8th European Conference on Radar in Meteorology and Hydrology, Garmisch-Partenkirchen, Germany, MIC.P07.

Ryzhkov, A., Zhang, P., Reeves, H., Kumjian, M., Tschallener, T., Troemel, S., et al. (2016). Quasi-vertical profiles – a new way to look at polarimetric radar data. *Journal of Atmospheric and Oceanic Technology, 33*, 551–562.

Schrom, R. S., Kumjian, M. R., & Lu, Y. (2015). Polarimetric radar signatures of dendritic growth zones within Colorado winter storms. *Journal of Applied Meteorology and Climatology, 54*, 2365–2388.

Schuur, T., Ryzhkov, A., & Clabo, D. (2005). *Climatological analysis of DSDs in Oklahoma as revealed by 2D-video disdrometer and polarimetric WSR-88D*. Preprints, 32nd Conference on Radar Meteorology., CD-ROM, 15R.4

Sinclair, V., Moisseev, D., & von Lerber, A. (2016). How dual-polarization radar observations can be used to verify model representation of secondary ice. *Journal of Geophysical Research – Atmospheres, 121*, 10954–10970.

Smith, P., Musil, D., Weber, S., Spahn, J., Johnson, G., & Sand, W. (1976). Raindrop and hailstone distributions inside hailstorms. In Preprints. *International Conference on Cloud Physics* (pp. 252–257), Boulder, CO.

Spahn, J., & Smith, P. (1976). Some characteristics of hailstone size distributions inside hailstorms. In Preprints. *17th Conference on Radar Meteorology* (pp. 187–191), Seattle, WA.

Straka, J. (2009). *Cloud and precipitation microphysics. Principles and parameterizations* (p. 392). Cambridge, UK: Cambridge University Press.

Szyrmer, W., & Zawadzki, I. (1999). Modeling of the melting layer. Part I: Dynamics and microphysics. *Journal of the Atmospheric Sciences, 56*, 3573–3592.

Takahashi, T., Endoh, T., Wakaham, G., & Fukuta, N. (1991). Vapor diffusional growth of free-falling snow crystals between -3 and -23°C. *Journal of the Meteorological Society of Japan, 69*, 15–30.

Tobin, D., & Kumjian, M. (2017). Polarimetric radar and surface-based precipitation-type observations of ice pellet to freezing rain transitions. *Weather and Forecasting, 32*, 2065–2082.

Troemel, S., Ryzhkov, A., Bick, T., Muhlbauer, K., & Simmer, C. (2017). Towards nowcasting of winter precipitation: The Black Ice Event in Berlin 2014. *Meteorologische Zeitschrift, 26*, 147–160.

Troemel, S., Ryzhkov, A., Zhang, P., & Simmer, C. (2014). Investigations of backscatter differential phase in the melting layer. *Journal of Applied Meteorology and Climatology, 53*, 2344–2359.

Ulbrich, C., & Atlas, D. (1982). Hail parameter relations: A comprehensive digest. *Journal of Applied Meteorology, 21*, 22–43.

Vivekanandan, J., Bringi, V., & Raghavan, R. (1990). Multiparameter radar modeling and observation of melting ice. *Journal of the Atmospheric Sciences, 47*, 549–563.

Vogel, J., Fabry, F., & Zawadzki, I. (2015). *Attempts to observe polarimetric signatures of riming in stratiform precipitation*. In 37th Conference on Radar Meteorology, Norman, OK, American Meteor Society, 6B.6. Retrieved from https://ams.confex.com/ams/37RADAR/webprogram/Paper275246.html

Wang, Y., Zhang, J., Ryzhkov, A., & Tang, L. (2013). C-band polarimetric radar QPE based on specific differential propagation phase for extreme typhoon rainfall. *Journal of Atmospheric and Oceanic Technology, 30*, 1354–1370.

Westbrook, C., & Heymsfield, A. (2011). Ice crystals growing from vapor in supercooled clouds between -2.5° and -22°C: Testing current parameterization methods using laboratory data. *Journal of the Atmospheric Sciences, 68*, 2416–2429.

Williams, E., Smalley, D., Donovan, M., Hallowell, R., Hood, K., Bennett, B., et al. (2015). Measurements of differential reflectivity in snowstorms and warm season stratiform systems. *Journal of Applied Meteorology and Climatology, 54*, 573–595.

Xie, X., Evaristo, R., Troemel, S., Saavedra, P., Simmer, C., & Ryzhkov, A. (2016). Radar observation of evaporation and implications for quantitative precipitation and cooling rate estimation. *Journal of Atmospheric and Oceanic Technology, 33*, 1779–1792.

Yokoyama, T., & Tanaka, H. (1984). Microphysical processes of melting snowflakes detected by two-wavelength radar. Part I: Principle of measurement based on model calculation. *Journal of the Meteorological Society of Japan, 62*, 650–666.

Zawadzki, I., Jung, E., & Lee, G. (2010). Snow studies. Part I: A study of natural variability of snow terminal velocity. *Journal of the Atmospheric Sciences, 67*, 1591–1604.

Zawadzki, I., Szyrmer, W., Bell, C., & Fabry, F. (2005). Modeling of the melting layer. Part III: The density effect. *Journal of the Atmospheric Sciences, 62*, 3705–3723.

Zhang, G., Luchs, S., Ryzhkov, A., Xue, M., Ryzhkova, L., & Cao, Q. (2011). Winter precipitation microphysics characterized by polarimetric radar and video disdrometer observations in central Oklahoma. *Journal of Applied Meteorology and Climatology, 50*, 1558–1570.

Chapter 8
Polarimetric Characteristics of Deep Convective Storms

In this chapter, we describe polarimetric signatures observed in mesoscale convective systems, hail-bearing storms, and supercell tornadic storms.

8.1 Mesoscale Convective Systems (MCSs)

The most common type of mesoscale convective system is an elongated line of severe thunderstorms (known as a squall line) followed by trailing stratiform precipitation (e.g., Bluestein 2013; Houze 1993; Houze et al. 1989; Markowski and Richardson 2010). A conceptual model of the kinematic, microphysical, and radar-echo structure of MCS viewed in vertical cross section oriented perpendicular to the convective line is illustrated in Fig. 8.1.

A typical composite PPI of the MCS observed with the KOUN WSR-88D is displayed in Fig. 8.2. A squall line is characterized by high reflectivity exceeding 50 dBZ and high differential reflectivity up to 4 dB. Differential reflectivity is particularly high at a leading edge of squall lines in areas of convective updrafts where intense size sorting of raindrops occurs (Sect. 7.1.3). If a convective cell comprising a squall line contains hail of significant size, then Z_{DR} can be reduced there. The correlation coefficient ρ_{hv} in a squall line is usually lower than in the trailing stratiform precipitation which is often separated from the squall line by a transition region of diminished values of Z. Differential phase Φ_{DP} is high in radial directions coincident with the orientation of squall line if the squall line passes over the radar. Rapid radial increase of Φ_{DP} signifies potential attenuation and differential attenuation which may significantly bias the estimates of Z and Z_{DR} even at S band. Ryzhkov and Zrnic (1995) report large negative biases of Z and Z_{DR} with magnitudes of 10 dB and 3–4 dB along a squall line of an MCS using differential phase measurements made with the Cimarron S-band polarimetric radar.

At longer distances from the radar, the beam intercepts the melting layer of the stratiform part of the MCS which causes increase of Z and Z_{DR} and decrease of ρ_{hv}.

A. V. Ryzhkov, D. S. Zrnic, *Radar Polarimetry for Weather Observations*,
Springer Atmospheric Sciences, https://doi.org/10.1007/978-3-030-05093-1_8

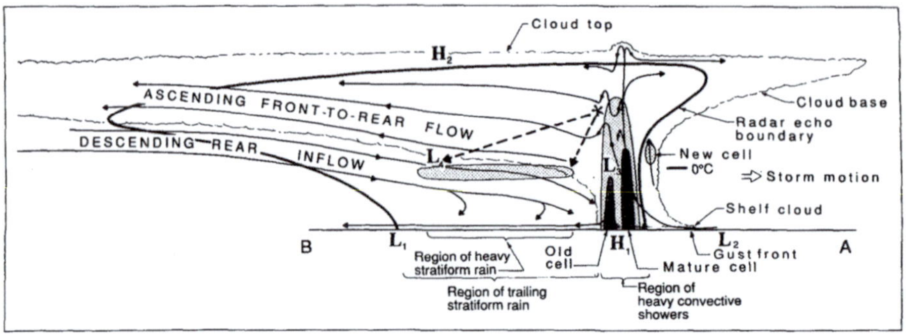

Fig. 8.1 Conceptual model of MCS. From Houze et al. (1989). © American Meteorological Society. Used with permission

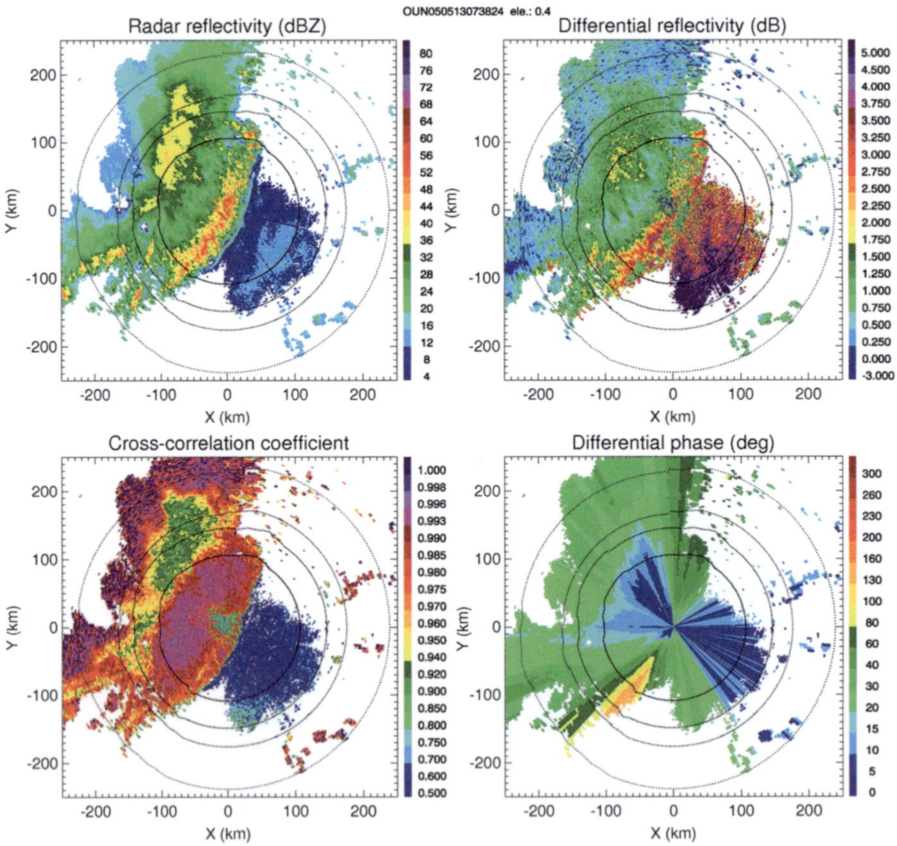

Fig. 8.2 Composite plot of Z, Z_{DR}, ρ_{hv}, and Φ_{DP} at 0.43° elevation angle measured by the KOUN WSR-88D on 13 May 2005 in central Oklahoma. Contours represent slant ranges associated with the bottom and top of the melting layer as functions of azimuth (see Chap. 9 for details)

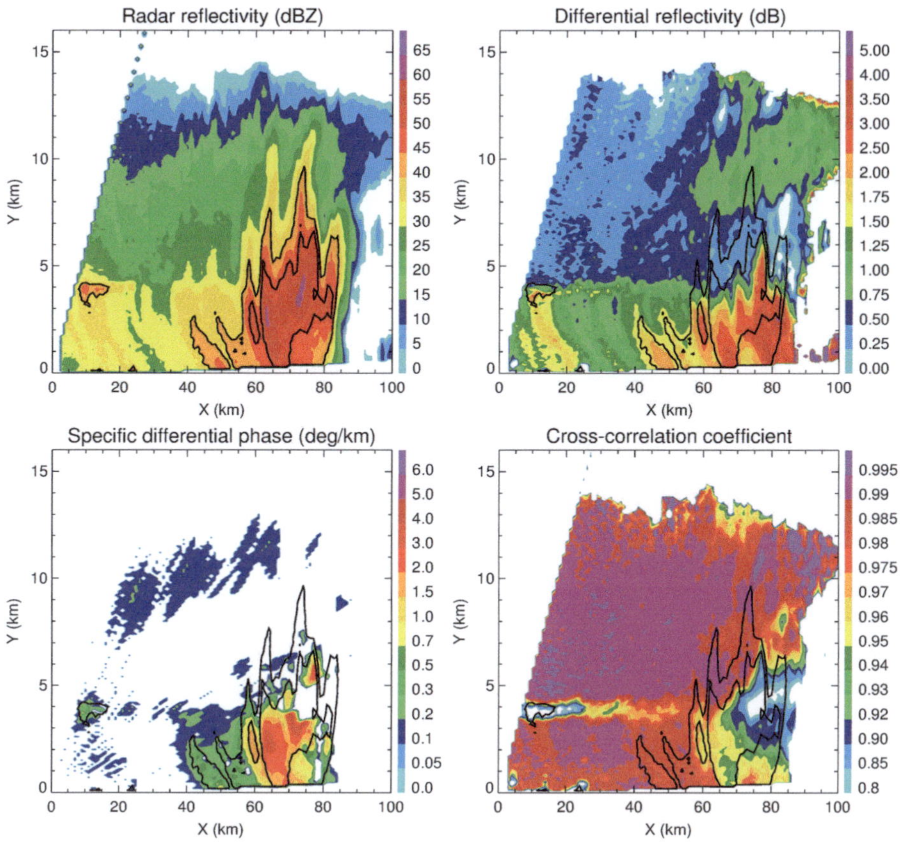

Fig. 8.3 Composite RHI of Z, Z_{DR}, K_{DP}, and ρ_{hv} measured by the KOUN WSR-88D on 17 May 2005 at 055651 UTC and $Az = 191°$. Z contours at 40 and 50 dBZ are overlaid on all panels

The correlation coefficient reverts to its high values close to 1 once the radar beam leaves the areas affected by the melting layer impact. Extremely high values of Z_{DR} and very low values of ρ_{hv} are observed ahead of the squall line in the southeast sector of the radar coverage area where biological scatterers such as insects and birds dominate the radar return.

A spatial distribution of Z, Z_{DR}, K_{DP}, and ρ_{hv} in the RHI along a vertical plane oriented perpendicular to the squall line of a typical MCS is shown in Fig. 8.3. The convective part of the MCS is located between 60 and 80 km from the radar and is characterized by high Z exceeding 60 dBZ signifying presence of hail aloft. One of the notable features is the Z_{DR} column extending up to 6.5 km height (i.e., 2.5 km above the environmental freezing level) at the distance of about 75 km. This Z_{DR} column has its base at $X > 77$ km in the inflow area ahead of the squall line where reflectivity near the surface is relatively low (generally less than 40 dBZ) which is a clear manifestation of the size sorting process. Higher than 0.75 dB values of Z_{DR} aloft (above 7 km AGL) manifest depolarization effects caused by oriented crystals near the tops of multiple convective cells (Sect. 6.9). These crystals are oriented quasi-vertically by electric fields generated in the proximity of convective updrafts. Slightly negative Z_{DR}s at midlevels on the rear side of the squall line are likely caused by differential attenuation in melting hail. Very low ρ_{hv} (below 0.8) at the heights between 3 and 5 km AGL at the rear side of the squall line is partially attributed to the presence of melting hail but is primarily caused by nonuniform beam filling (Sect. 6.8). Such drop of ρ_{hv} is observed in the area of the highest vertical gradient of differential phase Φ_{DP} (not shown) separating the altitudes where Φ_{DP} is high due to propagation in rain and low in ice particles above the freezing level.

Specific differential phase K_{DP} is positively correlated with Z in the convective region below the melting layer which is expected because the bulk of graupel/hail melts into raindrops producing high K_{DP}. K_{DP} cannot be reliably estimated in the areas of low ρ_{hv} (below 0.8–0.9), but some indication of K_{DP} higher than 1–2 km^{-1} is visible near the top of the Z_{DR} column ($X = 77$ km, $Y = 6$ km) which might be a manifestation of the so-called K_{DP} column (Hubbert et al. 1998; Loney et al. 2002; Van Lier-Walqui et al. 2016) (see Sect. 8.2). High Z_{DR} and K_{DP} above the freezing level indicate presence of water associated either with supercooled raindrops or with water-coated hail growing in a wet growth regime in a convective updraft.

The melting layer in a stratiform part of the MCS at the height of about 4 km and ranges less than 60 km from the radar is marked by well-pronounced ρ_{hv} minimum and abrupt drop of Z_{DR} with altitude. Note that in this vertical cross section, the reflectivity does not exhibit clearly defined "bright band" (except for very close proximity to the radar).

8.2 Hailstorms

Hail is formed when either graupel or large frozen raindrops grow by accreting supercooled cloud droplets in convective updrafts. Original hail embryos lofted by updraft may temporarily fall out of it and then recirculate back into the updraft where they continue to grow to larger size (Fig. 8.4) and their trajectories within a cloud can be quite complicated. Once a hailstone becomes large enough and its terminal velocity exceeds the one of the vertical updraft, it starts falling either against updraft or in the nearby downdraft. Hail can be effectively identified by the polarimetric radar and was a subject of numerous studies during the last three decades (Aydin et al. 1986; Aydin and Zhao 1990; Balakrishnan and Zrnic 1990; Bringi et al. 1986; Conway and Zrnic 1993; Holler et al. 1994; Hubbert et al. 1998; Illingworth et al. 1987; Kaltenboeck and Ryzhkov 2013; Kennedy et al. 2001, 2014; Ortega et al. 2016; Picca and Ryzhkov 2012; Ryzhkov et al. 2013a, b; Tessendorf et al. 2005 among others). Because hailstones have lower dielectric constant than water and are more randomly oriented compared to raindrops, the differential reflectivity of dry hailstones is usually much lower than the one of rain having the same reflectivity factor. Another polarimetric variable which historically

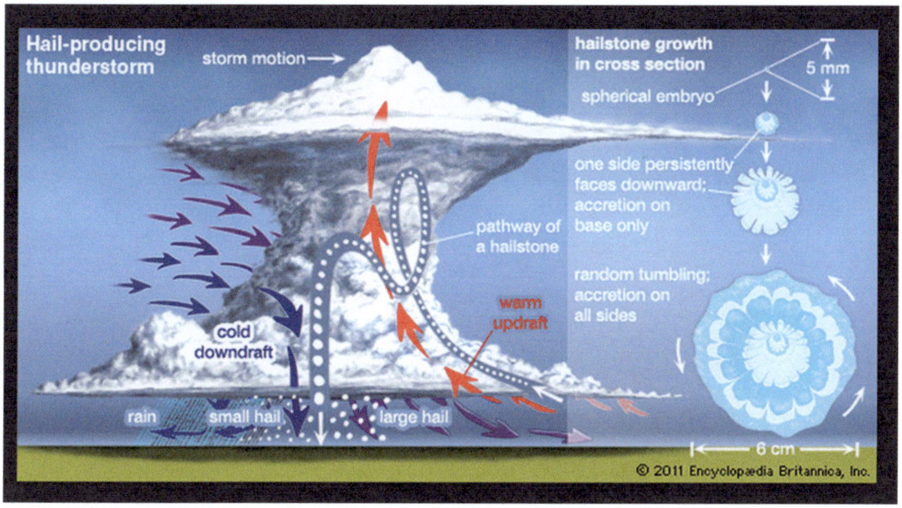

Fig. 8.4 Conceptual picture of a hailstorm. Reprinted with permission from *Encyclopedia Britannica*, © 2012 by Encyclopedia Britannica, Inc.

was first used to detect hail is the linear depolarization ratio *LDR* because it is usually significantly higher in hail than in rain with demarcation threshold of about −25 dB.

A typical vertical cross section of Z, Z_{DR}, ρ_{hv}, and depolarization ratio *DR* at S band in a hail-bearing storm is presented in Fig. 8.5. In this example, *DR* (serving as a proxy for a circular depolarization ratio *CDR*) was computed from the combination of Z_{DR} and ρ_{hv} (Ryzhkov et al. 2017) (see also Sect. 5.6). Very high reflectivity exceeding 65 dBZ combined with depolarization ratio larger than −10 dB near the top of the reflectivity core signifies hail aloft with the size likely over 5–6 cm (Fig. 5.20). Negative Z_{DR} associated with large Z and *DR* also confirms that hailstones may exceed the resonance size (Fig. 5.13). Very low ρ_{hv} below 0.8 is consistent with the presence of very large hail aloft. Slightly negative Z_{DR} behind the hail core in the region of moderate and weak reflectivity is attributed to the effect of differential attenuation which was not accounted for in the Z_{DR} data. Note that a "Z_{DR} hole" 2–3 km below the environmental freezing level (depicted by a dashed line) indicates that large partially melted hailstones characterized by low intrinsic Z_{DR} apparently dominate the radar return in the mixture with raindrops originating from smaller hailstones and graupel. Differential reflectivity's high value exceeding 2.5 dB (typical for rain) is restored where $Z > 50$ dBZ just 1 km above the surface.

A vertical cross section of Z, Z_{DR}, ρ_{hv}, and K_{DP} for the same storm but 20 min earlier (Fig. 8.6) exhibits classical signatures of the Z_{DR} column (enclosed in the white ellipse) and the K_{DP} column (in the yellow ellipse). A tilted Z_{DR} column is coincident with convective updraft where hail is growing. Due to intense size sorting in the updraft, very high Z_{DR} (above 4 dB) is associated with modest Z (about 30 dBZ). In this example, the Z_{DR} column stretches more than 2 km above the

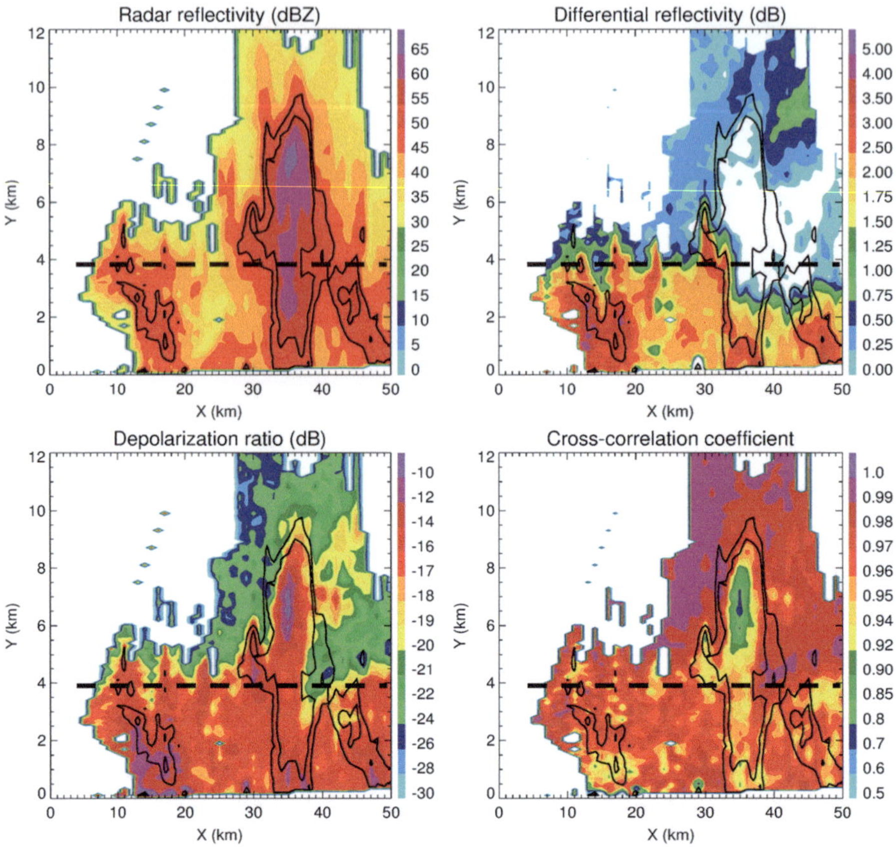

Fig. 8.5 Composite RHI of Z, Z_{DR}, ρ_{hv}, and depolarization ratio DR computed from Z_{DR} and ρ_{hv} in a hailstorm observed by the KOUN WSR-88D on 17 June 2005 at 0447 UTC and $Az = 305°$. Dashed black line denotes the environmental freezing level. Overlaid are contours of $Z = 45$, 50, and 55 dBZ

environmental freezing level. Hail that grows near the top of the Z_{DR} column falls along the trajectories farther away from the radar. Some of water-coated hailstones along with large supercooled raindrops produce high values of specific differential phase above the freezing level and comprise an upper part of the K_{DP} column. The base of the K_{DP} column consists mostly of raindrops and partially of melted hailstones falling in a downdraft (Loney et al. 2002). Large dry hail advected father away from the updraft is characterized by low (or even negative) Z_{DR} and creates a "Z_{DR} hole" below the freezing level at the distance of about 50 km from the radar. Hail mixed with rain is marked by a tilted column of very low ρ_{hv} (<0.9).

It is common for deep convective storms to have the Z_{DR} and K_{DP} columns separated at the lower levels but originated from the same location near the middle of

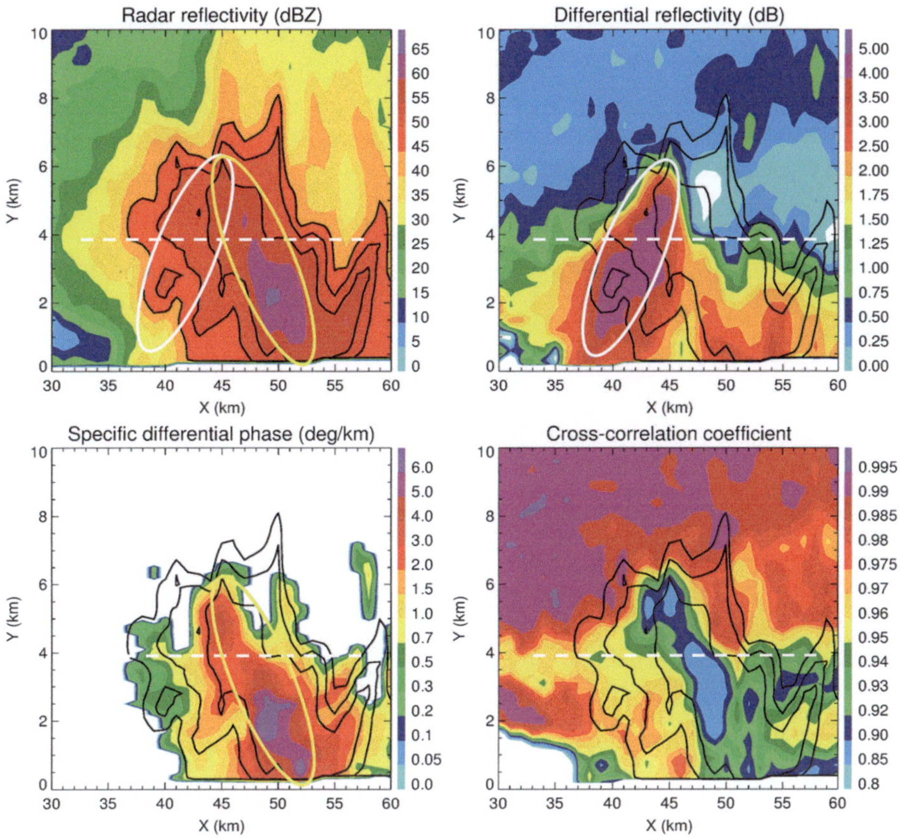

Fig. 8.6 Composite RHI of Z, Z_{DR}, ρ_{hv}, and K_{DP} in a hailstorm observed by the KOUN WSR-88D on 17 June 2005 at 0426 UTC and $Az = 300°$. The Z_{DR} column is enclosed in the white ellipse, and the K_{DP} column is enclosed in the yellow ellipse. Overlaid are contours of $Z = 45$, 50, and 55 dBZ

a convective updraft where partially frozen raindrops coexist with water-coated hailstones growing in the wet growth regime.

The Z_{DR} column indicates an updraft, whereas the K_{DP} column marks a major precipitation shaft in a downdraft. Note that K_{DP} can be extremely high (over 6–7° km^{-1} at S band) in a core of hailstorm as in the case illustrated in Fig. 8.6. The corresponding rain rate would reach unrealistic values (above 200–250 mm h^{-1}) if it is assumed that these K_{DP}s are attributed to pure rain (see Chap. 10). Definitely, melting hailstones/graupel also contribute to K_{DP} (Loney et al. 2002), and anomalously large K_{DP} in deep convective storms almost certainly indicates strong presence of hail mixed with rain (Ryzhkov et al. 2013b).

In the cases illustrated in Figs. 8.5 and 8.6, hail has reflectivity well above 55 dBZ justifying use of high Z for hail detection. However, there is growing evidence that giant hail may be observed in areas of modest Z (Blair and Leighton 2012). This happens in the proximity of very strong updrafts which can be almost free of

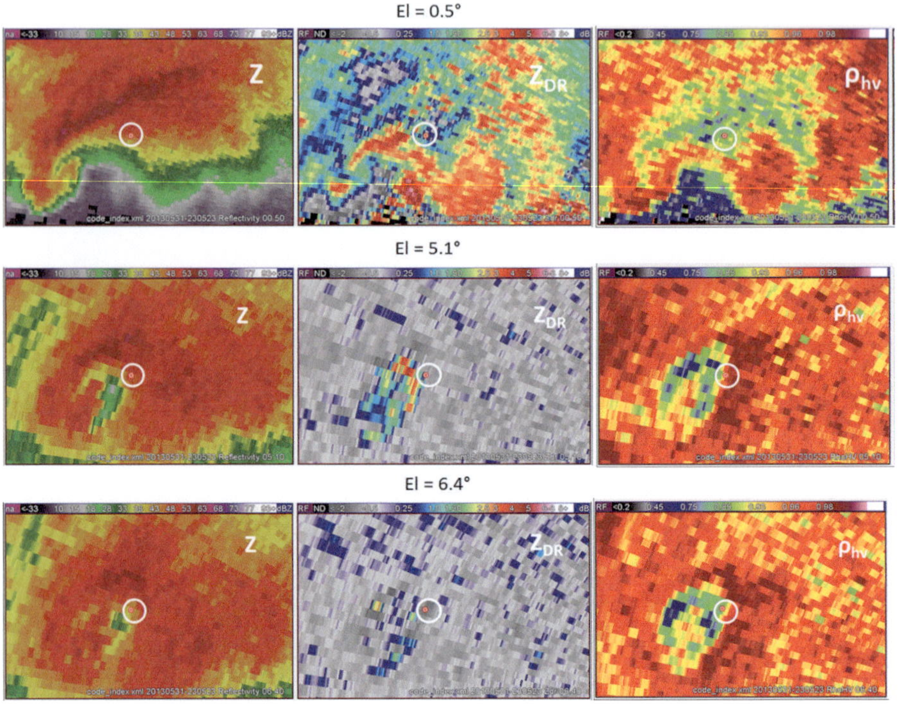

Fig. 8.7 Composite PPIs of Z, Z_{DR}, and ρ_{hv} at three antenna elevations for the storm on 31 May 2013 near Oklahoma City which produced 16 cm size hail. The bull's-eye marker enclosed in the white circle indicates the position of the largest hailstone reported at the ground

hydrometeors (i.e., bounded weak echo regions or BWER). Giant hail produced near the BWER boundary may fall just next to the updraft from which all smaller particles are blown away so that only a few very large hailstones remain and the corresponding reflectivity is relatively low.

A storm producing really giant hail (up to 16 cm in diameter) is illustrated in Fig. 8.7 which displays PPIs of Z, Z_{DR}, and ρ_{hv} at three elevation angles: 0.5°, 5.1°, and 6.4°. The storm occurred near Oklahoma City on 31 May 2013, and the data were collected by the KTLX (Oklahoma City OK) WSR-88D. The bull's-eye marker enclosed in the white circle indicates the position of hailstone measured to be 16 cm in diameter at about 2307 UTC. At the hail location and elevation 0.5°, reflectivity is modest 45–50 dBZ. At elevation 5.1°, the location of giant hail corresponds to the northeast edge of the BWER and along the edge of the Z_{DR} column. Note that the area of reduced ρ_{hv} (as low as 0.53) characterizes the northern and western areas of higher Z surrounding the BWER. At elevation 6.4°, the giant hail measurement occurred under the northeast edge of the BWER. The area along the periphery of the BWER has Z_{DR} occasionally less than -1.75 dB and ρ_{hv} as low as 0.51!

Hailstones are large scatterers with very pronounced resonance scattering effects; hence, polarimetric signatures of hail strongly depend on the radar wavelength. In some studies, hailstorms have been simultaneously observed with polarimetric radars operating at S and C bands (Borowska et al. 2011; Kaltenboeck and Ryzhkov 2013; Palmer et al. 2011; Picca and Ryzhkov 2012). Picca and Ryzhkov (2012) examined the differences between S-band and C-band polarimetric radar characteristics of a strong hailstorm which produced hailstones exceeding 10 cm and hail drifts in excess of 1.5 m in the Oklahoma City metropolitan area on 16 May 2010. The storm was observed by the S-band KOUN WSR-88D and C-band OU-PRIME radar (Palmer et al. 2011) separated by 6.9 km. At S band, the powerful nature of the storm is quite apparent, as 0.5° PPI Z values exceed 65 dBZ over sizeable area (Fig. 8.8; left top panel). Within the core, Z values reach a maximum 72.5 dBZ, which almost certainly indicates the presence of giant hail. In the area of $Z > 60$ dBZ, a large region of depressed Z_{DR} (<0.5 dB) and low ρ_{hv} (generally <0.95) is observed. Occasional patches of negative Z_{DR} down to -0.75 dB point to the hail size larger than 4–5 cm (Fig. 5.13, Chap. 5).

At C band, Z is noticeably lower than at S band by nearly 10 dB throughout the hail core, which suggests that numerous hailstones of at least marginally large sizes (2–2.5 cm) are present (Feral et al. 2003; Ryzhkov et al. 2013a). In areas of rain (outside of hail core), more expansive regions of $Z_{DR} > 5$ dB exist at C band than at S band. This difference is attributed to large drops (close to 6 mm) which cause resonance scattering at C band and greatly increase Z_{DR}. Even within the hail core, pockets of higher Z_{DR} (>2 dB) at C band are noted, similar to what has been observed in several other C-band studies (e.g., Anderson et al. 2011; Kaltenboeck and Ryzhkov 2013; Ryzhkov et al. 2007; Tabary et al. 2009). C-band Z_{DR} is 2–3 dB higher than S-band Z_{DR} between regions 1 and 2 of the hail core (overlaid circles in Fig. 8.8). Figure 8.8 also exhibits a ρ_{hv} difference between S and C bands of 0.2–0.4 throughout the hail core, owing to enhanced effects of resonance scattering at a shorter wavelength.

A more quantitative comparison between polarimetric radar variables at S and C bands for the hail case on 16 May 2010 is illustrated in Fig. 8.9 where radial profiles of Z, Z_{DR}, and ρ_{hv} are plotted for nearly coincident azimuths of KOUN and OU-PRIME beams (shown in Fig. 8.8). The first pair of azimuths (KOUN, 348°; OU-PRIME, 346°) passes through the inflow region where an abundance of large raindrops is not expected (solid curves in Fig. 8.9). Conversely, the second pair (KOUN, 353°; OU-PRIME, 351°) passes through a region producing extremely high attenuation and differential attenuation, likely characterized by a significant number of large drops and small hailstones (dashed curves in Fig. 8.9). At C band, both radials pass through areas of large hail with Z exceeding 60 dBZ, but only one (351°) exhibits anomalously high attenuation and differential attenuation causing negative Z and Z_{DR} biases of about 20 dB and 15 dB as the comparison between Z and Z_{DR} at the two wavelengths indicates. Gradual decrease of the cross-correlation coefficient at C band (right bottom panel in Fig. 8.9) is partially attributed to a nonuniform beam filling effect (Chap. 6) which is more pronounced at C band.

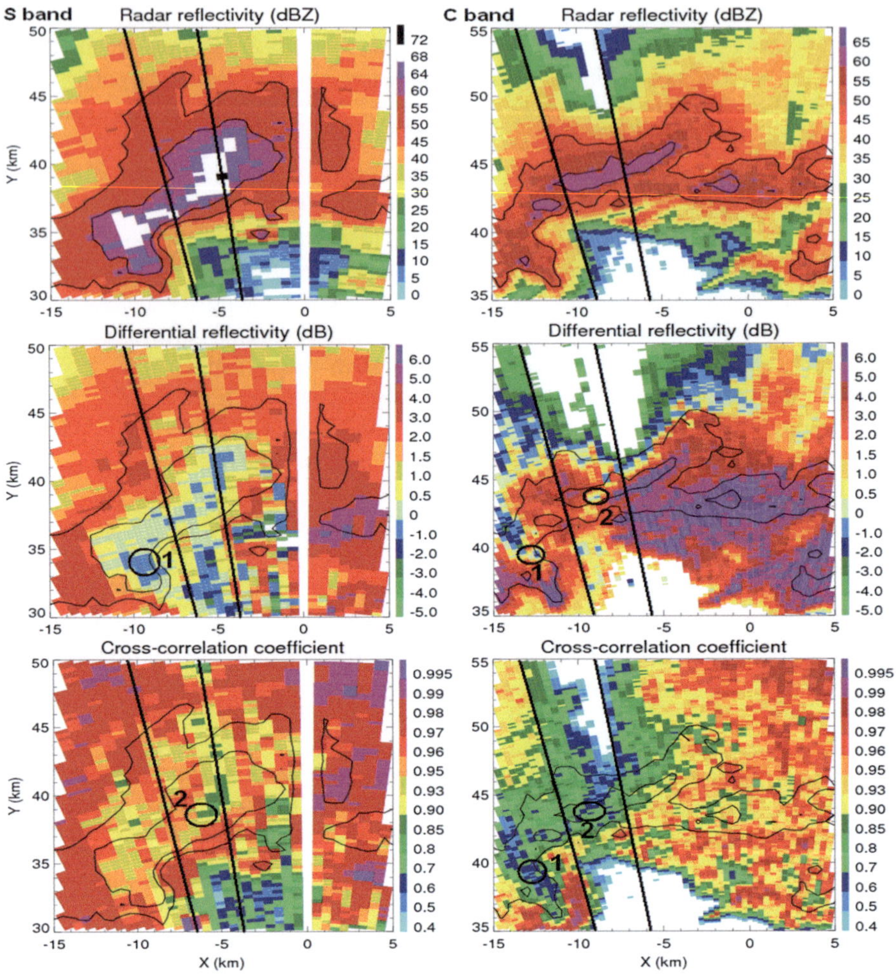

Fig. 8.8 KOUN (S band) and OU-PRIME (C band) 2059 UTC 0.5° and 2100 UTC 0.25° PPIs, of Z_H (top) from 16 May 2010 and corresponding PPIs of Z_{DR} (middle) and ρ_{hv} (bottom). Highlighted regions 1 and 2 represent areas in which giant hail is likely present deduced from the polarimetric variables. Black lines display the radials from which the polarimetric values are displayed in Fig. 8.9. Contours of $Z = 50$ and 60 dBZ are overlaid on all panels. The effects of strong attenuation/differential attenuation are evident in the fields of C-band Z and Z_{DR} which are not corrected for attenuation. From Picca and Ryzhkov (2012). © American Meteorological Society. Used with permission

A hotspot correction method described in Gu et al. (2011) (see Chap. 6) was applied for attenuation correction at C band, and the reflectivity is reasonably well restored (gray curve in the right top panel in Fig. 8.9). In the example shown in

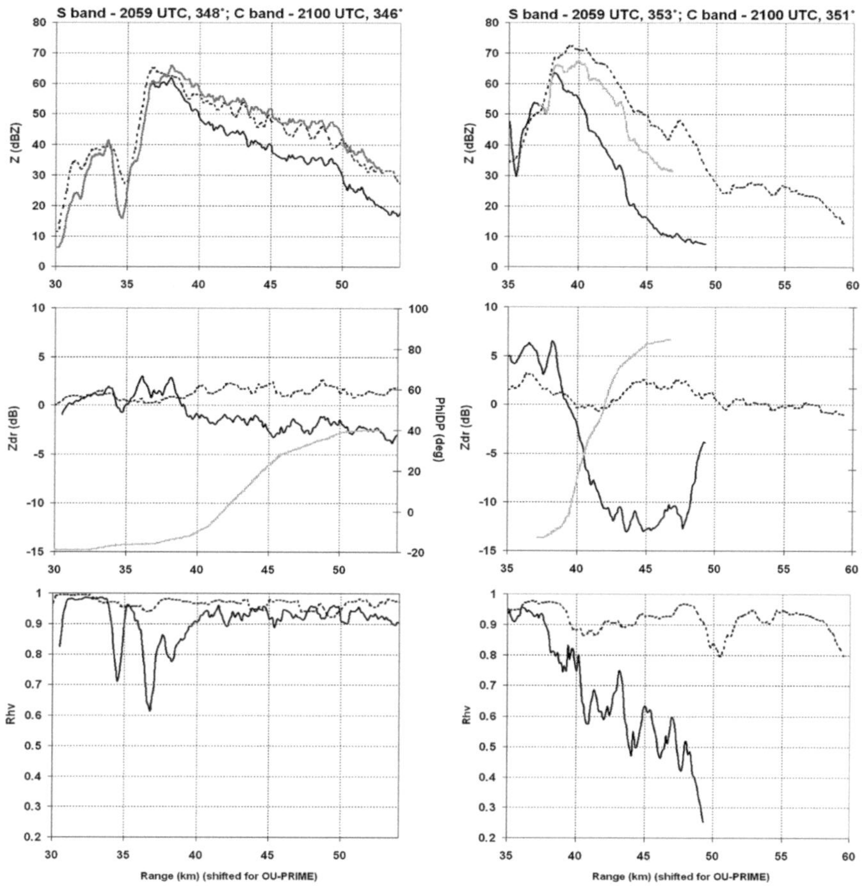

Fig. 8.9 Radial profiles of Z (top), Z_{DR} (middle), and ρ_{hv} (bottom) from KOUN at 2059 UTC (dashed curves) at 348° (left) and 353° (right) and from OU-PRIME at 2100 UTC (solid dark curves) at 346° (left) and 351° (right). Values are from the lowest elevation angle of each radar (0.5° on KOUN; 0.25° on OU-PRIME). $Z(C)$ corrected for attenuation and Φ_{DP} are plotted in the top and middle rows (solid light curves). C-band curves are shifted to the left by 6 km to account for the distance between the two radar sites and to facilitate comparison. Adapted from Picca and Ryzhkov (2012). © American Meteorological Society. Used with permission

Fig. 8.9, the attenuation-related biases of Z and Z_{DR} at S band are small compared to the ones at C band. However, the impact of attenuation in melting hail mixed with large raindrops can be significant even at S band. Ryzhkov et al. (2013b) reported the Z and Z_{DR} attenuation biases up to 15 dB and 6 dB in hail at S band (Fig. 6.26 in Chap. 6).

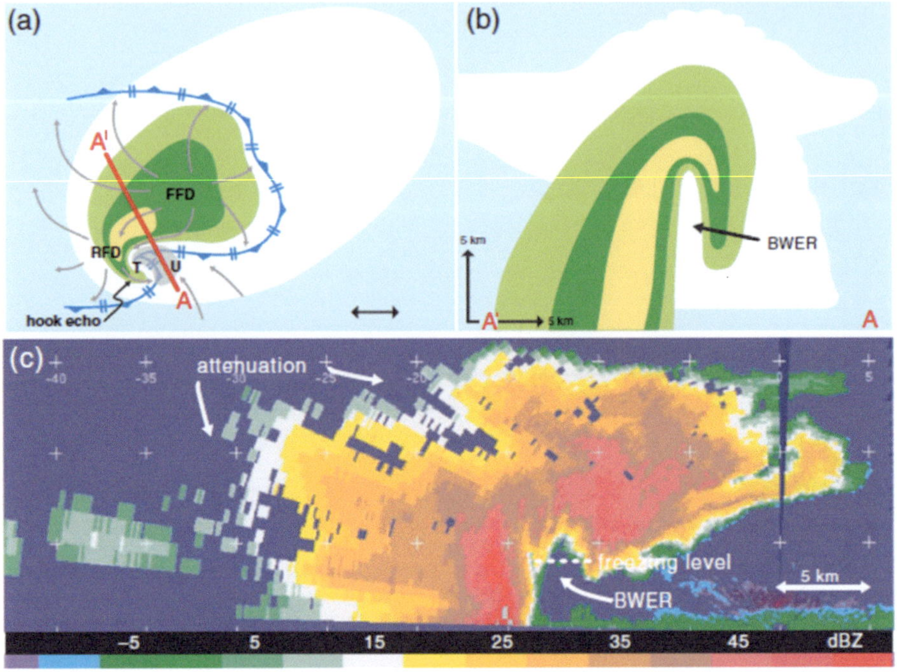

Fig. 8.10 (**a**, **b**) Conceptual drawing of a supercell tornadic storm. (**c**) Vertical cross section of a supercell storm. Adapted from Markowski and Richardson (2010), with permission

8.3 Supercell Tornadic Storms

Supercell storms are characterized by rotating updraft in the vicinity of which a tornado may develop. A conceptual plot of the tornadic supercell storm is shown in Fig. 8.10. Tornado (T) is typically formed at the inner side or a tip of a characteristic "hook echo" which separates areas of inflow updraft (U) and rear-flank downdraft (RFD) (Fig. 8.10a). A bulk of precipitation including hail falls in a forward-flank downdraft (FFD) region. A rotating updraft is commonly marked as a vault or BWER in a vertical cross section (Fig. 8.10b, c). Polarimetric radars were proven to be very efficient for recognizing various microphysical and kinematic features of supercell storms and the primary tools to detect tornado on the ground via identification of the "tornadic debris signature" (TDS). In this section, we provide a brief overview of supercell polarimetric signatures based on the studies by Ryzhkov et al. (2005), Kumjian and Ryzhkov (2008), and the review by Kumjian (2013).

A number of repetitive polarimetric signatures are found in the supercell storms. These features include the tornadic debris signature associated with tornado

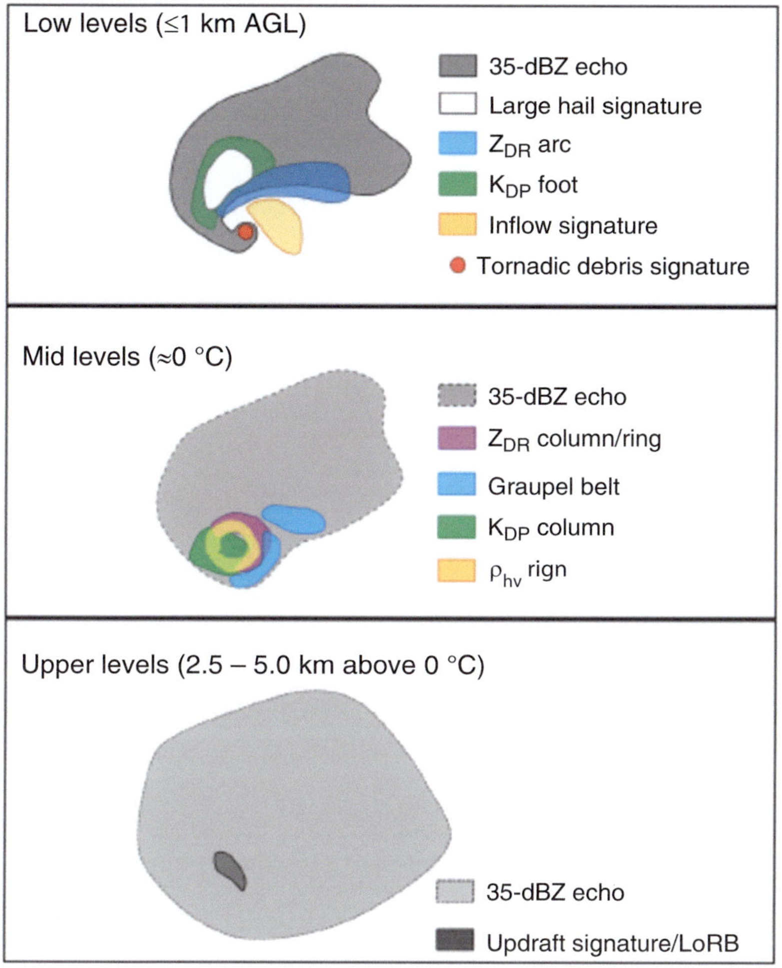

Fig. 8.11 Schematic of polarimetric signatures in a supercell at low level (<1 km), midlevel (approximately at the environmental 0 °C level), and upper level (2.5–5.0 km above the environmental 0 °C level). Adapted from Kumjian and Ryzhkov (2008), with updates based on Romine et al. (2008), Kumjian et al. (2010), and Snyder et al. (2013). © American Meteorological Society. Used with permission

touchdown, hail signatures near the ground, the "Z_{DR} arc" at low levels, reduced ρ_{hv} in the storm-inflow region at low levels and in the updraft, Z_{DR} and K_{DP} columns extending above the melting layer, and midlevel "rings" of enhanced Z_{DR} and decreased ρ_{hv}. These features are depicted in a conceptual diagram in Fig. 8.11.

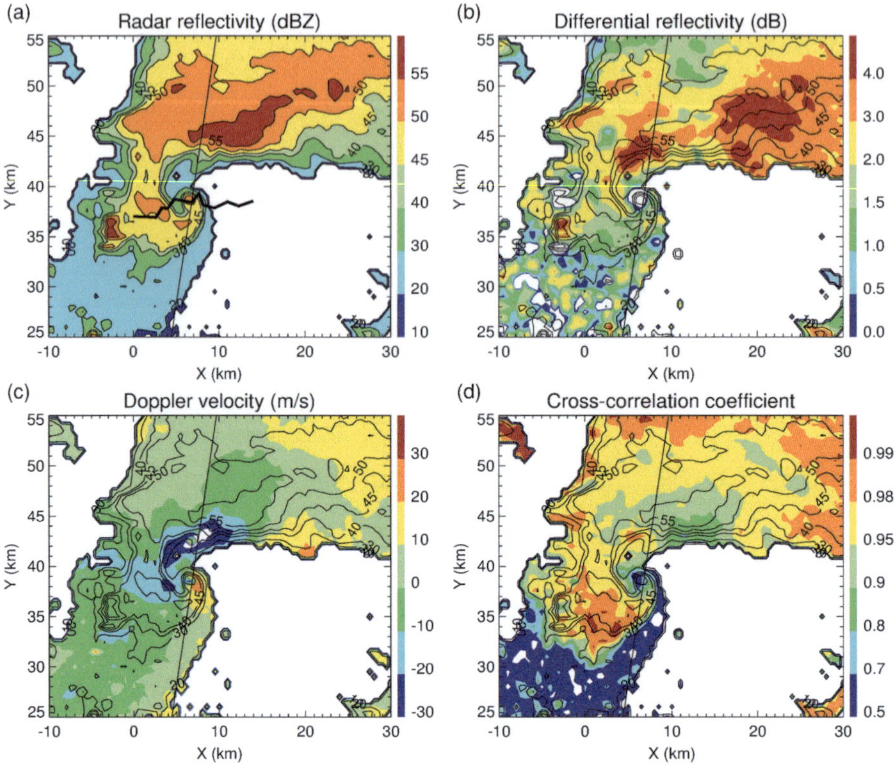

Fig. 8.12 Fields of (**a**) Z, (**b**) Z_{DR}, (**c**) v and (**d**) ρ_{hv} at the 0.5° scan in elevation on 10 May 2003, 0346 UTC. Thin solid lines indicate azimuthal direction of 10°, and the thick solid line in the Z panel depicts a part of observed tornado track on the ground. From Ryzhkov et al. (2005). © American Meteorological Society. Used with permission

8.3.1 Tornadic Debris Signature (TDS)

Ryzhkov et al. (2002, 2005) documented a pronounced polarimetric signature associated with tornado touchdown. Lofted tornadic debris typically has large size, irregular shape, random orientation, and high dielectric constant which produce high Z, low Z_{DR}, and particularly low ρ_{hv} which distinguish it from the meteorological scatterers. TDS is usually collocated with the tornado vortex signature (TVS) so named by Brown et al. (1978).

An example of the TDS is shown in Figs. 8.12 and 8.13. The storm on 9–10 May 2003 produced F3 tornado with a 29 km damage path (Ryzhkov et al. 2005). The fields of Z, Z_{DR}, ρ_{hv}, and Doppler velocity v at the lowest radar scan at El $= 0.5°$ and 0346 UTC are displayed in Fig. 8.12. At that time, a strong classical hook echo had developed with all of the indications of tornado occurrence at the tip of the hook ($X = 6.5$ km, $Y = 38.5$ km)—increased Z, TVS,

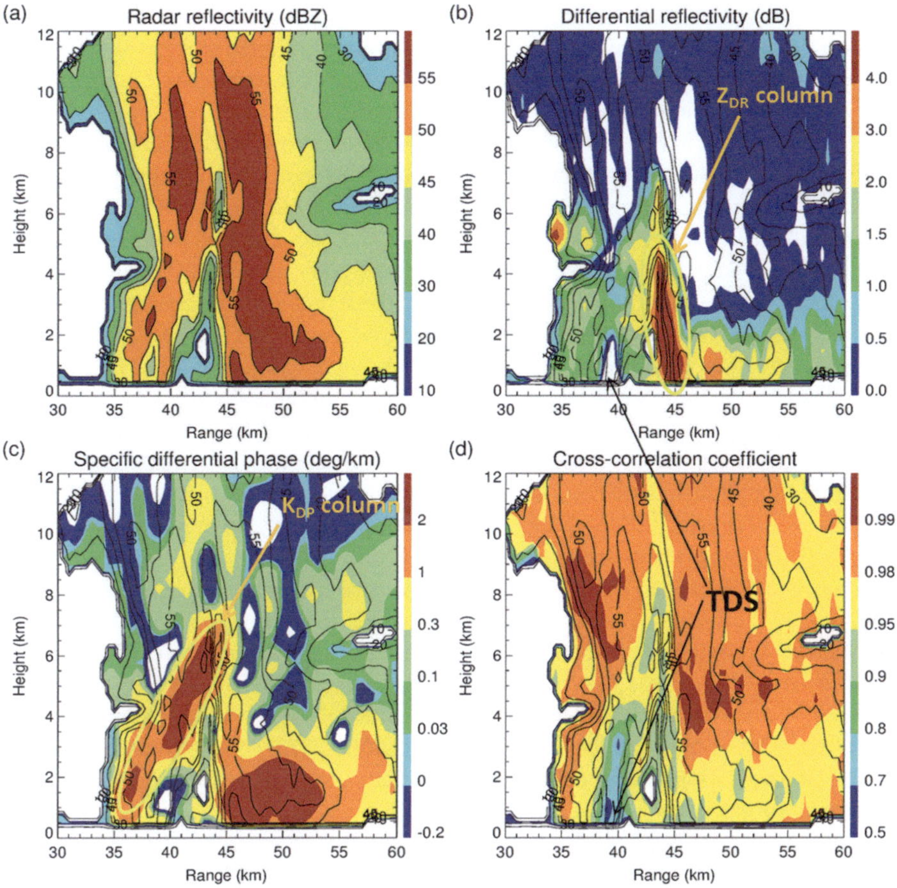

Fig. 8.13 Vertical cross section of (**a**) Z, (**b**) Z_{DR}, (**c**) K_{DP}, and (**d**) ρ_{hv} corresponding to azimuth of 10° (shown in Fig. 8.12) at 0346 UTC on 10 May 2003. The debris signature is centered at about 39.5 km from the radar. From Ryzhkov et al. (2005). © American Meteorological Society. Used with permission

anomalously low ρ_{hv}, and negative Z_{DR}. Note that tumbling of the debris particles commonly reduces Z_{DR} (sometimes down to negative values as in the 9–10 May 2003 case), although in some situations, precipitation entrained into the tornadic circulation can cause higher Z_{DR} values closer to what is observed in precipitation (Kumjian and Ryzhkov 2008). A remarkable tornadic debris signature at distances of 39–40 km from the radar is evident in the vertical cross section through the hook echo (Fig. 8.13). The columns of negative Z_{DR}, negative K_{DP}, and low ρ_{hv} extend vertically from the ground up to a height of 2 km (4 km for ρ_{hv}). The corresponding Z is between 50 and 55 dBZ. There is no doubt that the radar echo in this region is dominated by nonmeteorological scatterers, that is, debris. Negative values of Z_{DR} and K_{DP} might be attributed either to a certain degree of vertical common orientation of relatively small Rayleigh scatterers or large preferentially oriented scatterers (Bodine et al. 2014; Ryzhkov et al. 2005). In the resonance regime of scattering, both Z_{DR} and K_{DP} might have a negative sign even if the horizontal dimension of the scatterers is larger than vertical and the scatterers are horizontally oriented.

As can be seen in Fig. 8.13, two distinct columns of enhanced Z that are closely connected aloft and separated by a "vault" at altitudes below 7.5 km have strikingly different polarimetric attributes. The left column is associated with the rear-flank downdraft (RFD) and exhibits an impressive tilted column of high positive K_{DP}, low values of ρ_{hv} (in comparison with its right counterpart), and highly variable low to moderate Z_{DR}. The right column represents the main precipitation core, consisting of rain below 2–2.5 km and hail above, as can be inferred from Z_{DR}, K_{DP}, and ρ_{hv}. A well-defined Z_{DR} column with maximum values approaching 6.5 dB is observed in the weak echo region that is associated with the storm updraft.

In a series of studies (e.g., Bluestein et al. 2007, 2015; Bodine et al. 2013, 2014; Kumjian 2011; Kumjian and Ryzhkov 2008; Kurdzo et al. 2015; Palmer et al. 2011; Romine et al. 2008; Scharfenberg et al. 2005; Schultz et al. 2012a, b; Snyder et al. 2013; Snyder and Bluestein 2014; Van Den Broeke et al. 2008), the TDS signatures have been documented from a variety of tornadic storms at different radar wavelengths.

Another example of TDS for an EF5 tornadic storm is presented in Fig. 8.14. This tornado affected Moore, Oklahoma, and the surrounding area on 20 May 2013. It traveled 23 km and caused up to 1.7 km-wide damage. The tornado killed 24 people, injured over 200 others, and damaged many structures (Burgess et al. 2014). Because of its extreme intensity, tornado lifted heavy debris which is indicated by a "debris ball" of high reflectivity exceeding 60 dBZ. The corresponding dip in the cross-correlation coefficient exhibits anomalously low values (less than 0.45) and differential reflectivity is close to zero.

Tornadic debris signatures look very similar at different radar wavelengths as an example of the tornadic storm simultaneously observed with S-band and C-band polarimetric radars (Fig. 8.15) indicates. On 10 May 2010, a large tornado outbreak occurred in the extended part of Oklahoma. The strongest two tornadoes rated EF4 occurred in the proximity of the S-band KOUN WSR-88D and C-band OU-PRIME radars (Bodine et al. 2014; Palmer et al. 2011). In Fig. 8.15, the three top panels show Z, Z_{DR}, and ρ_{hv} measured by KOUN at 2246 UTC and the 0.87° elevation scan. Three bottom panels display the same polarimetric variables measured by OU-PRIME at the 1.0° elevation scan 1 min later. The two tornadoes marked as B1 and B2 are identified in the images from both radars by the combination of enhanced Z and depressed Z_{DR} and ρ_{hv}. The fine structure of the tornadic storm is better resolved by OU-PRIME because this radar has narrower antenna beam (0.48°) and is slightly closer to the tornado locations and data spacing in azimuth is about 0.5°. Outside the tornado areas, differences between the polarimetric variables at the two wavelengths are consistent with observations in Figs. 8.8 and 8.9. At the longer wavelength, Z and ρ_{hv} are higher, and Z_{DR} is lower in the areas not affected by differential attenuation which is attributed to the effects of resonance scattering from large raindrops and hailstones.

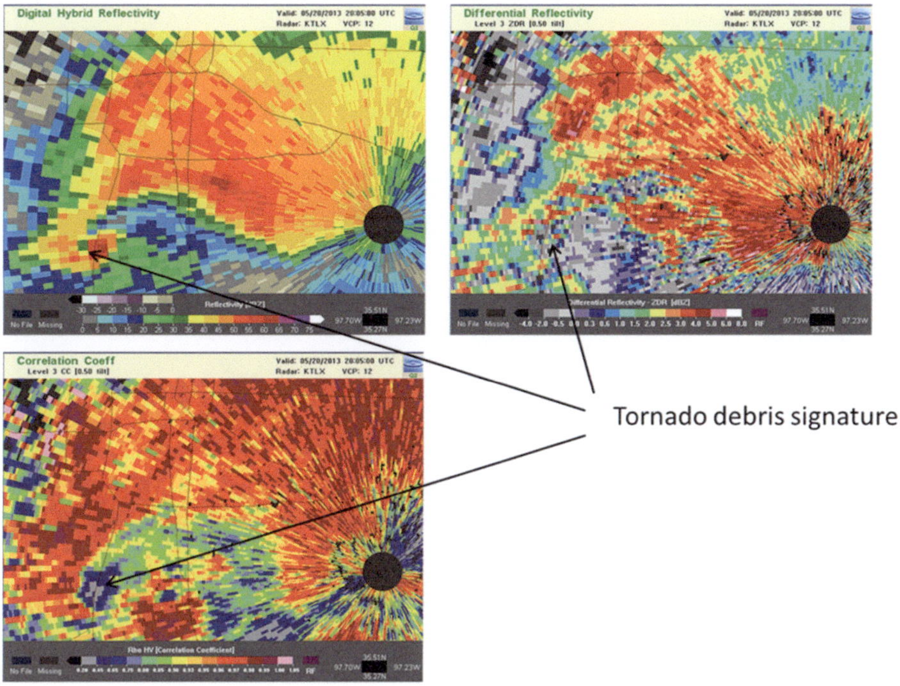

Fig. 8.14 Tornado debris signature in the fields of Z, Z_{DR}, and ρ_{hv} observed during the devastating tornado in Moore, OK, on 20 May 2013 at 2005 UTC. Antenna elevation is 0.5°

Although the TDS signatures are qualitatively similar, significant quantitative differences are observed at S and C bands. The S-band TDS exhibits higher Z and ρ_{hv} which is a consequence of non-Rayleigh scattering. The 90th percentile Z dual-wavelength differences of 10 dB are observed in this storm (Bodine et al. 2014) which is very similar to the dual-wavelength ratio in hailstorms (Picca and Ryzhkov 2012; see also Fig. 8.8). A larger dual-wavelength difference in median Z at S and C bands occurs while the tornado caused more damage suggesting a greater amount of lofted large debris. Comparisons reveal that S-band median ρ_{hv} is 0.12–0.15 higher than C-band ρ_{hv} (Bodine et al. 2014).

Tornado observations with mobile polarimetric radars are particularly valuable for investigation of fine-scale features because mobile radars can be deployed close to tornadoes and therefore obtain high spatial and temporal resolution of tornado signatures (Bluestein et al. 2007, 2015; Kurdzo et al. 2015; Snyder et al. 2013; Snyder and Bluestein 2014). Bluestein et al. (2007) report one of the first observations of TDS with a high-resolution X-band polarimetric radar (Fig. 8.16). The reflectivity pattern associated with F2 tornado is a ring of moderately intense echo (~40 dBZ) about 600 m in diameter, surrounding a weaker echo region (~15 to 25 dBZ) ~200 m in diameter; beyond the ring of intense echo, there are spiral bands of reflectivity of 40–50 dBZ (Fig. 8.16a). Such features are common in tornadoes

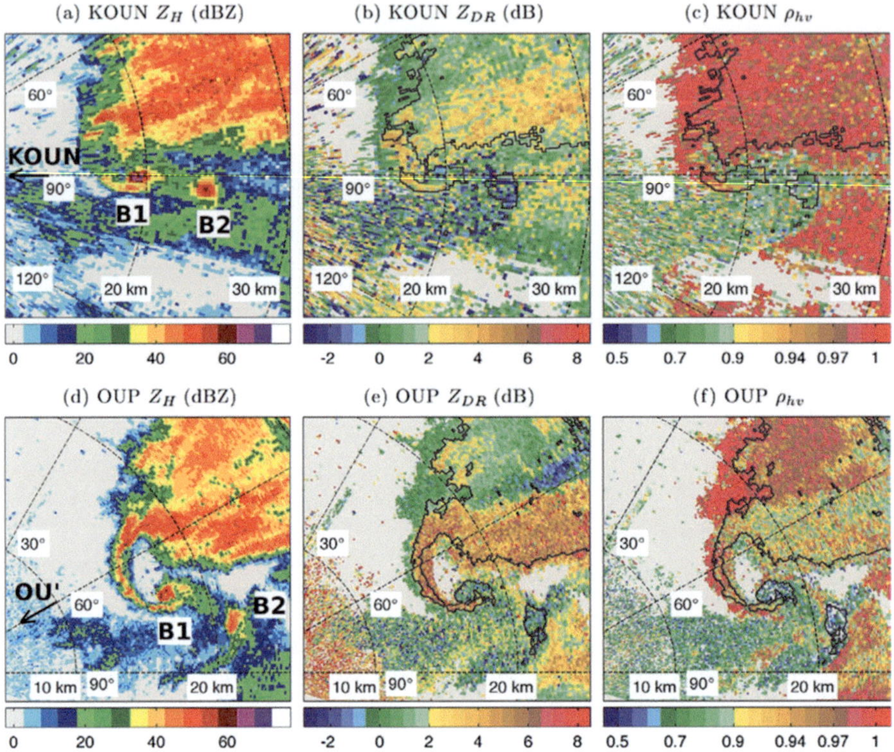

Fig. 8.15 Polarimetric radar measurements by the S-band KOUN WSR-88D (El = 0.87°) (panels **a**, **b**, and **c**) and C-band OU-PRIME radars (El = 1.0°) (panels **d**, **e**, and **f**) at 2247 UTC of the tornadic storm at 2246 UTC on 10 May 2010. The solid black lines indicate the 30 dBZ contour of *Z*, and tornadoes B1 and B2 are labeled on the *Z* panels. Range markers are 10 km apart, and the locations of KOUN and OU-PRIME (labeled here as OU') are denoted by the black arrows in the panels (**a**) and (**d**). From Palmer et al. (2011). © American Meteorological Society. Used with permission

viewed by radars at close range. The region encompassing the ring of moderately intense echo is collocated with the quasi-circular regions of relatively low Z_{DR} and ρ_{hv} (Fig. 8.16b, c). These regions of Z_{DR} (below 0.5 dB green and white colors) and ρ_{hv} (below 0.5 green color) are ~900 m in diameter. The tornado vortex is associated with the Doppler velocity couplet (i.e., TVS) over 300 m wide.

In severe tornadic cases, multiple tornadic vortices and TDS signatures can be resolved close to the radar. Such example in Fig. 8.17 is from the devastating El Reno, Oklahoma, tornado which occurred on 31 May 2013 and produced eight fatalities including three storm chasers (Wurman et al. 2014). The cyclonic tornado during its mature stage and the anticyclonic tornado were probed at a relatively close range of ~4 to 6 km from the X-band mobile RaXPol radar (Bluestein et al. 2015). This EF5 tornado produced a 4.2 km-wide damage path arguably tagging it as the widest tornado ever documented. The tornado (marked as a C for cyclonic) produced a ~5-km-wide ring of enhanced reflectivity (Fig. 8.17a), a polarimetric debris

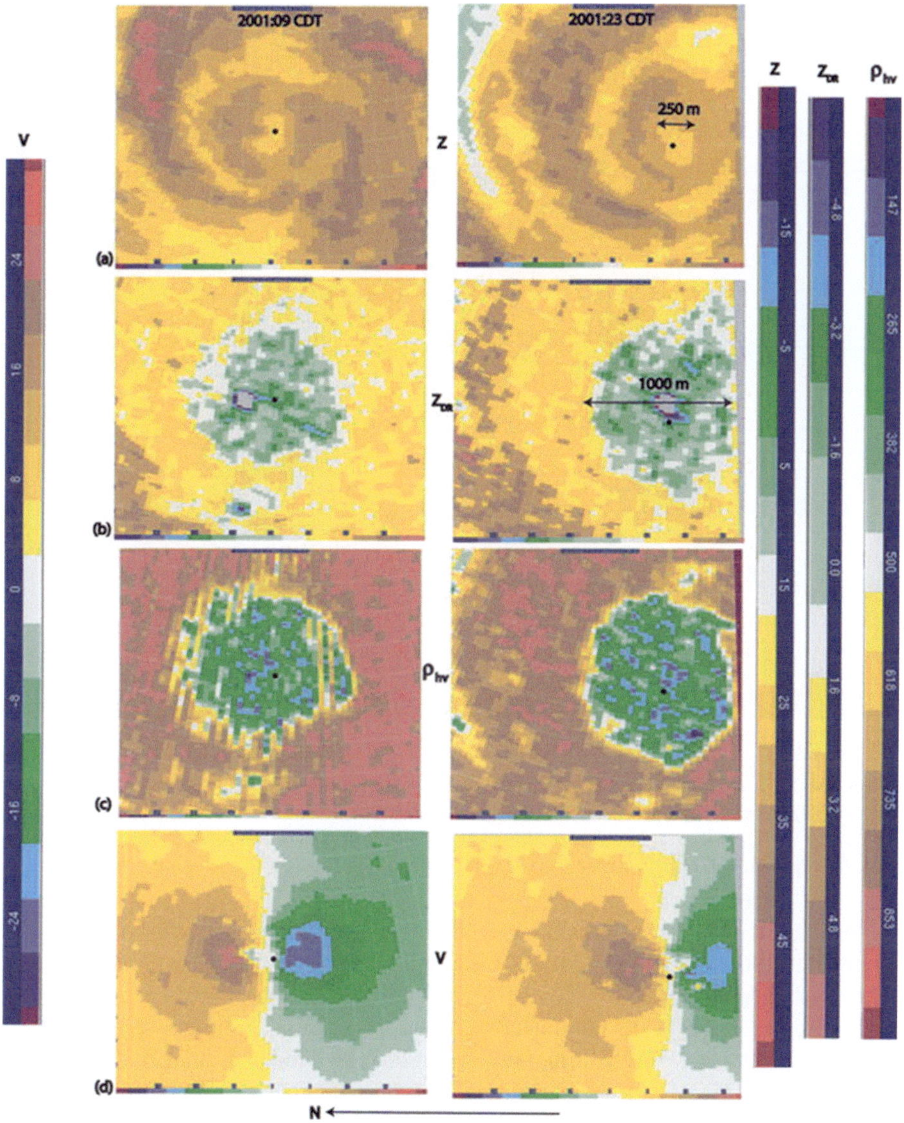

Fig. 8.16 Composite plot of (**a**) Z, (**b**) Z_{DR}, (**c**) ρ_{hv}, and (**d**) Doppler velocity v measured by the X-band UMass X-Pol radar in the close proximity of the tornado on 12 May 2004. Dot indicates the approximate location of the center of the tornado. From Bluestein et al. (2007). © American Meteorological Society. Used with permission

signature of low Z_{DR} (Fig. 8.17b) and ρ_{hv} (Fig. 8.17c), and a well-defined Doppler velocity couplet (Fig. 8.17d). A much smaller (less than 1 km in diameter) anticyclonic tornado exhibiting a TDS signature similar to the tornado C was simultaneously detected (marked as an A in Fig. 8.17).

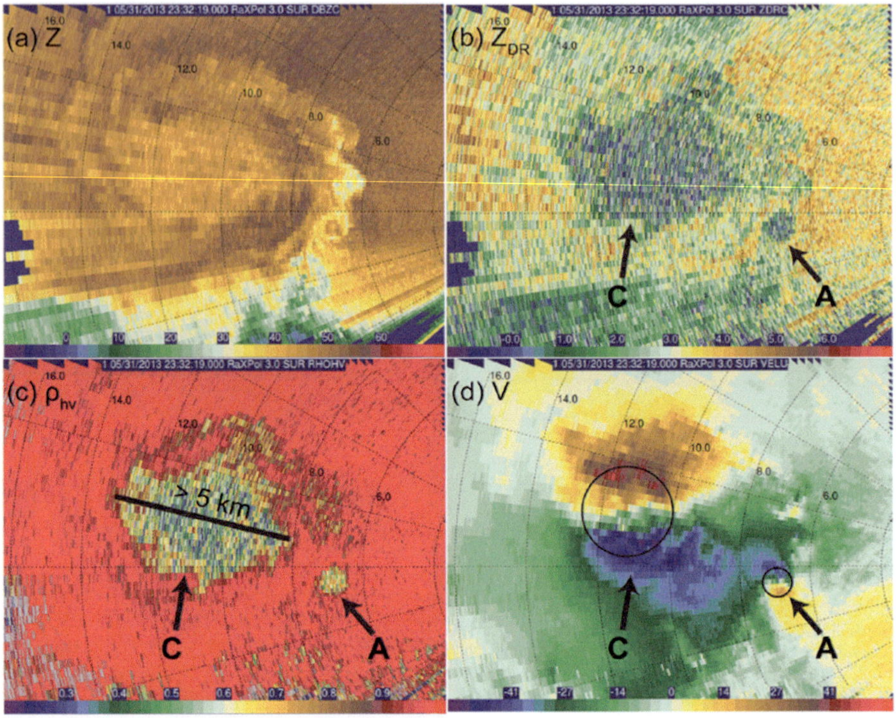

Fig. 8.17 Composite plot of (**a**) Z, (**b**) Z_{DR}, (**c**) ρ_{hv}, and (**d**) Doppler velocity (V) measured by X-band RaXPol radar during the El Reno, OK, devastating tornado on 31 May 2013 depicting the coexisting large cyclonic tornado (marked with a C) and smaller anticyclonic tornado (marked with an A) at a low elevation angle. Range rings are spaced 2 km apart. From Bluestein et al. (2015).

In addition to detecting tornadoes, dual-polarization radars can be used to roughly estimate the strength and severity of these vortices (Bodine et al. 2013; Snyder and Bluestein 2014). Bodine et al. (2013) studied correlations between the damage surveys and various TDS parameters such as maximal Z, minimal Z_{DR}, minimal ρ_{hv}, TDS height, and TDS volume. It was demonstrated that maximal Z, TDS height, and TDS volume increase while minimal values of Z_{DR} and ρ_{hv} decrease as tornado damage severity increases. On the basis of many observations, Entremont and Lamb (2014) present a strong correlation between the EF scale and maximal height of the TDS (Fig. 8.18) (see also Van Den Broeke and Jauernic (2014)). Such information provided by the TDS parameters could help forecasters identify changing trends in tornado damage severity and estimate the severity and extent of tornado damage in near-real time.

The TDS parameters are not a predictor of when a violent tornado will occur. However, they can enable forecasters to gauge tornado damage severity, particularly when spotter reports are unavailable (e.g., for rain-wrapped tornadoes or at night) or

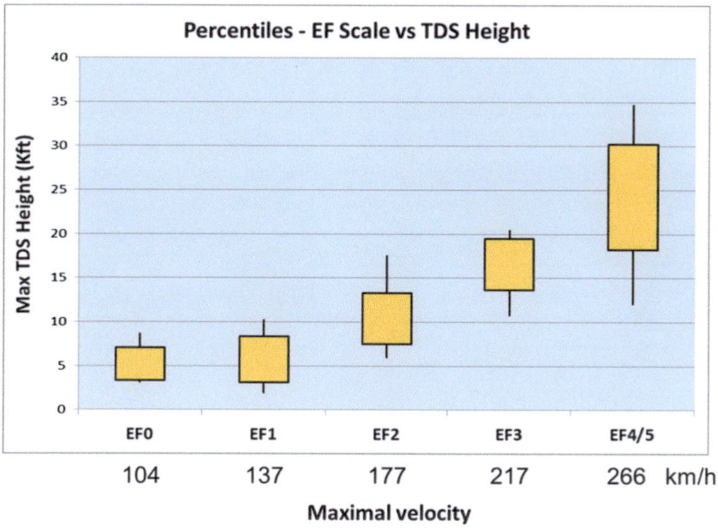

Fig. 8.18 Dependence of the maximal TDS height on the EF scale and maximal air velocity associated with tornado. From Entremont and Lamb (2014)

when low-level velocity data are not available (e.g., at long ranges). The TDS parameters can provide near-real time information about the severity of tornado damage to emergency managers and first responders to ensure that their efforts are immediately directed to the most severely affected areas.

8.3.2 Other Supercell Signatures

8.3.2.1 Z_{DR} Arc

One of the most common polarimetric signatures associated with supercell storms is the "Z_{DR} arc." It appears within the forward-flank reflectivity region at low levels (below about 1–2 km AGL) and is characterized by significant positive Z_{DR} collocated with low-to-moderate reflectivity (Fig. 8.19). Kumjian and Ryzhkov (2009) used a simplified sedimentation model to attribute the Z_{DR} arc to enhanced size sorting of rain associated with the strong low-level wind shear in the inflow environment of the supercell storm. They also demonstrated a positive correlation of the strength of the Z_{DR} arc with the magnitude of the low-level storm-relative helicity (SRH). Schematic depiction of how low-level veering winds in a supercell storm-relative frame cause enhancement of Z_{DR} along the right (inflow) edge of the forward-flank downdraft is shown in Fig. 8.20. Raindrops of three different sizes originating from the same source aloft follow very different trajectories so that the largest drop ends up at the surface just below the source producing very high Z_{DR} at

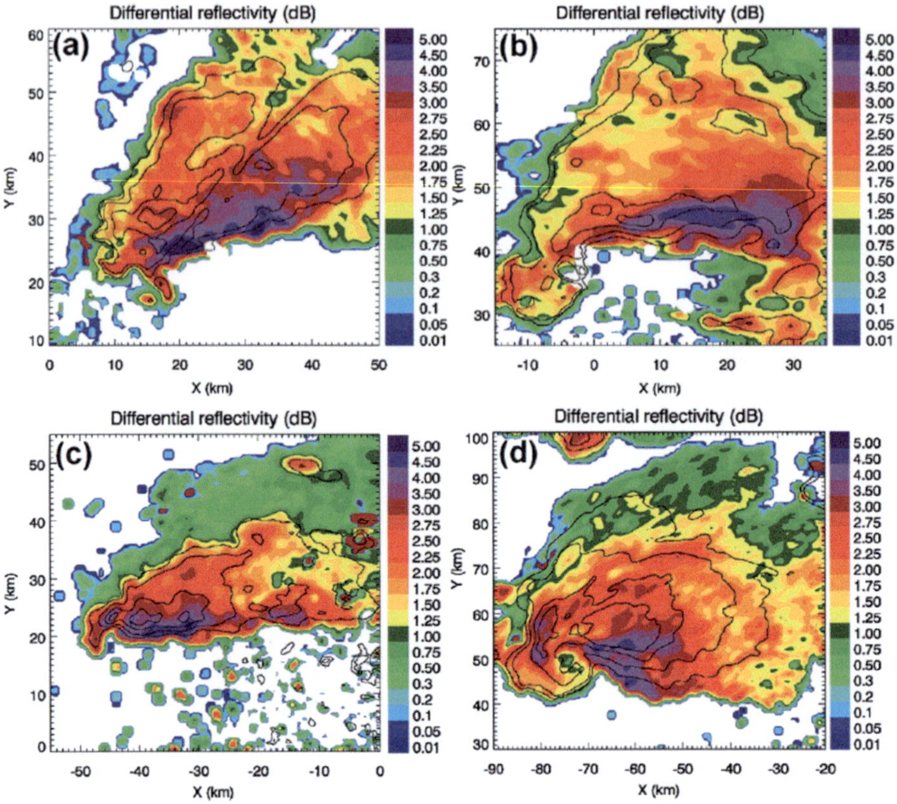

Fig. 8.19 The Z_{DR} arc signature in four supercell storms observed by the KOUN S-band polarimetric radar. The 30-, 40-, 50-, and 55-dBZ contours of Z are overlaid. The supercells were observed on (**a**) 2234 UTC 8 May 2003 at 1.5° elevation, (**b**) 0333 UTC 10 May 2003 at 0.5° elevation, (**c**) 2346 UTC 26 May 2004 at 0.4° elevation, and (**d**) 0044 UTC 30 May 2004 at 0.5° elevation. In each case, a strong enhancement of Z_{DR} (values in excess of 4 dB) is found along the Z gradient on the right forward flank (inflow side) of the storm. From Kumjian and Ryzhkov (2009). © American Meteorological Society. Used with permission

the surface near the edge of the radar echo, whereas smaller raindrops are blown further away inside the storm and follow longer trajectories.

Kumjian and Ryzhkov (2009) consider idealized rain shafts with prescribed initial distributions of liquid drops aloft. In typical supercell storms, most of rain is generated from graupel and hail, and it is a mixture of partially melted graupel/hail and completely melted graupel/hail (i.e., raindrops) that determines the magnitude of differential reflectivity below the freezing level (see Sect. 7.2.4.2). Dawson et al. (2014) used a more sophisticated 3D model with three-moment bulk microphysics which simulates both the genesis of precipitation aloft and its fallout. They conclude that both size sorting of raindrops and partially melted graupel/hail are responsible

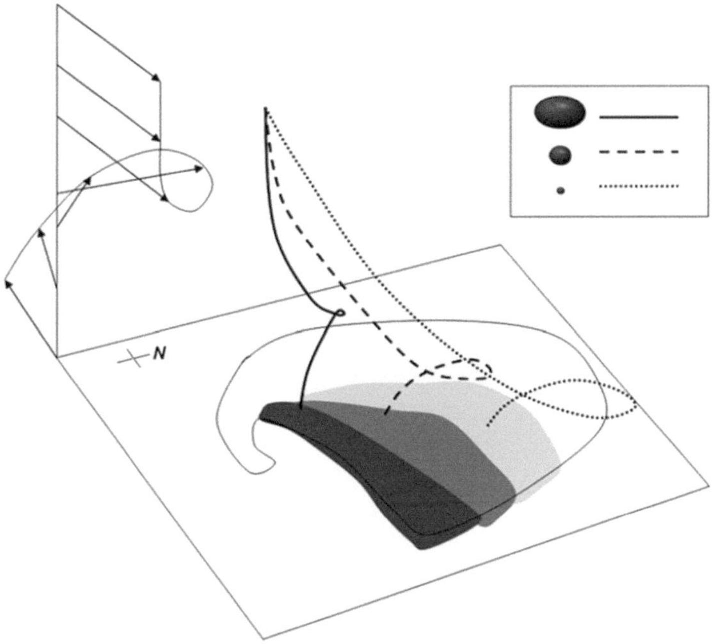

Fig. 8.20 Depiction of how low-level veering winds in a supercell storm-relative frame enhance Z_{DR} along the right (inflow) edge of the forward-flank downdraft region (outlined on the surface). The wind vectors indicate the veering flow. If projected onto a horizontal plane, the line connecting the wind vectors would represent the hodograph, the area of which is proportional to the low-level SRH. Large drops follow cyclonic trajectory depicted with the black solid curve, medium-size drops follow the dashed curve, and small drops follow the dotted curve. All three sizes start from the same point source. The shading represents the Z_{DR} enhancement, which is maximized at the edge of the forward-flank downdraft. From Kumjian and Ryzhkov (2009). © American Meteorological Society. Used with permission

for the apparent Z_{DR} arc signatures and size sorting of hail likely plays a bigger role. This is consistent with the general perception that Z_{DR} within a hail core or its close proximity is usually much lower than at its periphery where smaller-size graupel/hail completely melts into very large raindrops producing very high Z_{DR} values before spontaneous breakup.

The Z_{DR} arc is an important attribute of supercell storms. Nonsupercell storms usually do not exhibit pronounced Z_{DR} arc signatures, and the appearance of such signature may alert forecasters that the storm is transitioning into a severe supercell mode that has significant tornado potential. Kumjian et al. (2010) documented a cyclic behavior of the Z_{DR} arc apparently tied to the cyclic occlusion process of the low-level mesocyclone in a cyclic nontornadic supercell storm in Oklahoma. Leading up to the occlusion of the low-level mesocyclone, the arc extends back toward the inflow notch and increases in magnitude and areal extent. The arc tends to reach maximum strength and extent before weakening of the updraft. As the circulation occludes, the arc contracts and dissipates so that the region of very high Z_{DR} rapidly shrinks. Then, a new region of very high Z_{DR} may form at the edge of the FFD echo—a new Z_{DR} arc. According to Kumjian et al. (2010), the Z_{DR} arc cycle in their observations lasted from 7 to 20 min.

8.3.2.2 Rotating Updraft-Related Signatures

Both in supercell and nonsupercell storms, convective updrafts are indicated by the "Z_{DR} columns" extending well above the environmental freezing level (Figs. 7.10, 8.6, and 8.13). At lower levels (<1 km AGL), the updraft is connected to the "inflow signature" (Kumjian and Ryzhkov 2008), which appears as a significant reduction of ρ_{hv} (<0.85) in the inflow regions of supercells (Fig. 8.11). The low values of Z (<20 dBZ) and ρ_{hv} indicate nonmeteorological scatterers. These may be insects and/or other light debris such as grass and leaves ingested into storm updraft via strong near-surface winds occasionally exceeding 25 m s^{-1}. This debris is characterized by irregular shapes, random orientation, and relatively high dielectric constant all of which contribute to the drop in ρ_{hv}. The inflow ρ_{hv} signature is similar to the TDS signature (low ρ_{hv}) which is typically associated with heavier debris and is located in the RFD area of the storm.

The Z_{DR} column can be distorted into a ring or half-ring shape by the mesocyclone in supercells (e.g., Kumjian and Ryzhkov 2008; Palmer et al. 2011; Payne et al. 2010; Snyder et al. 2013, 2017b). Nearly collocated with the Z_{DR} ring is a circular or semicircular ring of depressed ρ_{hv} called the "ρ_{hv} ring." Midlevel rings of enhanced Z_{DR} and depressed ρ_{hv} wrapping convective updraft are visible at the 4.5 km horizontal constant altitude plan position indicator (CAPPI) composite map in Fig. 8.21. The updraft area is enclosed in a black circle. This storm produced tornado on the ground which is marked by the TDS signature at lower CAPPI level of 0.75 km (Fig. 8.22). The Z_{DR} and ρ_{hv} rings are commonly found above the environmental melting layer and most likely near the updraft-perturbed melting layer. Taken together, these signatures indicate mixed-phase and/or non-Rayleigh scatterers within or very near the updraft and mesocyclone. Payne et al. (2010) confirm the association between these signatures and the mesocyclonic vertical vorticity maximum and also document a U-shaped region of lightning activity associated with the signatures, which they call a "lightning ring."

Kumjian et al. (2010) discovered that the midlevel Z_{DR} ring may be closely tied to a feature dubbed as the "graupel belt" characterized by high ρ_{hv} (>0.98) and Z_{DR} near zero, indicative of ice-phase hydrometeors. The graupel belt wraps around the mesocyclone, and because of its proximity to the Z_{DR} ring, it is hypothesized that some of this graupel is entrained into the periphery of the warmer updraft, whereupon melting and/or accretion of liquid water contributes to the observed enhanced Z_{DR}.

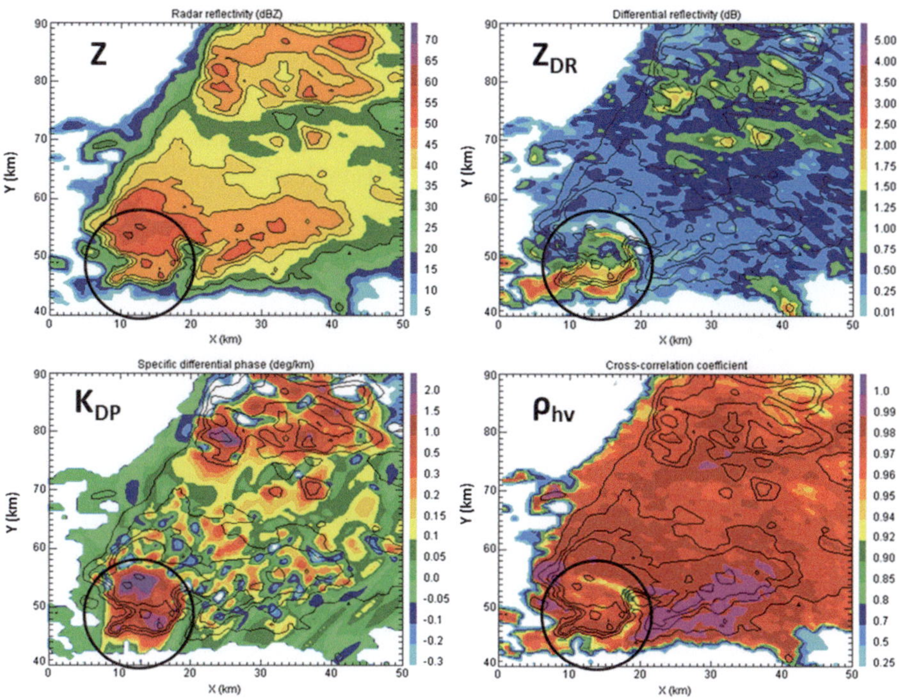

Fig. 8.21 Midlevel rings of enhanced Z_{DR} and depressed ρ_{hv} wrapping convective updraft within the black circle at the CAPPI level of 4.5 km. The data have been collected by the KOUN WSR-88D in central Oklahoma during a tornadic storm on 10 May 2009

8.4 Modeling Polarimetric Characteristics of Deep Convective Storms

Simulating deep convective storms in their full complexity requires sophisticated cloud models which consider a number of microphysical processes occurring simultaneously and interacting with each other in a complicated manner. Polarimetric radar variables have tremendous potential for improvement of existing storm-scale numerical weather prediction (NWP) models and their microphysical parameterization in particular. Two paradigms currently exist for doing this: retrievals and forward operators. Radar retrievals imply estimation of the prognostic variables in NWP models like mixing ratios of various habits of atmospheric particles and corresponding parameters of their size distributions such as total concentration and median size from the set of polarimetric variables (such as Z, Z_{DR}, K_{DP}, LDR, CDR, etc.) (Carlin et al. 2016). The estimated prognostic model variables are compared to the simulated ones or assimilated into the model to improve its performance. This is a significant step from utilizing only Z and Doppler velocity v in the past studies (e.g., Gao et al. 2004).

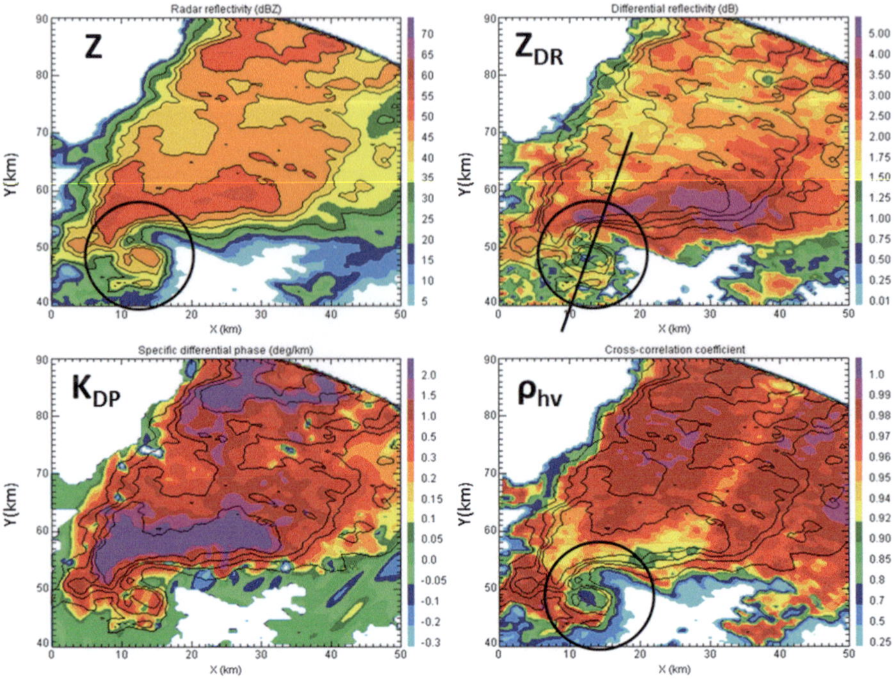

Fig. 8.22 The same as in Fig. 8.21 but at lower CAPPI level (0.5 km)

The second approach uses a polarimetric radar observation operator to convert the output of the NWP model into the fields of the observed polarimetric variables. The model is modified and optimized until it reproduces fields of Doppler velocities and polarimetric variables similar to the observed ones. Model simulations in conjunction with radar observations provide insight into the underlying process of cloud and precipitation development and help in the interpretation of the observed variables and their spatial/temporal structure.

Two types of cloud models for deep convective clouds are considered herein: models with bulk microphysics and models with spectral (bin) microphysics. Models with bulk parameterization of cloud microphysics prescribe a priori exponential or gamma size distributions of particles with different habits. Because of such simplification, the system of microphysical equations is commonly reduced to equations for one (mass content) or two (content and concentration) moments of the size distributions. Models with spectral microphysics explicitly treat the evolution of physical properties of particles in separate size intervals with no assumptions about their distributions. Thus these models represent better the physics of precipitation but are computationally more intense.

Polarimetric variables are very sensitive to the changes in the hydrometeor density, water content, and distributions across size spectra; these are not well captured by the bulk models. Most polarimetric variables (such as Z, Z_{DR}, and ρ_{hv})

are affected by hydrometeors in the tails (or high-order moments) of size distributions. Intense size sorting caused by updrafts and wind shear may create very unusual size distributions with shapes not adequately described in terms of exponential and gamma size distributions. It is clear that the polarimetric variables can hardly be reproduced realistically by only considering one- or two-moment bulk schemes.

Herein, we demonstrate that NWP models with spectral (bin) and bulk microphysics can reproduce meaningful polarimetric signatures in deep convective storms. One such model is the 2D nonhydrostatic mixed-phase spectral bin Hebrew University of Jerusalem Cloud Model (HUCM; e.g., Khain et al. 2004, 2011; Khain and Pinsky 2018). The model contains seven classes of hydrometeors, and each class is represented by size distribution functions in 43 size intervals (bins). The HUCM model is coupled with the polarimetric radar observation operator (Ryzhkov et al. 2011). Hydrometeors are considered as spheroids, with size distributions explicitly computed by HUCM. Scattering calculations involve a combination of Rayleigh formulas and T-matrix computations. The dielectric constant is specified according to the particle type (pure liquid for raindrops, uniform mixtures for melting snowflakes, two-layer spheroids for freezing drops and melting hail, etc.), and the model predicts variation of the liquid water mass fraction across the spectra. Because polarimetric variables depend crucially on the aspect ratios and orientation of hydrometeors which are not implicit in the model, these are prescribed in the forward operator (Ryzhkov et al. 2011).

A typical hail-bearing storm was simulated by HUCM with horizontal resolution of 300 m and vertical resolution of 100 m. The evolution of vertical cross sections of reflectivity Z, differential reflectivity Z_{DR}, and vertical air velocity w during the mature stage of the storm development is illustrated in Fig. 8.23. The fields of Z, Z_{DR}, and w are updated every 6 min during a 24-min period. A cyclic dependence of the Z and Z_{DR} spatial patterns and their close connection to updrafts is evident. Indeed, a developing convective cell marked as A in the Z_{DR} panels is associated with convective updraft at midlevels at 5280 s of the model simulation time. The updraft is marked by enhanced Z and a classical Z_{DR} column extending up to 5 km above ground, i.e., 2 km above the environmental freezing level at 3 km AGL. Six minutes later (at 5640 s), the Z_{DR} column associated with the cell A is transformed into the "Z_{DR} hole" stretching down to the surface and accompanied by rapid increase of Z at lower levels due to hail fallout. At the same time, a hint of another developing Z_{DR} column (cell B) is evident in the updraft area to the right of the decaying cell A. The cell B reaches its maximal development 6 min later (at 6000 s) and transforms into a Z_{DR} hole (at 6360 s). A third Z_{DR} column associated with the cell C marks another updraft pulse (at 6720 s). Figure 8.23 demonstrates that the model and the forward operator reproduce realistically looking Z_{DR} columns and holes with cyclic periodicity typical of deep convective storms containing hail (Kumjian et al. 2010). Moreover, the enhanced Z_{DR} below the freezing level (owing primarily to gravitational sedimentation) identifies the bottom of updraft even if the actual Z_{DR} column is shallow.

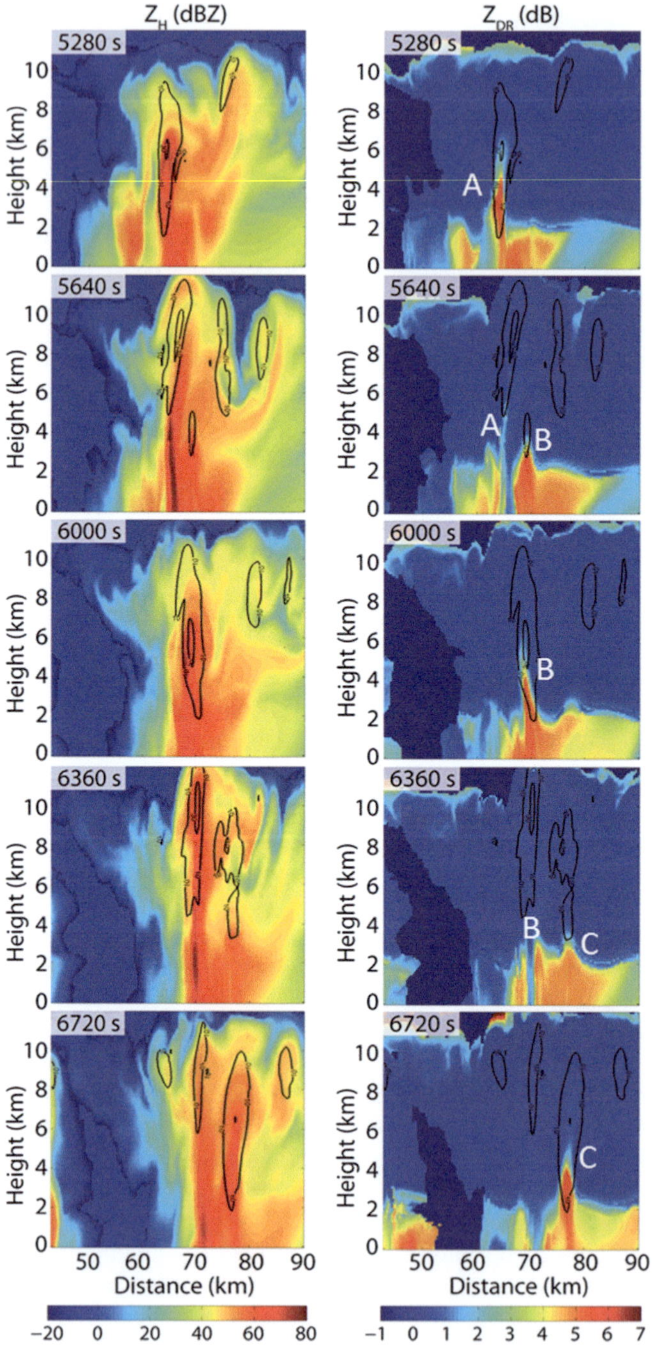

Fig. 8.23 Evolution of vertical cross section of Z (left panels) and Z_{DR} (right panels) throughout life cycle of a hailstorm simulated using HUCM. Vertical velocity values of 10 m s^{-1} and 30 m s^{-1}

Within the updrafts of deep moist convection, the freezing level is lifted with respect to the ambient air by latent heating effects. Because large raindrops do not freeze instantly upon entering a subfreezing environment, liquid or partly-liquid drops may exist well above the freezing level within updrafts of convective storms. Initial large raindrops at the base of the Z_{DR} column are likely produced by the recirculation process (Ilotoviz et al. 2016; Kumjian et al. 2014). That is, the initial Z_{DR} column forms by recirculation and insertion of small drops (grown in the first pass through the updraft) into the base of the updraft, whereupon they grow rapidly by collection of cloud droplets and smaller raindrops (Fig. 8.24).

The anatomy of the Z_{DR} column simulated by Kumjian et al. (2014) is illustrated in Fig. 8.25, and the mass distributions of the various hydrometeor species within the mature Z_{DR} column are in Fig. 8.26. It is evident (Fig. 8.25) that the base of the Z_{DR} column from the surface up to 1–1.5 km above the environmental freezing level is dominated by raindrops with size distribution skewed toward larger drops. Partially frozen supercooled raindrops and water-coated graupel/hail constitute the top of the Z_{DR} column. According to Fig. 8.26, at the surface and at 1 km AGL, only raindrops are present, and the distributions are dominated by large (>4 mm) sizes. At 2 km AGL, a second peak in the mass distribution indicative of tiny droplets (<0.05 mm in diameter) appears; it is associated with cloud water mass and implies the presence of an updraft. At 3 km AGL, the cumulative amount of cloud water mass increases, and we see the appearance of some hailstones (<1 cm in diameter). By 4 km, some larger hail is present, along with large partially frozen or freezing drops. By 5 km, most of the raindrops larger than 3 mm in diameter have disappeared from the mass distribution as they freeze and thus convert into substantial masses of hail and freezing drops. At 6 km AGL, very few raindrops larger than 1 mm exist, and we see massive amounts of freezing drops and hailstones.

Kumjian et al. (2014) and Ilotoviz et al. (2016, 2018) also quantify the relations between the height of the Z_{DR} column (determined as the maximal height of the 1 dB or 2 dB Z_{DR} contour above the environmental freezing level) and the updraft vertical velocity, revealing correlated behavior and evolution (Fig. 8.27). It is apparent that the strength of the updraft can be quantified (at least approximately) via vertical extension of the Z_{DR} column above the freezing level. This finding corroborates observational evidence of the link between the Z_{DR} column and storm updraft intensity. Further, the Z_{DR} column yields prognostic information about precipitation production, providing a strong positive correlation between the Z_{DR} column height and precipitation mass content at the ground ~10–15 min later. This finding motivated development of an operational algorithm for Z_{DR} column detection (Snyder et al. 2015).

Fig. 8.23 (continued) are contoured in black, and the time within the simulation is in the upper-left corner of each panel. The polarimetric emulator used to produce these images is described in Ryzhkov et al. (2011). White letters "A," "B," and "C" highlight three Z_{DR} columns. Courtesy of J. Snyder

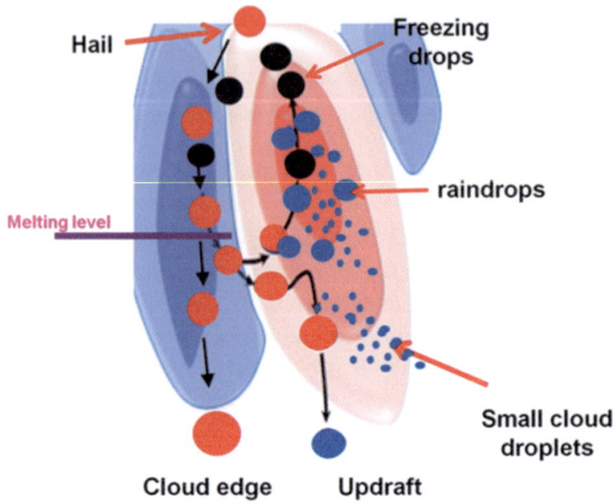

Fig. 8.24 A conceptual drawing of hail growth in an updraft-downdraft cloud zone at the mature stage of convective cell evolution in case of high concentration of cloud condensation nuclei (and cloud droplets). The zone of updraft is marked in red; the zone of downdrafts at cloud edge is marked in blue. Small blue circles denote small cloud droplets, large blue circles denote raindrops, black circles denote freezing drops, and red circles denote hail. From Ilotoviz et al. (2016). © American Meteorological Society. Used with permission

It is significant that the height and intensity of the Z_{DR} column also depend on the concentration of cloud condensation nuclei (CCN) near the surface (Ilotoviz et al. 2016, 2018). In a clean atmosphere with low concentration of CCN typical for maritime environment, cloud droplets start growing faster compared to the case of polluted atmosphere with high concentration of cloud concentration nuclei (CCN), whereby competition between larger number of cloud droplets for available water vapor is stronger. Therefore, the coalescence process at lower atmosphere levels is delayed in high CCN cases, and little rain is generated below the melting layer via warm-rain collision-coalescence mechanism as most smaller-size liquid droplets are transported by the updraft above the freezing level. These supercooled liquid droplets in high concentration favor wet growth of hail aloft via accretion of cloud water. Hailstones growing in the wet regime are larger in size and have a water shell which dramatically increases their differential reflectivity (Chap. 7). The contribution of wet hailstones to Z_{DR} aloft makes Z_{DR} columns taller and more intense. Such dependences of the Z_{DR} column volume and the heights of $Z_{DR} = 1$ dB and 2 dB contours on the CCN concentration are shown in Fig. 8.28.

Ilotoviz et al. (2016) also found that total amount of hail decreases, but its maximal size increases with increasing CCN concentration. Summarizing, one can conclude that the use of cloud model with spectral microphysics in conjunction with polarimetric radar measurements reveals a very important link between the CCN

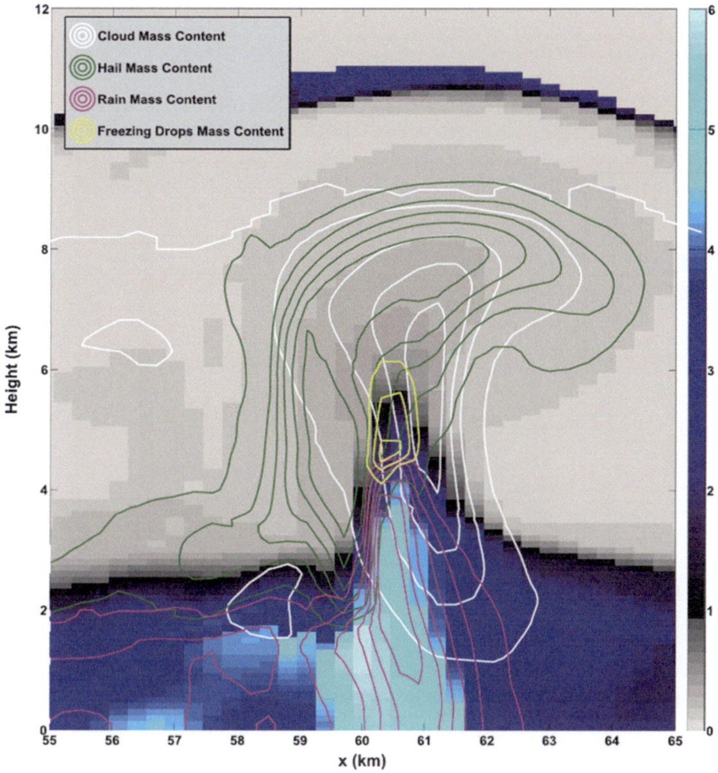

Fig. 8.25 Field of simulated Z_{DR} (color shading; dB) in a convective storm at 4680 s of simulation time. Overlaid are contours of hydrometeor mass content: cloud droplets (white), rain (magenta), hail (green), and freezing drops (yellow). Contour intervals are the same for each species, starting at 0.5 g m^{-3}, with increments of 0.5 g m^{-3}. From Kumjian et al. (2014). © American Meteorological Society. Used with permission

concentration, hail size and amount, and the characteristics of Z_{DR} columns in deep convective storms.

The presented simulation results are from a 2D version of the HUCM model. Polarimetric signatures in supercells have been simulated using 3D NWP models with bulk microphysics (e.g., Dawson et al. 2014; Johnson et al. 2016; Jung et al. 2010; Snyder et al. 2017a, b) using a version of the forward operator described in Jung et al. (2008) and further modified by Snyder et al. (2017a). The models with bulk microphysics do not explicitly treat mixed-phase hydrometeors (such as melting snow or hail) and commonly simulate mixing ratios and concentrations for pure liquid habits such as rain and pure ice habits (i.e., snow, hail, and graupel). Jung et al. (2008) and Dawson et al. (2014) suggest methodologies which partially mitigate inherent deficiencies of the bulk models by creating artificial classes of "melting snow" or "melting hail" by "borrowing" some amounts of water and ice from pure liquid and pure ice categories

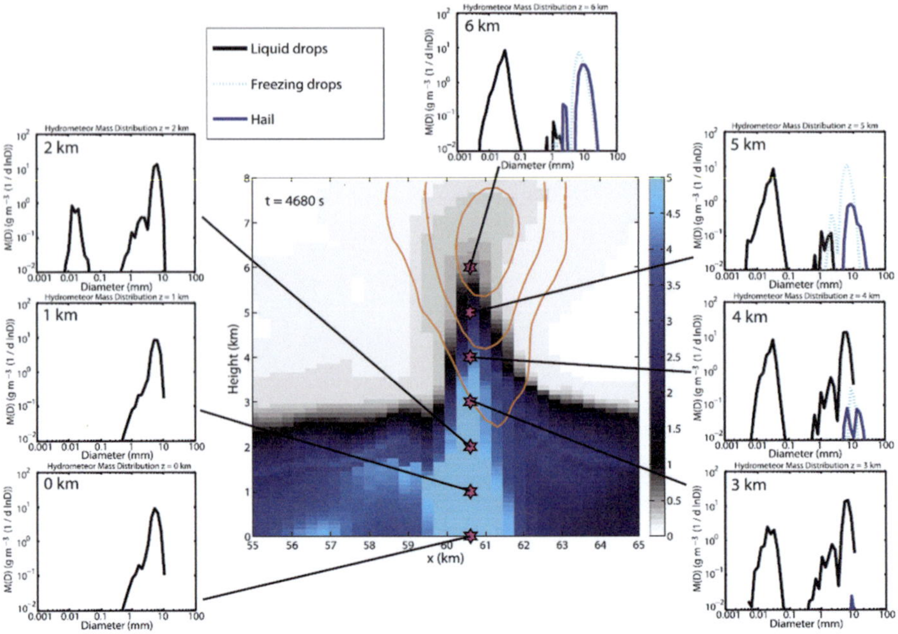

Fig. 8.26 Simulated Z_{DR} (color shading; dB) at 4680 s of simulation time, overlaid with updraft contours (brown, in 10 m s^{-1} increments starting with 10 m s^{-1}). Pink markers are at locations from which particle mass distributions are plotted in the outset panels on log-log scale for every 1 km in height, starting at the surface. The liquid-drop mass distribution is in black (this includes rain and cloud water), hail is in blue, and the freezing-drop distribution is in dashed cyan. From Kumjian et al. (2014). © American Meteorological Society. Used with permission

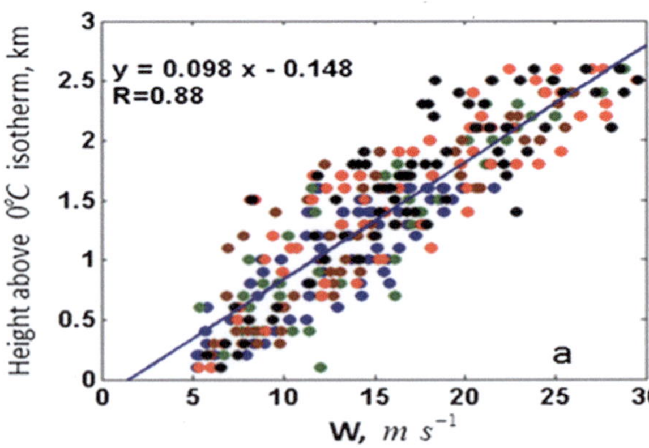

Fig. 8.27 Scatterplots of the 1 dB Z_{DR} contour height above the 0 °C level (km) versus the updraft speed (m s^{-1}) at that height. Different concentrations of CCN are marked by different colors: blue, 100 cm^{-3}; green, 400 cm^{-3}; brown, 1000 cm^{-3}; red, 2000 cm^{-3}; and black, 3000 cm^{-3}. Adapted from Ilotoviz et al. (2018). © American Meteorological Society. Used with permission

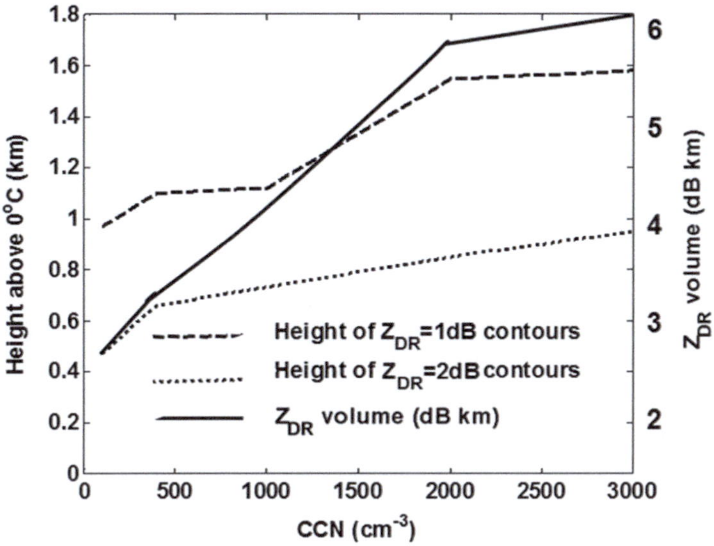

Fig. 8.28 Dependencies of the Z_{DR} column volume and the heights of Z_{DR} = 1 dB and 2 dB contours on the CCN concentration. Adapted from Ilotoviz et al. (2018). © American Meteorological Society. Used with permission

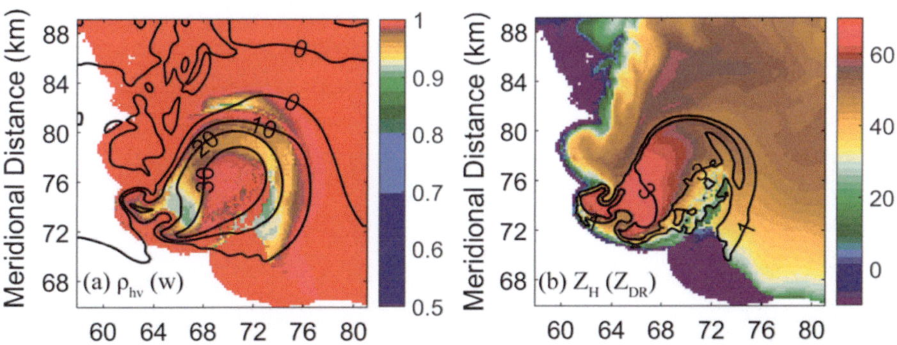

Fig. 8.29 The ρ_{hv} and Z_{DR} rings simulated in a supercell storm at ~4.6 km AGL as shown by (**a**) ρ_{hv} (colors) and w (contoured every 10 m s^{-1}), (**b**) Z (dBZ; colors), and Z_{DR} (dB; contoured every 1 dB). The polarimetric variables are calculated for X band. Adapted from Snyder et al. (2017a). © American Meteorological Society. Used with permission

in the model and mixing these. Presumably this would produce more realistic polarimetric signatures in the areas of mixed-phase hydrometeors.

Jung et al. (2010) simulated a supercell storm using the Advanced Regional Prediction System (ARPS) model (Xue et al. 2003) with single- and double-moment microphysics scheme. The forward operator described by Jung et al. (2008) was applied to examine its ability to simulate polarimetric signatures reported in observational

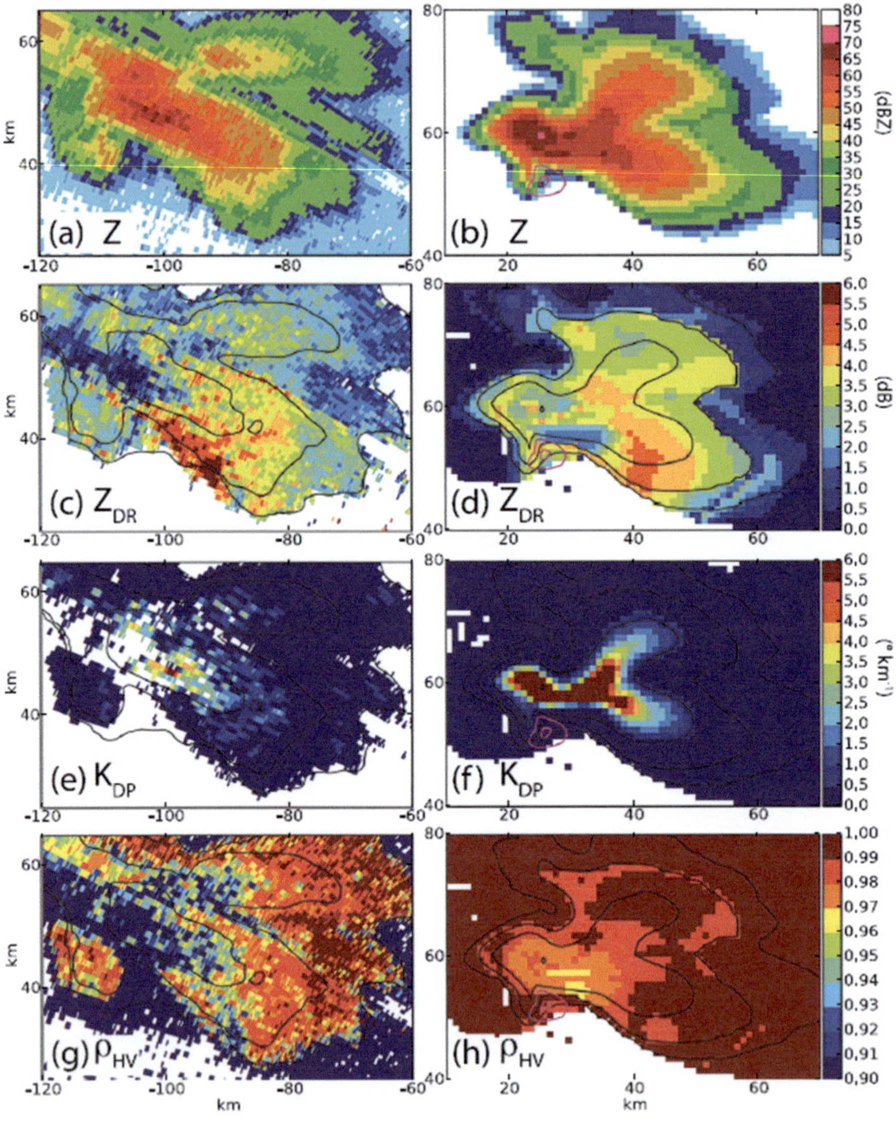

Fig. 8.30 Observed (left panels) and simulated (right panels) fields of Z, Z_{DR}, K_{DP}, and ρ_{hv} near the surface of the 1 June 2008 northwestern Oklahoma nontornadic supercell. Overlaid are the reflectivity contours in 20 dBZ increments, starting at 10 dBZ (black), and vertical velocity at about 3 km AGL in 10 m s^{-1} increments, starting at 10 m s^{-1}. From Dawson et al. (2014). © American Meteorological Society. Used with permission

studies (e.g., Kumjian and Ryzhkov 2008). The authors compared the simulations with double- and single-moment microphysics and found that certain signatures, such as Z_{DR} arc and midlevel Z_{DR} and ρ_{hv} rings, cannot be reproduced with the latter.

Snyder et al. (2017a, 2017b) used the same model with triple-moment microphysics of Mildbrand and Yau (2005) and a similar polarimetric operator to reproduce realistically looking Z_{DR} and K_{DP} columns as well as the Z_{DR} and ρ_{hv} midlevel rings in a simulated supercell. The results of simulations of Z, Z_{DR}, ρ_{hv}, and vertical velocity w at height level 4.6 km illustrated in Fig. 8.29 are generally consistent with radar observations in a typical supercell storm (e.g., Fig. 8.21).

Another modeling study of polarimetric signatures in the nontornadic supercell storm which occurred on 1 June 2008 in central Oklahoma was conducted by Dawson et al. (2014). It was primarily focused on the physical interpretation of the Z_{DR} arc. Various microphysical schemes were tested, and the triple-moment microphysics scheme developed by Mansel (2010) which takes into account variable density and terminal velocity of graupel and hail provided best agreement with radar observations (Fig. 8.30). The model reasonably well reproduces observational results of Kumjian et al. (2010) for this storm. These include reduction of Z_{DR} and ρ_{hv} in the hail area northeast of the convective updraft (indicated by magenta contours of vertical velocity in the right panels in Fig. 8.30) and Z_{DR} enhancement further away east of the updraft which can be qualified as the Z_{DR} arc. In this case, joint analysis of radar observations and the results of model simulations help to reveal and clarify important underlying mechanisms such as size sorting of melting graupel/hailstones and raindrops in the presence of strong veering of wind behind the formation of the Z_{DR} arc signature.

References

Anderson, M., Carey, L., Petersen, W., & Knupp, K. (2011). C-band dual-polarization radar signatures of hail. *Electronic Journal of Operational Meteorology*, 2011-EJ2. Retrieved from http://www.nwas.org/ej/pdf/2011-EJ2.pdf

Aydin, K., Seliga, T., & Balaji, V. (1986). Remote sensing of hail with a dual linear polarization radar. *Journal of Climate and Applied Meteorology, 25*, 1475–1484.

Aydin, K., & Zhao, Y. (1990). A computational study of polarimetric radar observables in hail. *IEEE Transactions on Geoscience and Remote Sensing, 28*, 412–422.

Balakrishnan, N., & Zrnic, D. (1990). Use of polarization to characterize precipitation and discriminate large hail. *Journal of the Atmospheric Sciences, 47*, 1525–1540.

Blair, S., & Leighton, J. (2012). Creating high-resolution hail datasets using social media and post-storm ground surveys. *Electronic Journal of Operational Meteorology, 13*, 32–45.

Bluestein, H. (2013). *Severe convective storms and tornadoes. Observations and dynamics* (p. 456). Berlin, Germany: Springer.

Bluestein, H., French, M., Tanamachi, R., Frasier, S., Hardwick, K., Junyent, F., & Pazmany, A. (2007). Close-range observations of tornadoes in supercells made with a dual-polarization, X-band, mobile Doppler radar. *Monthly Weather Review, 135*, 1522–1543.

Bluestein, H., Snyder, J., & Houser, J. (2015). A multiscale overview of the El Reno, Oklahoma, tornadic supercell of 31 May 2013. *Weather and Forecasting, 30*, 525–552.

Bodine, D., Kumjian, M., Palmer, R., Heinselman, P., & Ryzhkov, A. (2013). Tornado damage estimation using polarimetric radar. *Weather and Forecasting, 28*, 139–158.

Bodine, D., Palmer, R., & Zhang, G. (2014). Dual-wavelength polarimetric radar analyses of tornadic debris signatures. *Journal of Applied Meteorology and Climatology, 53*, 242–261.

Borowska, L., Ryzhkov, A., Zrnic, D., Simmer, C., & Palmer, R. (2011). Attenuation and differential attenuation of the 5 cm wavelength radiation in melting hail. *Journal of Applied Meteorology and Climatology, 50*, 59–76.

Bringi, V., Vivekanandan, J., & Tuttle, J. (1986). Multiparameter radar measurements in Colorado convective storms. Part II: Hail detection studies. *Journal of the Atmospheric Sciences, 43*, 2564–2577.

Brown, R. A., Lemon, L. R., & Burgess, D. W. (1978). Tornado detection by pulsed Doppler radar. *Monthly Weather Review, 106*, 29–38.

Burgess, D., et al. (2014). 20 May 2013 Moore, Oklahoma, Tornado: Damage survey and analysis. *Weather and Forecasting, 29*, 1229–1237.

Carlin, J., Ryzhkov, A., Snyder, J., & Khain, A. (2016). Hydrometeor mixing ratio retrievals for strom-scale radar data assimilation: Utility of current equations and potential benefits of polarimetry. *Monthly Weather Review, 144*, 2981–3001.

Conway, J., & Zrnic, D. (1993). A study of embryo production and hail growth using dual-Doppler and multiparameter radars. *Monthly Weather Review, 121*, 2511–2528.

Dawson, D., Mansell, E., Jung, Y., Wicker, L., Kumjian, M., & Xue, M. (2014). Low-level Z_{DR} signatures in supercell forward flanks: The role of size sorting and melting of hail. *Journal of the Atmospheric Sciences, 71*, 276–299.

Entremont, C., & Lamb, D. (2014). *Relationship between tornado debris signature (TDS) height and tornado intensity*. Training Material for Forecasters. https://slideplayer.com/slide/8421932/

Feral, L., Sauvageot, H., & Soula, S. (2003). Hail detection using S- and C-band radar reflectivity difference. *Journal of Atmospheric and Oceanic Technology, 20*, 233–248.

Gao, J., Xue, M., Brewster, K., & Droegemeier, K. (2004). A three-dimensional variational data analysis method with recursive filter for Doppler radars. *Journal of Atmospheric and Oceanic Technology, 21*, 457–469.

Gu, J.-Y., Ryzhkov, A., Zhang, P., Neilley, P., Knight, M., Wolf, B., & Lee, D.-I. (2011). Polarimetric attenuation correction in heavy rain at C band. *Journal of Applied Meteorology, 50*, 39–58.

Holler, H., Bringi, V., Hubbert, J., Hagen, M., & Meischner, P. (1994). Life cycle and precipitation formation in a hybrid-type hailstorm revealed by polarimetric and Doppler radar measurements. *Journal of the Atmospheric Sciences, 51*, 2500–2522.

Houze, R. (1993). *Cloud dynamics* (p. 573). Amsterdam, Netherlands: Academic Press.

Houze, R., Rutledge, S., Biggerstaff, M., & Smuul, B. (1989). Interpretation of Doppler weather radar displays of midlatitude mesoscale convective systems. *Bulletin of the American Meteorological Society, 70*, 608–619.

Hubbert, J., Bringi, V., Carey, L., & Bolen, S. (1998). CSU-CHILL polarimetric radar measurements from a severe hail storm in eastern Colorado. *Journal of Applied Meteorology, 37*, 749–775.

Illingworth, A., Goddard, J., & Cherry, S. (1987). Polarization radar studies of precipitation development in convective storms. *Quarterly Journal of the Royal Meteorological Society, 113*, 469–489.

Ilotoviz, E., Benmoshe, N., Khain, A., Phillips, V., & Ryzhkov, A. (2016). Effect of aerosols on freezing drops, hail, and precipitation in a mid-latitude storm. *Journal of the Atmospheric Sciences, 73*, 109–144.

Ilotoviz, E., Khain, A., Ryzhkov, A., & Snyder, J. (2018). Relationship between aerosols, hail microphysics, and Z_{DR} columns. *Journal of the Atmospheric Sciences, 75*, 1755–1781.

Johnson, M., Jung, Y., Dawson, D., & Xue, M. (2016). Comparison of simulated polarimetric signatures in idealized supercell storms using two-moment bulk microphysics schemes in WRF. *Monthly Weather Review, 144*, 971–996.

Jung, Y., Xue, M., & Zhang, G. (2010). Simulations of polarimetric radar signatures of a supercell storm using a two-moment bulk microphysics scheme. *Journal of Applied Meteorology and Climatology, 49*, 146–163.

Jung, Y., Zhang, G., & Xue, M. (2008). Assimilation of simulated polarimetric radar data for a convective storm using the Ensemble Kalman Filter. Part I: Observation operators for reflectivity and polarimetric variables. *Monthly Weather Review, 136*, 2228–2245.

Kaltenboeck, R., & Ryzhkov, A. (2013). Comparison of polarimetric signatures of hail at S and C bands for different hail sizes. *Atmospheric Research, 123*, 323–336.

Kennedy, P., Rutledge, S., Dolan, B., & Thaller, E. (2014). Observations of the 14 July 2011 Fort Collins hailstorm: Implications for WSR-88D-based hail detection and warnings. *Weather and Forecasting, 29*, 623–638.

Kennedy, P., Rutledge, S., Petersen, W., & Bringi, V. (2001). Polarimetric radar observations of hail formation. *Journal of Applied Meteorology, 40*, 1347–1366.

Khain, A., & Pinsky, M. (2018). *Physical processes in clouds and cloud modeling* (p. 686). Cambridge: Cambridge University Press.

Khain, A., Pokrovsky, A., Pinsky, M., Seifert, A., & Phillips, V. (2004). Simulation of effects of atmospheric aerosols on deep turbulent convective clouds using a spectral microphysics mixed-phase cumulus cloud model. Part I: Model description and possible applications. *Journal of the Atmospheric Sciences, 61*, 2963–2982.

Khain, A., Rosenfeld, D., Pokrovsky, A., Blahak, U., & Ryzhkov, A. (2011). The role of CCN in precipitation and hail in a midlatitude storm as seen in simulations using a spectral (bin) microphysics model in a 2D dynamic frame. *Atmospheric Research, 99*, 129–146.

Kumjian, M. (2011). Precipitation properties of supercell hook echoes. *Electronic Journal of Severe Storms Meteorology, 6*(5), 1–21 Retrieved from https://ejssm.org/ojs/index,php/ejssm/issue/view/30.

Kumjian, M. (2013). Principles and applications of dual-polarization weather radar. Part II: Warm- and cold-season applications. *Journal of Operational Meteorology, 1*(20), 243–264.

Kumjian, M., Khain, A., Benmoshe, N., Ilotoviz, E., Ryzhkov, A., & Phillips, V. (2014). The anatomy and physics of Z_{DR} columns: Investigating a polarimetric radar signature with a spectral bin microphysical model. *Journal of Applied Meteorology and Climatology, 53*, 1820–1843.

Kumjian, M., & Ryzhkov, A. (2009). Storm-relative helicity revealed from polarimetric radar measurements. *Journal of the Atmospheric Sciences, 66*, 667–685.

Kumjian, M., Ryzhkov, A., Melnikov, V., & Schuur, T. (2010). Rapid-scan super-resolution observations of a cyclic supercell with a dual-polarization WSR-88D. *Monthly Weather Review, 138*, 3762–3786.

Kumjian, M. R., & Ryzhkov, A. (2008). Polarimetric signatures in supercell storms. *Journal of Applied Meteorology and Climatology, 47*, 1940–1961.

Kurdzo, J., Bodine, D., Cheong, B.-L., & Palmer, R. (2015). High-temporal resolution polarimetric X-band Doppler radar observations of the 20 May 2013 Moore, Oklahoma, tornado. *Monthly Weather Review, 143*, 2711–2735.

Loney, M., Zrnic, D., Straka, J., & Ryzhkov, A. (2002). Enhanced polarimetric signatures above the melting level in a supercell storm. *Journal of Applied Meteorology, 41*, 1179–1194.

Mansel, E. (2010). On sedimentation and advection in multimoment bulk microphysics. *Journal of the Atmospheric Sciences, 67*, 3084–3094.

Markowski, P., & Richardson, Y. (2010). *Mesoscale meteorology in midlatitudes* (p. 407). Chichester, UK: Wiley.

Mildbrand, J., & Yau, M. (2005). A multimoment bulk microphysical parameterization. Part II: A proposed three-moment closure and scheme description. *Journal of the Atmospheric Sciences, 62*, 3065–3081.

Ortega, K., Krause, J., & Ryzhkov, A. (2016). Polarimetric radar characteristics of melting hail. Part III: Validation of the algorithm for hail size discrimination. *Journal of Applied Meteorology and Climatology, 55*, 829–848.

Palmer, R., Bodine, D., Kumjian, M., Cheong, B., Zhang, G., Cao, Q., Bluestein, H., Ryzhkov, A., Yu, T., & Wang, Y. (2011). Observations of the 10 May 2010 tornado outbreak using

OU-PRIME: Potential for new science with high-resolution polarimetric radar. *Bulletin of the American Meteorological Society, 92*, 871–891.

Payne, C., Schuur, T., MacGorman, D., Biggerstaff, M., Kumjian, M., & Rust, W. (2010). Polarimetric and electrical characteristics of a lightning ring in a supercell storm. *Monthly Weather Review, 138*, 2405–2425.

Picca, J., & Ryzhkov, A. (2012). A dual-wavelength polarimetric analysis of the 16 May 2010 Oklahoma City extreme hailstorm. *Monthly Weather Review, 140*, 1385–1403.

Romine, G., Burgess, D., & Wilhelmson, R. (2008). A dual-polarization-radar -based assessment of the 8 May 2003 Oklahoma City area tornadic supercell. *Monthly Weather Review, 136*, 2849–2870.

Ryzhkov, A., Burgess, D., Zrnic, D., Smith, T., & Giangrande, S. (2002). *Polarimetric analysis of a 3 May 1999 Tornado*. Preprints, 21 Conference on Severe Local Storms, San Antonio, TX, Amer. Meteor. Soc., 14.2. Retrieved from https://ams.confex.com/ams/pdfpapers/47348.pdf

Ryzhkov, A., Kumjian, M., Ganson, S., & Khain, A. (2013a). Polarimetric radar characteristics of melting hail. Pt I: Theoretical simulations using spectral microphysical modeling. *Journal of Applied Meteorology and Climatology, 52*, 2849–2870.

Ryzhkov, A., Kumjian, M., Ganson, S., & Zhang, P. (2013b). Polarimetric radar characteristics of melting hail. Pt II: Practical implications. *Journal of Applied Meteorology and Climatology, 52*, 2871–2886.

Ryzhkov, A., Matrosov, S., Melnikov, V., Zrnic, D., Zhang, P., Cao, Q., Knight, M., Simmer, C., & Troemel, S. (2017). Estimation of depolarization ratio using radars with simultaneous transmission/reception. *Journal of Applied Meteorology and Climatology, 56*, 1797–1816.

Ryzhkov, A., Pinsky, M., Pokrovsky, A., & Khain, A. (2011). Polarimetric radar observation operator for a cloud model with spectral microphysics. *Journal of Applied Meteorology and Climatology, 50*, 873–894.

Ryzhkov, A., Schuur, T., Burgess, D., & Zrnic, D. (2005). Polarimetric tornado detection. *Journal of Applied Meteorology, 44*, 557–570.

Ryzhkov, A., & Zrnic, D. (1995). Precipitation and attenuation measurements at a 10 cm wavelength. *Journal of Applied Meteorology, 34*, 2121–2134.

Ryzhkov, A., Zrnic, D., Zhang, P., Krause, J., Park, H., Hudak, D., et al. (2007). *Comparison of polarimetric algorithms for hydrometeor classification at S and C bands*. In Extended Abstracts, 33rd Conference Radar Meteorology, Cairns, Australia, Amer. Meteor. Soc., 10.3. Retrieved from http://ams.confex.com/ams/pds/papers/123109.pdf

Scharfenberg, K., Miller, D., Schuur, T., Schlatter, P., Giangrande, S., Melnikov, V., et al. (2005). The Joint Polarization Experiment: Polarimetric radar in forecasting and warning decision making. *Weather and Forecasting, 20*, 775–788.

Schultz, C., Nelson, S., Carey, L., Belanger, C., Carcione, B., Darden, C., et al. (2012a). Dual-polarization tornadic debris signatures. Part I: Examples and utility in an operational setting. *Electronic Journal of Operational Meteorology, 13*(9), 120–137.

Schultz, C., Nelson, S., Carey, L., Belanger, L., Carcione, B., Darden, C., et al. (2012b). Dual-polarization tornadic debris signatures. Part II: Comparisons and caveats. *Electronic Journal of Operational Meteorology, 13*(10), 138–150.

Snyder, J., & Bluestein, H. (2014). Some considerations for the use of high-resolution mobile radar data in tornado intensity determination. *Weather and Forecasting, 29*, 799–827.

Snyder, J., Bluestein, H., Dawson, D., & Jung, Y. (2017a). Simulations of polarimetric, X-band radar signatures in supercells. Part I: Description of experiment and simulated ρ_{hv} rings. *Journal of Applied Meteorology and Climatology, 56*, 1977–1999.

Snyder, J., Bluestein, H., Dawson, D., & Jung, Y. (2017b). Simulations of polarimetric, X-band radar signatures in supercells. Part II: Z_{DR} columns and rings and K_{DP} columns. *Journal of Applied Meteorology and Climatology, 56*, 2001–2026.

Snyder, J., Bluestein, H., Venkatesh, V., & Frasier, S. (2013). Observations of polarimetric signatures in supercells by an X-band mobile Doppler radar. *Monthly Weather Review, 141*, 3–29.

Snyder, J., Ryzhkov, A., Kumjian, M., Picca, J., & Khain, A. (2015). Developing a Z_{DR} column detection algorithm to examine convective storm updrafts. *Weather and Forecasting, 30*, 1819–1844.

Tabary, P., Fradon, B., Illingworth, A.J., & Vulpiani, G. (2009). *Hail detection and quantification with a C-band polarimetric radar: Challenges and promises*. In Extended Abstracts, 34th Conference on Radar Meteorology, Williamsburg, VA, Amer. Meteor. Soc., 10A.4. Retrieved from http://ams.confex.com/ams/34Radar/techprogram/paper_155530.htm

Tessendorf, S., Miller, J., Wiens, K., & Rutledge, S. (2005). The 29 June 2000 supercell observed during STEPS. Part I: Kinematics and microphysics. *Journal of the Atmospheric Sciences, 62*, 4127–4150.

Van Den Broeke, M., & Jauernic, S. (2014). Spatial and temporal characteristics of polarimetric tornadic debris signatures. *Journal of Applied Meteorology and Climatology, 53*, 2217–2231.

Van Den Broeke, M., Straka, J., & Rasmussen, E. (2008). Polarimetric radar observations at low levels during tornado life cycles in a small sample of classic southern plains supercells. *Journal of Applied Meteorology and Climatology, 47*, 1232–1247.

Van Lier-Walqui, M., Fridlind, A., Helmus, A., MacGorman, D., North, K., Kollias, P., & Posselt, D. (2016). On polarimetric radar signatures of deep convection for model evaluation: Columns of specific differential phase observed during MC3E. *Monthly Weather Review, 144*, 737–758.

Wurman, J., Kosiba, K., Robinson, P., & Marshall, T. (2014). The role of multiple-vortex tornado structure in causing storm researcher fatalities. *Bulletin of the AMS, 95*, 31–45.

Xue, G., Wang, D., Gao, J., Brewster, K., & Droegemeier, K. (2003). The Advanced Regional Prediction System (ARPS), storm-scale numerical weather prediction and data assimilation. *Meteorology and Atmospheric Physics, 82*, 139–170.

Chapter 9
Polarimetric Classification of Radar Echo

9.1 General Principles of Classification

Dual-polarization radar is uniquely suited for the classification of radar echoes. Having dual-polarization capability is essential for discriminating between meteorological and nonmeteorological scatterers and for distinguishing between different hydrometeor types, such as rain, hail, graupel, and snow of various habits. Furthermore, knowledge of the hydrometeor type within the radar resolution volume is required for choosing an appropriate polarimetric relation to accurately quantify the amount of precipitation.

There are several methods that can be potentially used for hydrometeor identification, such as decision tree, statistical decision theory, neural networks, and fuzzy logic. But as of this writing, there is no automatic (artificially intelligent) method that can outperform human experts. The first reported manual classification by humans using the polarimetric variables and knowledge of storm structure was made on vertical cross sections of hail storms (Zrnic et al. 1993). But manual classification is very slow and hard to make in real time especially if several storms are present at the same time. This deficiency doesn't apply to automatic methods. Although "artificially intelligent" methods are not perfect, they can quickly point to forecasters' regions of interest where precipitation hazard may loom. Some of these methods are described next.

The decision tree methodology is based on thresholds or hard boundaries and Boolean logic to distinguish between different types of radar echo. The decision tree method is not efficient for hydrometeor classification because the measurement set for different hydrometeor types is not mutually exclusive. In other words, the probability density functions of a certain radar variable, say, radar reflectivity of rain and snow, may overlap heavily. In addition, the decision tree technique does not provide allowance for measurement errors. The statistical decision theory is an optimum approach. It associates regions in the space of polarimetric variables with classes so that the probability of correctly identifying these is maximized for a given

A. V. Ryzhkov, D. S. Zrnic, *Radar Polarimetry for Weather Observations*,
Springer Atmospheric Sciences, https://doi.org/10.1007/978-3-030-05093-1_9

probability of wrong classification. This requires a comprehensive statistical information that is not yet available. Another candidate method is based on neural networks. Although powerful, this technique needs a verification training set of considerable size that currently does not exist. The fuzzy logic technique is well suited for classifying radar returns because it is easy to implement and can identify classes for which the domains in the space of polarimetric variables overlap and the variables may be contaminated by measurement noise. Furthermore, in the fuzzy logic approach, it is easy to assign confidence factors to various types of measurements.

Utilization of the principles of fuzzy logic for hydrometeor classification was explored by Straka and Zrnic (1993) and Straka (1996). Their classification scheme became the basis of several subsequent attempts for hydrometeor identification (Zrnic and Ryzhkov 1999; Vivekanandan et al. 1999). A further step in the development was made by Liu and Chandrasekar (2000) who used a feedback scheme to adjust the weights of the fuzzy classifier; they also gave a comprehensive set of values for one-dimensional membership functions. Meanwhile, Starka et al. (2000) have presented an extensive set of tables and two-dimensional graphs specifying typical ranges of variability of various polarimetric radar variables and their relations with the reflectivity factor. These or similar graphs have been used for building the membership functions in more sophisticated fuzzy logic classification schemes (Zrnic et al. 2001; Schuur et al. 2003; Keenan 2003; Lim et al. 2005; Marzano et al. 2007; Park et al. 2009; Dolan & Rutledge 2009; Dolan et al. 2013; Snyder et al. 2010; Chandrasekar et al. 2013; Al-Sakka et al. 2013; Thompson et al. 2014; Bechini and Chandrasekar 2015).

The need for fuzzy logic approach is primarily driven by strong overlap of the ranges of variability of various radar variables associated with different classes of radar echo in a multiparameter space. This is illustrated in Fig. 9.1 where the scatterplots of Z_{DR} vs Z for three types of snow are presented in the Z–Z_{DR} plane. These scatterplots overlap significantly among themselves and even more so with the typical region for rain which is located between the two green curves in Fig. 9.1.

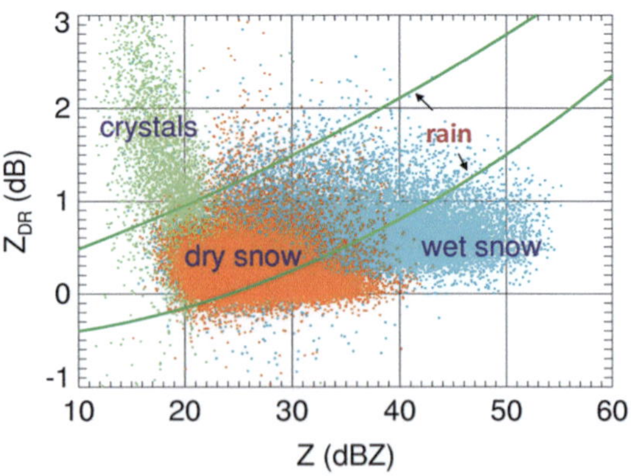

Fig. 9.1 Scatterplots of Z_{DR} vs. Z for different snow types. The region of expected data for rain is between the two green curves. Data were obtained with the S-band Cimarron dual-polarization radar in Oklahoma. From Ryzhkov et al. (2005). © American Meteorological Society. Used with permission

In the fuzzy logic methodology, different classes of radar echo (or hydrometeors) are described by the sets of the membership functions $P^{(i)}(V_j)$ characterizing distributions of polarimetric variables V_j for the i-th class. One of the commonly used shapes of the membership function is trapezoidal described by four parameters: x_1, x_2, x_3, and x_4 (Fig. 9.2). We illustrate the essence of the classification algorithm using a simple example of discrimination between three classes of returns associated with (1) meteorological scatterers, (2) biological scatterers, and (3) ground clutter/anomalous propagation. Assume that the following five polarimetric variables are available: Z, Z_{DR}, ρ_{hv}, and two texture parameters—for Z (SD(Z)) and for Φ_{DP} (SD(Φ_{DP})). The two latter variables are especially suitable for discrimination between meteorological and nonmeteorological radar returns. The parameters SD(Z) and SD(Φ_{DP}) characterize the intensity of small-scale spatial fluctuations of Z and Φ_{DP} along the beam. To obtain SD(Z), we average Z data (sampled every 0.25 km on the WSR-88D radars) along the radial using 1-km-width running average window and subtract the smoothed estimates of Z from the original values as shown in Fig. 9.3.

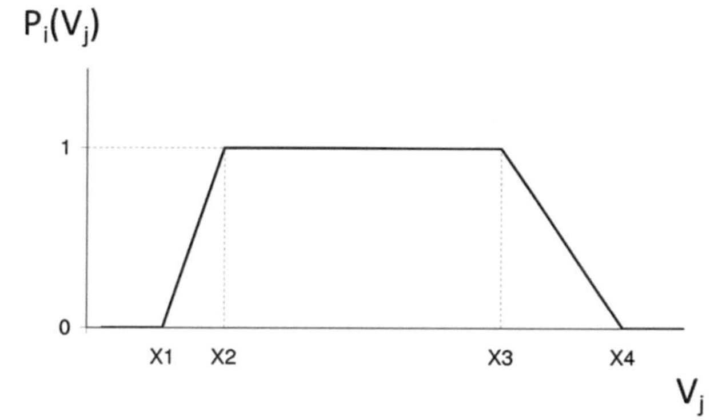

Fig. 9.2 Trapezoidal membership function where V_j is an arbitrary polarimetric variable

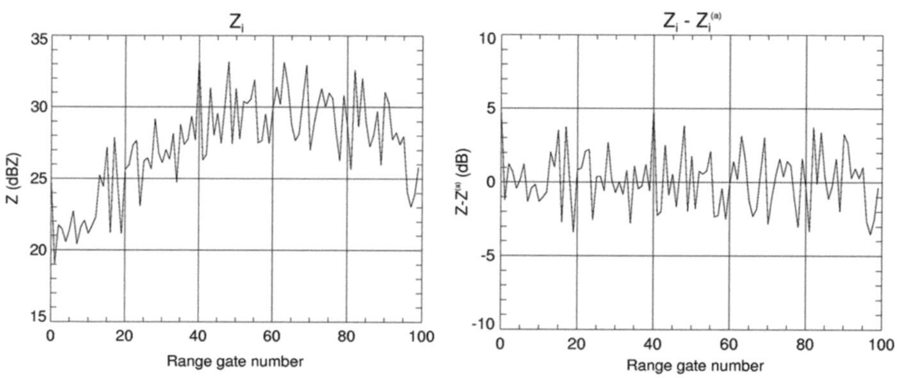

Fig. 9.3 Determination of the texture parameter of Z (SD(Z)). Data were obtained with the S-band Cimarron radar

A similar procedure is used for computing the parameter SD(Φ_{DP}), but with an averaging window that is two times larger.

An example of the membership functions for the three classes is presented in Fig. 9.4. Assume that at a given range location $Z = 25$ dBZ, $Z_{DR} = 3.0$ dB, $\rho_{hv} = 0.55$, SD(Z) = 5 dB, and SD(Φ_{DP}) = 35°. The corresponding values of the membership functions for the three classes are in Table 9.1. If only reflectivity is used for classification, then rain and AP have higher scores than biota (second column in Table 9.1). If the texture of *Z* is used, then AP has the highest score.

However, utilization of all radar variables via the aggregation process leads to a different conclusion. In this simple case, aggregation means summation of the values of all membership functions for each class, i.e., summation of the numbers in the three horizontal rows in Table 9.1. The results of this summation are in the column marked as *A*. The highest aggregation of all five measurements per class is for the biota (Bio) which tells that the most likely source of the radar return at this particular range location is biota.

This simple version of a fuzzy logic classifier effectively discriminates between the three types of radar return in the whole radar coverage area (Fig. 9.5). In the case on 24 August 2002, the three types of returns received by the S-band polarimetric radar are mixed, and the reflectivity field has no clues for identifying these. The classification algorithm identifies extended regions of ground clutter caused by anomalous propagation embedded in precipitation. The values of Z_{DR} are generally negative, and ρ_{hv} is quite low in areas where ground clutter/AP dominates

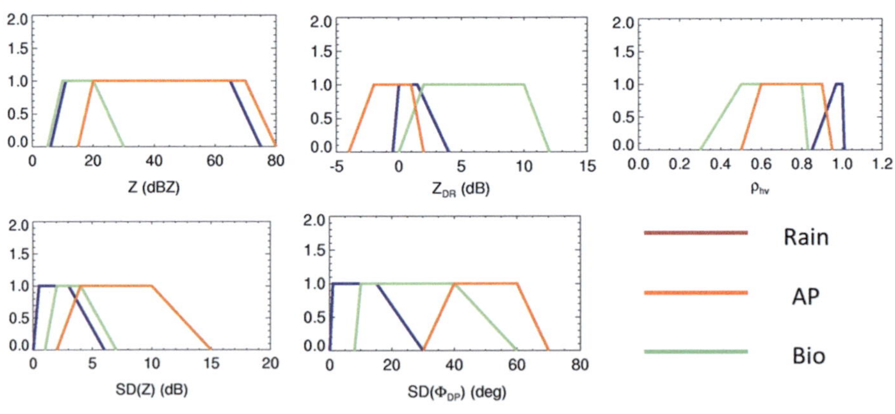

Fig. 9.4 The membership functions of Z, Z_{DR}, ρ_{hv}, and texture parameters SD(Z) and SD(Φ_{DP}) for rain, ground clutter/anomalous propagation (AP), and biota (Bio)

Table 9.1 Values of the membership functions and aggregation values in the example

Class	*P* (*Z*)	*P* (Z_{DR})	*P* (ρ_{hv})	*P* (SD(Z))	*P* (SD(Φ_{DP}))	*A*
Rain	1.0	0.5	0.0	0.33	0.0	1.83
AP	1.0	0.0	0.5	1.0	0.5	3.00
Bio	0.5	1.0	1.0	0.67	1.0	4.17

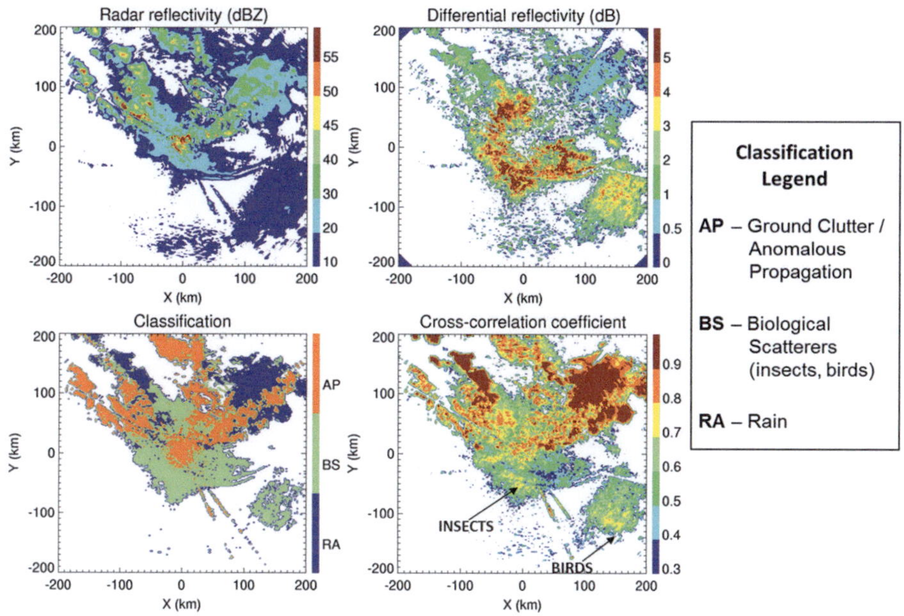

Fig. 9.5 The fields of Z, Z_{DR}, ρ_{hv}, and results of classification at elevation 0.5° obtained with the KOUN radar in three types of returns for the case on 24 August 2002. Adapted from Ryzhkov et al. (2005). © American Meteorological Society. Used with permission

the radar signal. The biota is characterized by very high Z_{DR} and low ρ_{hv}. Both nonmeteorological returns are characterized by much higher values of the texture parameters SD(Z) and SD(Φ_{DP}) than weather return. Animation of the classification fields for this particular case shows high degree of spatial/temporal "coherency" of these fields which increases confidence in the performance of the classification scheme. Note the difference between regions likely associated with insects (south of the radar) and birds in the SE sector of the image. "Birds" produce lower Z_{DR} and higher Φ_{DP} (attributed to large backscatter differential phase) than "insects."

The fuzzy logic classification generally contains three steps: (1) fuzzification, (2) aggregation or inference, and (3) defuzzification. In the first step (fuzzification), the values of the membership functions for various classes and different variables are obtained. In the second step, the values of the membership functions are summed (as in a previous example), or multiplied, or combined in other ways for each class. A simplest aggregation scheme prescribes computing the aggregation value A_i for i-th class of radar echo via summation of the membership functions:

$$A_i = \sum_j P^{(i)}(V_j). \tag{9.1}$$

Different radar variables may have different classification potential which can be reflected by the vector of weights **W** in the aggregation rule:

$$A_i = \frac{\sum_j W_j P^{(i)}(V_j)}{\sum_j W_j}. \tag{9.2}$$

If the informative content of a particular polarimetric variable depends on the class, then the vector of weights $\mathbf{W}$ can be replaced by a matrix of weights $\hat{\mathbf{W}}$ and

$$A_i = \frac{\sum_j W_{ij} P^{(i)}(V_j)}{\sum_j W_{ij}}. \tag{9.3}$$

Finally, any degradation in the quality of radar measurements by bias or noise can be addressed through the confidence vector $\boldsymbol{Q}$:

$$A_i = \frac{\sum_j W_{ij} Q_j P^{(i)}(V_j)}{\sum_j Q_j W_{ij}}. \tag{9.4}$$

That is, at each range location, the j-th polarimetric variable is supplemented with its confidence index Q_j depending on vulnerability to several factors which can compromise its measurement. These factors include (1) radar miscalibration, (2) attenuation, (3) nonuniform beam filling (NBF), (4) partial beam blockage (PBB), (5) the magnitude of ρ_{hv} (which determines the statistical errors of all polarimetric variables), and (6) receiver noise. The definition of the confidence vector is in Park et al. (2009), where the hydrometeor classification algorithm (HCA) deployed on the WSR-88Ds is described in detail. The last step in the classification process (defuzzification) ranks all aggregation values A_i and finds a maximal one which designates the most likely class of the radar return.

9.2 Hydrometeor Classification Algorithms (HCA)

9.2.1 HCA on the WSR-88D Network

The operational version of the hydrometeor classification algorithm (HCA) implemented on the US network of polarimetrically upgraded WSR-88Ds is described in Park et al. (2009). The HCA discriminates between ten classes of radar returns: (1) ground clutter including the one due to anomalous propagation (GC/AP), (2) biological scatterers (BS), (3) dry aggregated snow (DS), (4) wet snow (WS), (5) crystals of various orientations (CR), (6) graupel (GR), (7) "big drops" (BD), (8) light and moderate rain (RA), (9) heavy rain (HR), and (10) a mixture of rain and hail (RH). The "big drops" category designates rain with a drop size distribution skewed toward large sizes. This usually implies the presence of drops bigger than 3 mm and the deficit of smaller drops. "Big drop" signatures are generally caused by size sorting associated with convective updrafts often exemplified with Z_{DR} enhancement.

Six polarimetric variables are directly utilized for discrimination. These are (1) reflectivity factor at horizontal polarization Z, (2) differential reflectivity Z_{DR}, (3) cross-correlation coefficient ρ_{hv} between horizontally and vertically polarized radar returns, (4) specific differential phase K_{DP}, (5) texture parameter SD(Z) of the Z field, and (6) texture parameter SD(Φ_{DP}) of the field of differential phase Φ_{DP}. Due to the range of K_{DP}, it is more convenient to use its values in logarithmic scale rather than in linear scale. Hence, the parameter LKdp determined as

$$\begin{array}{lll} \text{LKdp} = 10\log(K_{DP}) & \text{if} & K_{DP} > 10^{-3}\text{deg km}^{-1} \\ \text{LKdp} = -30 & \text{if} & K_{DP} \leq 10^{-3}\text{deg km}^{-1} \end{array} \tag{9.5}$$

is used.

The membership functions $P^{(i)}(V_j)$ are assumed to have trapezoidal shapes (Fig. 9.2), and the parameters x_1–x_2 of these functions are presented in Table 9.2. Trapezoids adequately approximate the symmetric shapes of the probability distributions of polarimetric variables as well as the asymmetric ones which is not the case for other popular membership functions like Gaussian or beta (e.g., Liu and Chandrasekar 2000; Lim et al. 2005).

The parameters x_1–x_4 of the membership functions of Z_{DR} and LKdp depend on reflectivity for all four categories of rain (BD, RA, HR, and RH). The dependency is chosen to describe a nonrectangular region in the Z, Z_{DR} space where the typical pair of these values resides for each of the four rain categories (similar to Fig. 2a in Straka et al. (2000)). Thus functions f_1–f_3 and g_1–g_2 determine the values x_1–x_4 in Table 9.2 and are given by

$$\begin{aligned} f_1(Z) &= -0.50 + 2.50 \times 10^{-3}Z + 7.50 \times 10^{-4}Z^2, \\ f_2(Z) &= 0.68 - 4.81 \times 10^{-2}Z + 2.92 \times 10^{-3}Z^2, \\ f_3(Z) &= 1.42 + 6.67 \times 10^{-2}Z + 4.85 \times 10^{-4}Z^2, \end{aligned} \tag{9.6}$$

and

$$g_1(Z) = -44.0 + 0.8Z \quad \text{and} \quad g_2(Z) = -22.0 + 0.5Z, \tag{9.7}$$

where the reflectivity Z is expressed in dBZ.

The matrix of weights $\hat{\mathbf{W}}$ in (9.4) characterizes the discriminating efficiency of each variable with respect to a particular class. The elements of the matrix $\hat{\mathbf{W}}$ are presented in Table 9.3. As an example, ρ_{hv} is the most important variable for identification of biological scatterers ($W_{23} = 1.0$) followed by the texture parameters of Z and Φ_{DP} ($W_{25} = W_{26} = 0.8$), Z_{DR} ($W_{22} = 0.6$), and LKdp, which does not have

Table 9.2 Parameters of the membership functions for ten classes

$P(Z(\text{dBZ}))$										
	GC/AP	BS	DS	WS	CR	GR	BD	RA	HR	RH
x_1	15	5	5	25	0	25	20	5	40	45
x_2	20	10	10	30	5	35	25	10	45	50
x_3	70	20	35	40	20	50	45	45	55	75
x_4	80	30	40	50	25	55	50	50	60	80
$P(Z_{DR}(\text{dB}))$										
	GC/AP	BS	DS	WS	CR	GR	BD	RA	HR	RH
x_1	−4	0	−0.3	0.5	0.1	−0.3	$f_2 - 0.3$	$f_1 - 0.3$	$f_1 - 0.3$	−0.3
x_2	−2	2	0.0	1.0	0.4	0.0	f_2	f_1	f_1	0.0
x_3	1	10	0.3	2.0	3.0	f_1	f_3	f_2	f_2	f_1
x_4	2	12	0.6	3.0	3.3	$f_1 + 0.3$	$f_3 + 1.0$	$f_2 + 0.5$	$f_2 + 0.5$	$f_1 + 0.5$
$P(\rho_{hv})$										
	GC/AP	BS	DS	WS	CR	GR	BD	RA	HR	RH
x_1	0.5	0.3	0.95	0.88	0.95	0.90	0.92	0.95	0.92	0.85
x_2	0.6	0.5	0.98	0.92	0.98	0.97	0.95	0.97	0.95	0.90
x_3	0.9	0.8	1.00	0.95	1.00	1.00	1.00	1.00	1.00	1.00
x_4	0.95	0.83	1.01	0.985	1.01	1.01	1.01	1.01	1.01	1.01
$P(\text{LKdp})$										
	GC/AP	BS	DS	WS	CR	GR	BD	RA	HR	RH
x_1	−30	−30	−30	−30	−5	−30	$g_1 - 1$	$g_1 - 1$	$g_1 - 1$	−10
x_2	−25	−25	−25	−25	0	−25	g_1	g_1	g_1	−4
x_3	10	10	10	10	10	10	g_2	g_2	g_2	g_1
x_4	20	10	20	20	15	20	$g_2 + 1$	$g_2 + 1$	$g_2 + 1$	$g_1 + 1$
$P(\text{SD}(Z)(\text{dB}))$										
	GC/AP	BS	DS	WS	CR	GR	BD	RA	HR	RH
x_1	2	1	0	0	0	0	0	0	0	0
x_2	4	2	0.5	0.5	0.5	0.5	0.5	0.5	0.5	0.5
x_3	10	4	3	3	3	3	3	3	3	3
x_4	15	7	6	6	6	6	6	6	6	6
$P(\text{SD}(\Phi_{DP})(\text{deg}))$										
	GC/AP	BS	DS	WS	CR	GR	BD	RA	HR	RH
x_1	30	8	0	0	0	0	0	0	0	0
x_2	40	10	1	1	1	1	1	1	1	1
x_3	50	40	15	15	15	15	15	15	15	15
x_4	60	60	30	30	30	30	30	30	30	30

Reflectivity-dependent parameters f_1–f_3 and g_1–g_2 are defined by Eqs. (9.6) and (9.7), respectively

any diagnostic value for biological scatterers ($W_{24} = 0$). As in Lim et al. (2005), the matrix $\hat{\mathbf{W}}$ is determined subjectively, but further refinement based on objective criteria could be made. One possible criterion to accomplish this refinement is the methodology described by Park et al. (2007). It considers the degree of overlap between probability density functions of each polarimetric variable for different

Table 9.3 Matrix of weights

	Variable					
Class	Z	Z_{DR}	ρ_{hv}	LKdp	SD(Z)	SD(Φ_{DP})
GC/AP	0.2	0.4	1.0	0.0	0.6	0.8
BS	0.4	0.6	1.0	0.0	0.8	0.8
DS	1.0	0.8	0.6	0.0	0.2	0.2
WS	0.6	0.8	1.0	0.0	0.2	0.2
CR	1.0	0.6	0.4	0.5	0.2	0.2
GR	0.8	1.0	0.4	0.0	0.2	0.2
BD	0.8	1.0	0.6	0.0	0.2	0.2
RA	1.0	0.8	0.6	0.0	0.2	0.2
HR	1.0	0.8	0.6	1.0	0.2	0.2
RH	1.0	0.8	0.6	1.0	0.2	0.2

GC/AP stands for ground clutter including the one due to anomalous propagation, BS for biological scatterers, DS for dry snow, WS for wet snow, CR for crystals, GR for graupel, BD for big drops, RA for light rain and moderate rain, HR for heavy rain, and RH for rain/hail mixture

classes of radar return. More overlapping usually requires lowering the values of weight. A distinctive feature of the HCA implemented on WSR-88Ds is that it utilizes the confidence vector **Q** characterizing the quality of radar measurements (Park et al. 2009).

Because the membership functions of some classes of liquid and frozen hydrometeors (such as rain and dry snow) heavily overlap, the classification of these can be improved significantly by taking into account the location of the radar resolution volume with respect to the melting layer. The top and bottom of the melting layer can be determined from an upper air sounding, numerical model output, or polarimetric data (if echoes cover a sufficiently large area). The melting layer detection algorithm (MLDA) developed at the National Severe Storms Laboratory (NSSL) determines the top and bottom of the melting layer as functions of azimuth for any given radar scan (Giangrande et al. 2008).

Checking consistency of the designated class with the location of the melting layer is an important part of the classification routine. The geometry of the radar beam with respect to the melting layer is illustrated in Fig. 9.6; the slant ranges R_b and R_t correspond to geometrical projections of the melting layer onto the cone generated by the beam center at the lowest tilt. In fact, mixed-phase hydrometeors partially fill the radar resolution volume in a much broader interval of ranges (R_{bb}, R_{tt}) due to the effects of beamwidth. Once the boundaries of the melting layer are determined, the fuzzy logic aggregation (9.3) and classification procedures are applied with two restrictions: snow is not allowed below the ML bottom, and pure rain is prohibited above the ML top except for the category of "big drops"; these are commonly transported by convective updrafts well above the environmental freezing level (Z_{DR} columns).

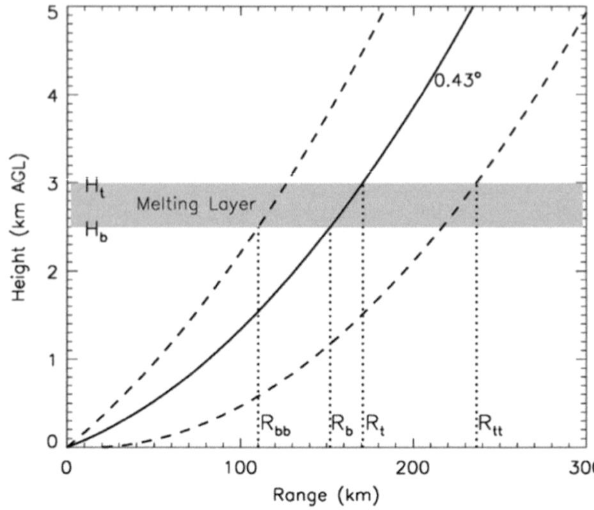

Fig. 9.6 Geometry of the radar beam with respect to the melting layer. The beam axis is drawn with the thick solid line, whereas the locations of the $\pm 0.5°$ beam extent (3 dB beamwidth) are indicated with the dashed curves. Dotted lines are at slant ranges R_{bb}, R_b, R_t, and R_{tt}, corresponding to geometrical projections of the melting layer from the radar beam onto the base scan. The heights H_b and H_t represent the melting layer bottom and top heights

The following subsets of classes are allowed within five slant range intervals:

$$\begin{array}{cc} 0 < R < R_{bb} & \text{GC/AP, BS, BD, RA, HR, RH} \\ R_{bb} < R < R_b & \text{GC/AP, BS, WS, GR, BD, RA, HR, RH} \\ R_b < R < R_t & \text{GC/AP, BS, DS, WS, GR, BD, RH} \\ R_t < R < R_{tt} & \text{GC/AP, BS, DS, WS, CR, GR, BD, RH} \\ R > R_{tt} & \text{DS, CR, GR, RH} \end{array} \tag{9.8}$$

To check if the results are reasonable, a set of "hard thresholds" is used, thereby reducing the number of clearly wrong class designations, for example, hail is considered to be misclassified if $Z < 40$ dBZ. In such cases the algorithm accepts the class with the next highest aggregation value.

An example of the classification product at the lowest beam elevation scan for the case of MCS system in central Oklahoma observed with the KOUN WSR-88D on 13 May 2005 is displayed in Fig. 9.7 along with the fields of Z, Z_{DR}, and ρ_{hv}. Biological scatterers and ground clutter (two shades of blue) are identified ahead of the squall line. The leading edge of the squall line is marked by a thin line of "big drops" associated with size sorting caused by updrafts and wind shear (see Chap. 7). It is followed by heavy rain (dark shade of red) and occasional regions of hail (pink). A gradual transition from rain to wet snow (green), dry snow (blue), and crystals (dark shade of green) is clear in the stratiform part of MCS once the radar beam penetrates and overshoots the melting layer. Elevated areas of graupel (yellow) are evident in the convective part of the storm at its southern flank.

Classification is performed at each antenna elevation, and the vertical cross sections of Z, Z_{DR}, ρ_{hv}, and classification results constructed from the series of conical scans are displayed in Fig. 9.8. The bright band is identified as wet snow (light green) and separates dry snow and rain in the stratiform part of the storm. In the storm's convective part, graupel is shown in yellow and hail in pink. Hail is

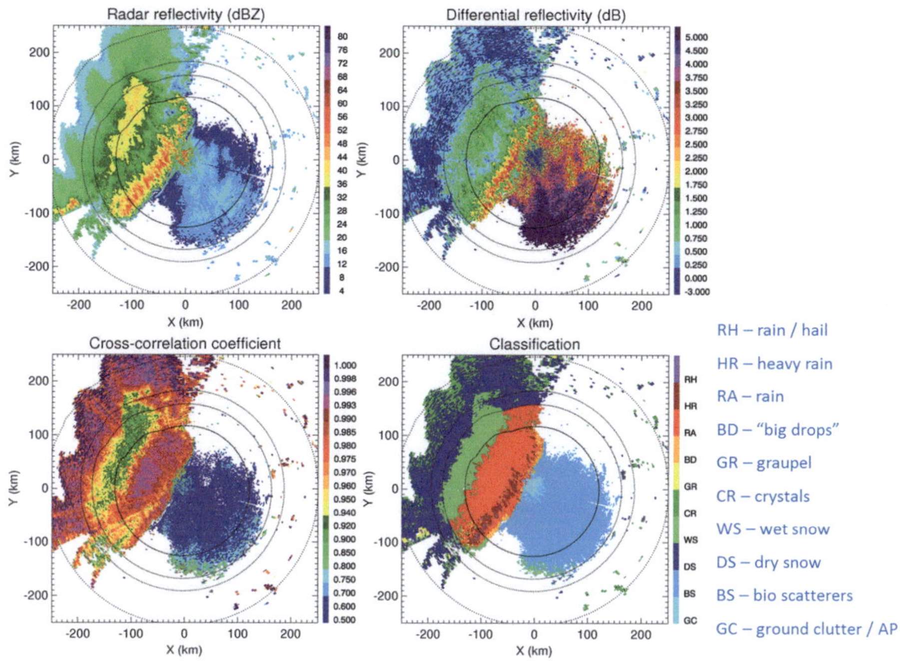

Fig. 9.7 Composite plot of Z, Z_{DR}, ρ_{hv}, and classification results at a 0.43° elevation angle measured by the KOUN WSR-88D at 0646 UTC 13 May 2005. Contours represent slant ranges R_{bb}, R_b, R_t, and R_{tt} (associated with the melting layer) as functions of the azimuth. Adapted from Park et al. (2009). © American Meteorological Society. Used with permission

detected aloft (two small areas) and likely melts completely before reaching the ground. Biological scatterers are recognized in the boundary layer ahead of the storm at the distances beyond 95 km from the radar.

9.2.2 Other Classification Algorithms

A series of alternative fuzzy logic hydrometeor classifiers has been devised at Colorado State University (CSU) (e.g., Liu and Chandrasekar 2000; Lim et al. 2005; Keranen et al. 2007; Dolan and Rutledge 2009; Dolan et al. 2013; Thompson et al. 2014; Bechini and Chandrasekar 2015). There are significant commonality and some differences between the NSSL and CSU algorithms which we describe herein.

9.2.2.1 Membership Function Approximations

The CSU schemes utilize a beta function approximation for the membership function as opposed to the trapezoidal function determined in Fig. 9.2. The beta membership function is defined as

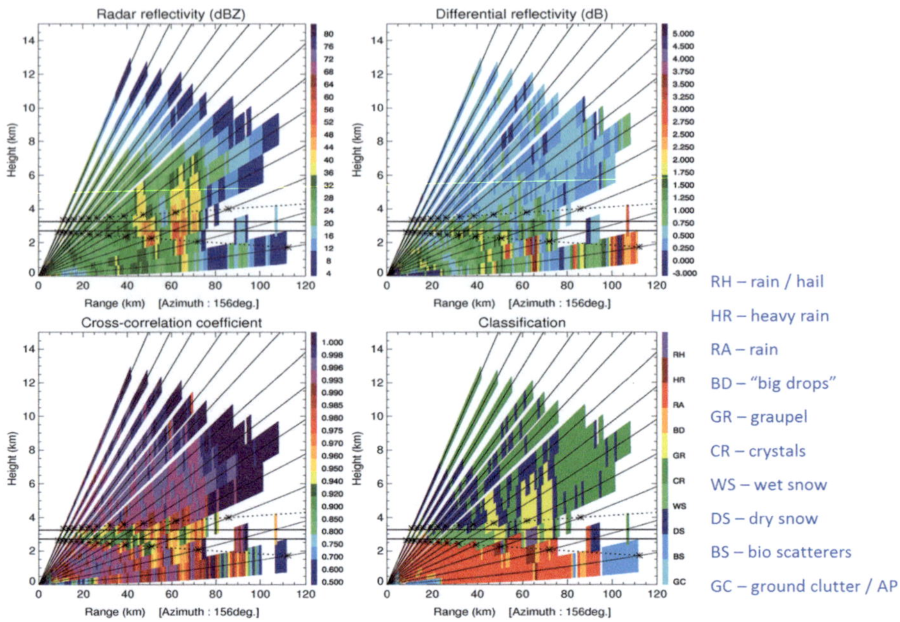

Fig. 9.8 Composite plot of Z, Z_{DR}, ρ_{hv}, and the results of classification in the RHI cross section at 1013 UTC on 13 May 2005. The azimuth angle is 156°. Dotted lines indicate slant ranges R_{bb}, R_b, R_t, and R_{tt} (associated with the melting layer) at the azimuth. From Park et al. (2009). © American Meteorological Society. Used with permission

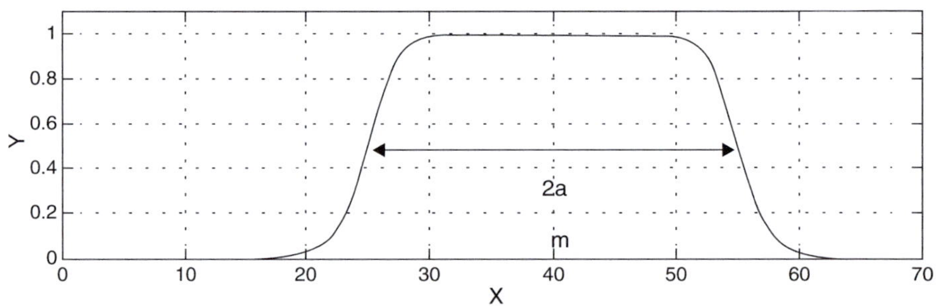

Fig. 9.9 Beta function

$$\mathrm{beta}(x, m, a, b) = \frac{1}{1 + \left[\left(\frac{x-m}{a}\right)^2\right]^b}, \tag{9.9}$$

where x is the value of the input variable and three parameters define the shape of a beta function, namely, the center of the function m, the width a, and the slope b (Fig. 9.9). The beta function (9.9) has symmetric shape. To mimic possible asymmetry of the probability distributions of polarimetric variables, Bechini and Chandrasekar (2015) define an asymmetric beta function given by

$$\mathrm{beta}(x, m, a_{r,l}, b_{r,l}) = \begin{cases} \dfrac{1}{1 + \left(\frac{x-m}{a_l}\right)^{2b_l}}; & x \leq m \\ \dfrac{1}{1 + \left(\frac{x-m}{a_r}\right)^{2b_r}}; & x > m \end{cases} \tag{9.10}$$

In (9.10), a_l and a_r are the left- and right-side half-widths, respectively; and b_l and b_r are left- and right-side slopes. An example of the values of the parameters used to define the asymmetric beta function for the polarimetric radar variables at S band used in the study of Bechini and Chandrasekar (2015) is shown in Table 9.4. The values w in Table 9.4 have the meaning of the elements of the matrix of weights similar to the one in Table 9.3.

9.2.2.2 Variables Used for Classification

Most of the classification algorithms described in the literature use the following set of polarimetric variables: Z, Z_{DR}, K_{DP}, and ρ_{hv} as well as temperature information or the heights of the top and bottom of the melting layer for better separation of liquid, frozen, and mixed-phase hydrometeors. Liu and Chandrasekar (2000) and Lim et al. (2005) also utilize linear depolarization ratio *LDR* in their classification schemes.

9.2.2.3 Hydrometeor Classes

The number of discriminated hydrometeor categories varies in different versions of classifiers. For example, Liu and Chandrasekar (2000) consider two categories of graupel (wet and dry) and hail (small and large). Dolan and Rutledge (2009), Dolan et al. (2013) distinguish between low-density graupel (density $\rho < 0.55$ g cm^{-3}) and high-density graupel ($\rho > 0.55$ g cm^{-3}) and treat vertically aligned ice as a separate crystal category. Vertically aligned crystals characterized by negative Z_{DR} and K_{DP} indicate the regions of possible strong electric fields. Thompson et al. (2014) and Bechini and Chandrasekar (2015) consider dendrites as a separate hydrometer class. A category of rimed snow is identified as a special class of dry snow by Lim et al. (2013).

9.2.2.4 Determination of the Membership Functions

The parameters of the membership functions can be fine-tuned in a three-step process. First, microphysical properties of the various hydrometeor habits are postulated on the basis of existing observations and/or models. Then, scattering

Table 9.4 Values of the parameters used to define the asymmetric beta function for the polarimetric radar variables at S band. Adapted from Bechini and Chandrasekar (2015)

	Z_h				Z_{dr}				ρ_{HV}				K_{dp}			
Class	m	a_l/a_r	b_l/b_r	w	m	a_l/a_r	b_l/b_r	w	m	a_l/a_r	b_l/b_r	w	m	a_l/a_r	b_l/b_r	w
Large drops	40	15	8	1.0	$0.4 + 0.08\tilde{Z}_h$	$0.5 + 0.02\tilde{Z}_h$	2	0.8	0.99	0.03	2	1.0	0.3	0.3/1.0	2	0.5
Drizzle	0	25	8	1.0	0.5	0.7	2	0.8	1.00	0.02	2	0.8	0.01	0.2	3	0.5
Rain	34	14	8	1.0	$-0.4 + 0.055\tilde{Z}_h$	$0.8 + 0.025\tilde{Z}_h$	2	0.8	0.99	0.03	2	1.0	0.3	0.3/1.0	2	0.5
Heavy rain	50	10	8	1.0	$-0.4 + 0.055\tilde{Z}_h$	$0.8 + 0.025\tilde{Z}_h$	2	0.8	0.99	0.05	2	1.0	0.5	0.2/10.0	2	0.8
Rain plus hail	65	15	4	1.0	0.5	1.5	3	1.0	0.95	0.07	2	0.8	5.0	5.0	8	1.0
Hail	65	13	4	1.0	−2.0	3.0	3	1.0	1.00	0.06	2	0.8	0.0	1.0	2	1.0
Graupel	42	14	5	1.0	1.2	2.0	2	0.8	1.00	0.06	2	1.0	0.8	1.5	2	0.5
Wet ice	20	25	4	1.0	1.5	1.7	3	1.5	0.88	0.08	2	1.0	1.0	1.2	2	0.5
Dry ice	22.5	15	5	1.0	0.2	1.0	2	0.4	1.00	0.05	2	1.0	0.2	0.3	2	1.0
Crystals	0	22.5	5	1.0	1.5	3.5	2	0.5	0.98	0.06	2	1.0	0.5	0.7	2	1.0
Dendrites	20	15	5	1.0	1.0	0.5/1.5	2	1.0	1.00	0.07	3	0.5	0.8	0.4/2.5	2	2.0

coefficients of a specific class are computed and integrated over their postulated distributions thus generating the electric fields to which the radar signals are proportional. Finally, these signals are transformed into the polarimetric variables. The membership functions corresponding to the classes of large hydrometeors such as hail or wet snow (and even rain) are noticeably different at S, C, and X bands due to the effects of resonance scattering. The X-band versions of the hydrometeor classification algorithm are described in the papers by Dolan and Rutledge (2009) and Snyder et al. (2010). Keenan (2003), Keranen et al. (2007), Marzano et al. (2007), and Dolan et al. (2013) specify the hydrometeor identifiers at C band, whereas the classifiers at all three microwave frequency bands, S, C, and X, are explored by Al-Sakka et al. (2013), Thompson et al. (2014), and Bechini and Chandrasekar (2015).

One-dimensional membership functions parameterized by the reflectivity Z are used in most classifiers. The WSR-88D HCA utilizes these types of the membership functions for $P^{(i)}(Z_{\mathrm{DR}})$ and $P^{(i)}(K_{\mathrm{DP}})$ with their parameters depending on Z. Bivariate or two-dimensional membership functions $P^{(i)}(Z, Z_{\mathrm{DR}})$, $P^{(i)}(Z, K_{\mathrm{DP}})$, and $P^{(i)}(Z, \rho_{\mathrm{hv}})$ are sometimes used instead (Al-Sakka et al. 2013).

Grazioli et al. (2015), Wen et al. (2015, 2016), and Besic et al. (2016) suggest using cluster analysis for an objective determination of the distinguishable classes of radar echo and for derivation of the membership functions of polarimetric variables and the temperature. This approach, called unsupervised or semi-supervised classification, avoids scattering simulations conducted over an arbitrary defined number of hydrometeor classes.

9.2.2.5 Inference Rules

Several versions of the Colorado State University algorithm use a "hybrid" inference rule rather than the additive ones given by Eqs. (9.1)–(9.4). This rule combines additive and product methods of inference so that the equations for aggregation values A_i have a form (Lim et al. 2005)

$$\begin{aligned} A_i = \; & P^{(i)}(Z)P^{(i)}(H) \\ & \times\left[w_i(Z_{\mathrm{DR}})P^{(i)}(Z_{\mathrm{DR}}) + w_i(K_{\mathrm{DP}})P^{(i)}(K_{\mathrm{DP}})\right. \\ & \left.+w_i(\rho_{\mathrm{hv}})P^{(i)}(\rho_{\mathrm{hv}}) + w_i(LDR)P^{(i)}(LDR)\right] \end{aligned} \tag{9.11}$$

or (Al-Sakka et al. 2013)

$$A_i = P^{(i)}(T)P^{(i)}(\mathrm{BB})\left[P^{(i)}(Z, Z_{\mathrm{DR}}) + P^{(i)}\left(Z, K_{\mathrm{DP}}\right) + P^{(i)}\left(Z, \rho_{\mathrm{hv}}\right)\right] \tag{9.12}$$

In (9.11), the multiplicative term $P^{(i)}(H)$ is the membership function characterizing the height H of the hydrometeors belonging to the i-th class with respect to the melting layer. In (9.12), a similar membership function $P^{(i)}(T)$ uses the air temperature T. The membership function $P^{(i)}(\mathrm{BB})$ delineates the top (BB top) and the bottom of the bright band (BB bottom).

9.2.2.6 Bin- Versus Region-Based Classifiers

All hydrometeor classification algorithms discussed so far belong to the category of "bin" classifiers. These designate a hydrometeor class on the basis of the polarimetric data collected at a particular spatial location (range, azimuth, and elevation). The result of classification at a given point is independent of results at a neighboring location. Because of the measurement errors and algorithm imperfections, this often leads to a noisy classification product and apparently wrong class designations which may not have physical meaning and thus appear as outliers. Bechini and Chandrasekar (2015), utilizing ideas of cluster analysis, introduced region-based classification to mitigate the "speckliness" in the classification product. The method involves several processing steps. In the first step, a guess classification is produced using a conventional bin-type fuzzy logic. Then the cluster analysis is performed to identify and connect regions populated with adjacent bins of the same hydrometeor class. Finally, the set of connected regions is passed to the fuzzy logic algorithm for the final classification. An example comparing bin-based and region-based classification results in the vertical cross section is presented in Fig. 9.10 for a deep convective storm observed with the S-band CHILL polarimetric radar. The results of region-based classification look much smoother and exhibit better delineation of solid and mixed-phase precipitation although the classification products may differ depending on the parameters λ and α of the clustering routine (compare panels *b* and *f* in Fig. 9.10). More details are provided in Bechini and Chandrasekar (2015).

9.3 Melting Layer Detection

9.3.1 *Existing Melting Layer Detection Algorithms*

Effective performance of the hydrometeor classification algorithms is contingent on the accurate melting layer designation which is a crucial module of any classification scheme. The melting layer signature in stratiform clouds is pronounced at higher elevation PPIs as a depression of the cross-correlation coefficient ρ_{hv} (Fig. 9.11). A circular-shape Z_{DR} enhancement also indicates the melting layer very well. A less discernible is the classical bright band signature in the Z field as it may disappear at certain azimuthal directions (Fig. 9.11). Obviously, the ρ_{hv} signature is the most effective discriminator of melting hydrometeors. The melting layer (ML) appearances at lower elevation angles are smeared by beam broadening and

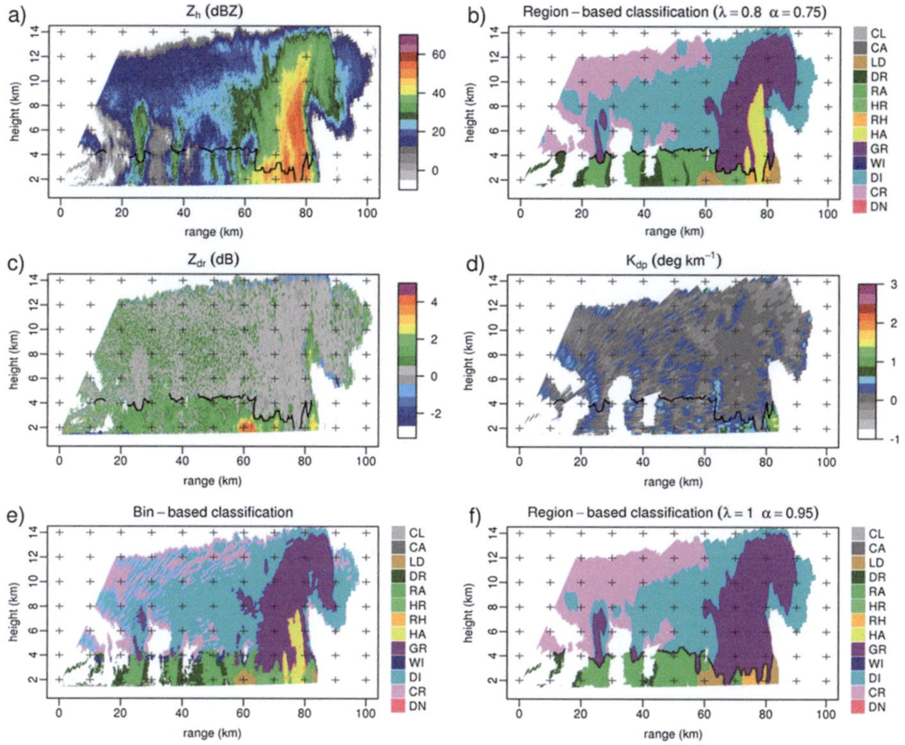

Fig. 9.10 Composite RHI for the storm observed by S-band CHILL radar. (**a**) Reflectivity and (**b**) final region-based hydrometeor classification. Panels (**c–f**) are Z_{DR}, K_{DP}, the bin-based hydrometeor classification after temperature adjustment, and the region-based classification. The overlaid solid black line in (**a–d**) marks the separation between solid and liquid-mixed phase particles in the final classification of (**b**). Class identificators in the legend are the following: *CL* clutter, *CA* clear air, *LD* large drops, *DR* drizzle, *RA* rain, *HR* heavy rain, *RH* rain/hail mixture, *HA* hail, *GR* graupel, *WI* wet ice, *DI* dry ice, CR crystals, *DN* dendrites. From Bechini and Chandrasekar (2015). © American Meteorological Society. Used with permission

may be contaminated by nonuniform beam filling or partial beam blockage. At elevation angles larger than 10°, the slant range extent of the ML decreases rapidly. In addition, at higher elevation angles, the ML signature is at close slant ranges where contamination by ground clutter is very likely. This is the reason why the first version of the automatic melting layer detection algorithm (MLDA) implemented on the WSR-88D network used data collected at elevation angles between 4° and 10° (Giangrande et al. 2008) (Fig. 9.12).

The MLDA searches for range locations exhibiting polarimetric signatures of melting snow along slant radial profiles of Z, Z_{DR}, and ρ_{hv} (Fig. 9.12). For each radial, the gates where ρ_{hv} falls between 0.90 and 0.97 are identified. The ρ_{hv} ML signature should be consistent with Z and Z_{DR} maxima in the vicinity of the identified ρ_{hv} drops. The algorithm then looks for Z and Z_{DR} maxima in a 500 m window above the locations matching the ML ρ_{hv} signature. If the maximum Z falls

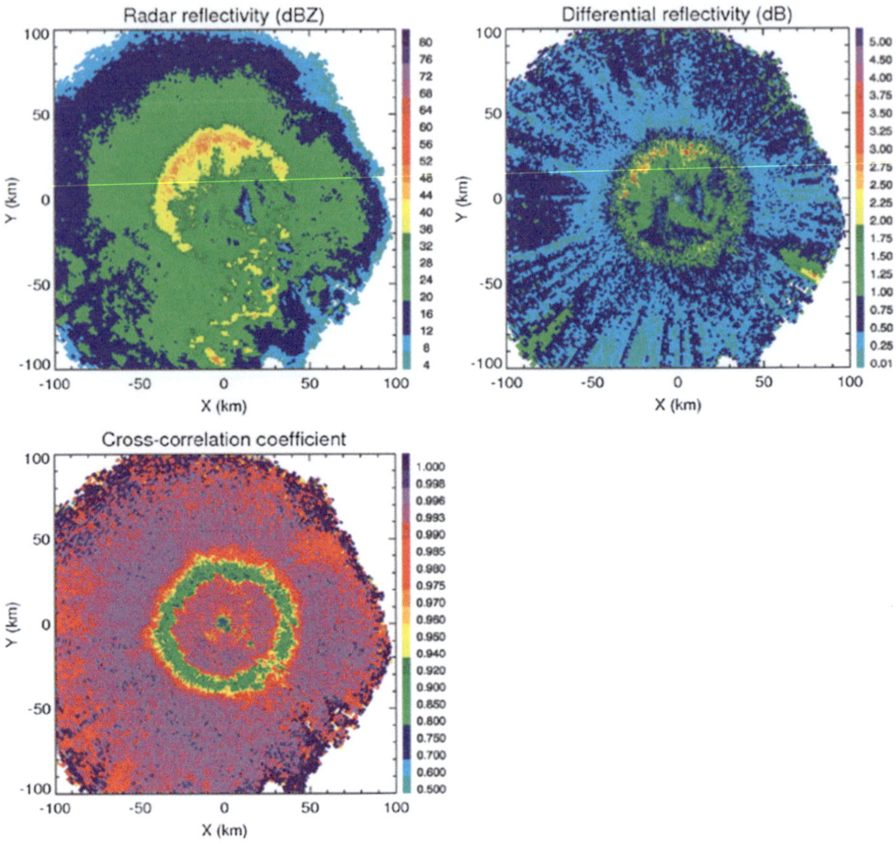

Fig. 9.11 Polarimetric signature of the melting layer in a composite PPI of Z, Z_{DR}, and ρ_{hv} from the KOUN WSR-88D observations at the 4.5° elevation angle; time is 1016 UTC 13 May 2005

between 30 and 47 dBZ and the maximum Z_{DR} is within the interval 0.8–2.5 dB, then the location of the ρ_{hv} signature is identified as an ML point. The corresponding azimuth and height of the ML points are stored in a two-dimensional array which contains the total number of ML points in a height-azimuth grid. Figure 9.13 provides an example of the ML point locations in a height-azimuth plane after all azimuths and elevations between 4° and 10° are examined in a single radar coverage volume.

Once the heights that encompass a majority of the ML points are determined, the MLDA algorithm estimates the ML top as the height below which 80% of ML points reside. Similarly, the ML bottom is determined as the height below which 20% of the ML points reside. The heights of the ML top and bottom vary with azimuth as depicted by solid lines in Fig. 9.13. The contours corresponding to the top and bottom of the melting layer are overlaid on the composite PPI of the polarimetric variables and the results of classification for the case of MCS are in Fig. 9.7.

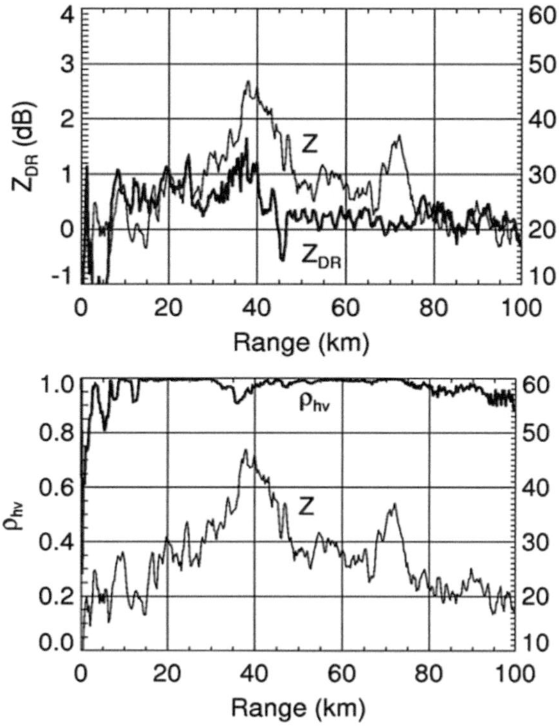

Fig. 9.12 Example of radial dependencies of Z, Z_{DR}, and ρ_{hv} through the melting layer

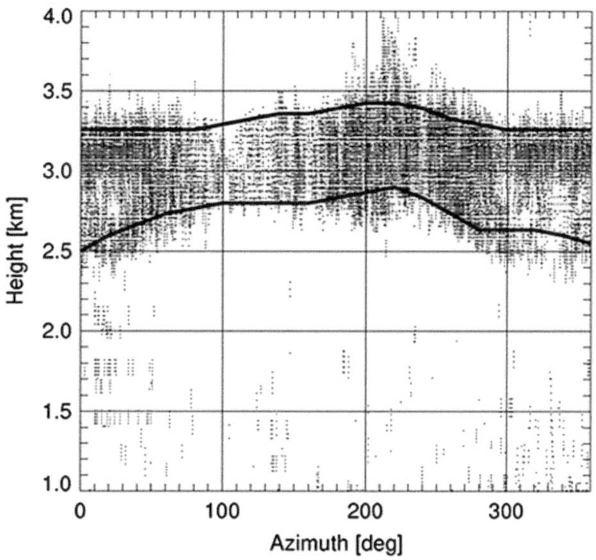

Fig. 9.13 The ML points mapped on the height-azimuth plane. The 80% (ML top) and 20% (ML bottom) height contours are overlaid on the image in solid lines. Adapted from Giangrande et al. (2008). © American Meteorological Society. Used with permission

Of the four quasi-circular contours, the two internal ones depict the ML top and bottom and approximately correspond to the ranges R_t and R_b according to their definition in Fig. 9.6. Note that these two internal contours represent physical boundaries of the melting layer. However, due to the beam broadening, the ML appears in the wider area between the two external quasi-circular contours.

The Giangrande et al. (2008) MLDA was validated using sounding data and the output of the Rapid Update Cycle (RUC) model for 18 events encompassing 136 h of observations in central Oklahoma. It was found that the temperature of the bottom of the melting layer varies between 2° and 6 °C. According to the MLDA retrievals, the depth of the ML changed from 100 to 900 m, and the median value was about 400 m in the dataset used for validation. It was also shown that in 85% of cases, the MLDA yields unbiased estimates of the height of the melting layer with an RMS error of 0.22 km. The remaining 15% of the cases were primarily associated with mature warm-season mesoscale convective systems for which local radar-derived ML heights were occasionally 1 km lower than the RUC model prediction. This might be attributed to a strong modulation of the ML by precipitation or the effects of ML "sagging" described in Chap. 7. In such situations, the radar may provide a more reliable designation of the ML than the RUC or similar models.

The S-band MLDA described in Giangrande et al. (2008) was later adapted to C-band radar data from southern Canada by Boodoo et al. (2010) and tested for two frontal systems over this region using the aircraft measurements of the height of the 0° isotherm. The Canadian study demonstrated the algorithm's ability to detect rapidly changing melting layer height during frontal passages. In a separate validation effort, Boodoo et al. (2010) ran the algorithm on 530 days during a 2-year period and compared its output with the wet-bulb temperature profiles retrieved from the GEM model. They obtained results similar to the US study with the bias in the ML top determination of <0.1 km and the RMS error of 0.26 km.

There are alternative techniques for detecting the melting layer based on slightly different principles. The Brandes and Ikeda (2004) method matches observed polarimetric radar measurement profiles with idealized model profiles of Z, LDR, and ρ_{hv} expected in the ML, whereas Tabary et al. (2006) present a similar technique for operational ML identification at C band but use solely profiles of ρ_{hv} as well as Matrosov et al. (2007) at X band. Wolfensberger et al. (2016) utilize the product of Z (in linear scale) and (1-ρ_{hv}) for detection of the melting layer at X band.

9.3.2 *Hybrid Melting Layer Detection Algorithm*

Existing ML detection algorithms function well for slow-moving warm-season stratiform events. However, the performance of these algorithms, including the MLDA, may be compromised in intense winter storms that are characterized by low, rapidly changing, and/or spatially complex MLs. One of the primary reasons that the existing fuzzy logic-based HCA fails during transitional winter weather events is that the presence/absence of the ML over the radar domain is determined by the projection of the ML detections and heights made near the radar (due to the use of high elevation angles from 4° to 10°) to more distant ranges.

The "hybrid" melting layer detection algorithm (HMLDA) suggested by Schuur et al. (2015) seeks to eliminate this deficiency by combining radar observations at all elevation angles with numerical model output to explicitly map ML location over the radar domain. The HMLDA takes advantage of the radar's ability to provide accurate ML designations at locations near the radar with a numerical model's ability to extend that coverage to more distant ranges where radar detections may be unreliable due to either beam broadening or beam overshooting the ML entirely. The HMLDA has been developed as part of a surface Hydrometeor Classification Algorithm (sHCA) that combines polarimetric radar observations with thermodynamic output from a numerical model to produce a surface-based precipitation type which complements the basic NEXRAD HCA. A more detailed description of HMLDA is presented in Sect. 9.7.

9.4 Detection of Hail and Determination of its Size

Hail is among the most damaging atmospheric hazards, and its detection is a very important aspect of the hydrometeor classification algorithm. In early studies, hail was identified if reflectivity factor Z exceeds a certain threshold, typically 53–55 dBZ at S band. Another reflectivity-centric approach employed vertically integrated liquid water (VIL) to detect hail (Greene and Clark 1972). The magnitude of VIL is computed using vertical profile of Z. Waldvogel et al. (1979) discovered that the height of the 45-dBZ contour above the freezing level correlates well with hail occurrence at the surface. Later, Witt et al. (1998) used data presented by Waldvogel et al. (1979) to develop a relation according to which larger probabilities of hail correspond to higher altitudes above the melting layer of the maximum reflectivity. Another approach for hail detection during the era preceding introduction of dual-polarization radars uses dual-wavelength radar measurements (e.g., Eccles and Atlas 1973; Feral et al. 2003). Atlas and Ludlam (1961) showed that, due to effects of resonance scattering, the ratio of reflectivities at S and X bands for hail is much larger than for rain and hence can be used for hail detection. Two distinct limitations of this dual-wavelength technique are the strong attenuation of radar returns at the shorter wavelength and the need to match antenna patterns at the two wavelengths.

First polarimetric methods for hail detection utilize combined measurements of Z and Z_{DR} or Z and K_{DP}. Rain and hail are separated in the Z–Z_{DR} plane by a boundary defined with relations $Z_{DR} = f(Z)$ as specified by Leitao and Watson (1984) and Aydin et al. (1986)—see Fig. 9.14. Similar relations in the Z–K_{DP} plane were suggested by Balakrishnan and Zrnic (1990). Linear depolarization ratio LDR was used for identification of hail by Bringi et al. (1986), Holler et al. (1994), Kennedy et al. (2001), and Zeng et al. (2001), among others.

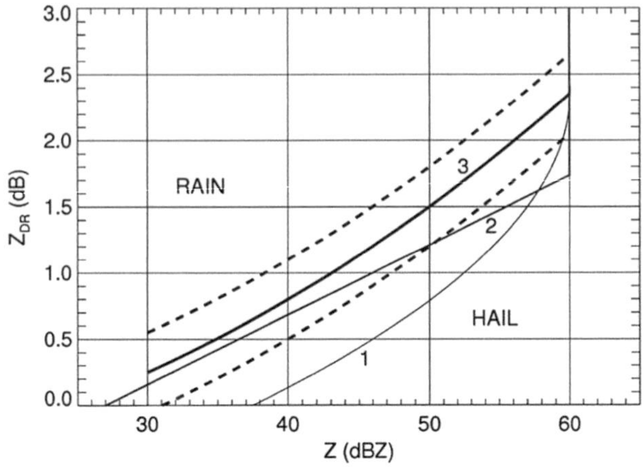

Fig. 9.14 Boundaries between rain and hail in the Z–Z_{DR} plane proposed by Leitao and Watson (1984) (curve 1), Aydin et al. (1986) (curve 2), and in the NSSL's fuzzy logic classification algorithm (Heinselman and Ryzhkov 2006) (curve 3). Dashed lines encompass the area where hail and rain overlap in the fuzzy logic algorithm. From Heinselman and Ryzhkov (2006). © American Meteorological Society. Used with permission

A fuzzy logic approach for hydrometeor classification quickly demonstrated its advantages for hail detection compared to the traditional Z-based methods or polarimetric techniques utilizing "hard thresholds" or decision tree methodology. Ryzhkov et al. (2005) and Heinselman and Ryzhkov (2006) reported significant improvement compared to the earlier single-polarization version of the WSR-88D hail detection algorithm (HDA) of Witt et al. (1998) in the course of a limited validation study during the Joint Polarization Experiment (JPOLE) in 2003. A simplified version of HCA was tested for 4 hail events with 47 rain and hail reports collected by storm-intercept teams. The superior performance of the HCA resulted primarily from the algorithm's lack of false alarms compared to HDA. Because the conventional HDA quantifies severe hail probability and maximum hail size in a storm as a whole, it does not specify the actual location of hail. In contrast, the HCA determines the precise regions of hail within a storm cell because the polarimetric variables it uses depend directly on the properties of hydrometeors in a radar resolution volume, and these differ significantly for hail compared to other hydrometeors.

It is hail of sufficiently large size and high density that inflicts substantial damage so that in addition to hail detection, discrimination of its sizes and densities is required. According to the US National Weather Service standard, hail exceeding 2.5 cm (1.0 in.) is considered high-impact and dangerous (i.e., "severe"), and hail in excess of 5.0 cm (2.0 in.) is considered "significantly severe." In the majority of research studies, the term "small hail" is used for size smaller than 2.5 cm, and the terms "large hail" and "giant hail" are synonymous with "severe" and "significantly severe." The first version of the WSR-88D HCA algorithm described in Sect. 9.2.1 detects "rain mixed with hail" (Park et al. 2009) and makes no distinction between small and large hail. The Colorado State University HCA distinguishes between "graupel/small hail" and "hail" without specifying the borderline size between "small hail" and "hail" (e.g., Lim et al. 2005). Depue et al. (2007) recommend using the hail differential reflectivity parameter (H_{DR}) which incorporates both Z and Z_{DR} (Aydin et al. 1986) for estimating maximal size at the ground. But this method does not consider the hail melting process, which has a very strong impact on the vertical profile of Z_{DR} (Picca and Ryzhkov 2012).

Ryzhkov et al. (2013b) and Ortega et al. (2016) propose an extension of the Park et al. (2009) algorithm by splitting the class "rain mixed with hail" (RH) into three separate size categories of small, large, and giant hail. The corresponding module in the WSR-88D HCA algorithm is called "Hail Size Discrimination Algorithm" (HSDA). The HSDA primarily capitalizes on the fact that the magnitudes and vertical profiles of Z, Z_{DR}, and ρ_{hv} in the rain/hail mixture below the freezing level depend on the dominant size of hail undergoing melting. Because the relations between Z, Z_{DR}, and ρ_{hv} depend strongly on the degree of melting and on the radar wavelength, the methods for hail detection and determination of its size should be wavelength-specific and must take into account the height of the radar resolution volume with respect to the freezing level.

The HSDA operates after the HCA and only on pixels identified as "rain/hail." The HSDA uses Z, Z_{DR}, and ρ_{hv} within a fuzzy logic scheme to calculate an aggregation value A_i for three hail classes using equation similar to (9.4):

$$A_i = \frac{\sum_j W_j Q_j P^{(i)}(V_j)}{\sum_j Q_j W_j}, \tag{9.13}$$

where $V_1 = Z$, $V_2 = Z_{DR}$, and $V_3 = \rho_{hv}$. In (9.13), the confidence vector $\mathbf{Q}$ is identical to the one in Park et al. (2009). It is important that the membership functions vary for each of the three polarimetric variables depending on the height of the radar sampling volume with respect to six height intervals. The six height intervals are defined by the location of the center of the radar resolution volume H with respect to the heights of wet-bulb temperature $H(T_w)$ at T_w equal to 0° and −25 °C:

$$\begin{aligned}
&(a)\ H > H(T_w = -25^\circ \text{C}),\\
&(b)\ H(T_w = 0^\circ \text{C}) < H < H(T_w = -25^\circ \text{C}),\\
&(c)\ H(T_w = 0^\circ \text{C}) - 1\text{km} < H < H(T_w = 0^\circ \text{C}),\\
&(d)\ H(T_w = 0^\circ \text{C}) - 2\text{km} < H < H(T_w = 0^\circ \text{C}) - 1\text{km},\\
&(e)\ H(T_w = 0^\circ \text{C}) - 3\text{km} < H < H(T_w = 0^\circ \text{C}) - 2\text{km, and}\\
&(f)\ H < H(T_w = 0^\circ \text{C}) - 3\text{km}.
\end{aligned} \tag{9.14}$$

The membership functions for these intervals are depicted in Fig. 9.15 and specified in Table 9.5. The parameters of the membership functions of HSDA are determined on the basis of the output of the model studies in Ryzhkov et al. (2013a) as well as radar observations. At $H > H\,(T_w = -25\ ^\circ\text{C})$ (interval (a)), Z is the major

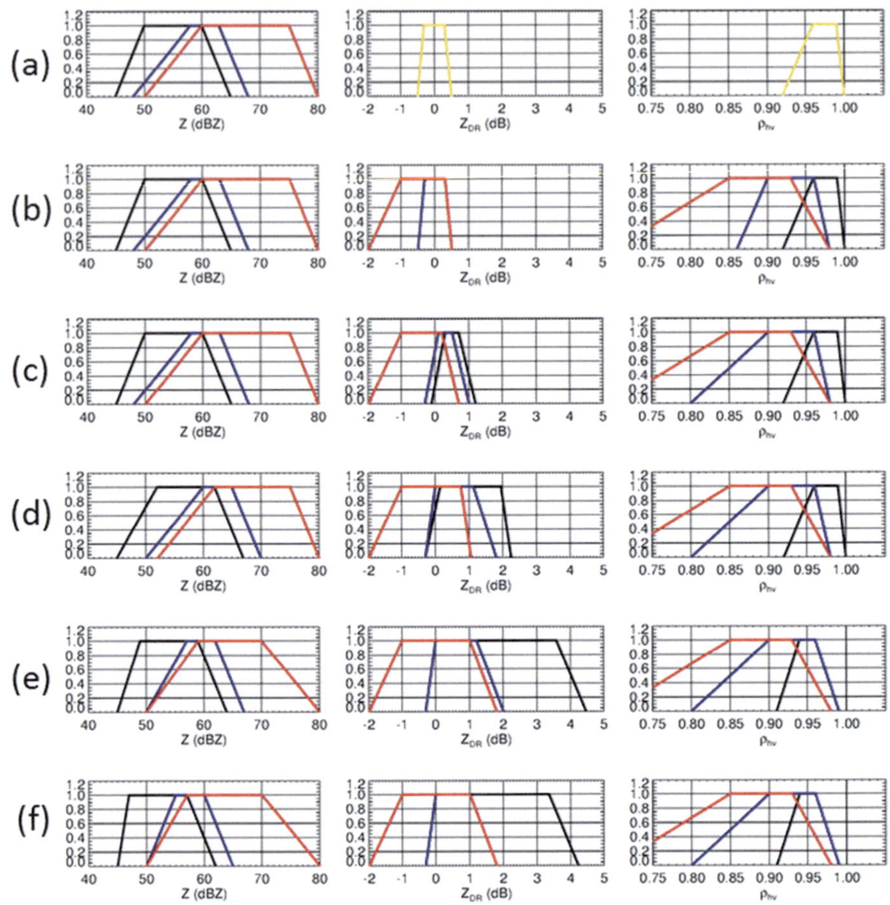

Fig. 9.15 Membership functions for small hail (black), large hail (blue), and giant hail (red) for different heights H of the radar sampling volume. (**a**) $H > H(T_w = -25\ ^\circ\text{C})$, (**b**) $H(T_w = 0\ ^\circ\text{C}) < H < H(T_w = -25\ ^\circ\text{C})$, (**c**) $H(T_w = 0\ ^\circ\text{C}) - 1$ km $< H < H(T_w = 0\ ^\circ\text{C})$, (**d**) $H(T_w = 0\ ^\circ\text{C}) - 2$ km $< H < H(T_w = 0\ ^\circ\text{C}) - 1$ km, (**e**) $H(T_w = 0\ ^\circ\text{C}) - 3$ km $< H < H(T_w = 0\ ^\circ\text{C}) - 2$ km, (**f**) $H < H(T_w = 0\ ^\circ\text{C}) - 3$ km. Orange represents overlapped membership functions for small, large, and giant hail

discriminator between the three categories of hail because Z_{DR} and ρ_{hv} vary very little with hail size at higher altitudes. Therefore, the membership functions of Z_{DR} and ρ_{hv} are the same for all three hail size categories (Fig. 9.15a). In the altitude range (b) between the 0 °C and −25 °C wet-bulb temperature isotherms, Z and ρ_{hv} have major discriminative potency, whereas Z_{DR}may become negative for giant hail in wet growth (Fig. 9.15b). All three variables possess strong classification capability below the freezing level, and the discriminative power of Z_{DR} increases toward the ground (Fig. 9.15c–f). The discrimination efficiency of each variable at each height layer is characterized by the vector of weights **W** defined in Table 9.6.

Table 9.5 Membership functions (MF) for height layers and hail classes of HSDA

Height layer	MF bound	Small hail			Large hail			Giant hail		
		Z	Z_{DR}	ρ_{hv}	Z	Z_{DR}	ρ_{hv}	Z	Z_{DR}	ρ_{hv}
$H \geq H(T_w = -25\ °C)$	x_1	45	−0.50	0.92	48	−0.50	0.92	50	−0.50	0.92
	x_2	50	−0.30	0.96	58	−0.30	0.96	60	−0.30	0.96
	x_3	60	0.30	0.99	63	0.30	0.99	75	0.30	0.99
	x_4	65	0.50	1.00	68	0.50	1.00	80	0.50	1.00
$H(T_w = -25\ °C) > H > H(T_w = 0\ °C)$	x_1	45	−0.50	0.92	48	−0.50	0.86	50	−2.0	0.70
	x_2	50	−0.30	0.96	58	−0.30	0.90	60	−1.0	0.85
	x_3	60	0.30	0.99	63	0.30	0.96	75	0.20	0.93
	x_4	65	0.50	1.00	68	0.50	0.98	80	0.50	0.98
$H(T_w = 0\ °C) > H > H(T_w = 0\ °C) - 1$ km	x_1	45	−0.10	0.93	48	−0.30	0.80	50	−2.0	0.70
	x_2	50	0.30	0.96	58	0.10	0.91	60	−1.0	0.85
	x_3	60	0.70	0.99	63	0.50	0.97	75	0.20	0.94
	x_4	65	1.20	1.00	68	1.00	0.98	80	0.70	0.98
$H(T_w = 0\ °C) - 1$ km $> H > H(T_w = 0\ °C) - 2$ km	x_1	45	$g_2 - 0.3$	0.94	50	$g_3 - 0.3$	0.80	52	−2.0	0.70
	x_2	52	g_2	0.96	60	g_3	0.91	62	−1.0	0.85
	x_3	62	g_1	0.98	65	g_2	0.97	75	g_3	0.93
	x_4	67	$g_1 + 0.3$	1.00	70	$g_2 + 0.3$	0.98	80	$g_3 + 0.3$	0.98
$H(T_w = 0\ °C) - 2$ km $> H > H(T_w = 0\ °C) - 3$ km	x_1	45	$f_2 - 0.3$	0.91	50	$f_3 - 0.3$	0.80	50	−2.0	0.70
	x_2	49	f_2	0.94	57	f_3	0.90	59	−1.0	0.85
	x_3	59	f_1	0.96	62	f_2	0.96	100	g_3	0.93
	x_4	64	$f_1 + 0.3$	0.99	67	$f_2 + 0.3$	0.99	101	$g_3 + 0.3$	0.98
$H < H(T_w = 0\ °C) - 3$ km	x_1	45	$f_2 - 0.3$	0.91	50	$f_3 - 0.3$	0.80	50	−2.0	0.70
	x_2	47	f_2	0.94	55	f_3	0.90	57	−1.0	0.85
	x_3	57	f_1	0.96	60	f_2	0.96	100	g_3	0.93
	x_4	62	$f_1 + 0.3$	0.99	65	$f_2 + 0.3$	0.99	101	$g_3 + 0.3$	0.98

In table $f_1 = -0.5 + 2.5 \times 10^{-3} Z + 7.5 \times 10^{-4} Z^2$; $f_2 = 0.1 \times (Z - 50.0) > 0$; $f_3 = 0.1 \times (Z - 60.0) > 0$; $g_1 = -0.9 + 1.5 \times 10^{-2} Z + 5.0 \times 10^{-4} Z^2$; $g_2 = 0.075 \times (Z - 50.0) > 0$; $g_3 = 0.075 \times (Z - 60.0) > 0$, i.e., negative values of f_2, f_3, g_2, and g_3 should be replaced by zero

Table 9.6 Weights of polarimetric variables for each height layer used in the **W** vector (Ortega et al. 2016)

Height layer	Z	Z_{DR}	ρ_{hv}
$H \geq H(T_w = -25\ °C)$	1.0	0.3	0.6
$H(T_w = -25\ °C) > H > H(T_w = 0\ °C)$	1.0	0.3	0.6
$H(T_w = 0\ °C) > H > H(T_w = 0\ °C) - 1$ km	0.8	0.5	0.6
$H(T_w = 0\ °C) - 1$ km $> H > H(T_w = 0\ °C) - 2$ km	0.7	0.8	0.6
$H(T_w = 0\ °C) - 2$ km $> H > H(T_w = 0\ °C) - 3$ km	0.7	1.0	0.6
$H < H(T_w = 0\ °C) - 3$ km	0.7	1.0	0.6

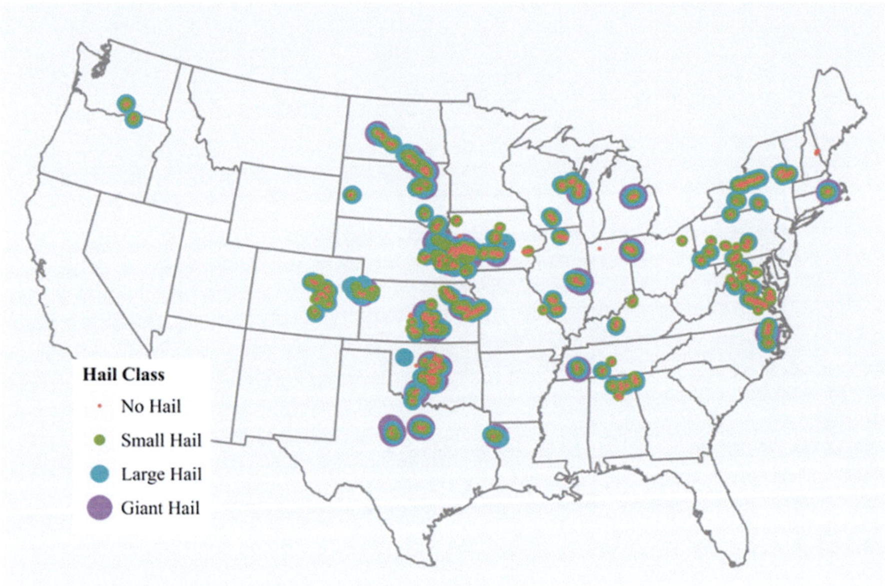

Fig. 9.16 Map of SHAVE hail reports. From Ortega et al. (2016). © American Meteorological Society. Used with permission

The HSDA algorithm was tested on a large number of hailstorms observed by the dual-polarization WSR-88D systems, with ground truth collected as part of the NSSL Severe Hazards Analysis and Verification Experiment (SHAVE) spanning the years 2010 through 2014 (Ortega et al. 2009, 2016). The primary goal of SHAVE was to collect hail reports—including information on maximum and average hail sizes and the start and end time of the hail fall—via telephone surveys in the wake of thunderstorms at spacing near 1.6 km, including reports of small hail and no hail along and near the thunderstorm's path. In the Ortega et al. (2016) study, 79 SHAVE operations were used. These operations yielded 3257 total reports (1115 no-hail reports, 1150 small-hail reports, 786 large-hail reports, and 206 giant-hail reports) across the contiguous USA (Fig. 9.16). An example of the composite PPI of Z, Z_{DR}, ρ_{hv}, and classification results which include three categories of hail size, small hail

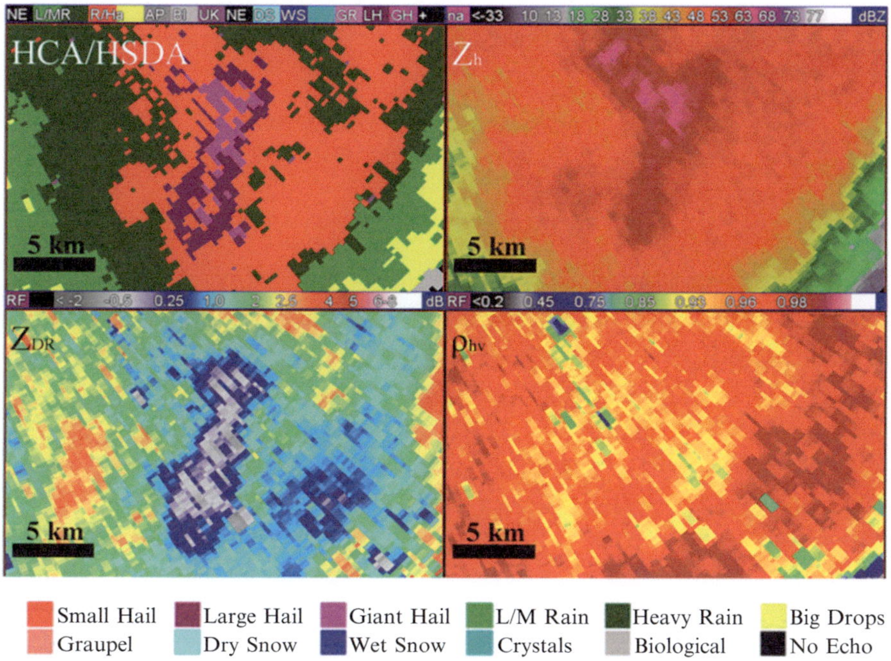

Fig. 9.17 HCA and HSDA output (top left), Z (top right), Z_{DR} (bottom left), and ρ_{hv} (bottom right) for the storm producing giant hail approximately 47 km from the KDMX (Des Moines, IA) WSR-88D at 1856 UTC 30 June 2014. Adapted from Ortega et al. (2016). © American Meteorological Society. Used with permission

(SH or RH), large hail (LH), and giant hail (GH), in the storm that produced giant hail according to the SHAVE observations is presented in Fig. 9.17.

Giant hail recognized by HSDA was supported by the SHAVE surface report. This storm produced giant hail approximately 47 km from the KDMX (Des Moines, IA) WSR-88D at 1856 UTC 30 June 2014. The HSDA designations are made at all antenna elevation scans to see the transition from completely frozen hydrometeors above the freezing level to rain and hail of different size at lower altitudes. In Fig. 9.18, the results of classification are displayed at four elevation scans for the same hailstorm as in Fig. 9.17. The area of large and giant hail is smaller at the lowest elevation of 0.5° compared to the elevation of 1.45° due to gradual melting of hail below the freezing level. The largest area of giant hail is indicated at the beam height of about 10 km ASL.

The SHAVE validation technique limitation is the absence of accurate timing of hail occurrence at the surface—the residents of the area affected by hail were able to correspond the size category of hail and its location at the surface but not precise time of occurrence. Various matching techniques were employed in the Ortega et al. (2016) study to obtain vertical "matching" profiles of Z, Z_{DR}, and ρ_{hv} corresponding to the given SHAVE hail report. These techniques rely on maximum Z at the exact point location of the report but within a 4×4 km search window. Vertical matching

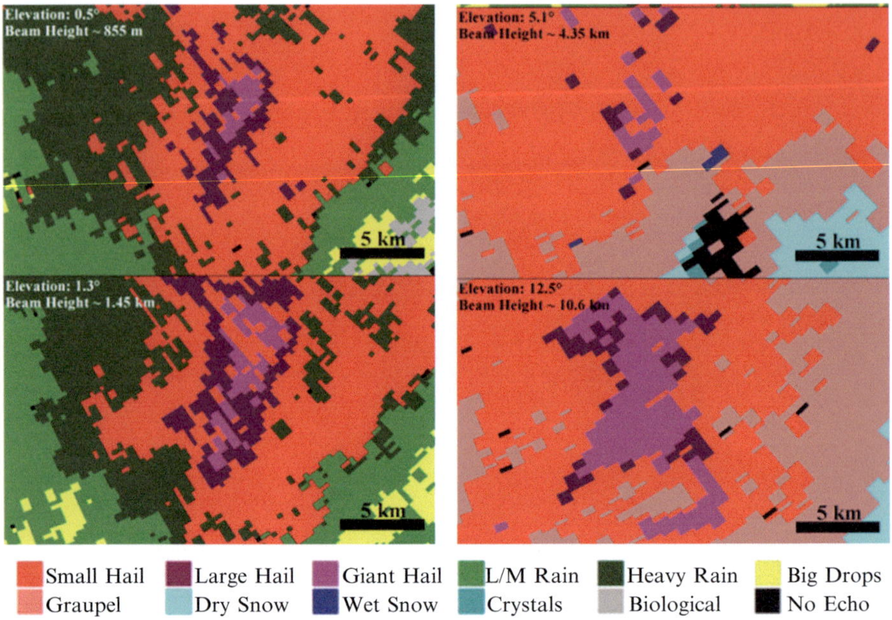

Fig. 9.18 HCA and HSDA output at multiple antenna elevations for the same volume as in Fig. 9.17. $H(T_w = 0\ °\mathrm{C})$ is 3.82 km, and $H(T_w = -25\ °\mathrm{C})$ is 8.23 km. The storm is approximately 47 km from the radar. Adapted from Ortega et al. (2016). © American Meteorological Society. Used with permission

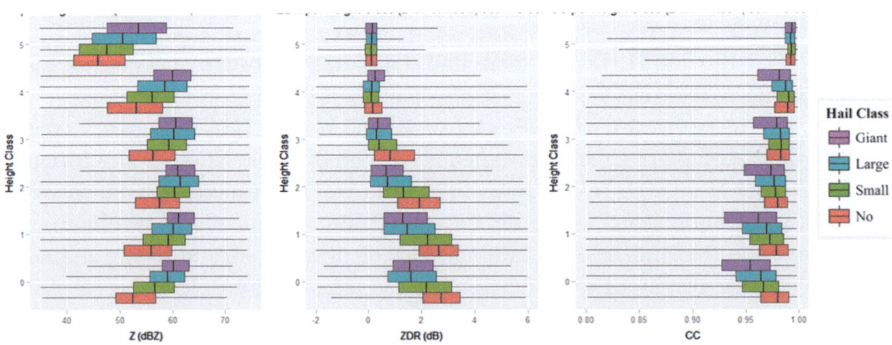

Fig. 9.19 Statistical distributions of Z, Z_{DR}, and ρ_{hv} at different height intervals for no hail, small hail, large hail, and giant hail for all SHAVE cases. The whiskers mark the 95th percentile, the boxes mark the interquartile range, and the vertical lines mark the median value. Height class layers correspond to the height intervals specified by (9.14). Adapted from Ortega et al. (2016). © American Meteorological Society. Used with permission

profiles of the three polarimetric variables are created for each hail report, and statistical distributions of Z, Z_{DR}, and ρ_{hv} are generated for each of the six height intervals and four hail size classes (no hail, small, large, and giant in Fig. 9.19). The statistical distributions of the polarimetric variables and their dependencies on the height with respect to the freezing level obtained from the observations are consistent with the membership functions for different hail sizes and altitudes displayed in Fig. 9.15 and Table 9.5.

Median Z for giant hail is 4–5 dBZ higher than the median for small hail. The vertical profiles of Z (left panel in Fig. 9.19) reveal very little change in Z with height for giant hail, while there is a 3–4 dB increase of Z with height up to the level where shedding of raindrops from the surface of melting hailstones starts, which for large hail is in the layer between 2 and 3 km below the melting level. This is in agreement with theory (Ryzhkov et al. 2013a). There is a significant overlap of the Z distributions for large and giant hail within the layers at or below the melting level (height intervals 1–4 in Fig. 9.19). Some overlap of the Z distributions for small and large hail is also present; however, small hail distributions are shifted toward smaller Z compared to those of large hail.

The vertical profiles and ranges of the Z_{DR} distributions (middle panel in Fig. 9.19) correspond well to the theoretical predictions for all hail size categories. The largest changes in Z_{DR} with decreasing height are for smaller hail sizes. As was observed with the Z distributions, the biggest separation is between "no hail" and giant hail reports. At the lowest height interval (marked as 0 in Fig. 9.19), the median Z_{DR} changes from 2.1 dB for small hail to 0.9 dB for giant hail. These are very close to the centers of the trapezoids representing the membership functions for Z_{DR} (2.3 dB and 1.2 dB for small and giant hail, respectively). Such consistency of the observations and theory attests to the validity of basic models of melting hail in Ryzhkov et al. (2013a) used for designing the membership functions of the HSDA algorithm. The Z_{DR} differences between different hail sizes gradually decrease with height and practically disappear at the melting layer and above. Above the melting layer, Z_{DR} loses its discrimination power (except the situations with giant hail in wet growth) and, thus, justifies minimization of the weight W for Z_{DR} at and above the melting layer (Table 9.5).

Hail with larger sizes exhibits smaller ρ_{hv} compared to smaller sizes at all heights (right panel in Fig. 9.19) as predicted by Balakrishnan and Zrnić (1990). An exception is for "no hail" reports near and above the melting level. "No hail" reports tend to be associated with ρ_{hv} lower than those with small hail and similar to large hail. The distributions also show that $\rho_{hv} < 0.9$ tend to be within the tails of the distributions, even for giant hail. A count of the number of pixels within the search windows with $\rho_{hv} < 0.9$ in the height layer $H(T_w = 0\ °C) - 3\ \text{km} \leq H_b < H(T_w = 0\ °C) - 2\ \text{km}$ reveals only 10% of all pixels are associated with such values. Comparison of the observed distributions to the membership functions shows that the observed ρ_{hv} tends to be higher.

The validation technique based on the matching of the most common HSDA hail size designation at lowest scan within the 4 × 4 km area surrounding SHAVE hail report provides an objective metric of the algorithm skill. The corresponding probability of detection (POD) is 0.594, the false-alarm ratio (FAR) is 0.136, and the critical success index (CSI) is 0.543 (Ortega et al. 2016). The HSDA outperforms the traditional single-polarization algorithm (with CSI equal to 0.324) confirming the ability of polarimetric radars to better quantify hail size.

The HSDA algorithm described herein does not capitalize on the polarimetric signature of giant hail growing in the wet regime which is associated with negative Z_{DR} and anomalously low ρ_{hv} near the tops of Z_{DR} columns (Sect. 8.2). Because of the strong resonance scattering effects in hail, the S-band version of HSDA has to be modified for application at shorter wavelengths. Moreover, stronger effects of attenuation/differential attenuation at C and X bands have to be accounted for before determination of hail size can be reliably made.

9.5 Automated Tornado Detection

The original WSR-88D HCA algorithm (Park et al. 2009) was modified by Snyder and Ryzhkov (2015) to add a tornado class or "tornadic debris signature" (TDS). The TDS described in detail in Sect. 8.3.1 provides very valuable information about a tornado on the ground. Of the 745 tornadoes examined in Van den Broeke and Jauernic (2014), ~62% of the tornadoes that were rated 2 on the enhanced Fujita (EF) scale (i.e., EF2) or greater were associated with a TDS, and 100% of the tornadoes that were rated EF4 or EF5 were associated with a TDS.

Identification of tornado requires some estimate of vertical vorticity, which is accomplished by adding azimuthal wind shear (*AS*) as an input variable to the classification scheme. Because debris from tornadoes can be advected away from the maximal *AS* typically found near the center of the tornado, the *AS* field is "dilated" by determining the 95th percentile value of valid *AS* in a 4 radial by 8 range gate neighborhood around each range location. This algorithm is not designed to classify regions of debris fallout that can occur well away from the tornado in the forward-flank region of a supercell—the primary intention of the TDS classification is to identify TDSs that are associated with the debris field near tornadoes.

For TDS classification, the HCA uses the following five inputs: Z, Z_{DR}, ρ_{hv}, SD(Φ_{DP}), and dilated *AS*. Membership functions for TDS classification (Fig. 9.20) were developed using the TDS characteristics reported in the literature with some modification based on internal testing of the algorithm. In developing the TDS classification routine, Snyder and Ryzhkov (2015) tried to minimize false TDS designations while still properly classifying all range locations composing a TDS. For this purpose, a set of additional "sanity checks" are applied after the HCA determines the TDS as a most likely class. The following criteria must be met: (1) $\rho_{hv} < 0.92$, (2) $Z > 25$ dBZ, (3) $AS > 0.005$ s^{-1}, (4) TDS aggregation value >0.8, and (5) the center of the radar beam must be below the melting layer. The results of classifications are spatially filtered using a 5 × 5 mode filter to reduce speckliness.

An example of automated TDS detection is shown in Fig. 9.21. A series of supercells produced tornadoes across central and eastern Oklahoma on the afternoon and evening of 10 May 2010. At 2250 UTC, four tornadoes were ongoing simultaneously to the east and northeast of the KOUN WSR-88D. A hook echo is associated with the southern supercell producing tornadoes B and C. Collocated with each of

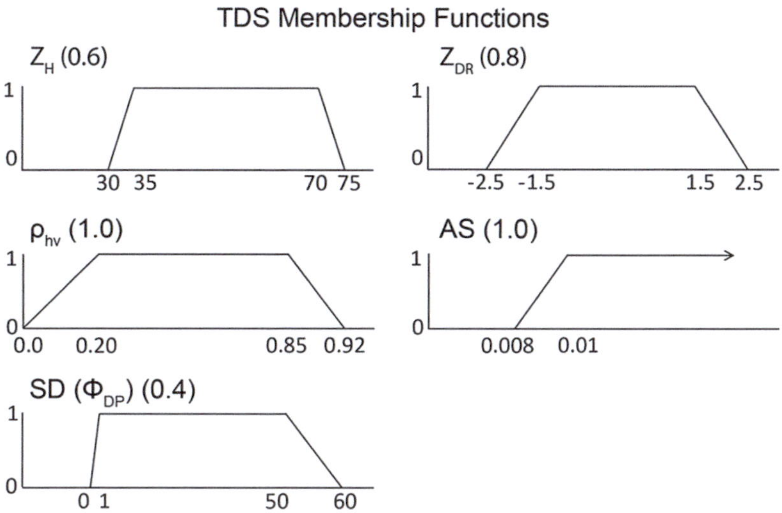

Fig. 9.20 The Z, Z_{DR}, ρ_{hv}, AS, and SD(ΦDP) membership functions for the TDS class. Numbers in parentheses following the name of each variable represent the weight of the variable. The four numbers listed along the abscissa represent the four points at which the trapezoidal membership function is defined, except for the AS membership function, which has no upper bound. From Snyder and Ryzhkov (2015). © American Meteorological Society. Used with permission

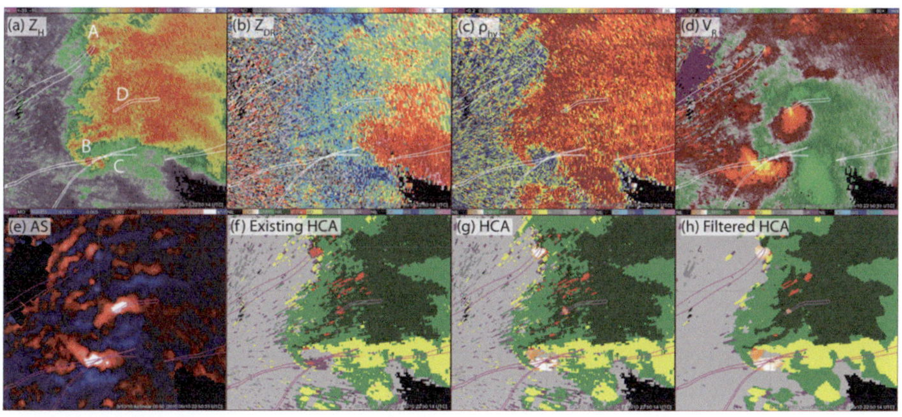

Fig. 9.21 Images of (**a**) Z (dBZ), (**b**) Z_{DR} (dB), (**c**) ρ_{hv}, (**d**) v (m s^{-1}), (**e**) AS (s^{-1}), and results from the (**f**) existing HCA and the (**g**) new HCA at near 2250 UTC 10 May 2010 from the KOUN WSR-88D. (**h**) New HCA results filtered with the 5 × 5 mode filter to reduce speckling. Tornado paths as determined by the Norman NWS Forecast Office are shown as white polygons in (**a–d**) or magenta polygons in (**e–h**). Letters A, B, C, and D denote the paths of four tornadoes that were present at this time. HCA output classes are labeled as tornadic debris signature (TDS), light/moderate rain (LMR), heavy rain (HR), rain-hail mixture (RH), big drops (BD), anomalous propagation (AP), biological (BI), unknown (UK), and no echo (NE). From Snyder and Ryzhkov (2015). © American Meteorological Society. Used with permission

the three tornadoes (A–C), Z_{DR} is between -1 dB and 1 dB (panel b), ρ_{hv} is generally <0.6 (panel c), and v reveals strong cyclonic rotation (panel d). A fourth tornado, labeled D is beginning at this time—the primary indication of this tornado is reduced ρ_{hv} and strong cyclonic rotation. Azimuthal wind shear is highest for tornadoes B, C, and D; the smaller size and weaker magnitude of the cyclonic couplet associated with tornado A results in its much lower *AS*. Each tornado appears to be associated with a TDS, although the close proximity of tornadoes B and C makes it hard to tell that two different tornadoes are present. The original version of HCA (without TDS class) labeled as "existing HCA" in the panel f classifies tornadoes A and D mostly as RH (rain/hail mixture), whereas tornadoes B and C are generally classified as UK (unknown) because the maximal aggregation value falls below the threshold 0.4. A modified HCA version (labeled as "HCA" in panel g) classifies each of the tornadoes as TDS (colored white in panel g). This is even better displayed in the panel h showing mode-filtered classification results.

Three other examples of TDS identification and echo classification are shown in Fig. 9.22. Violent tornadoes in central Oklahoma on 20 May 2013 (left column of Fig. 9.22) and in central Arkansas on the evening of 27 April 2014 (center column) each are associated with very prominent TDSs with quasi-circular areas of locally maximized Z, strong cyclonic shear evident in v, and $\rho_{hv} < 0.5$. The modified HCA correctly classifies those locations as TDS; aggregation values of 1.0 near the center of each TDS generally decrease to ~0.8 to 0.85 near the periphery of each TDS, where debris may be sparser and may be mixing with meteorological scatterers. On 16 June 2014, several intense tornadoes occurred in eastern Nebraska; at 2119.

UTC (right column of Fig. 9.22), two tornadoes were present, and each is associated with a TDS. The modified HCA properly classifies both TDSs.

The modified HCA (Snyder and Ryzhkov 2015) can be used to create "TDS tracks" and plot the paths of tornadoes in near-real time. Six cases featuring TDSs are presented in Fig. 9.23, which shows all locations where a TDS was identified by the algorithm over some time period (in this specific case, it includes the duration of each event). Discontinuities of the TDS designations along the tracks are caused by a 5–6 min update interval of the volume coverage on the WSR-88D network.

9.6 Convective Updraft Detection

A convective updraft is commonly associated with the Z_{DR} column, at least in the early stage of the updraft development (Chaps. 7 and 8). Kumjian et al. (2014), Snyder et al. (2015), and Ilotoviz et al. (2016) combined spectral bin microphysics in a high-resolution numerical models with an advanced forward operator and showed that the location and depth of the simulated Z_{DR} columns match quite well with the location and intensity of convective storm updrafts; changes in the maximum vertical velocity within an updraft tend to be seen after changes in the height of a Z_{DR} column. Moreover, Picca et al. (2010) and Kumjian et al. (2014) demonstrate that an increase in updraft strength (manifested as an increase in the Z_{DR} column

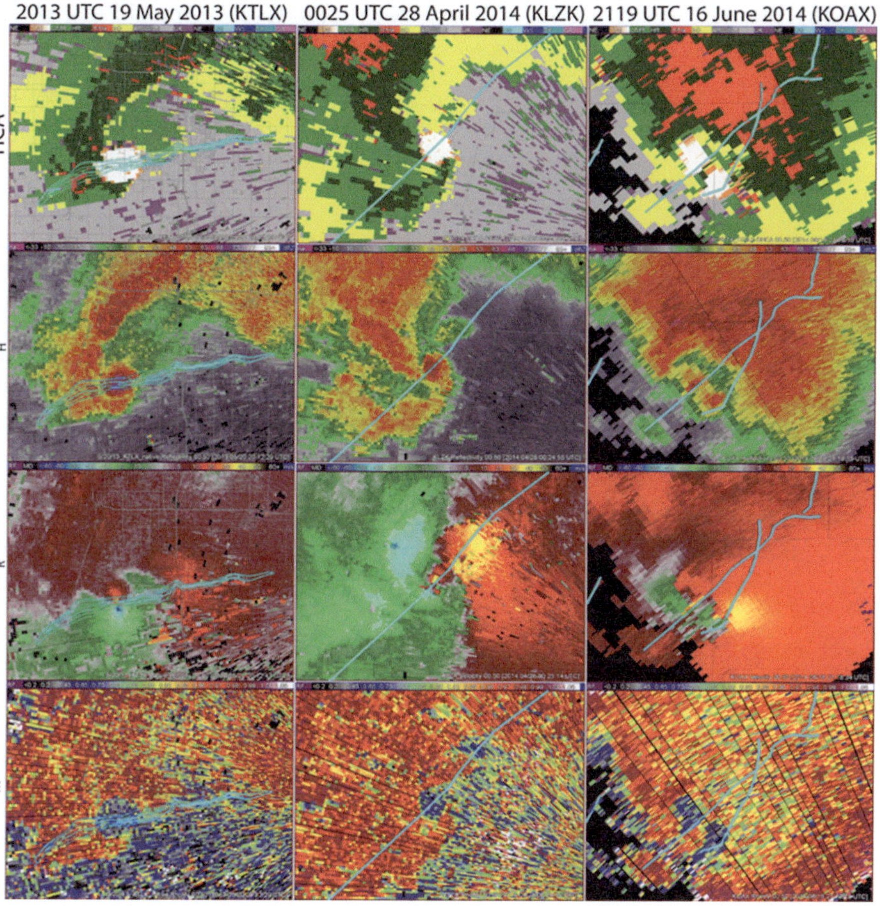

Fig. 9.22 (top) HCA output, (upper middle) Z (dBZ), (lower middle) ρ_{hv}, and (bottom) V_R (m s^{-1}) valid (left) 2012 UTC 20 May 2013 from the Twin Lakes radar (KTLX) in central Oklahoma, (center) 0025 UTC 28 April 2014 from the Little Rock radar (KLZK) in central Arkansas, and (right) 2119 UTC 16 June 2014 from the Omaha radar (KOAX) in eastern Nebraska. The cyan polygons or curves in each panel represent the paths of tornadoes produced by these supercells. All data are from the 0.5° elevation angle. HCA output classes are as in Fig. 9.21. From Snyder and Ryzhkov (2015). © American Meteorological Society. Used with permission

height or volume above the environmental freezing level) precedes an increase in surface precipitation rate and/or hail intensity by 10–15 min. Thus, the Z_{DR} column signature can be effectively used for nowcasting subsequent storm development.

This motivates development of automated Z_{DR} column detection algorithm for the operational WSR-88D (Snyder et al. 2015). According to the algorithm, the filtered Z_{DR} data are objectively analyzed onto a three-dimensional latitude-longitude-height grid with a grid spacing of 0.0025° × 0.0025° in latitude and longitude and 250 m in height. The algorithm requires the height of the

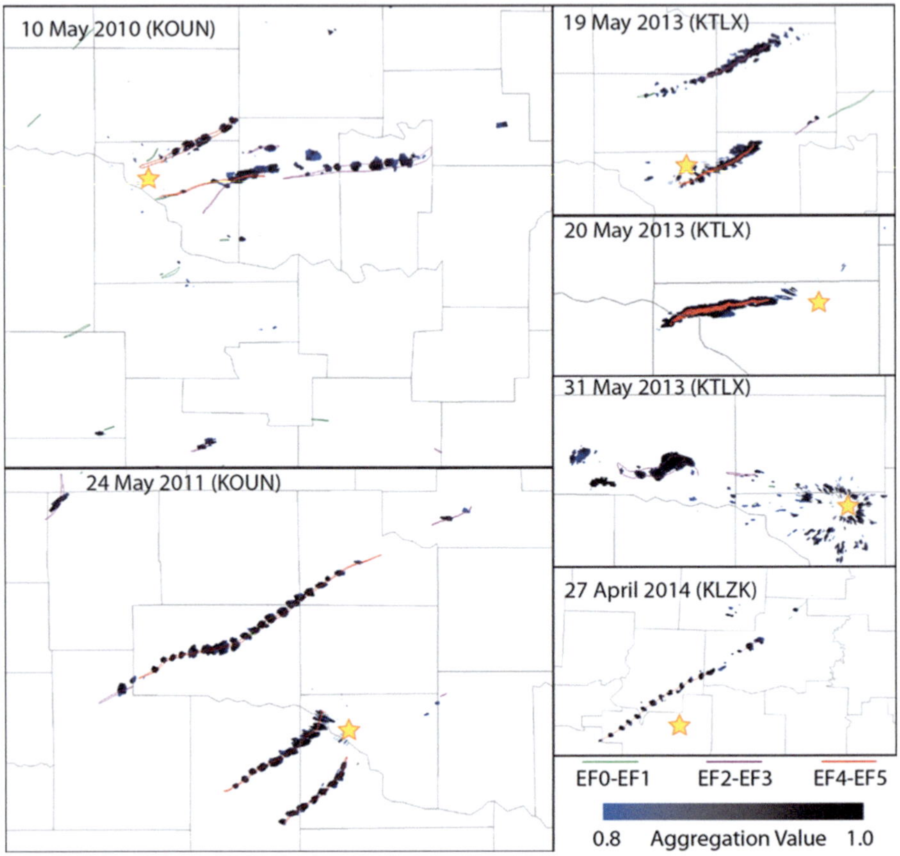

Fig. 9.23 TDS tracks for several tornado events across the south-central USA (marked by date and radar). The tracks trace only those locations for which the HCA selected TDS. County borders are in gray, and the maximum EF scale rating assigned to each tornado is marked in green for "weak" tornadoes of EF0–EF1, purple for "strong" tornadoes of EF2–EF3, or red for "violent" tornadoes of EF4–EF5. The aggregation values for the TDS classification range from 0.8 (blue) to 1.0 (black). Yellow stars represent the location of the indicated radar. Only HCA output from the 0.5°-elevation-angle scans are shown. From Snyder and Ryzhkov (2015). © American Meteorological Society. Used with permission

environmental 0 °C level which can be obtained from the numerical weather prediction models such as the Rapid Update Cycle (RUC) or Rapid Refresh (RAP) models. At each horizontal grid point, the number of vertically consecutive grid points with $Z_{DR} > 1$ dB above the 0 °C level is counted such that the result is the vertically continuous depth of the Z_{DR} column (as defined by $Z_{DR} > 1$ dB). The algorithm provides a filtered, two-dimensional map of the depth of the Z_{DR} column above the model-provided environmental 0 °C level.

An example of the algorithm performance is illustrated next. The algorithm was applied to a powerful supercell storm which hit the Norman, Oklahoma area on the afternoon and evening of 24 May 2011. Figure 9.24 shows a view of the supercell at

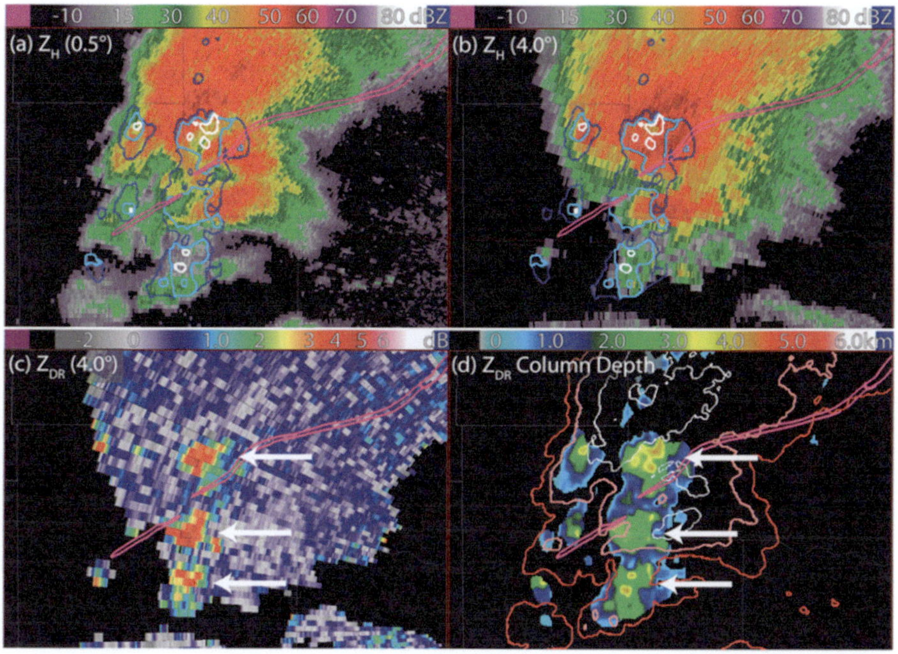

Fig. 9.24 Data from a supercell near El Reno at ~2055 UTC close to the genesis of a violent tornado: (**a**) Z at 0.5° elevation angle (dBZ), (**b**) Z at 4.0° elevation angle (dBZ), (**c**) filtered Z_{DR} at 4.0° elevation angle (dB), and (**d**) Z_{DR} column depth generated by the Z_{DR} column algorithm (km; above the environmental 0 °C level). Three primary Z_{DR} columns are identified by the algorithm and are noted by white arrows in (**c**) and (**d**). The Z_{DR} column depth contours at 1, 2.3, and 3.7 km are marked in dark blue, light blue, and white contours in (**a**) and (**b**). Contours of Z at 10, 30, and 50 dBZ are marked in red, pink, and white in (**d**). The purple swaths are tornado tracks from NWSFO Norman. From Snyder et al. (2015). © American Meteorological Society. Used with permission

2055 UTC, near the time of the genesis of a violent tornado. Although someone familiar with the general structure of supercells may be able to infer the position of updrafts based upon Z data (two upper panels in Fig. 9.24), details such as updraft intensity and structure typically are not captured by Z, particularly at low elevation angles. Turning to Z_{DR}, however, one can identify three primary regions of enhanced Z_{DR} above the environmental freezing level (e.g., at the 4.0° elevation angle in the panel c). At this particular time, the Z_{DR} column detection algorithm identifies three large Z_{DR} columns aligned in nearly a north-south fashion (panel d). The deepest Z_{DR} column exceeding 3.7 km in height above the freezing level is located near the notch of the hook echo associated with the supercell producing the tornado, whereas the other two Z_{DR} columns are associated with convective storms developing south of and merging into the primary supercell.

A nowcasting value of the Z_{DR} column product is illustrated in Fig. 9.25 which exhibits the ~30-min evolution of a multicell convective complex over western Florida on 24 June 2013. A time evolution of Z at 0.5° elevation angle and Z_{DR}

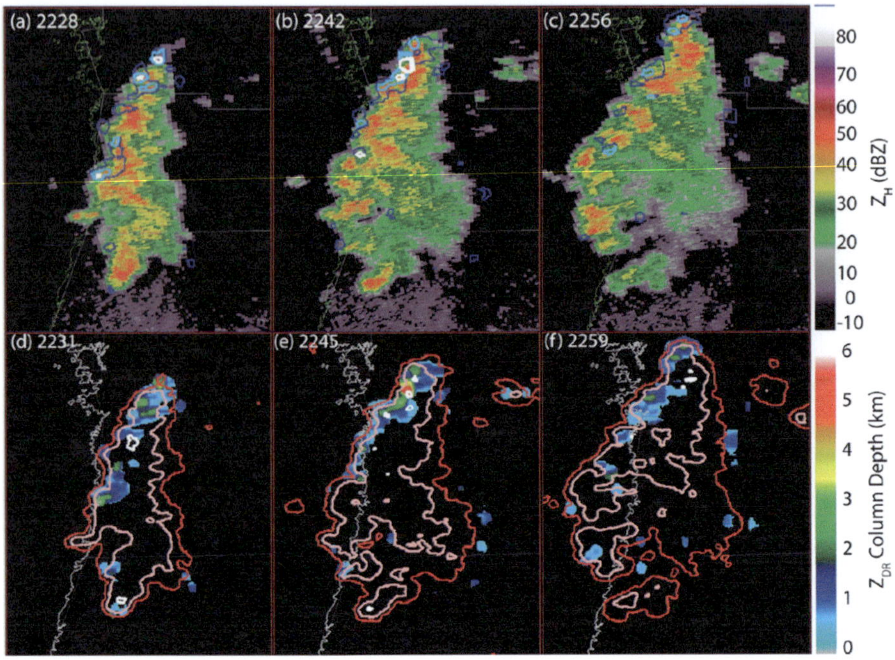

Fig. 9.25 Time series (with time increasing from left to right) of (**a–c**) Z at 0.5° elevation angle and (**d–f**) Z_{DR} column depth from the WSR-88D in Tampa Bay, Florida (KTBW), on the afternoon of 24 June 2013. Contours of Z_{DR} column depth of 1, 2, and 3 km are shown in dark blue, light blue, and white, respectively, in (**a–c**); Z contours at 10, 30, and 50 dBZ are dark red, pink, and white, respectively, in (**d–f**). From Snyder et al. (2015). © American Meteorological Society. Used with permission

column depth retrieved from the data collected by the KTBW WSR-88D in Tampa Bay, Florida is displayed. At 2228 UTC, the strongest Z is evident at the southern flank of the convective cluster, whereas the Z_{DR} column algorithm output shows that the deepest Z_{DR} columns are located along the northern periphery of the convective complex. This indicates that this area may be the preferred location of continued convective development. About 30 min later, the southern part of the cluster has weakened, leaving higher Z on the northern edge of the convection, where the deepest Z_{DR} columns were located at the previous time. This example clearly demonstrates that the depth of the Z_{DR} columns provides useful short-term prognostic information relevant to operational meteorologists.

9.7 Classification of Winter Precipitation

The initial HCA algorithm implemented on the WSR-88D (Park et al. 2009) was developed for warm-season weather and therefore requires modification to address classification issues related to transitional winter weather, such as detection of freezing rain and discrimination between rain, ice pellets/sleet, and different types of snow. This algorithm precludes any ice-phase precipitation types below the melting layer, thereby excluding precipitation types such as sleet. Second, it is essentially "local." That is, it provides class designations at every elevation scan, using only radar information collected at that scan rather than in a full 3D volume. Hence, precipitation type is determined only from data observed on conical surfaces and may not represent hydrometeors near the ground. This is particularly a problem at large distances from the radar where even a 0.5° elevation scan may be several kilometers above ground. Third, the initial WSR-88D HCA algorithm is entirely radar based; no thermodynamic information is utilized in the classification process. This latter point is critical for the detection of freezing rain because radar observations alone cannot distinguish between rain and freezing rain.

A new classification paradigm for surface winter precipitation has been introduced by Schuur et al. (2012). Figure 9.26 presents a schematic that summarizes the classification procedures of the winter surface HCA (WsHCA). The first step of the process is the fuzzy logic-based classification (Park et al. 2009). The algorithm then projects the fuzzy logic-based classifications from the lowest elevation scan to the surface as snow or ice crystals for cold-season events at locations above which the entire atmospheric column has $T < -5$ °C. The projection is rain, big drops, or hail for warm-season events where the surface temperature at a location is higher than 5 °C. For intermediate conditions typical of transitional winter weather events, the algorithm uses vertical profiles of model wet-bulb temperature to provide a background precipitation classification type. Polarimetric radar observations are then used to either confirm or reject the background classification. For example, if a radar bright band (suggesting an elevated warm layer) is observed immediately above an area with background classification of dry snow (suggesting the absence of an elevated warm layer in the model output), the background classification is declared inconsistent with observations and is modified according to a set of empirical rules. The polarimetric data are also used to provide further refinement of precipitation type categories if the observations are found to be consistent with the background classification.

The existing operational NWP models poorly describe phase transitions of winter precipitation and do not produce reliable discrimination between rain and snow,

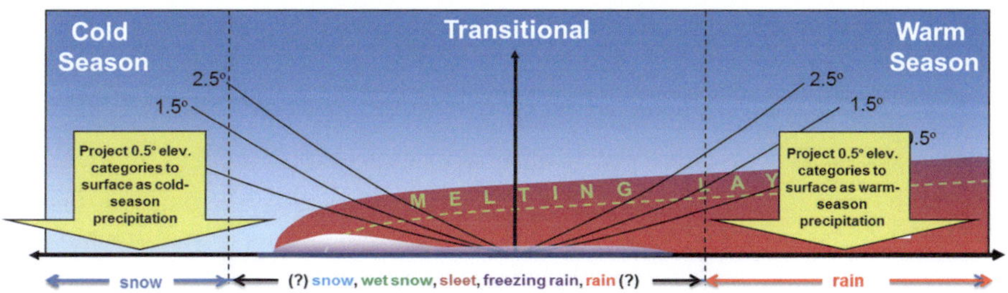

Fig. 9.26 Schematic showing the classification process of the surface-based HCA. Courtesy Dr. T. Schuur

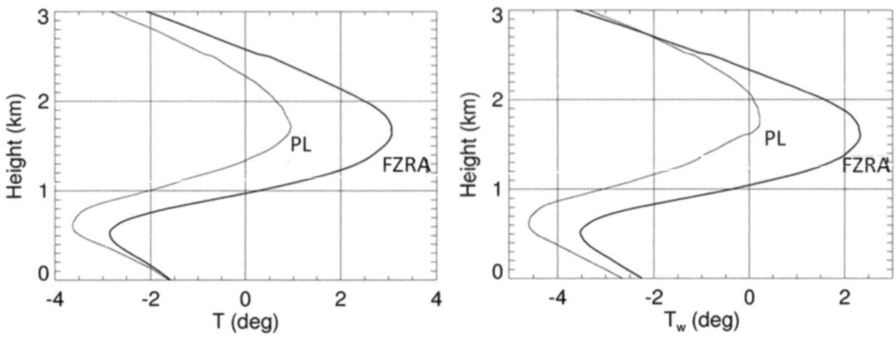

Fig. 9.27 Median low-altitude profiles of temperature (T) and wet-bulb temperature (T_w) for freezing rain (FZRA) and ice pellets (PL) retrieved from large dataset in the USA

especially between freezing rain and ice pellets. This is a difficult problem because relatively minor changes in a vertical profile of temperature may cause dramatic transformation of dangerous and high-impact freezing rain into relatively harmless ice pellets or sleet. Figure 9.27 shows median vertical profiles of temperature and wet-bulb temperature for freezing rain and ice pellet events obtained from a 10-year dataset of soundings and ASOS observations (Reeves et al. 2014). Surface temperatures associated with these two precipitation types are indistinguishable, and the only difference is in the temperature of elevated inversion which is about 2° higher for freezing rain.

One of the major deficiencies of the models with bulk microphysics is inadequate treatment of the processes of melting and refreezing of hydrometeors. Even the most sophisticated models use primarily the vertical profiles of temperature and humidity to identify "warm" atmospheric layers where melting of snowflakes occurs and "cold" layers where completely or partially melted snowflakes undergo refreezing. The degree of melting/refreezing depends crucially on the residence time of the particles in these layers. The residence times vary dramatically depending on the terminal velocity of hydrometeors which is a strong function of particle size, density, shape, and phase composition. The models with bulk microphysics essentially ignore such dependencies.

A background classifier based on a simple 1D spectral bin model which uses a vertical profile of wet-bulb temperature T_w was suggested by Ryzhkov et al. (2014) and described in full detail by Reeves et al. (2016). It uses explicit microphysical modeling of melting and refreezing (Sects. 7.2.4.1 and 7.2.5.2). Consider the three thermal profile types in Fig. 9.28. The simplest are the completely subfreezing profiles shown (Fig. 9.28a). The surface precipitation type for these profiles is dictated by whether the cloud-top wet-bulb temperature ($T_{\text{w-ct}}$) is less than the ice nucleation temperature (T_{ice}). T_{ice} is defined as the wet-bulb temperature at which the ice nucleus in a water droplet becomes activated (usually <-5 °C; e.g., Meyers et al. 1992). Herein, a $T_{\text{ice}} = -6$ °C is assumed. If $T_{\text{w-ct}} \leq T_{\text{ice}}$, ice crystals form, and the resulting surface precipitation type will be snow (SN). Otherwise, supercooled liquid water droplets form resulting in freezing rain (FZRA).

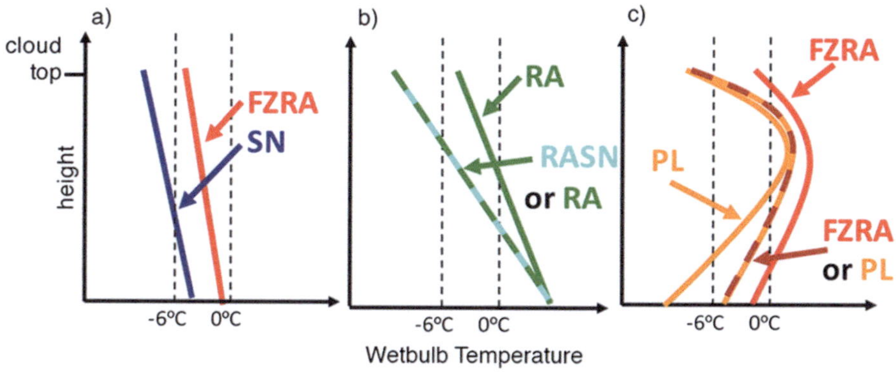

Fig. 9.28 Idealized profiles of wet-bulb temperature assuming (**a**) no, (**b**) one, and (**c**) two crossings of the 0 °C isotherm. The possible precipitation types for each profile are indicated. From Reeves et al. (2016). © American Meteorological Society. Used with permission

The profiles in Fig. 9.28b cross the 0 °C isotherm one time and have surface T_ws that are higher than 0 °C. If $T_{\text{w-ct}} > T_{\text{ice}}$, only liquid water droplets form aloft and warm to above-freezing temperatures as they enter the melting layer yielding rain (RA) at the surface. If not, the precipitation type will depend on whether the frozen particles passing through the melting layer have sufficient time to completely melt before reaching the surface. The larger and more heavily rimed a particle is, the more time it will take to completely melt. Therefore, liquid drops (i.e., RA) and/or partially melted ice particles can reach the surface. This latter class is sometimes referred to as wet snow (WS) or rain/snow mix (RASN).

The third profile type has an elevated warm layer capping a surface-based subfreezing layer (Fig. 9.28c). Assuming the cloud top is above the highest crossing of the 0 °C isotherm, there are two possible precipitation types. If all hydrometers completely melt in the melting layer or $T_{\text{w-ct}} > T_{\text{ice}}$, the subfreezing layer will require $T_w \leq T_{\text{ice}}$ for ice nucleation and refreezing to occur. If this condition is met and the droplets refreeze before reaching the ground, ice pellets (PL) will be observed. Otherwise, the precipitation type will be FZRA. If the melting layer is sufficiently cool and shallow or the hydrometeors sufficiently large and rimed before entering it, they may not completely melt. In this case, the hydrometeors will have active ice nuclei, and refreezing will instantaneously occur when the hydrometeors enter the subfreezing layer resulting in PL. It is possible, given a spectrum of drop sizes, for multiple phases (some frozen, some liquid) to occur in the same volume yielding a mixture of FZRA, RA, and PL. This mixture is referred to as FZRAPL if surface temperature is negative or RAPL if it is positive.

A one-dimensional spectral bin model assumes a certain size distribution of particles above the freezing level and uses the temperature and humidity vertical profiles retrieved from the Rapid Refresh (RAP; Brown et al. 2011) analyses to calculate the liquid water fraction f_w for a spectrum of hydrometeor sizes as a function of height above ground and melted hydrometeor diameter. The algorithm

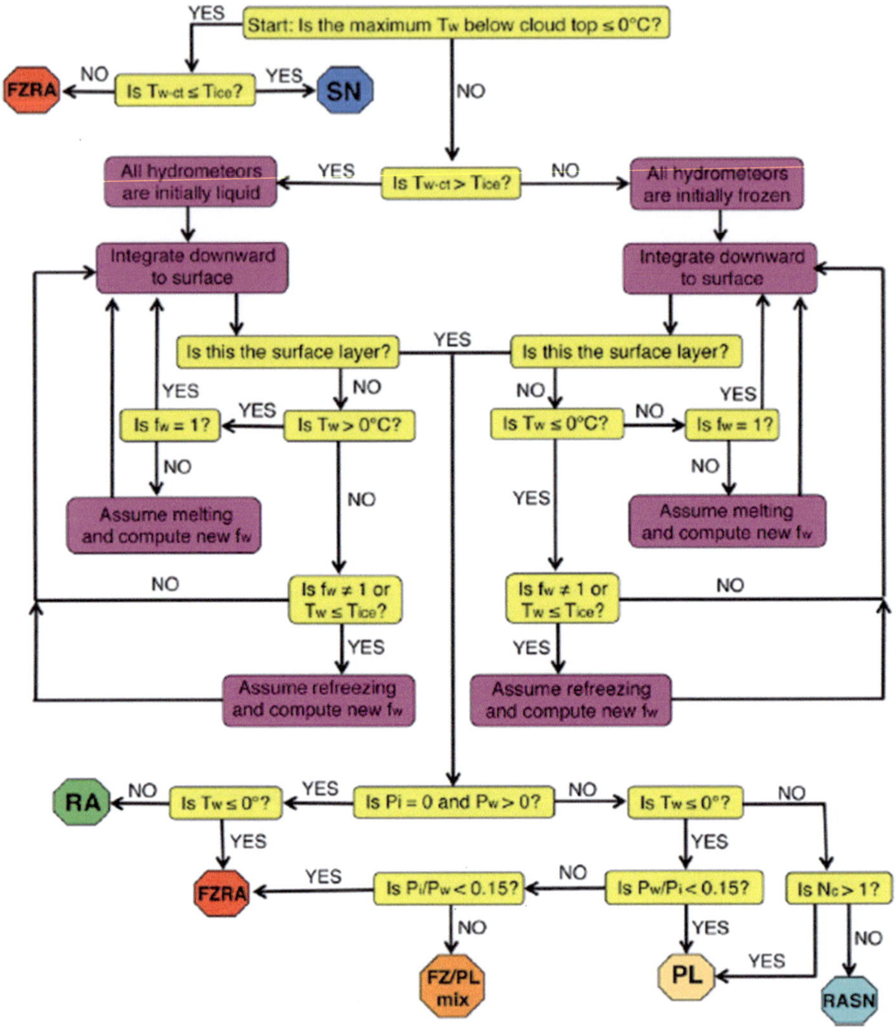

Fig. 9.29 A flowchart showing the logic used by the SBC. In the SBC, T_w, T_{ice}, $T_{\text{w-ct}}$, f_w, P_i, P_w, and N_c are the wet-bulb temperature; ice-nucleation temperature; wet-bulb temperature at cloud top; liquid water fraction; precipitation rates of ice and liquid water, respectively; and the number of crossings that the T_w profile makes with the 0 °C isotherm. From Reeves et al. (2016). © American Meteorological Society. Used with permission

then follows the flow chart depicted in Fig. 9.29. At the surface, the liquid water and ice precipitation rates P_w and P_i are computed as

$$P_w = \frac{3600}{\rho_w} \int m(D) f_w(D) V_t(D) N(D) dD \tag{9.15}$$

and

$$P_i = \frac{3600}{\rho_i} \int m(D)[1 - f_w(D)]V_t(D)N(D)dD, \tag{9.16}$$

where ρ_w and ρ_i are the densities of water and solid ice (g cm^{-3}) and $m(D)$, $f_w(D)$, $V_t(D)$, and $N(D)$ are the hydrometeor mass (kg), mass water fraction, terminal velocity (m s^{-1}), and size distribution (m^{-3} mm^{-1}). In (9.15) and (9.16), P_w and P_i are expressed in mm h^{-1}. If $P_i = 0$ and $P_w > 0$, the precipitation is classified either as RA or FZRA (if surface $T_w < 0$ °C). The precipitation is classified as FZRA, PL, FZRAPL, RAPL, or RASN, depending on the surface temperature, the ratio P_i/P_w, and the number of crossings the T_w profile makes with the 0 °C isotherm, as specified in Fig. 9.29.

The performance of the SBC background classifier is illustrated in Fig. 9.30 where vertical profiles of T_w combined with the mass water fraction f_w presented as a function of height and particle melted diameter D_w is shown for eight transitional winter events. Pure rain (RA) was diagnosed in the panels (i - j) because the temperature at the surface T_s is positive and surface $f_w = 1$ for all particle sizes. Freezing rain (FZRA) was identified if $T_s < 0$ °C and surface $f_w = 1$ either for all sizes (panels (e–f)) or for the majority of sizes so that $P_i/P_w < 0.15$ (panels (g–h) and (k–l)). Similarly, ice pellets (PL) were classified as a dominant precipitation type near the ground if surface $f_w = 0$ either for all sizes (panels (a–b) and (o–p)) or for the majority of sizes so that $P_w/P_i < 0.15$ (panels (c–d)). In one case, a mixture of rain and ice pellets (RAPL) was diagnosed because both liquid and ice particles coexist in comparable amounts (panels (m–n)). In all eight cases, the SBC designations match ASOS measurements at the ground.

The SBC algorithm was applied nationally, and the fields of designated precipitation types were compared to the observed types by public and recorded on the meteorological Phenomena Identification Near the Ground (mPING; Elmore et al. 2014) system (Fig. 9.31). The mPING observations are made by citizens in multiple locations across the USA (left panels in Fig. 9.31). The mPING reports and SBC analyses (right panels in Fig. 9.31) are very consistent for the three selected cases of the transitional winter weather.

Polarimetric measurements can be used for initialization of the SBC model by providing information about the presence or absence of the radar return in a given location, the height of the radar echo top, and, possibly, the size distribution and the degree of riming of snow aloft. Checking the consistency of the SBC designations with radar measurements is an essential part of the winter surface hydrometeor classification algorithm (WsHCA). An example of the synergetic use of the cloud modeling and radar observations is presented next.

Figures 9.32 and 9.33 show the results of the application of WsHCA to a winter weather event sampled by the KVWX (Evansville, IN) WSR-88D at 170143 UTC on 5 January 2014. The Z, Z_{DR}, and ρ_{hv} at this time (Fig. 9.32) exhibit a clear delineation line in the polarimetric fields (i.e., Z_{DR} enhancement, ρ_{hv} decrease) between cold-season precipitation to the N-NW and warm-season precipitation

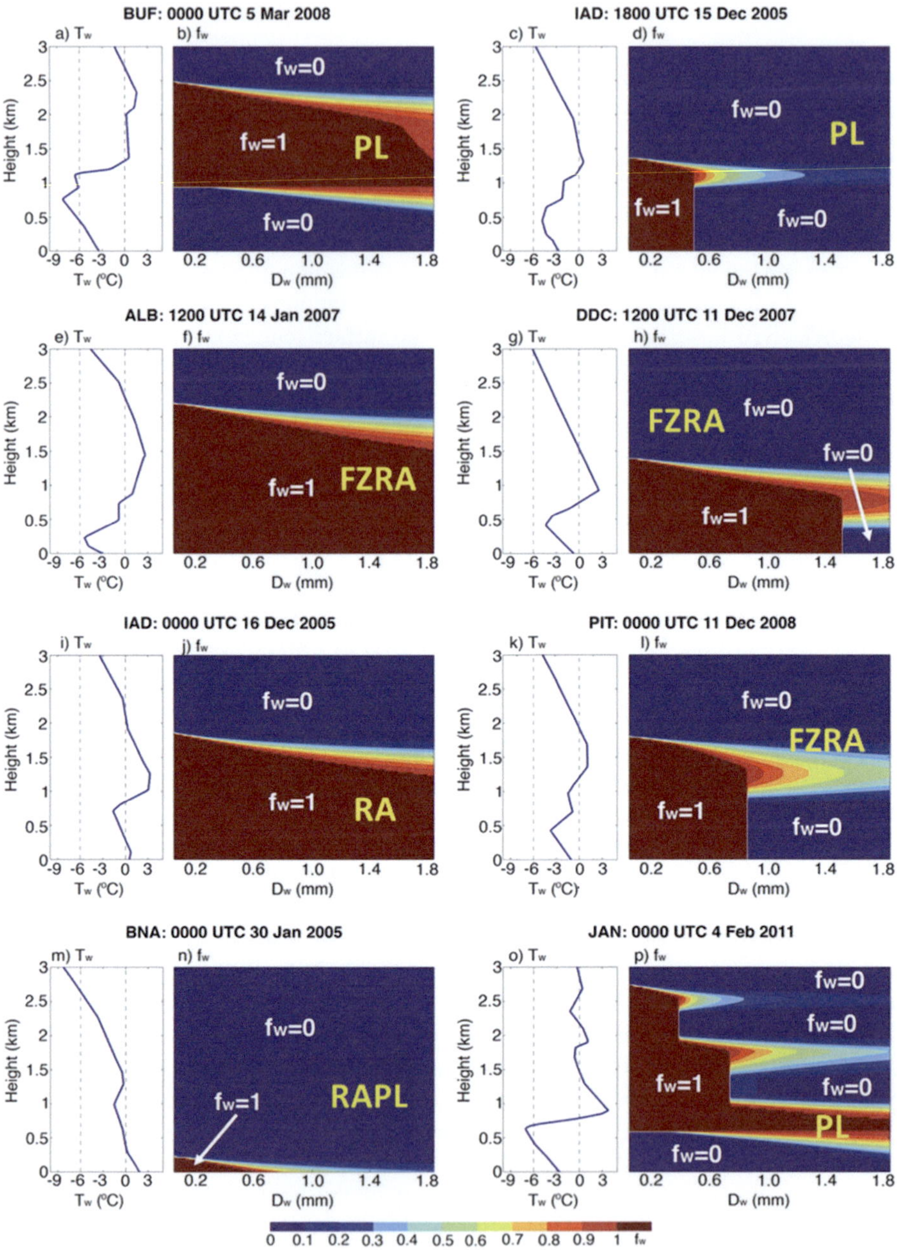

Fig. 9.30 Vertical profiles of wet-bulb temperature T_w and vertical distributions of mass water fraction f_w as a function of the hydrometeor melting diameter D_w retrieved from the SBC model for eight transitional winter events. The symbols RA (rain), FZRA (freezing rain), PL (ice pellets), and RAPL (mixture of rain and ice pellets) indicate the dominant type of precipitation near the surface determined by the SBC algorithm. Adapted from Reeves et al. (2016). © American Meteorological Society. Used with permission

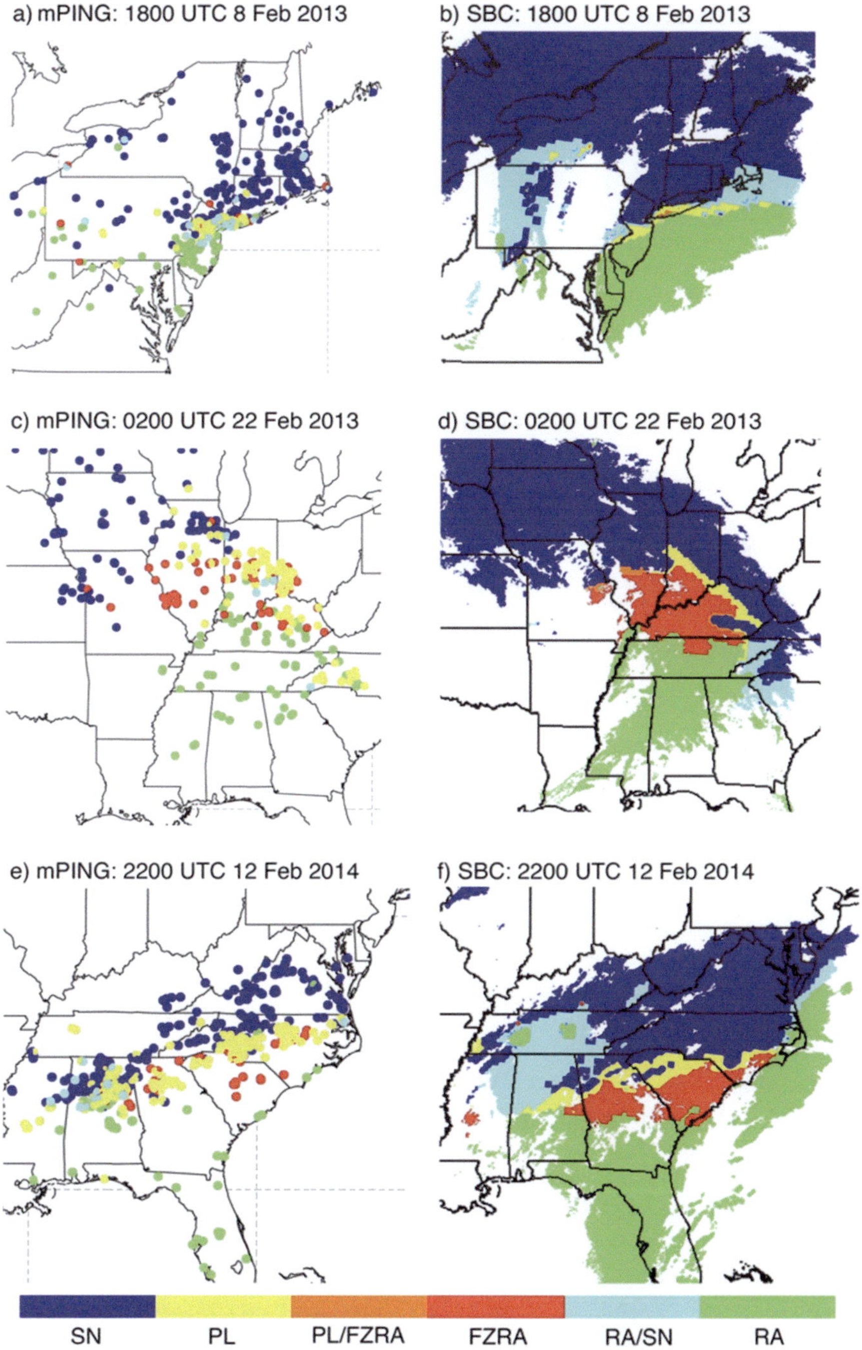

Fig. 9.31 The (**a**), (**c**), and (**e**) mPING distributions of observed precipitation type for the 2 h period surrounding the time indicated at the top of each panel and (**b**), (**d**), and (**f**) the corresponding SBC analyses. In (**b**), (**d**), and (**f**), only the areas with observed composite reflectivity larger than 0 dBZ are shaded. From Reeves et al. (2016). © American Meteorological Society. Used with permission

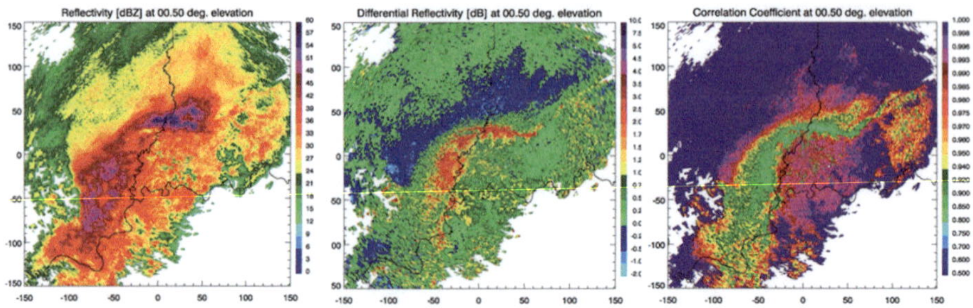

Fig. 9.32 KVWX (Evansville, IN) reflectivity (Z), differential reflectivity (Z_{DR}), and correlation coefficient (ρ_{hv}) at El = 0.5° on 5 January 2014 at 1702 UTC. Figure courtesy of Dr. T. Schuur and J. Krause

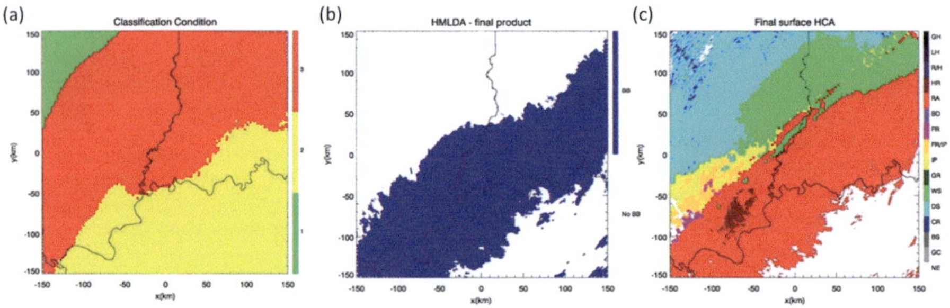

Fig. 9.33 Results of (**a**) the environmental "classification," (**b**) hybrid melting layer detection algorithm (HMLDA), and (**c**) WsHCA classification at 1702 UTC on 5 January 2014. Figure courtesy of Dr. T. schuur and J. Krause

to the S-SE. Surface observations from the mPING at this time show mostly snow to the N-NW of this line, rain to the S-SE, and occasional reports of freezing rain and ice pellets near the boundary.

In Fig. 9.33a, plotted are the "classification conditions" used to combine the WSR-88D HCA described in Sect. 9.2.1 with the SBC-derived surface precipitation type. Green color in Fig. 9.33a indicates locations where ice-phase categories from the fuzzy logic HCA at KVWX's 0.5° elevation scan are projected to the surface as winter precipitation; yellow color indicates locations where ice-phase categories are projected as rain, and red color indicates locations where the SBC-based surface classification, which is later modified by radar observations, is used. The results of the hybrid melting layer detection algorithm (HMLDA) and WsHCA at this time are presented in Fig. 9.33b, c. Comparing the HMLDA results to Fig. 9.32, it is seen that the location of the combined radar and model detected ML agrees quite well with the ML boundary suggested by the polarimetric variables in Fig. 9.32. Furthermore, the final classification by the WsHCA (Fig. 9.33) is also in good agreement with the polarimetric signatures in Fig. 9.32, as well as the mPING surface observations.

9.8 Discrimination Between Meteorological and Nonmeteorological Radar Returns

9.8.1 Identification of Land and Sea Clutter

Ground clutter has polarimetric characteristics very distinct from the ones of meteorological scatterers. Therefore, even the very simple fuzzy logic classification algorithm described in Sect. 9.1 (Fig. 9.5) can identify areas of returns dominated by clutter or anomalous propagation. It is very common that the returns attributed to ground clutter and weather scatterers are mixed, and ground clutter filtering is needed to separate the contributions from clutter and weather. Siggia and Passarelli Jr (2004) suggested Gaussian model adaptive processing (GMAP) to do that. Various signal processing techniques have been proposed to identify samples contaminated by ground clutter. These include the clutter mitigation decision (CMD) algorithm (Hubbert et al. 2009a, b), spectrum-time estimation and processing (STEP) methodology (Cao et al. 2012), spectrum clutter identification (SCI) technique (Li et al. 2013), and clutter environment analysis using adaptive processing (CLEAN-AP) filter (Torres and Warde 2014). A good overview of the modern techniques for clutter detection and cancelling can be found in the book of Zhang (2016) (Sects. 5.4, 5.5, and 5.6).

In contrast to the land clutter, detection and mitigation of the sea clutter have not been studied as much. Conventional removal of the zero Doppler frequency component is not effective because the return from sea waves has nonzero Doppler velocity. Weather radars were not traditionally used for studying sea clutter although there is continuous demand from the navy and defense communities to distinguish between sea and weather returns. Very few studies of sea clutter with polarimetric weather radars are reported in the literature (e.g., Ryzhkov et al. 2002; Rico-Ramirez and Cluckie 2008; Hannessen and Weipert 2011a).

Next, we provide an example of polarimetric data collected by a C-band radar located at the Makung Island in the Taiwan Strait about 50 km west of Taiwan (Fig. 9.34). At the lowest elevation of 0.5°, the radar returns from sea, land, and weather are often observed simultaneously. A composite PPI of Z, Z_{DR}, ρ_{hv} and Doppler velocity v in a case the radar echo is dominated by sea clutter is in Fig. 9.35. At typical antenna elevations of weather surveillance radars, sea surface is illuminated via lower part of the main radar beam and/or sidelobes which reduce the reflected radar signal. Nonetheless, depending on the sea roughness, the corresponding Z may exceed 40 dBZ and can be comparable with the reflectivity of clouds and precipitation. Different elements of sea waves make different contributions to H and V components of radar returns; hence, the cross-correlation coefficient ρ_{hv} of sea clutter is significantly lower than that of weather. Additional reduction in ρ_{hv} may occur if the H and V copolar antenna patterns are not perfectly matched. Differential reflectivity of sea clutter exhibits very strong azimuthal dependence which correlates with dominant direction of the ocean waves driven by wind. Z_{DR} of sea clutter can vary from −3 to 4 dB as a function of azimuth (Fig. 9.35). This is in striking contrast to the backscatter of land clutter which has no pronounced

Fig. 9.34 Location of Makung Island and Taiwan main island

azimuthal dependence. As in the case of ground (land) clutter, the texture parameters of Z, Z_{DR}, and Φ_{DP} associated with sea clutter are considerably larger than those associated with weather.

Ocean waves (swell or wind-generated) have nonzero Doppler velocity that is often a scaled-down Doppler velocity of the wind. The magnitude of Doppler velocity associated with sea clutter rarely exceeds 5 ms^{-1}. Contamination from sea clutter may cause discontinuities in the fields of Doppler velocity if the weather and sea returns are mixed, and then these can be wrongly identified as areas of strong wind shear.

If the returns from sea clutter and rain overlap, discrimination between the two becomes very challenging (Fig. 9.36). It is impossible to distinguish between rain and sea clutter based on Z and v only, although Doppler velocity may provide some clues. By itself, the cross-correlation coefficient may not give clear picture either. But the combination of polarimetric variables and Doppler spectral moments enables clear distinction.

A simple fuzzy logic algorithm to discriminate between sea clutter (SC), ground (land) clutter (GC), and rain (RA) using seven radar variables is described next. The variables are Z, Z_{DR}, ρ_{hv}, v, SD(Z), SD(Z_{DR}), and SD(Φ_{DP}). The membership functions have trapezoidal shapes and are determined by the four parameters, x_1, x_2, x_3, and x_4, as in Park et al. (2009). The parameters of 21 membership functions

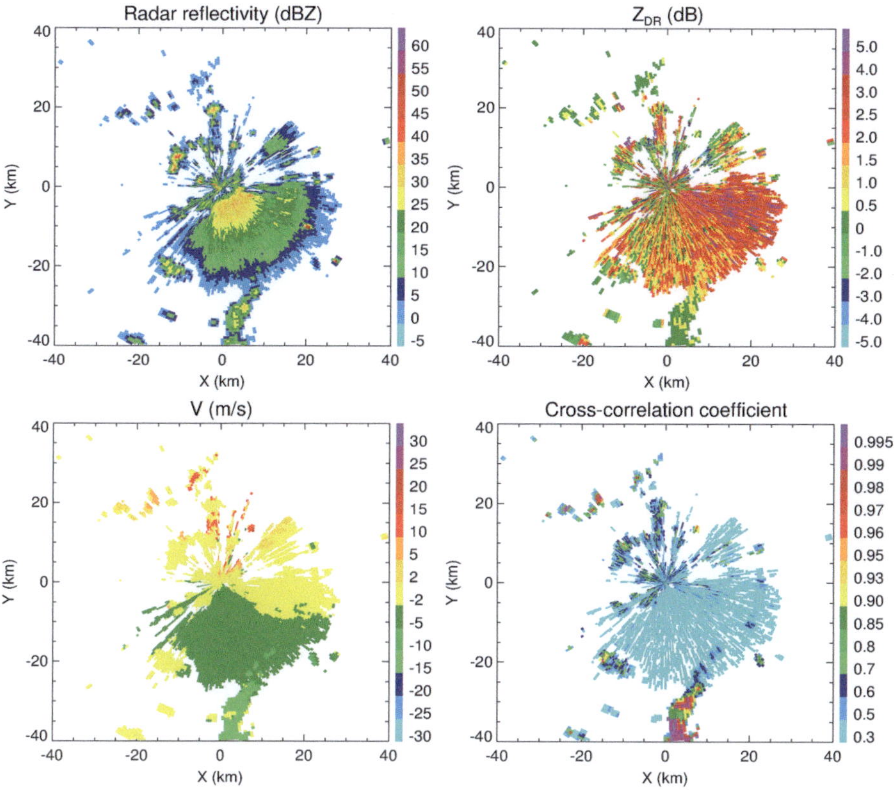

Fig. 9.35 Composite plot of Z, Z_{DR}, v, and ρ_{hv} of the radar return from sea clutter at El = 0.5°. The sea clutter is not mixed with the return from rain. Data courtesy of the Weather Wing of the Republic of China Air Force and Central Weather Bureau

are presented in Table 9.7. Each variable has a weight with respect to a particular class, and the matrix of these weights is in Table 9.8.The results of classification are displayed in the bottom right panel in Fig. 9.36. Sea clutter (SC) is painted blue, ground clutter (GC) is green, and rain is red (RA). The classifier distinguishes between rain and sea clutter mainly at relatively close distances east of the radar. Although the main island of Taiwan is located 50 km east from the radar, the central mountain range causing ground clutter is at the distance of about 80 km. The algorithm correctly identifies ground clutter (GC) there (green color) with some precipitation overlaid.

9.8.2 Biological Scatterers

Biological scatterers produce spectacular polarimetric signatures which are easy to identify and distinguish from meteorological returns. This is because insects, birds,

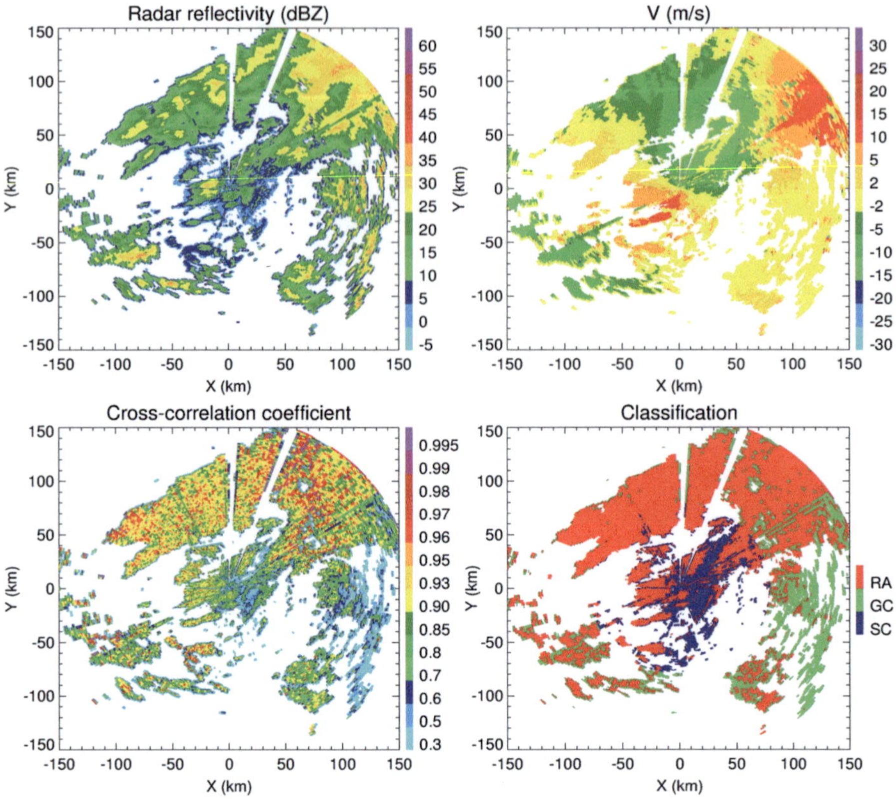

Fig. 9.36 Composite plot of Z, v_r, ρ_{hv}, and classification results of classification of radar returns from rain, sea clutter, and land clutter. *RA* rain, *GC* ground clutter, *SC* sea clutter. Data courtesy of the Weather Wing of the Republic of China Air Force and Central Weather Bureau

and bats have very irregular shape, a dielectric constant close to the one for water, and may exhibit high degree of common orientation, especially during massive migration. Polarimetric weather radars provide tremendous opportunities for ornitologists and enthomologists to study behavioral patterns of these flying creatures and advance aeroecology in general (Chilson et al. 2012). A recent review of dual-polarization radar for biological applications can be found in Stepanian et al. (2016). For meteorological applications, it is important to discriminate between clear air returns caused by passive tracers of air flows such as small insects and strong flyers such as birds. The latter can bias Doppler wind estimates and, therefore, need to be censored.

First measurements of polarimetric characteristics of biota using weather radars have been made by Mueller and Larkin (1985), Achtemeier (1991), and Zrnic and Ryzhkov (1998, 1999). In these studies, the authors note very high values of Z_{DR} (7–9 dB) which they attribute to insects. Zrnic and Ryzhkov (1998) found that birds are characterized by lower values of Z_{DR} (mainly between -1 and 3 dB) but much

Table 9.7 Parameters of the membership functions for discrimination between rain, ground clutter, and sea clutter

	Z	Z_{DR}	ρ_{hv}	v	SD(Z)	SD(Z_{DR})	SD(Φ_{DP})
Sea clutter (SC)							
x_1	−5	−3	0.15	−6	0	0	0
x_2	0	−2	0.25	−5	1	0.7	40
x_3	30	5	0.45	5	4	2.5	120
x_4	40	6	0.7	6	6	4	160
Ground clutter (GC)							
x_1	5	−4	0.1	−2	0	−1	−1
x_2	10	−2	0.15	−1	0.5	0	0
x_3	70	2	0.9	1	4.5	1.3	140
x_4	80	3	0.95	2	6	4	160
Rain (RA)							
x_1	5	$f_1 - 0.3$	0.9	−40	0	−1	−1
x_2	10	f_1	0.95	−35	0.5	0	0
x_3	45	f_2	1.00	35	2.5	0.4	10
x_4	50	$f_2 + 0.5$	1.01	40	5	1	40

Functions f_1 and f_2 are defined by Eq. (9.6)

Table 9.8 Matrix of weights in the fuzzy logic algorithm for discrimination between rain, ground clutter, and sea clutter

Class	Z	Z_{DR}	ρ_{hv}	v	SD(Z)	SD(Z_{DR})	SD(Φ_{DP})
SC	0.4	0.4	1.0	0.8	0.4	0.5	0.6
GC	0.2	0.4	1.0	0.8	0.4	0.5	0.6
RA	0.4	0.8	1.0	0.4	0.4	0.5	0.6

higher backscatter differential phase δ than insects (Fig. 9.37). However, the scatterplots for birds and insects in the Z_{DR}–δ plane (Fig. 9.37) overlap significantly, particularly at relatively low Z_{DR} and δ.

Another important feature of clear air echo produced by biota is strong azimuthal dependencies of all polarimetric variables (Figs. 9.38 and 9.39). In the composite plot of Z, Z_{DR}, δ, and ρ_{hv} (Fig. 9.38), the reflectivity field of migrating songbirds hints to the presence of anisotropic scatterers but by itself is inconclusive. The fields of the polarimetric variables exhibit strong azimuthal dependence indicating that songbirds fly with their axes oriented in the SW–NE direction so that Z_{DR} and δ are maximal in the orthogonal directions. Similar azimuthal patterns are typical for insects filling a convective boundary layer (Fig. 9.39). Again, one can conclude that insects are oriented along the wind in the NNW–SSE direction as indicated by the fields of Doppler velocity and differential reflectivity. In Fig. 9.38, it is clear that both Z_{DR} and δ are symmetric with respect to the 45° azimuth but are not circularly symmetric. This is because the bird's head view is different from its tail view.

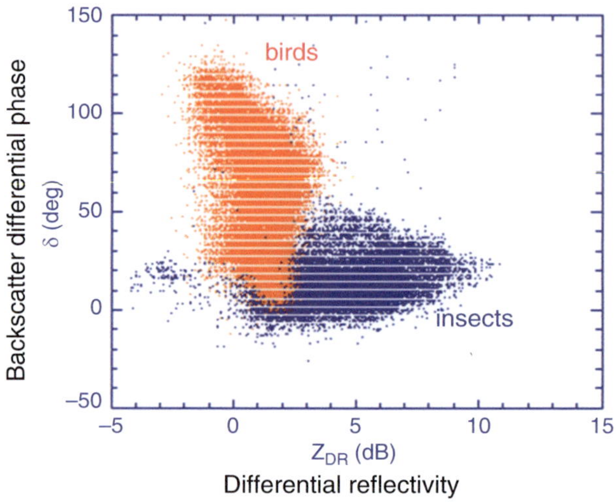

Fig. 9.37 Z_{DR} vs δ scatterplots for insects and birds. Data for insects were collected in central Oklahoma by the S-band NSSL Cimarron research radar on 10 August 1993 and for birds on 7 October 1996. Adapted from Zrnic and Ryzhkov (1999). © American Meteorological Society. Used with permission

Zrnic and Ryzhkov (1998) tried to reproduce azimuthal dependencies of Z_{DR} and δ of birds using scattering computations for prolate spheroids with different sizes and aspect ratios. They matched the observed azimuthal dependencies of Z_{DR} and δ in a case of migratory birds (Fig. 9.40a) and the corresponding theoretical dependencies assuming a prolate spheroid with equivolume diameter $D_e = 4$ cm and aspect ratio of 2.5 (Fig. 9.40b). The nonmonotonic azimuthal changes of Z_{DR} and δ are attributed to resonance scattering by the scatterers with typical bird sizes.

In the case illustrated in Fig. 9.39, the patterns of δ (indicated as Φ_{DP} in the plot) and ρ_{hv} are asymmetric with respect to the wind direction (and possibly mean orientation of insects). Melnikov et al. (2015) suggest that such asymmetry might be related to a strong coupling and depolarization of orthogonally polarized waves of horizontal and vertical polarizations in the simultaneous transmission/reception mode of operation. The effects of nonuniform beam filling can also contribute to the observed patterns. Indeed, the sharp drops of ρ_{hv} (Fig. 9.38) are observed at azimuthal directions where the azimuthal gradients of the backscatter differential phase δ are particularly high in agreement with Ryzhkov (2007).

The radar signatures of roosting birds are very different from the ones of migratory birds (Fig. 9.41). A ring signature west of the radar corresponds to a dispersing roost of purple martins as documented by Van den Broeke (2013). In either case, the cross-correlation coefficient ρ_{hv} of the bird returns (usually between 0.5 and 0.8) is much lower than the one of meteorological scatterers. The ρ_{hv} of insects is somewhat higher than that of birds but is still lower than of hydrometeors (Fig. 9.39).

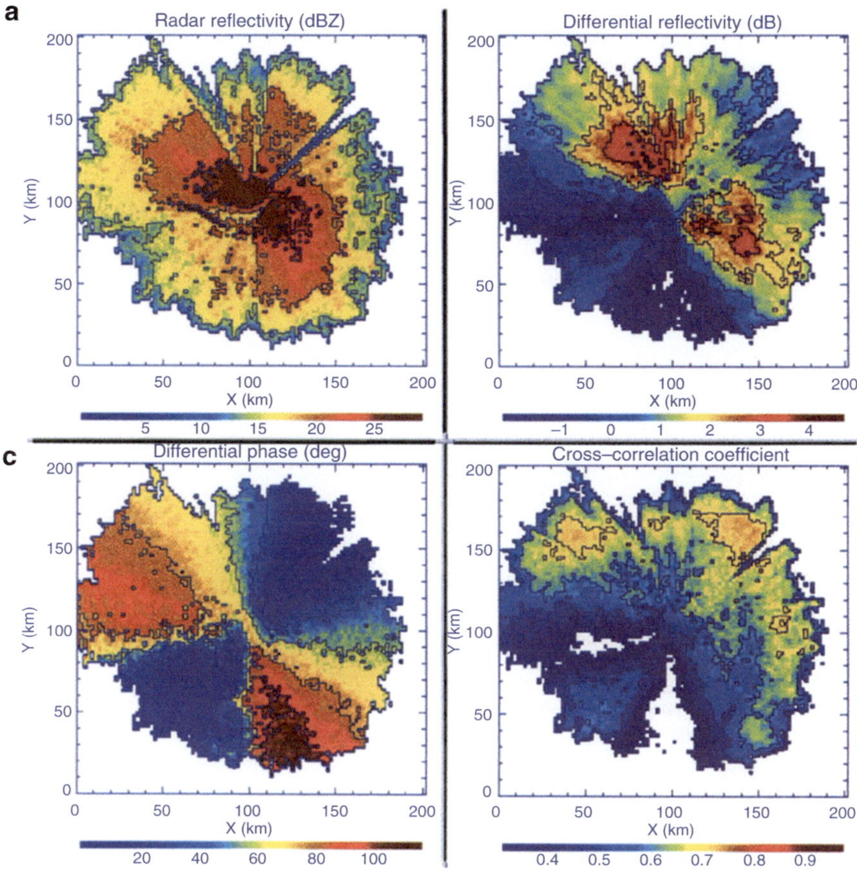

Fig. 9.38 Fields of (**a**) reflectivity factor, (**b**) differential reflectivity, (**c**) differential phase, and (**d**) correlation coefficient at elevation 0.5° from migrating songbirds (passerines). The measurements have been made in central Oklahoma by the S-band NSSL Cimarron research radar on 13 August 1997 at 0236 UTC. From Zrnic and Ryzhkov (1999). © American Meteorological Society. Used with permission

9.8.3 Chaff, Smoke Plumes, Dust Storms, and Volcanic Ash

There are several types of scatterers in the atmosphere which are neither of meteorological nor biological origin. These include military chaff, smoke plumes, dust, and volcanic ash which produce returns that are almost impossible to identify in the fields of reflectivity alone. Polarimetric radars can be used efficiently to recognize these types of clear air echo.

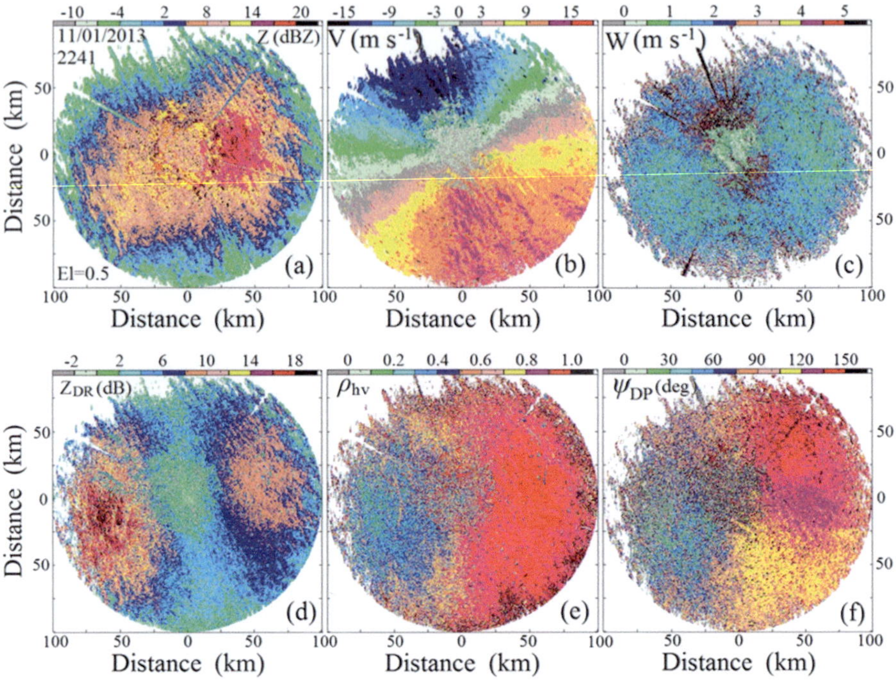

Fig. 9.39 Composite plot of (**a**) reflectivity Z, (**b**) Doppler velocity v, (**c**) spectrum width σ_v, (**d**) differential reflectivity Z_{DR}, (**e**) cross-correlation coefficient ρ_{hv}, and (**f**) total differential phase Φ_{DP} at antenna elevation 0.5°. The returns are caused by insects flying in a convective boundary layer. The measurements were made in central Oklahoma by the KOUN WSR-88D on 1 November 2013 at 2241 UTC. From Melnikov et al. (2015). © American Meteorological Society. Used with permission

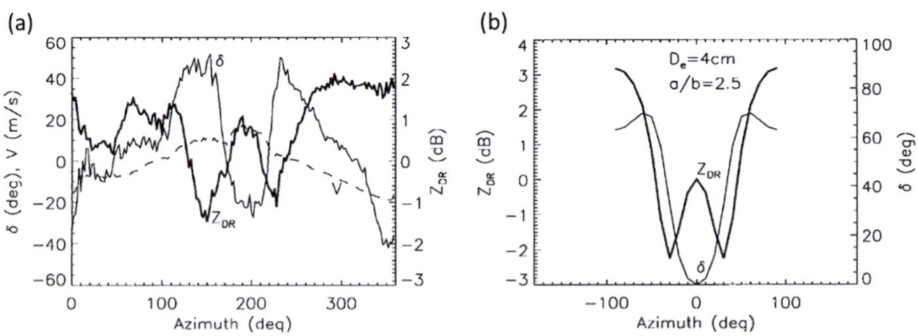

Fig. 9.40 (**a**) Observed azimuthal dependencies of differential reflectivity Z_{DR}, backscatter differential phase δ, and Doppler velocity V at elevation 1.5° from returns caused by migratory birds in central Oklahoma. The measurements have been made by the S-band NSSL Cimarron radar after midnight on 7 October 1996. (**b**) Theoretical azimuthal dependence of Z_{DR} and δ for a prolate water spheroid with equivolume diameter 4 cm and aspect ratio of 2.5. The azimuth angle of 0° corresponds to the direction of the symmetry axis. From Zrnic and Ryzhkov (1998).© IEEE. Used with permission

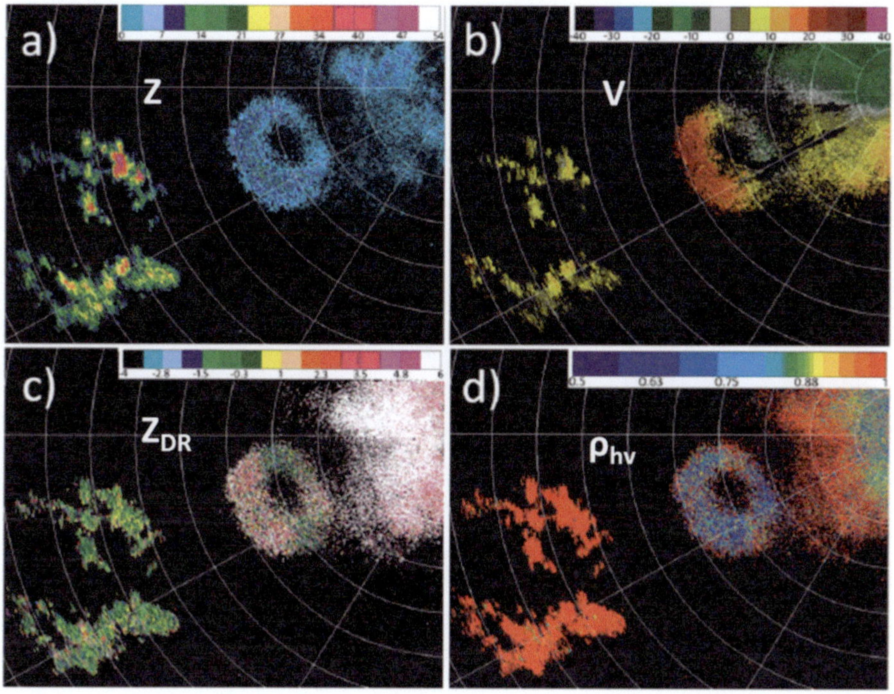

Fig. 9.41 Composite PPI of (**a**) reflectivity Z, (**b**) Doppler velocity V, (**c**) differential reflectivity Z_{DR}, and (**d**) cross-correlation coefficient ρ_{hv} illustrating polarimetric signatures of roosting birds (purple martins). Data have been collected by the KHTX WSR-88D at 1057 UTC 5 July 2012. Range rings are spaced 25 km apart. From Van den Broeke (2013). © American Meteorological Society. Used with permission

9.8.3.1 Chaff

Chaff is used in radar countermeasure applications and consists of scatterers which are commonly released from aircraft. These scatterers are composed of individual metallic strands and reflect radiation at certain wavelengths. Chaff presents multiple issues for aviation and weather service including false weather identification and contamination of weather returns because it may have reflectivity comparable to the one from clouds. An analysis of polarimetric properties of chaff is in Zrnic and Ryzhkov (2004) who discriminate chaff from ground clutter and snow in the fields of Z_{DR} and ρ_{hv}. These measurements in Oklahoma demonstrate that Z_{DR} from chaff is widely distributed between 0 and 6 dB, whereas ρ_{hv} is very low (0.2–0.5) and can serve as a primary discriminator between chaff, ground clutter, and weather.

A more recent and comprehensive study of polarimetric properties of chaff obtained from the network of WSR-88D has been carried out by Kurdzo et al. (2018). From 22 cases of chaff release, Kurdzo et al. (2018) found that the distributions of Z_{DR} and ρ_{hv} from chaff are wider than reported by Zrnic and Ryzhkov

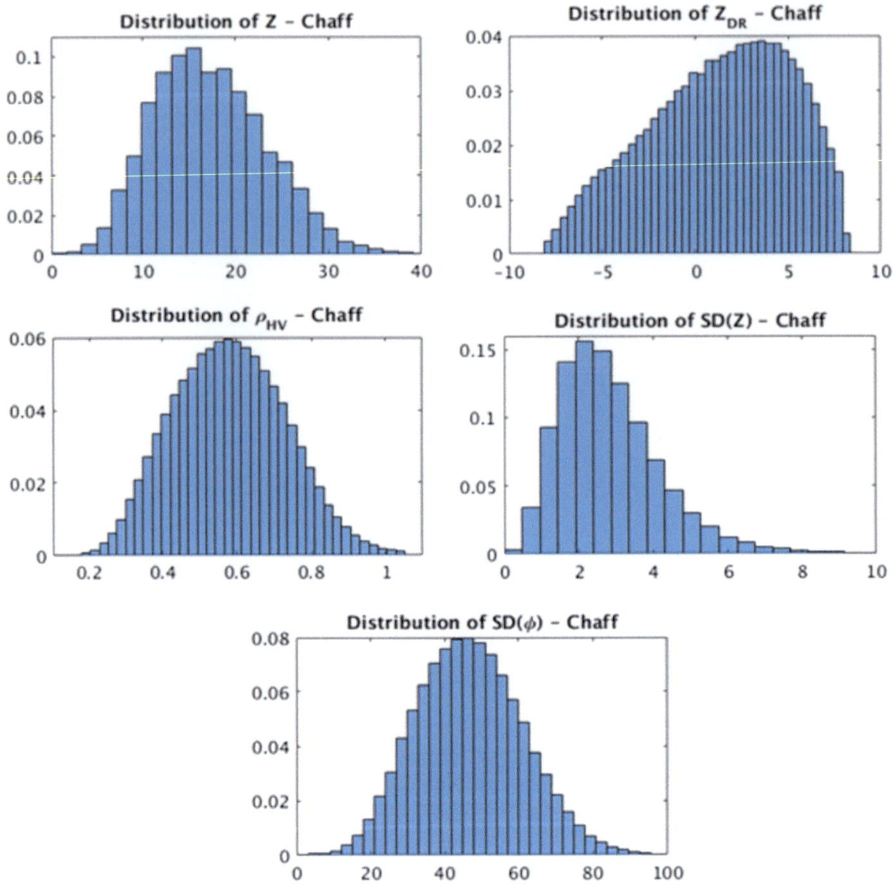

Fig. 9.42 Distributions of Z, Z_{DR}, ρ_{hv}, and texture parameters of Z and Φ_{DP} at S band in chaff cases across the USA. From Kurdzo et al. (2018). © American Meteorological Society. Used with permission

(2004). Specifically, the proportion of negative Z_{DR} (down to -8 dB) is larger and ρ_{hv} is bigger (Fig. 9.42). This suggests, for example, that a fuzzy logic classification method with all polarimetric variables should be used for reliable identification of chaff. Such algorithm (Chaff Detection Algorithm or CDA) is described in Kurdzo et al. (2018). Similar to the TDS classification algorithm described in Sect. 9.5, the basic hydrometeor classification algorithm (HCA) implemented on the WSR-88D operational network is modified to add a chaff class (CHF), and an example of discrimination between chaff and rain is in Fig. 9.43. Because chaff is released locally from aircrafts and falls slowly while drifting with the wind, its returns stretch into narrow trails (painted black in Fig. 9.43).

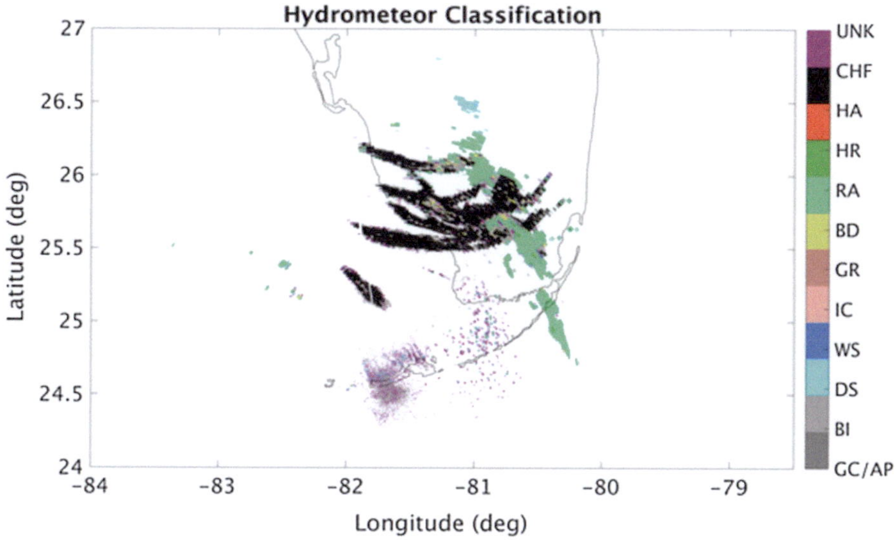

Fig. 9.43 Results of radar echo classification including chaff (CHF) using data collected by the KBYX WSR-88D on 23 February 2016 at the Florida Keys. From Kurdzo et al. (2018). © American Meteorological Society. Used with permission

9.8.3.2 Smoke Plumes

Ash and other small debris can be ejected into the atmosphere by fires and volcanic eruptions. A vertical plume of ash is usually clearly visible at its point of origin, i.e., location of a fire. The ash disperses by slowly drifting with the wind while falling. The associated debris (such as incompletely burned tree leaves or pieces of grass) may have significant size and produce returns with reflectivity of up to 40 dBZ (Fabry 2015). Polarimetric radar observations of smoke plumes in Oklahoma are described in Melnikov et al. (2008). Figure 9.44 displays the fields of Z, Z_{DR}, Φ_{DP}, and ρ_{hv} that represent relatively uniform radar echo from insects within the distances to 50 km from the radar and three very distinct bands of ash echoes originated at the locations of three forest fires south of the radar and aligned with the SE wind direction. Reflectivity of ash in the proximity of fires is generally higher (by 8–15 dB) than the reflectivity of biota, but the ρ_{hv} is definitely lower (mostly <0.4 as opposed to 0.5–0.9 in biota). Differential reflectivity of ash is broadly distributed between -2 and 5 dB and heavily overlaps with typical $Z_{DR}$of weather echoes; hence, ρ_{hv} is a primary polarimetric variable for identification of smoke plumes.

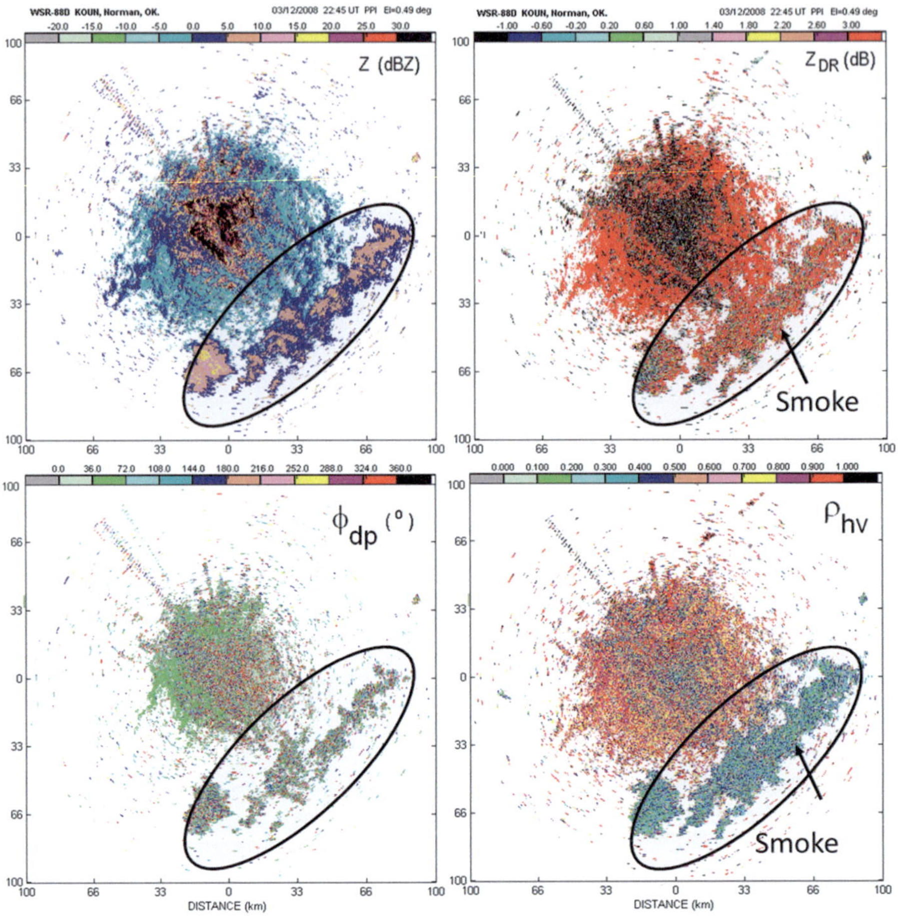

Fig. 9.44 Fields of Z, Z_{DR}, Φ_{DP}, and ρ_{hv} at antenna elevation 0.5° measured by the KOUN WSR-88D during wildfire in central Oklahoma at 2245 UTC 12 March 2008. From Melnikov et al. (2008). Copyright 2008; with permission from Elsevier

9.8.3.3 Dust Storms

Wind-lofted dust and sand may produce returns with reflectivities up to 25 dBZ (Van den Broeke and Alsarraf 2016). The value of ρ_{hv} of dust is much lower than of meteorological scatterers (usually well below 0.8). Differential reflectivity varies in a wide range including large positive and negative values. Very interesting observations of a powerful sandstorm ("haboob") that hit the Phoenix Sky Harbor International Airport in Arizona at 0704 MDT 4 July 2014 are reported by Zhang et al. (2015). The Z and Z_{DR} measurements with the KIWA (Phoenix AZ) WSR-88D are presented in Fig. 9.45. This haboob was closely associated with nearby thunderstorms which produced strong cold outflow spawning updrafts along the gust front.

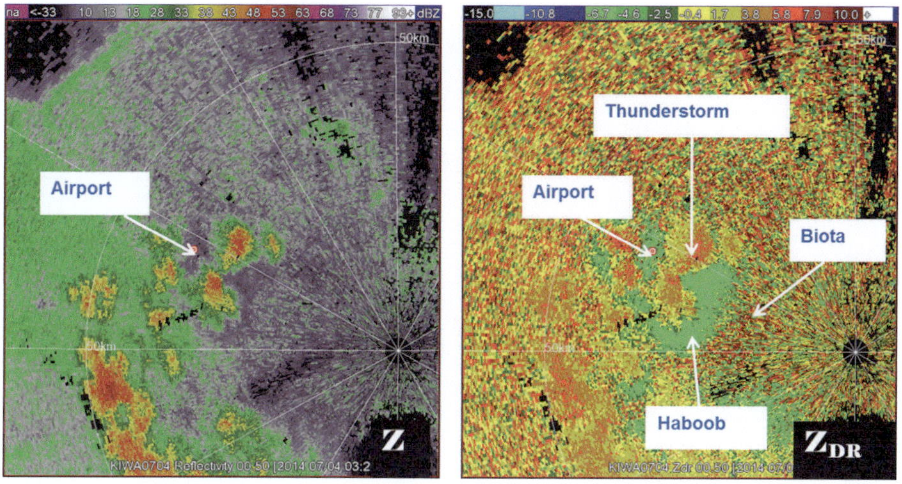

Fig. 9.45 Composite PPI of Z and Z_{DR} measured by the KIWA WSR-88D near Phoenix Airport during the haboob event at 0321 UTC 4 July 2014. Figure courtesy of Dr. P. Zhang

Updrafts lofted dust, sand, and other debris from the ground up to 3 km above the surface.

The most stunning signature is observed in the field of differential reflectivity which exhibits large negative values (down to −8 dB) in the haboob area (Fig. 9.45). Z_{DR} is within the range between 2 and 3 dB in the thunderstorm core and is highly positive (up to 5–7 dB) in the biota farther away from the storm. A possible reason for large negative Z_{DR} within the core of the dust storm might be a vertical orientation of scatterers caused by electrification as documented by Williams et al. (2009). In that study, an upward directed electrostatic field is reported within the haboobs in the Sahel region of Africa. Most of lofted dust and sand is composed of clay particles with sizes between 1 and 100 μm and other silicate minerals in the size range of 0.1 mm to several mm. However, all these scatterers are mainly isotropic and cannot produce such magnitudes of Z_{DR} regardless of their orientation. Much larger anisotropic scatterers like pieces of grass or insects caught in the gust front updraft and mixed with much smaller dust and sand particles are probably responsible for anomalous Z_{DR}. That is, dust and sand act as charge separators creating vertically oriented electric field which aligns much bigger anisotropic particles, and these produce large negative Z_{DR}. Note that farther away from the thunderstorm and its gust front and the corresponding updraft, Z_{DR} resumes its commonly high positive values associated with biota.

Zhang et al. (2015) also observed scattered negative Z_{DR} along a leading edge of a squall line outflow boundary in Oklahoma and Van den Broeke and Alsarraf (2016) confirmed the presence of small patches of $Z_{DR} < -3$ dB in their studies of dust storms in the US Southern Plains and in the Arabian Peninsula. These patches are commonly observed close to the ground (mostly below 0.5 km) and typically do not extend through a large portion of the dust storm. The same authors claim that the prevalent Z_{DR} in dust and sandstorms is positive but still 2–4 dB lower than in biota.

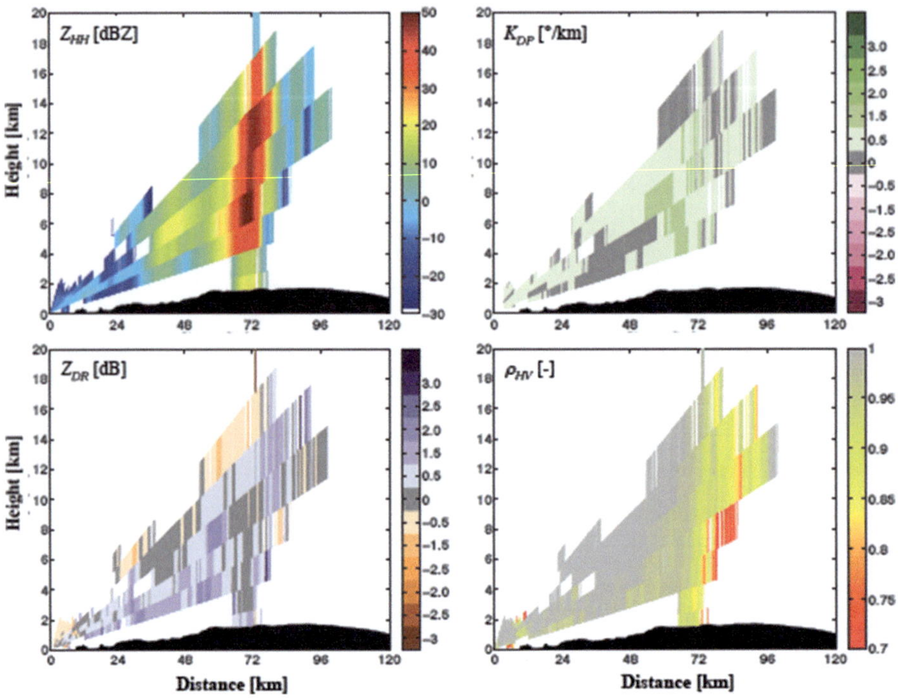

Fig. 9.46 Composite RHI of Z, Z_{DR}, K_{DP}, and ρ_{hv} of the volcanic ash released during the Grimsvotn eruption at 0712 UTC 22 May 2011. The measurements were made by an X-band dual-polarization radar. From Montopoli et al. (2013)

9.8.3.4 Volcanic Ash

Radar observations of volcanic ash are infrequent and inconclusive. Herein, we illustrate an example of polarimetric measurements of volcanic ash released by the Grimsvotn volcano in Iceland on 22 May 2011. It was observed by a X-band dual-polarization radar (Hannessen and Weipert 2011a, b; Montopoli et al. 2013). A composite RHI plot of Z, Z_{DR}, K_{DP}, and ρ_{hv} of the volcanic ash is in Fig. 9.46. A tower of enhanced Z just above the volcano vent is clearly seen stretching up to the height of 20 km. The lower part of the tower is marked by Z_{DR} close to 0 dB. The cross-correlation coefficient ρ_{hv} is reduced within the column of high Z and tends to decrease in the rear side of the column most likely due to propagation effects.

The corresponding composite PPI at El $= 4.8°$ is presented in Fig. 9.47. This PPI scan intercepts the core of volcanic eruption at the height between 5 and 6 km. Notable is a sharp reduction of ρ_{hv} down to 0.8–0.9 in the middle of the core and its rear side. The specific differential phase K_{DP} is high, up to $1.5°\ \text{km}^{-1}$ in the area surrounding the volcano. Low Z_{DR} coincident with high K_{DP} indicates that principal contributors to Z_{DR} are large ash particles with irregular shape which may exhibit

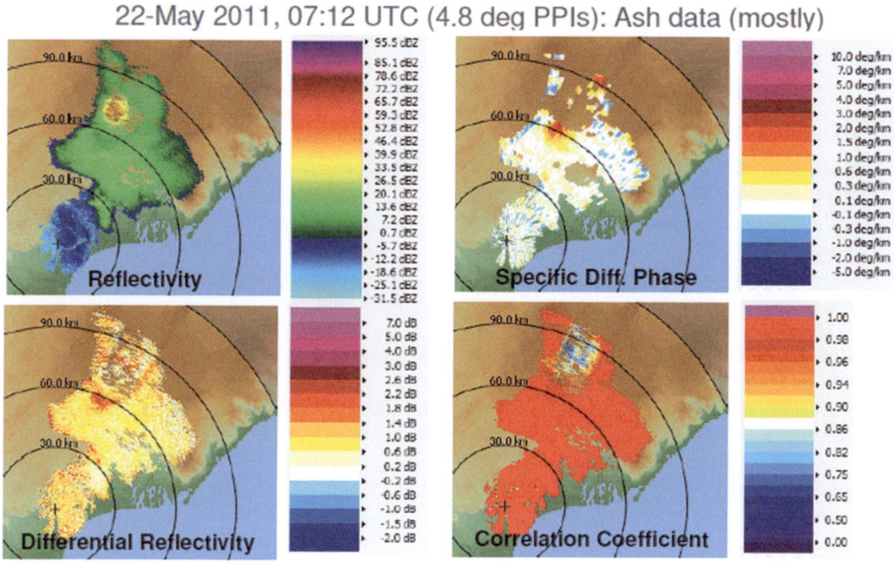

Fig. 9.47 Composite PPI of Z, Z_{DR}, K_{DP}, and ρ_{hv} of the volcanic ash released during the Grimsvotn eruption at 0712 UTC 22 May 2011. The measurements were made at elevation 4.8° by an X-band dual-polarization radar. The volcano is located at the distance ~70 km NE of the radar. From Hannessen and Weipert (2011b)

tumbling behavior, whereas high K_{DP} is caused by smaller but more nonspherical particles in high concentration. Smaller-size ash particles may act as ice nuclei for nonspherical ice crystals. Ash generally nucleates ice early after emplacement in the high troposphere because magmatic water and entrained lower tropospheric water are carried upward so that they condense and freeze. In some ash clouds, this ice rapidly sublimates and disappears in a few hours. But in early stages of ash cloud formation when coarse ash is present, there is likely to be a lot of ice. Once largest-size ash and ice particles sediment, reflectivity decreases, and differential reflectivity increases likely because now smaller-size, less spherical particles (possibly covered with ice and/or indistinguishable from ice crystals) dominate the radar signal.

References

Achtemeier, G. (1991). The use of insects as tracers for "clear air" boundary-layer studies by Doppler radar. *Journal of Atmospheric and Oceanic Technology, 8*, 746–765.

Al-Sakka, H., Boumahmoud, A., Fradon, B., Frasier, S., & Tabary, P. (2013). A new fuzzy logic hydrometeor classification scheme applied to the French X-, C-, and S-band polarimetric radars. *Journal of Applied Meteorology and Climatology, 52*, 2328–2344.

Atlas, D., & Ludlam, F. (1961). Multi-wavelength radar reflectivity of hailstorms. *Quarterly Journal of the Royal Meteorological Society, 87*, 523–534.

Aydin, K., Seliga, T., & Balaji, V. (1986). Remote sensing of hail with a dual linear polarization radar. *Journal of Applied Meteorology, 25*, 1475–1484.

Balakrishnan, N., & Zrnic, D. (1990). Use of polarization to characterize precipitation and discriminate large hail. *Journal of the Atmospheric Sciences, 47*, 1525–1540.

Bechini, R., & Chandrasekar, V. (2015). A semisupervised robust hydrometeor classification method for dual-polarization radar applications. *Journal of Atmospheric and Oceanic Technology, 32*, 22–47.

Besic, N., Figueras i Ventura, J., Grazioli, J., Gabella, M., Germann, U., & Berne, A. (2016). Hydrometeor classification through statistical clustering of polarimetric radar measurements: A semi-supervised approach. *Atmospheric Measurement Techniques, 9*, 4425–4445.

Boodoo, S., Hudak, D., Donaldson, N., & Leduc, M. (2010). Application of dual-polarization radar melting-layer detection algorithm. *Journal of Applied Meteorology and Climatology, 49*, 1779–1793.

Brandes, E., & Ikeda, K. (2004). Freezing-level estimation with polarimetric radar. *Journal of Applied Meteorology, 43*, 1541–1553.

Bringi, V., Vivekanandan, J., & Tuttle, J. (1986). Multiparameter radar measurements in Colorado convective storms. Part II: Hail detection studies. *Journal of the Atmospheric Sciences, 43*, 2564–2577.

Brown, J. M., Smirnova, T. G., Olson, J. B., Grell, G. A., Dowell, D. C., Benjamin, S., et al. (2011). *Improvement and testing of WRF physics options for application to rapid refresh and high resolution rapid refresh*. In Preprints, 14th Conf. on Mesoscale Processes/15th Conf. on Aviation, Range, and Aerospace Meteorology, Los Angeles, CA, Amer. Meteor. Soc. 5.5. Retrieved from https://ams.confex.com/ams/14Meso15ARAM/webprogram/Paper191234.html

Cao, Q., Zhang, G., Palmer, R., Knight, M., May, R., & Stafford, R. (2012). Spectrum – Time estimation and processing (STEP) for improving weather radar data quality. *IEEE Transactions on Geoscience and Remote Sensing, 50*, 4670–4683.

Chandrasekar, V., Keranen, R., Lim, S., & Moisseev, D. (2013). Recent advances in classification of observations from dual-polarization weather radars. *Atmospheric Research, 119*, 97–111.

Chilson, P., Frick, W., Kelly, J., Howard, K., Larkin, R., Diehl, R., Westbrook, J., Kelly, T., & Kunz, T. (2012). Partly cloudy with a chance of migration: Weather, radars, and aeroecology. *Bulletin of the American Meteorological Society, 93*, 669–686.

Depue, T., Kennedy, P., & Rutledge, S. (2007). Performance of the hail differential reflectivity (H_{DR}) polarimetric radar hail indicator. *Journal of Applied Meteorology and Climatology, 46*, 1290–1301.

Dolan, B., & Rutledge, S. (2009). A theory-based hydrometeor identification algorithm for X-band polarimetric radars. *Journal of Atmospheric and Oceanic Technology, 26*, 20171–22088.

Dolan, B., Rutledge, S., Lim, S., Chandrasekar, V., & Thurai, M. (2013). A robust C-band hydrometeor identification algorithm and application to a long-term polarimetric radar dataset. *Journal of Applied Meteorology and Climatology, 52*, 2162–2186.

Eccles, P., & Atlas, D. (1973). A dual-wavelength radar hail detector. *Journal of Applied Meteorology, 12*, 847–854.

Elmore, K., Flamig, Z., Lakshmanan, V., Kaney, B., Reeves, H., Farmer, V., & Rothfusz, L. (2014). mPING: Crowd-sourcing weather reports for research. *Bulletin of the American Meteorological Society, 95*, 1335–1342.

Fabry, F. (2015). *Radar meteorology. Principles and practice* (p. 256). Cambridge, UK: Cambridge University Press.

Feral, L., Sauvageot, H., & Soula, S. (2003). Hail detection using S- and C-band radar reflectivity difference. *Journal of Atmospheric and Oceanic Technology, 20*, 233–248.

Giangrande, S. E., Krause, J. M., & Ryzhkov, A. V. (2008). Automatic designation of the melting layer with a polarimetric prototype of the WSR-88D radar. *Journal of Applied Meteorology and Climatology, 47*, 1354–1364.

Grazioli, J., Turia, D., & Berne, A. (2015). Hydrometeor classification from polarimetric radar measurements: A clustering approach. *Atmospheric Measurement Techniques, 8*, 149–170.

Greene, D., & Clark, R. (1972). Vertically integrated liquid water – A new analysis tool. *Monthly Weather Review, 100*, 548–552.

Hannessen, R., & Weipert, A. (2011a). A probability-based sea clutter suppression method for polarimetric weather radar systems. In *Weather radar and hydrology, proceedings of a symposium held in Exeter, UK, April 2011* (pp. 52–57).

Hannessen, R., & Weipert, A. (2011b). *An algorithm to detect and quantify volcanic eruptions using polarimetric X-band radar data*. In International workshop on X-band Weather Radar, Delft, The Netherlands.

Heinselman, P., & Ryzhkov, A. (2006). Validation of polarimetric hail detection. *Weather and Forecasting, 21*, 839–850.

Holler, H., Brini, V., Hubbert, J., Hagen, M., & Meischner, P. (1994). Life cycle and precipitation formation in hybrid-type hailstorm revealed by polarimetric and Doppler radar measurements. *Journal of the Atmospheric Sciences, 51*, 2500–2522.

Hubbert, J., Dixon, M., & Ellis, S. (2009b). Weather radar ground clutter. Part II: Real-time identification and filtering. *Journal of Atmospheric and Oceanic Technology, 26*, 1181–1197.

Hubbert, J., Dixon, M., Ellis, S., & Meymaris, G. (2009a). Weather radar ground clutter. Part I: Identification, modeling, and simulation. *Journal of Atmospheric and Oceanic Technology, 26*, 1165–1180.

Ilotoviz, E., Benmoshe, N., Khain, A., Phillips, V., & Ryzhkov, A. (2016). Effect of aerosols on freezing drops, hail, and precipitation in a mid-latitude storm. *Journal of the Atmospheric Sciences, 73*, 109–144.

Keenan, T. (2003). Hydrometeor classification with a C-band polarimetric radar. *Australian Meteorological Magazine, 52*, 23–31.

Kennedy, P., Rutledge, S., Petersen, W., & Bringi, V. (2001). Polarimetric radar observations of hail formation. *Journal of Applied Meteorology, 40*, 1347–1366.

Keranen, R., Saltikoff, E., Chandrasekar, V., Lim, S., Holmes, J., & Selzler, J. (2007). *Real-time hydrometeor classification for the operational forecasting environment*. In Preprints, 33rd Conf. on Radar Meteorology, Cairns, Australia, Amer. Meteor. Soc. P11B.11. Retrieved from https://ams.confex.com/ams/pdfpapers/123476.pdf

Kumjian, M., Khain, A., Benmoshe, N., Ilotoviz, E., Ryzhkov, A., & Phillips, V. (2014). The anatomy and physics of Z_{DR} columns: Investigating a polarimetric radar signature with a spectral bin microphysical model. *Journal of Applied Meteorology and Climatology, 53*, 1820–1843.

Kurdzo, J., Williams, E., Smalley, D., Bennett, B., Patterson, D., Veillette, M., & Donovan, M. (2018). Polarimetric observations of chaff using the WSR-88D network. *Journal of Applied Meteorology and Climatology, 57*, 1063–1081.

Leitao, M., & Watson, P. (1984). Application of dual linearly polarized radar data to prediction of microwave path attenuation at 10 – 30 GHz. *Radio Science, 19*, 209–221.

Li, Y., Zhang, G., Doviak, R., Lei, L., & Cao, Q. (2013). A new approach to detect ground clutter mixed with weather signals. *IEEE Transactions on Geoscience and Remote Sensing, 51*, 2373–2387.

Lim, S., Chandrasekar, V., & Bringi, V. (2005). Hydrometeor classification system using dual-polarization radar measurements: Model improvements and in-situ verification. *IEEE Transactions on Geoscience and Remote Sensing, 42*, 792–801.

Lim, S., Moisseev, D., Chandrasekar, V., & Lee, D.-R. (2013). Classification and quantification of rimed snow based on spatial variability of radar reflectivity. *Journal of the Meteorological Society of Japan, 91*, 763–774.

Liu, H., & Chandrasekar, V. (2000). Classification of hydrometeors based on polarimetric radar measurements: Development of fuzzy logic and neuro-fuzzy systems, and in situ verification. *Journal of Atmospheric and Oceanic Technology, 17*, 140–167.

Marzano, F., Scaranari, D., & Vulpiani, G. (2007). Supervised fuzzy-logic classification of hydrometeors using C-band weather radars. *IEEE Transactions on Geoscience and Remote Sensing, 45*, 3784–3799.

Matrosov, S., Clark, K., & Kingsmill, D. (2007). A polarimetric radar approach to identify rain, melting layer, and snow regions for applying corrections to vertical profiles of reflectivity. *Journal of Applied Meteorology and Climatology, 46*, 154–166.

Melnikov, V., Istok, M., & Westbrook, J. (2015). Asymmetric radar echo patterns from insects. *Journal of Atmospheric and Oceanic Technology, 32*, 659–674.

Melnikov, V., Zrnic, D., Rabin, R., & Zhang, P. (2008). Radar polarimetric signatures of fire plums in Oklahoma. *Geophysical Research Letters, 35*, L14815.

Meyers, M., DeMott, P., & Cotton, W. (1992). New primary ice-nucleation parameterization in an explicit cloud model. *Journal of Applied Meteorology, 31*, 708–721.

Montopoli, M., Vulpiani, G., Cimini, D., Piccotti, E., & Marzano, F. (2013). Interpretation of observed microwave signatures from ground dual-polarization radar and space multi-frequency radiometer for the 2011 Grimsvotn volcanic eruption. *Atmospheric Measurement Techniques Discussions, 6*, 6215–6248.

Mueller, E., & Larkin, R. (1985). Insects observed using dual-polarization radar. *Journal of Atmospheric and Oceanic Technology, 2*, 49–54.

Ortega, K., Krause, J., & Ryzhkov, A. (2016). Polarimetric radar characteristics of melting hail. Part III: Validation of the algorithm for hail size discrimination. *Journal of Applied Meteorology and Climatology, 55*, 829–848.

Ortega, K. L., Smith, T. M., Manross, K. L., Kolodziej, A. G., Scharfenberg, K. A., Witt, A., & Gourley, J. (2009). The severe hazards analysis and verification experiment. *Bulletin of the American Meteorological Society, 90*, 1519–1530.

Park, H.-S., Ryzhkov, A., Zrnic, D., & Kim K.-E. (2007). *Optimization of the matrix of weights in the polarimetric algorithm for classification of radar echoes*. In Preprints, 33rd conf. on radar meteorology, Cairns, Australia, Amer. Meteor. Soc. P11B.4. Retrieved from http://ams.confex.com/ams/pdfpapers/123123.pdf

Park, H.-S., Ryzhkov, A., Zrnic, D., & Kim, K.-E. (2009). The hydrometeor classification algorithm for the polarimetric WSR-88D. Description and application to an MCS. *Weather and Forecasting, 24*, 730–748.

Picca, J., Kumjian, M., & Ryzhkov, A. (2010). *ZDR columns as a predictive tool for hail growth and storm evolution*. In Preprints, 25th conf. on severe local storms, Denver, CO, Amer. Meteor. Soc. 113. Retrieved from https://ams.confex.com/ams/pdfpapers/175750.pdf

Picca, J., & Ryzhkov, A. (2012). A dual-wavelength polarimetric analysis of the 16 May 2010 Oklahoma City extreme hailstorm. *Monthly Weather Review, 140*, 1385–1403.

Reeves, H., Elmore, K., Ryzhkov, A., Schuur, T., & Krause, J. (2014). Source of uncertainty in precipitation-type forecasting. *Weather and Forecasting, 29*, 936–953.

Reeves, H., Ryzhkov, A., & Krause, J. (2016). Discrimination between winter precipitation types based on spectral-bin microphysical modeling. *Journal of Applied Meteorology and Climatology, 55*, 1747–1761.

Rico-Ramirez, A., & Cluckie, I. (2008). Classification of ground clutter and anomalous propagation using dual-polarization weather radar. *IEEE Transactions on Geoscience and Remote Sensing, 46*, 1892–1904.

Ryzhkov, A., Kumjian, M., Ganson, S., & Khain, A. (2013a). Polarimetric radar characteristics of melting hail. Pt I: Theoretical simulations using spectral microphysical modeling. *Journal of Applied Meteorology and Climatology, 52*, 2849–2870.

Ryzhkov, A., Kumjian, M., Ganson, S., & Zhang, P. (2013b). Polarimetric radar characteristics of melting hail. Pt II: Practical implications. *Journal of Applied Meteorology and Climatology, 52*, 2871–2886.

Ryzhkov, A., Reeves, H., Krause, J., & Burcham, H. (2014). *Discrimination between winter precipitation types based on explicit microphysical modeling of melting and refreezing in the polarimetric hydrometeor classification algorithm*. In Eighth European conf. on radar meteorology, Garmisch-Partenkirchen, Germany. Europeran Meteorological Society. MIC.P07. Retrieved from http://www.pa.op.dlr.de/erad2014/programme/ExtendedAbstracts/198_Ryzhkov.pdf

Ryzhkov, A., Zhang, P., Doviak, R., & Kessinger, C. (2002). *Discrimination between weather and sea clutter using Doppler and dual-polarization weather radars*. In Proc. 27th general assembly of the international union of radio science, Maastricht, Netherlands. F1.P.3. Retrieved from https://www.ursi.org/Proceedings/ProcGA02/papers/p1383.pdf

Ryzhkov, A. V. (2007). The impact of beam broadening on the quality of radar polarimetric data. *Journal of Atmospheric and Oceanic Technology, 24*, 729–744.

Ryzhkov, A. V., Schuur, T. J., Burgess, D. W., Giangrande, S., & Zrnic, D. S. (2005). The joint polarization experiment: Polarimetric rainfall measurements and hydrometeor classification. *Bulletin of American Meteorological Society, 86*, 809–824.

Schuur, T., Krause, J., & Ryzhkov A. (2015). *A new melting layer detection algorithm that combines polarimetric radar-based detection with thermodynamic output from numerical models*. In 37th Conference on Radar Meteorology. 7B.1.

Schuur, T., Park, H.-S., Ryzhkov, A., & Reeves, H. (2012). Classification of precipitation types during transitional winter weather using the RUC model and polarimetric radar retrievals. *Journal of Applied Meteorology and Climatology, 51*, 763–779.

Schuur, T., Ryzhkov, A., & Heinselman, P. (2003). *Observations and classification of echoes with the polarimetric WSR-88D radar*. NOAA/National Severe Storms Laboratory report (46pp).

Siggia, A., & Passarelli, R., Jr. (2004). Gaussian model adaptive processing (GMAP) for improved ground clutter cancellation and moment calculation. In *Proceedings of ERAD (2004), Visby, Island of Gotland, Sweden* (pp. 63–73).

Snyder, J., Bluestein, H., Zhang, G., & Frasier, S. (2010). Attenuation correction and hydrometeor classification of high resolution, X-band, dual-polarized mobile radar measurements in severe convective storms. *Journal of Atmospheric and Oceanic Technology, 27*, 1979–2001.

Snyder, J., & Ryzhkov, A. (2015). Automated detection of polarimetric tornado debris signatures. *Journal of Applied Meteorology and Climatology, 54*, 1861–1870.

Snyder, J., Ryzhkov, A., Kumjian, M., Picca, J., & Khain, A. (2015). Developing a Z_{DR} column detection algorithm to examine convective storm updrafts. *Weather and Forecasting, 30*, 1819–1844.

Stepanian, P., Horton, K., Melnikov, V., Zrnic, D., & Gauthreaux, S., Jr. (2016). Dual-polarization radar products for biological applications. *Ecosphere, 7*(11), e01539. https://doi.org/10.1002/ecs2.1539.

Straka, J. (1996). Hydrometeor fields in a supercell storm as deduced from dual-polarization radar. In *Preprints, 18th Conf. on Severe Local Storms, San Francisco, CA, Amer. Meteor. Soc.* (pp. 551–554).

Straka, J., & Zrnic, D. (1993). An algorithm to deduce hydrometeor types and contents from multiparameter radar data. In *Preprints, 26th Int. conf. on radar meteorology, Norman, OK, Amer. Meteor. Soc.* (pp. 513–516).

Straka, J. M., Zrnic, D. S., & Ryzhkov, A. V. (2000). Bulk hydrometeor classification and quantification using multiparameter radar data. Synthesis of relations. *Journal of Applied Meteorology, 39*, 1341–1372.

Tabary, P., Le Henaff, A., Vulpiani, G., Parent-du-Chatelet, J., & Gourley, J. J. (2006). Melting layer characterization and identification with a C-band dual-polarization radar: A long-term analysis. In *Preprints, Fourth European conf. on radar in meteorology and hydrology (ERAD 2006), Barcelona, Spain* (pp. 17–20).

Thompson, E., Rutledge, S., Dolan, B., Chandrasekar, V., & Cheong, B.-L. (2014). A dual-polarization radar hydrometeor classification algorithm for winter precipitation. *Journal of Atmospheric and Oceanic Technology, 31*, 1457–1481.

Torres, S., & Warde, D. (2014). Ground clutter mitigation for weather radars using the autocorrelation spectral density. *Journal of Atmospheric and Oceanic Technology, 31*, 2049–2066.

Van den Broeke, M. (2013). Polarimetric radar observations of biological scatterers in hurricanes Irene (2011) and Sandy (2012). *Journal of Atmospheric and Oceanic Technology, 30*, 2754–2767.

Van den Broeke, M., & Alsarraf, H. (2016). Polarimetric radar observations of dust storms at C and S bands. *Journal Of Operational Meteorology, 4*(9), 123–131.

Van den Broeke, M., & Jauernic, S. (2014). Spatial and temporal characteristics of polarimetric tornadic debris signatures. *Journal of Applied Meteorology and Climatology, 53*, 2217–2231.

Vivekanandan, J., Zrnic, D. S., Ellis, S. M., Oye, D., Ryzhkov, A. V., & Straka, J. (1999). Cloud microphysics retrieval using S-band dual-polarization radar measurements. *Bulletin of the American Meteorological Society, 80*, 381–388.

Waldvogel, A., Federer, B., & Grimm, P. (1979). Criteria for the detection of hail cells. *Journal of Applied Meteorology, 18*, 1521–1525.

Wen, G., Protat, A., May, P., Moran, W., & Dixon, M. (2016). Cluster-based method for hydrometeor classification using polarimetric variables. Part II: Classification. *Journal of Atmospheric and Oceanic Technology, 33*, 45–59.

Wen, G., Protat, A., May, P., Wang, X., & Moran, W. (2015). Cluster-based method for hydrometeor classification using polarimetric variables. Part I: Interpretation and analysis. *Journal of Atmospheric and Oceanic Technology, 32*, 1320–1330.

Williams, E., Nathou, E., Hicks, E., Pontikis, C., Russell, B., Miller, M., & Bartholomew, M. (2009). The electrification of dust-lofting gust fronts ('haboobs') in the Sahel. *Atmospheric Research, 91*, 292–298.

Witt, A., Eilts, M., Stumpf, G., Johnson, J., Mitchell, E., & Thomas, K. (1998). An enhanced hail detection algorithm for the WSR-88D. *Weather and Forecasting, 13*, 286–303.

Wolfensberger, D., Scipion, D., & Berne, A. (2016). Detection and characterization of the melting layer based on polarimetric radar scans. *Quarterly Journal of the Royal Meteorological Society, 142*, 108–124.

Zeng, Z., Yuter, S., House, R., & Kingsmill, D. (2001). Microphysics of the rapid development of heavy convective precipitation. *Monthly Weather Review, 129*, 1882–1904.

Zhang, G. (2016). *Weather radar Polarimetry* (p. 304). Boca Raton, FL: CRC Press.

Zhang, P., Zrnic, D., & Ryzhkov, A. (2015) Observations of negative Z_{DR} in sandstorms and storm outflows. In *Preprints, 37th conference on Radar Meteorology, Norman, OK, Amer. Meteor. Soc.* (p. 259). Retrieved from https://ams.confex.com/ams/37RADAR/webprogram/Handout/Paper275678/37thPoster_PZhang.pdf

Zrnic, D., & Ryzhkov, A. (1998). Observations of insects and birds with a polarimetric radar. *IEEE Transactions on Geoscience and Remote Sensing, 36*, 661–668.

Zrnic, D., & Ryzhkov, A. (1999). Polarimetry for weather surveillance radars. *Bulletin of the American Meteorological Society, 80*, 389–406.

Zrnic, D. S., Bringi, V. N., Balakrishnan, N., Aydin, K., Chandrasekar, V., & Hubbert, J. (1993). Polarimetric measurements in a severe hailstorm. *Monthly Weather Review, 121*, 2221–2238.

Zrnic, D. S., & Ryzhkov, A. V. (2004). Polarimetric properties of chaff. *Journal of Atmospheric and Oceanic Technology, 21*, 1017–1024.

Zrnic, D. S., Ryzhkov, A. V., Straka, J., Liu, Y., & Vivekanandan, J. (2001). Testing a procedure for the automatic classification of hydrometeor types. *Journal of Atmospheric and Oceanic Technology, 18*, 892–913.

Chapter 10
Polarimetric Measurements of Precipitation

Improvement of quantitative precipitation estimation (QPE) is one of the primary benefits of a dual-polarization radar. For decades, rainfall rate R was estimated from the reflectivity factor Z via a multitude of R (Z) relations prone to the variability of drop-size distributions (DSD) and other artifacts. The sources of errors in radar rainfall estimation are listed in Table 10.1. The errors caused by at least half of the listed factors can be reduced significantly using polarimetric radar measurements (right column in Table 10.1).

10.1 Precipitation Rate

Rainfall rate R is the depth of water accumulated at the surface during a specified time period and is usually expressed in millimeters per hour (mm h^{-1}). Rain rate depends on the raindrop size distribution N (D) in a unit volume of air and the terminal velocity of raindrops V_t (D), and both depend on the raindrop's equivolume diameter D. Imagine a bucket with an area A; the rain rate is the depth ΔH of water accumulated at the bottom of the bucket over a time interval T (Fig. 10.1). The total number of raindrops N_t (D) with diameters between D and $D + \Delta D$ reaching the bottom of the bucket during the time interval T is

$$N_t(D) = N(D)\Delta D V_t(D) T A \qquad (10.1)$$

In (10.1), $N(D)$ is in m^{-3}mm^{-1}, D and ΔD are in mm, V_t is in m s^{-1}, T is in s, and A is in m^2. The volume Vol (D) of these drops is

$$Vol(D) = \frac{\pi}{6} D^3 N_t(D), \qquad (10.2)$$

A. V. Ryzhkov, D. S. Zrnic, *Radar Polarimetry for Weather Observations*,
Springer Atmospheric Sciences, https://doi.org/10.1007/978-3-030-05093-1_10

Table 10.1 Sources of errors in radar rainfall estimation

Factors	Addressed using radar polarimetry?
1. Variability of raindrop size distributions	Yes
2. Contamination from mixed-phase and frozen hydrometeors	Yes
3. Radar miscalibration	Yes
4. Attenuation in precipitation	Yes
5. Partial beam blockage	Yes
6. Nonuniform beam filling	
7. Sampling errors (spatial and temporal)	
8. Overshooting precipitation	
9. Evaporation	Yes
10. Vertical air motions	

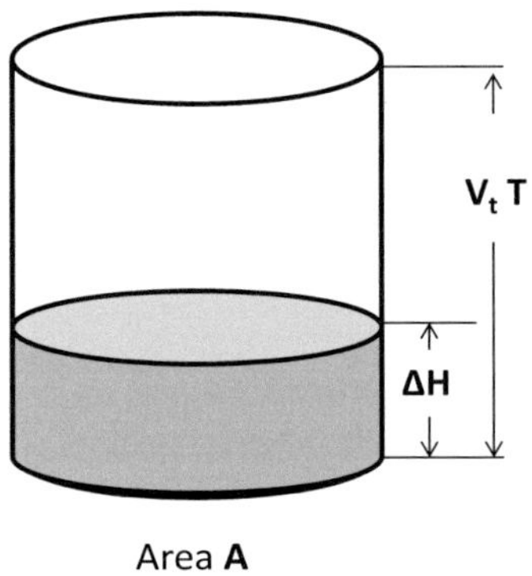

Fig. 10.1 Conceptual plot illustrating definition of rain rate

where the unit of $Vol(D)$ is mm^3. The total volume of all raindrops collected in a bucket during an hour ($T = 3600$ s) is

$$VOL = 600\pi A \int D^3 V_t(D) N(D) dD, \tag{10.3}$$

and the rain rate is obtained as a ratio of VOL and A

$$R = \Delta H = VOL/A = 6 \times 10^{-4} \pi \int D^3 V_t(D) N(D) dD. \tag{10.4}$$

In (10.4), the rain rate R is in mm h^{-1}, D in mm, $V_t(D)$ in m s^{-1}, and $N(D)$ in m^{-3}mm^{-1}.

If precipitation is snow, then the snow water equivalent (SWE) rate S is defined as a depth of melted water from the accumulated snow in the bucket. The derivation of the precipitation rate is similar to the one for rain, but the total volume of snow at any given size bin should be converted into equivalent volume of water using the ratio of the snowflake density ρ_s (a function of its size) and the density of water ρ_w. Therefore, (10.4) for rain rate is modified to

$$S = 6 \times 10^{-4}\pi \int \frac{\rho_s(D)}{\rho_w} D^3 V_t(D) N(D) dD. \tag{10.5}$$

The dependence of the terminal velocity of raindrops V_t on the equivolume diameter D can be well approximated with the following relations given by Atlas and Ulbrich (1977):

$$V_t = 3.78D^{0.67}, \tag{10.6}$$

Atlas et al. (1973)

$$V_t = 9.65 - 10.3\exp(-0.6D), \tag{10.7}$$

and Brandes et al. (2002)

$$V_t = \begin{cases} -0.102 + 4.93D - 0.955D^2 + 0.0793D^3 - 0.00236D^4 & D < 6\,\text{mm} \\ 9.17 & D > 6\,\text{mm} \end{cases}. \tag{10.8}$$

In all three formulas, V_t is in m s^{-1} and D is in mm.

There are many relations for terminal velocity of ice and snow particles due to a huge variability of their habits (e.g., Heymsfield et al., 2013; Mitchell and Heymsfield 2005). Herein, we include the one from Brandes et al. (2007) which is valid for aggregated snow:

$$V_t = 0.768D^{0.142}, \tag{10.9}$$

where D is the equivalent volume diameter of snowflake. The size dependencies of the terminal velocity given by (10.6), (10.7), and (10.9) are plotted in Fig. 10.2.

The plots (Fig. 10.2) are valid at the standard atmospheric pressure and air density near the surface. Terminal velocities of hydrometeors are larger at higher altitudes where the air density is lower, and such dependence is quantified as

$$V_t(h) = \left[\frac{\rho_a(0)}{\rho_a(h)}\right]^{1/2} V_t(0). \tag{10.10}$$

The variability of DSD is the major factor affecting the accuracy of radar rainfall measurements and the algorithms based on the use of the $R(Z)$ relations. This is because the reflectivity factor is a 6th moment of the raindrop size distribution (in the Rayleigh approximation):

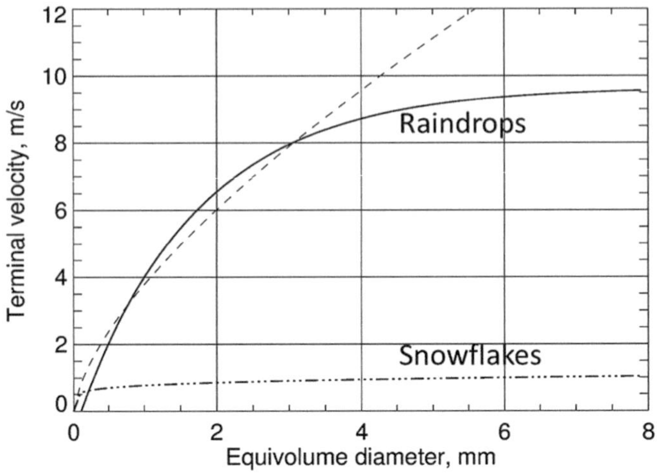

Fig. 10.2 Terminal velocity of raindrops and snowflakes as a function of the equivolume diameter (10.6) (dashed line), (10.7) (solid line), and (10.9) (dotted-dashed line) at the atmospheric pressure of 1013 mb

$$Z = \int D^6 N(D) dD, \tag{10.11}$$

whereas the rain rate is approximately proportional to a 3.67th moment of the size distribution as can be seen after introducing (10.6) into (10.4), so that

$$R \sim \int D^{3.67} N(D) dD. \tag{10.12}$$

Hence, Z is mostly determined by a higher end of the raindrop spectrum, while the largest contribution to R is made by smaller-size raindrops. Indeed, a single 5 mm raindrop produces the same reflectivity as a million of 0.5 mm drops, whereas the corresponding rain rate for smaller drops is 267 times larger assuming that $V_t(5\text{ mm}) = 9.0\text{ m s}^{-1}$ and $V_t(0.5\text{ mm}) = 2.4\text{ m s}^{-1}$. The DSD variability causes incredibly broad variety of the $R(Z)$ relations cited in the literature. Battan (1973) lists 69 $R(Z)$ relations (Fig. 10.3). Figure 10.3 shows that the rain rate can change more than an order of magnitude for a given Z depending on the $R(Z)$ relation chosen.

The -Marshall and Palmer (1948) relation

$$Z = 200R^{1.6} \tag{10.13}$$

is among the most commonly used $R(Z)$ relations. In (10.13), rain rate R is in mm h^{-1} and Z is in linear units (mm^6 m^{-3}). The choice of a particular $R(Z)$ depends on the type of rain (convective vs. stratiform, tropical vs. continental, etc.), season,

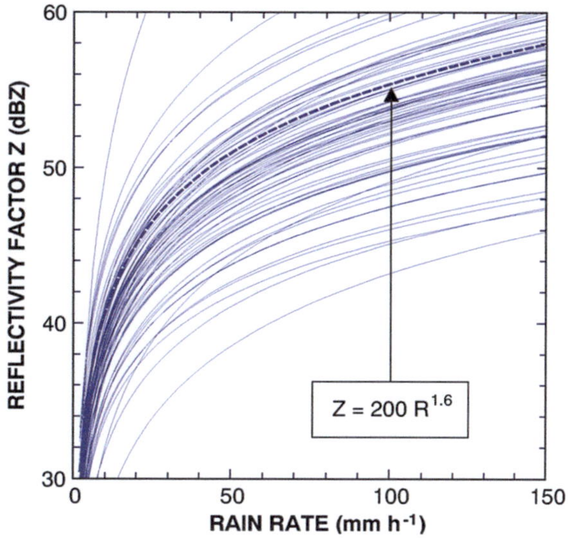

Fig. 10.3 Reflectivity as a function of rain rate according to 69 $R(Z)$ relations. From Battan (1973)

and climate region. For example, the following five $R(Z)$ relations are utilized on the US network of WSR-88Ds:

$$\begin{aligned} &Z = 300R^{1.4} && \text{in convective rain,} \\ &Z = 250R^{1.2} && \text{in tropical,} \\ &Z = 200R^{1.6} && \text{in summer stratiform rain,} \\ &Z = 130R^{2.0} && \text{in winter stratiform rain (Eastern USA),} \\ &Z = 75R^{2.0} && \text{in winter stratiform rain (Western USA).} \end{aligned} \tag{10.14}$$

The choice of a particular $R(Z)$ relation is at the discretion of a radar operator/meteorologist, and there are no objective criteria for choosing an optimal one for a given time and location.

10.2 The Impact of the Raindrop Size Distribution Variability on the Radar Rainfall Estimates

The impact of the raindrop size distribution variability on the radar rainfall estimates can be quantified using disdrometer measurements. In this chapter, we utilize a large dataset of disdrometer observations in Oklahoma which encompass sufficiently wide range of different rain types from tropical to continental (Schuur et al. 2005). The key parameters of DSD (rain rate R, median volume diameter D_0, and normalized concentration N_w—see Chap. 4 for definition of N_w) and the polarimetric variables consisting of reflectivity Z, differential reflectivity Z_{DR}, specific differential phase K_{DP}, and specific attenuation A have been computed for every DSD. The polarimetric variables at S ($\lambda = 11$ cm), C ($\lambda = 5.3$ cm), and X ($\lambda = 3.2$ cm) wavelengths have been simulated using the T-matrix code. In computations of polarimetric variables, the raindrop shape-diameter dependence (4.29) from Brandes et al. (2002) was used, and the width of the canting angle distribution $\sigma = 10^0$ was assumed.

A scatterplot of log(R) versus reflectivity factor at S band is displayed in Fig. 10.4. Consistent with Fig. 10.3, this plot illustrates inherent uncertainty of rain rate estimates from Z. Because the reflectivity is heavily weighted by the largest raindrops, it is natural to expect that the $R(Z)$ relation optimized for a whole disdrometer dataset would tend to underestimate "warm" rain dominated by high concentration of smaller drops and overestimate "cold" rain that mostly originates from ice aloft and is characterized by abundance of large drops and relative deficit of smaller drops (see discussion in Sect. 4.1.2, Chap. 4). To quantify the performance of radar rainfall relations for different rain types, we segregate all measured size distributions into several categories based on Z and Z_{DR}. Discrimination between different rain types and drop-size distributions can be made using the dependence of Z_{DR} on Z in a given spatial/temporal domain in the storm. The reflectivity Z depends on both D_0 and N_w, whereas Z_{DR} is a function of D_0 only. Higher values of Z_{DR} for a given Z indicate a more continental rain type.

Schuur et al. (2005) simulate average Z-Z_{DR} dependencies from the Oklahoma DSD data separately for rain events associated with (1) mesoscale convective systems, (2) isolated storms and showers, (3) large synoptic-scale systems, (4) hailstorms, and (5) tropical systems. These dependencies are illustrated in Fig. 10.5. As expected, for a given Z, the magnitude of Z_{DR} is highest in rain from hail-bearing storms because these are characterized by relatively low total concentrations of raindrops (i.e., low N_w) and abundance of large raindrops originated from melting hail. Tropical rain events have the lowest Z_{DR} for a fixed Z due to generally smaller

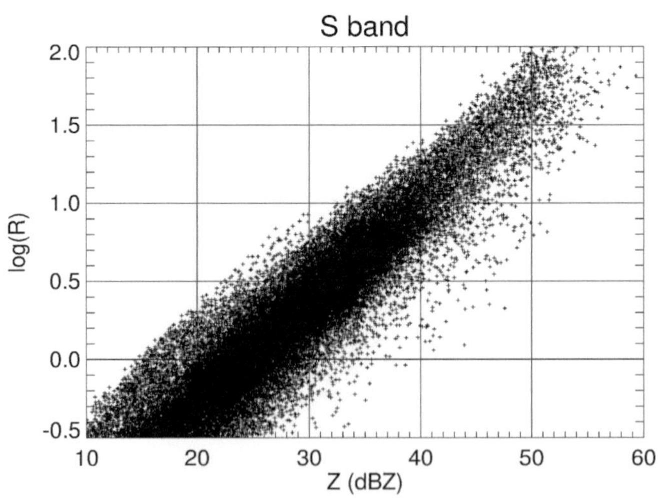

Fig. 10.4 Scatterplot of log(R) vs Z at S band for the Oklahoma disdrometer dataset

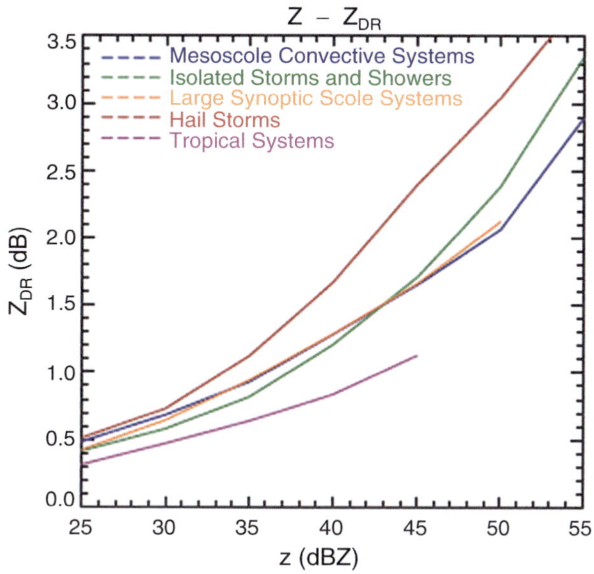

Fig. 10.5 Average Z-Z_{DR} dependencies retrieved from disdrometer data in Oklahoma for different rain types (from Schuur et al. 2005)

drop sizes and very high total concentrations of raindrops. The simulated Z-Z_{DR} dependencies are consistent with the average ones estimated from S-band radar data collected in different climate regions: Oklahoma, Florida, and Taiwan (Fig. 6.5 in Chap. 6). Oklahoma is characterized by the highest Z_{DR}, whereas Taiwan has the lowest Z_{DR} reflecting the difference between continental and tropical precipitation regimes.

The range of Z_{DR} variability is also demonstrated in Fig. 10.6 (identical to Fig. 6.3 from Chap. 6 and repeated herein for convenience) where the curves corresponding to different percentiles of the Z_{DR} distributions for a given Z from the Oklahoma DSD dataset are plotted. Figure 10.6 indicates that, for example, at $Z = 40$ dBZ, Z_{DR} varies from 0.4 to 4.2 dB and about 40% of Z_{DR} values are below 1.0 dB. The curves of constant $\log(N_w)$ in the Z-Z_{DR} plane are in Fig. 10.7. Note that the curves of constant $\log(N_w)$ follow closely the Z_{DR} percentile curves within the range of Z between 30 and 50 dBZ in Fig. 10.6. Therefore, the magnitude of $\log(N_w)$ characterizes rain type, and the knowledge of N_w may be very helpful in optimizing any of the radar relations for rainfall estimation (see discussion in the next section).

The value of $\log(N_w)$ can be roughly estimated from polarimetric radar measurements, either using a combination of Z and Z_{DR} (Illingworth and Thompson 2005; Ryzhkov et al. 2014; Tabary et al. 2011) or the ZPHI method (Testud et al. 2000, see Chap. 6). Herein we present the examples of the fields of $\log(N_w)$ retrieved from Z and Z_{DR} in three storms producing heavy rain. The storms were observed by the US WSR-88Ds. One of the storms produced a flash flood in Oklahoma City on 14 June 2010. Another was associated with Hurricane Irene on 27 August 2011 and caused massive flash floods along the East Coast of the USA. The third storm (20 May 2011) represents an MCS typical of the US Great Plains. The fields of

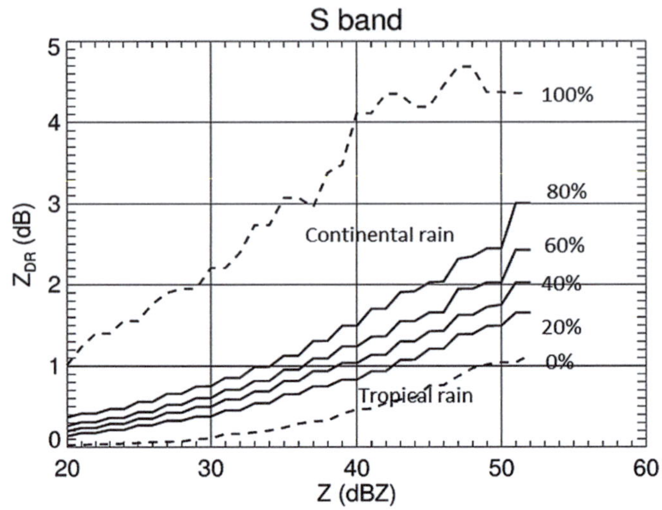

Fig. 10.6 Lines of different percentiles of the Z_{DR} distributions for a given Z obtained from the Oklahoma DSD measurements

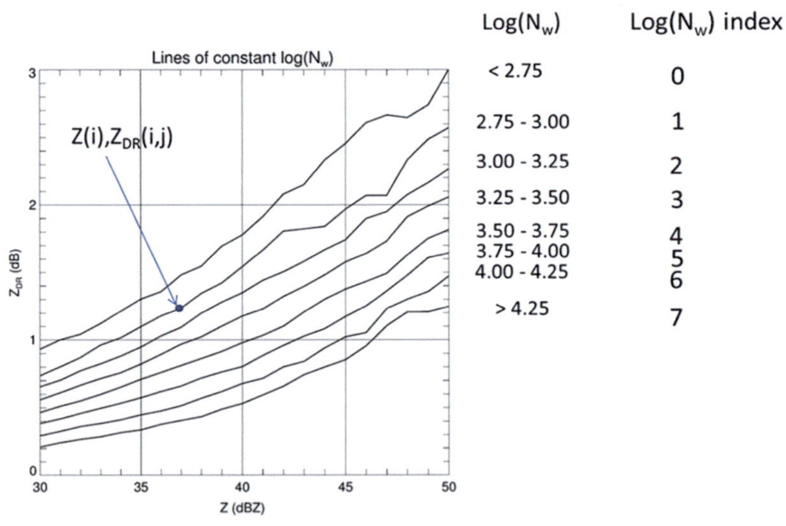

Fig. 10.7 Curves of constant log(N_w) in the Z-Z_{DR} plane

log(N_w) from the three storms are represented in Fig. 10.8 together with the corresponding Z maps.

The lowest values of log(N_w) are in the squall line on 20 May 2011 which represents typical continental rain (middle panels in Fig. 10.8). The N_w is almost two orders of magnitude higher in the Hurricane Irene (left panels). It is interesting that the case of the OKC flash flood (14 June 2010) exhibits features of both

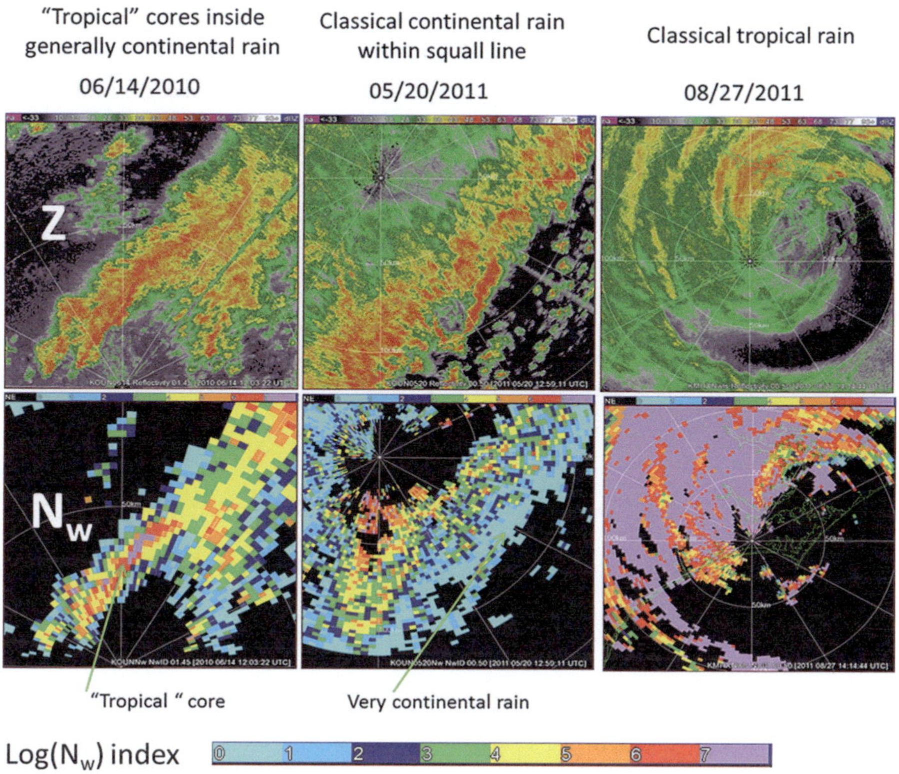

Fig. 10.8 PPIs of Z and $\log(N_w)$ from three storms producing heavy rain and observed with three different WSR-88Ds. The color coding in the lower panels covers a range in $\log(N_w)$ from less than 2.625 (index 0) to larger than 4.375 (index 7) (see Fig. 10.7). Low values of $\log(N_w)$ represent "continental rain" characterized by high concentrations of large raindrops, whereas high values of $\log(N_w)$ correspond to "tropical rain" with high concentrations of smaller raindrops. N_w is expressed in $\mathrm{m}^{-3}\mathrm{mm}^{-1}$. Adapted from Ryzhkov et al. (2014). © American Meteorological Society. Used with permission

continental and tropical rain with the cores of "tropical" rain immersed within generally "continental" rain. In addition to large differences in $\log(N_w)$ between the three storms, Fig. 10.8 also illustrates its significant variability within a particular storm. Robust retrieval of N_w requires well-calibrated measurements of Z and Z_{DR} with the accuracy of 1 dB and 0.1 dB, respectively. This is easier to implement at S band because attenuation-related biases of Z and Z_{DR} are small compared to biases at C and X bands.

We segregate all drop-size distributions into five categories using the Z-Z_{DR} percentile lines (Fig. 10.6) and present the $\log(R)$ vs Z scatterplots separately for the most tropical rain category and Z_{DR} belonging to the 0–20% percentile range (between two lowest curves in Fig. 10.6) and the most continental rain category within the 80–100% percentile range of Z_{DR} (between two highest curves in Fig. 10.6). We designate these two families of DSDs as belonging to "very tropical"

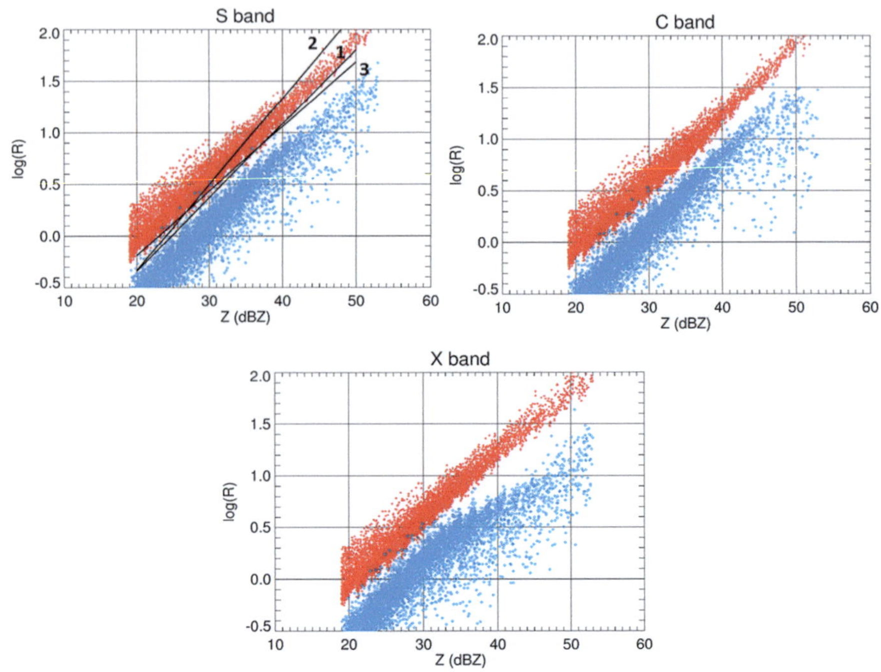

Fig. 10.9 Scatterplots of log(R) vs Z at S, C, and X bands for "very tropical" rain (red dots) and "very continental" rain (blue dots). Black lines corresponding to the $R(Z)$ relations $Z = 300\ R^{1.4}$, $Z = 250\ R^{1.2}$, and $Z = 200\ R^{1.6}$ are overlaid in the S-band plot (lines 1, 2, and 3). Rain rate is in mm h^{-1}

and "very continental" rain. In Fig. 10.9, the log(R) vs Z scatterplots corresponding to very tropical rain at S, C, and X bands are marked by red symbols, whereas the very continental scatterplots are presented in blue symbols. Recall that these simulations were made at the radar wavelengths 11.0 cm (S band), 5.3 cm (C band), and 3.2 cm (X band). Overlaid on the S-band panel are the lines corresponding to the following three $R(Z)$ relations: $Z = 300\ R^{1.4}$, $Z = 250\ R^{1.2}$, and $Z = 200\ R^{1.6}$. It is obvious that the red and blue scatterplots are well separated and all three $R(Z)$ relations tend to overestimate very continental rain and underestimate very tropical rain although the "tropical" $R(Z)$ relation $Z = 250\ R^{1.2}$ may be better for tropical rain in the reflectivity range 30–40 dBZ. Such pattern with large separation of the "tropical" and "continental" clusters of the R–Z pairs is common among each of the three frequency bands.

Various polarimetric radar rainfall relations have been suggested to overcome the inherent uncertainty of the Z-based QPE techniques. One of the first ideas was to use differential reflectivity Z_{DR} in combination with Z to reduce the impact of the DSD variability (Seliga et al. 1981). Indeed, Z_{DR} is a good measure of the median volume raindrop diameter (Chap. 5) and helps to reduce the ambiguity if combined with Z. Whereas the R(Z, Z_{DR}) relations definitely improve the quality of rainfall estimation, some limitations of the Z_{DR} use became evident. First, Z_{DR} is even more than Z weighted by largest raindrops in the size spectrum, while the bulk of rain is produced by drops in the lower end of the spectrum. Second, differential reflectivity should be well calibrated to yield QPE improvement. Required calibration accuracy of 0.1–0.2 dB is not easy to achieve on polarimetric radars with simultaneous transmission and reception of horizontally and vertically polarized waves (Chap. 6). Additionally, Z_{DR} is biased by differential attenuation which can be prohibitive at shorter wavelengths (e.g., *C* and *X* bands). Potentially better improvement can be achieved with the polarimetric variables that have dependence on rain rate closer to linear and are immune to the effects of miscalibration and attenuation. Such variables are specific differential phase K_{DP} and specific attenuation A.

Sachidananda and Zrnic (1986, 1987) noted that the relation between the specific differential phase, K_{DP}, and rain rate R is almost linear because K_{DP} is a lower-order moment of DSD than Z. Therefore the uncertainty in the K_{DP}-based estimates of rain rate due to the DSD variability is smaller than the uncertainty in the Z-based estimates. This is illustrated in Fig. 10.10 where the scatterplots of log(R) versus log(K_{DP}) for very tropical and very continental rain at *S*, *C*, and *X* bands are plotted. The scatterplots for the two rain types are narrower and closer to each other compared to the similar plots in Fig. 10.9 for all three frequency bands. The dispersion is particularly small for heavier rain containing larger raindrops. Note

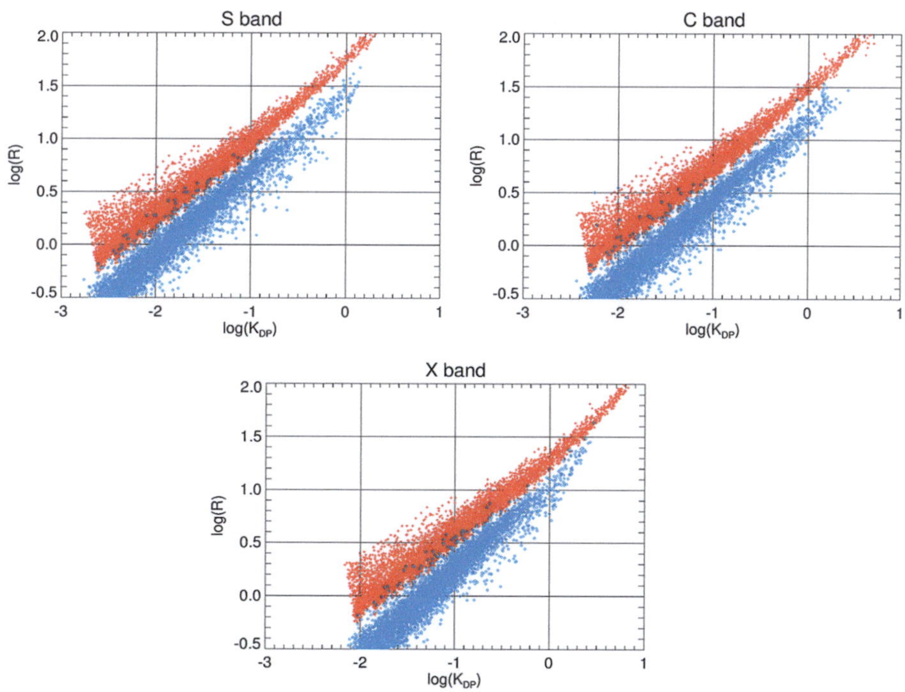

Fig. 10.10 Scatterplots of log(R) vs log(K_{DP}) at *S*, *C*, and *X* bands for "very tropical" rain (red dots) and "very continental" rain (blue dots). K_{DP} is expressed in deg km^{-1}

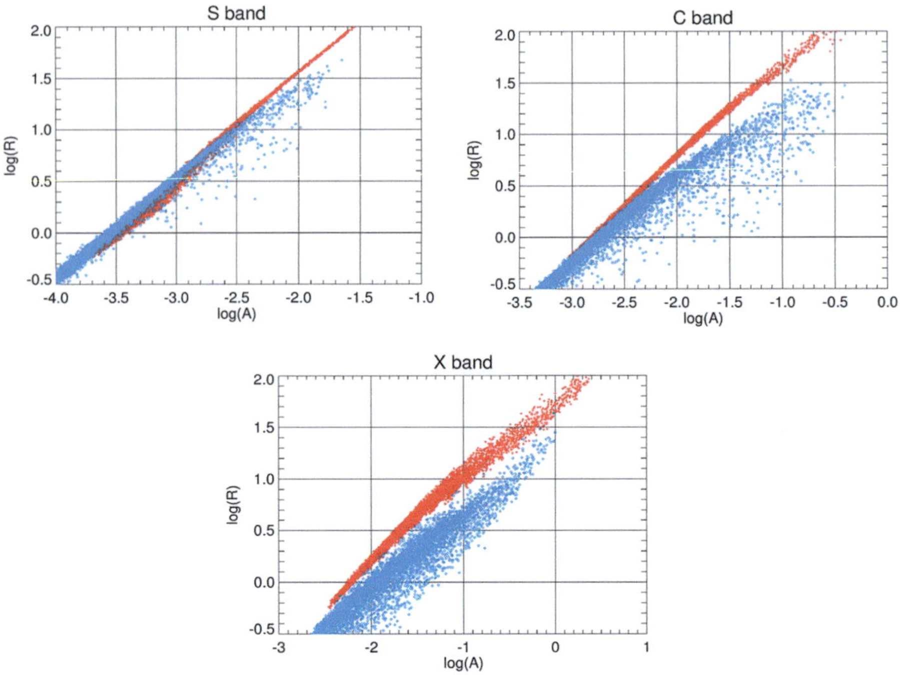

Fig. 10.11 Scatterplots of log(R) vs log(A) at S, C, and X bands for "very tropical" rain (red dots) and "very continental" rain (blue dots). Specific attenuation is computed for the temperature 20 °C

that the advantage of the $R(K_{DP})$ estimators in lighter rain is less pronounced because it is commonly dominated by almost spherical small drops.

Specific attenuation A is suitable for rainfall estimation because it almost linearly depends on rain rate (Table 5.2). The scatterplots of log(R) vs A for very tropical and very continental rain (Fig. 10.11) use values of A computed for the temperature 20 °C. It is remarkable that the scatterplots of log(R) vs log(A) are much narrower than the ones for Z and K_{DP} (particularly at S and C bands) and the clusters from tropical and continental rain almost coincide at S band. Clearly the A-based radar rainfall relations are least sensitive to the DSD variability at all three frequency bands and at S band in particular. The $R(A)$ relations are especially beneficial for lower rain rates and for tropical rain of any intensity. This is because the smaller raindrops dominate lighter rain and tropical rain, and for these the specific attenuation is very close to the 3.67th moment of drop-size distribution. The presence of the resonance-size raindrops that is more common at shorter wavelengths or in deep continental convection would make the $R(A)$ relations more sensitive to the DSD variability.

Thompson et al. (2018) used large disdrometer dataset obtained from the west Pacific area dominated by tropical rain (Indo-Pacific warm pool) to generate similar log(R) vs log(K_{DP}) and log(R) vs log(A) scatterplots at S, C, and X bands for convective and stratiform tropical rain. It is interesting that their simulation results for convective tropical rain perfectly match the scatterplots for "very tropical" rain displayed in Figs. 10.10 and 10.11, whereas tropical stratiform rain results better align with "very continental" rain in the Oklahoma disdrometer dataset. This is not surprising because coalescence is a primary microphysical process driving "warm" tropical rain either in the tropics or in the continental regions. Coalescence requires high concentration of small raindrops. Although "very continental" rain in the midlatitudes may have much higher reflectivity than stratiform rain in the tropics, they are similar in the sense that both are dominated by "cold" rain process in which coalescence does not play a significant role.

10.3 Polarimetric Radar Rainfall Estimators

A number of radar rainfall relations which involve Z, Z_{DR}, K_{DP}, A, and their combinations can be derived using disdrometer data. Additional relations to be explored utilize combinations of Z and N_w and K_{DP} and N_w because rain rate normalized by N_w (R/N_w) is almost uniquely determined by the values of Z and K_{DP} normalized by N_w (Testud et al. 2001; Illingworth and Blackman 2002). We utilize the Oklahoma disdrometer dataset of 47,144 DSDs for simulations and assume that the mean canting angle of raindrops is 0° and the width of the canting angle distribution is 10°.

Herein, we present the power-law $R(Z)$, $R(Z, Z_{DR})$, $R(K_{DP})$, $R(K_{DP}, Z_{DR})$, $R(A)$, $R(Z, N_w)$, and $R(K_{DP}, N_w)$ relations optimized for the Oklahoma DSD dataset at S ($\lambda = 11$ cm), C ($\lambda = 5.3$ cm) , and X ($\lambda = 3.2$ cm) frequencies and illustrate their sensitivity to the DSD variability. For example, the power-law relation $R(Z, Z_{DR})$ has a form $R(Z, Z_{DR}) = aZ^b Z_{dr}^c$ where factors a, b, and c are constant and Z and Z_{dr} are expressed in linear scale, i.e., $Z = 10^{0.1Z(dBZ)}$ and $Z_{dr} = 10^{0.1Z_{DR}(dB)}$. The scatterplots of the $R(Z)$, $R(Z, Z_{DR})$, $R(K_{DP})$, and $R(K_{DP}, Z_{DR})$ estimates versus R are presented in Figs. 10.12, 10.13, and 10.14, whereas the scatterplots $R(A)$, $R(Z, N_w)$, and $R(K_{DP}, N_w)$ estimates versus R are in Figs. 10.15, 10.16, and 10.17. The polarimetric estimates $R(Z, Z_{DR})$, $R(K_{DP})$, and $R(K_{DP}, Z_{DR})$ are evidently less sensitive to the DSD variability than the conventional $R(Z)$ estimate for heavier rain, and the corresponding scatter (Figs. 10.12, 10.13, and 10.14) is narrower for polarimetric relations at higher rain rates. The $R(K_{DP})$ estimate is least sensitive to the variability of drop-size distributions for rain rates above 10 mm h^{-1}. As expected from previous considerations, the relations $R(A)$, $R(Z, N_w)$, and $R(K_{DP}, N_w)$ are much less affected by the DSD variability in a very broad range of rain intensities including light rain.

The sensitivity of different radar rainfall estimates to the DSD variations can be quantified by comparing biases in the cases of "very tropical" and "very continental" rain. All rain rates and their radar estimates in each of the two categories are summed up for all DSDs from the Oklahoma dataset (Σ_g and Σ_r, respectively), and the ratio Σ_r/Σ_g is the bias measure of all listed rainfall estimators separately for the categories of the "very tropical" and "very continental" rain. All radar rainfall relations listed in

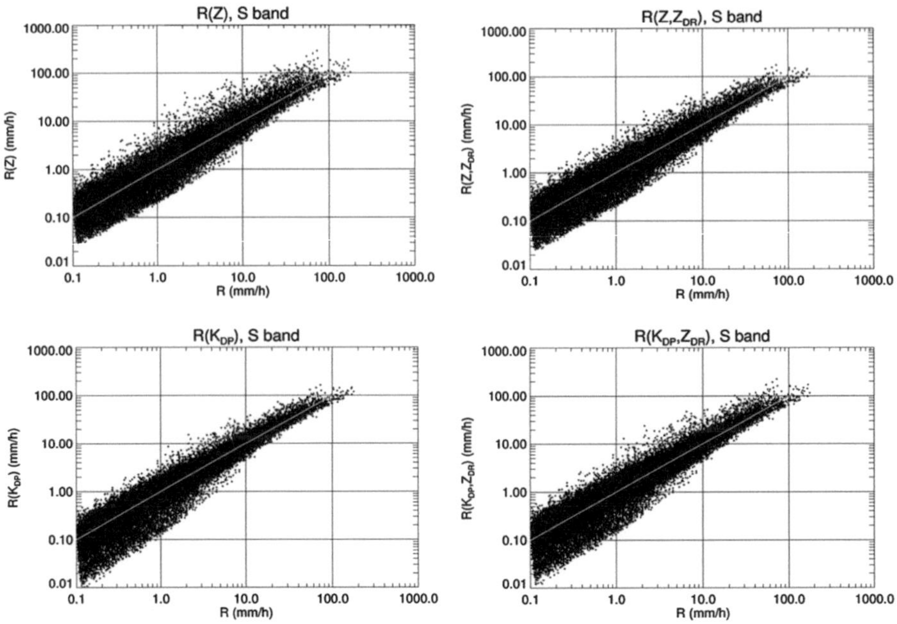

Fig. 10.12 Scatterplots of rain rates versus their estimates from relations $R(Z) = 1.94 \times 10^{-2} Z^{0.694}$, $R(Z, Z_{\mathrm{DR}}) = 1.66 \times 10^{-2} Z^{0.738} Z_{dr}^{-0.979}$, $R(K_{\mathrm{DP}}) = 47.1\, K_{\mathrm{DP}}^{0.790}$, and $R(K_{\mathrm{DP}}, Z_{\mathrm{DR}}) = 39.4\, K_{\mathrm{DP}}^{0.754} Z_{dr}^{0.537}$ optimized for the Oklahoma DSD dataset at *S* band ($\lambda = 11.0$ cm)

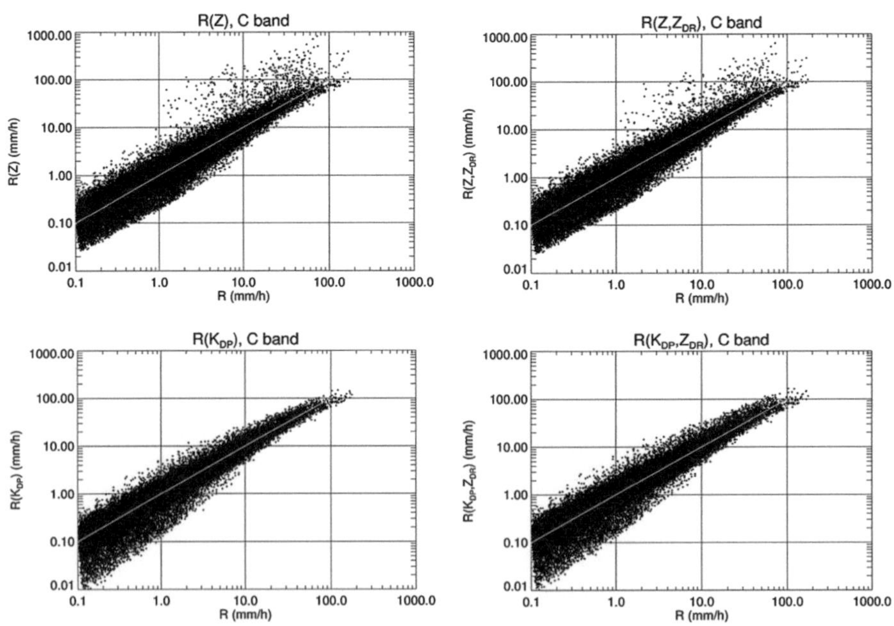

Fig. 10.13 Scatterplots of rain rates versus their estimates from relations $R(Z) = 1.77 \times 10^{-2} Z^{0.712}$, $R(Z, Z_{\mathrm{DR}}) = 1.70 \times 10^{-2} Z^{0.731} Z_{dr}^{-0.459}$, $R(K_{\mathrm{DP}}) = 25.3\, K_{\mathrm{DP}}^{0.776}$, and $R(K_{\mathrm{DP}}, Z_{\mathrm{DR}}) = 23.5\, K_{\mathrm{DP}}^{0.756} Z_{dr}^{0.276}$ optimized for the Oklahoma DSD dataset at *C* band ($\lambda = 5.3$ cm)

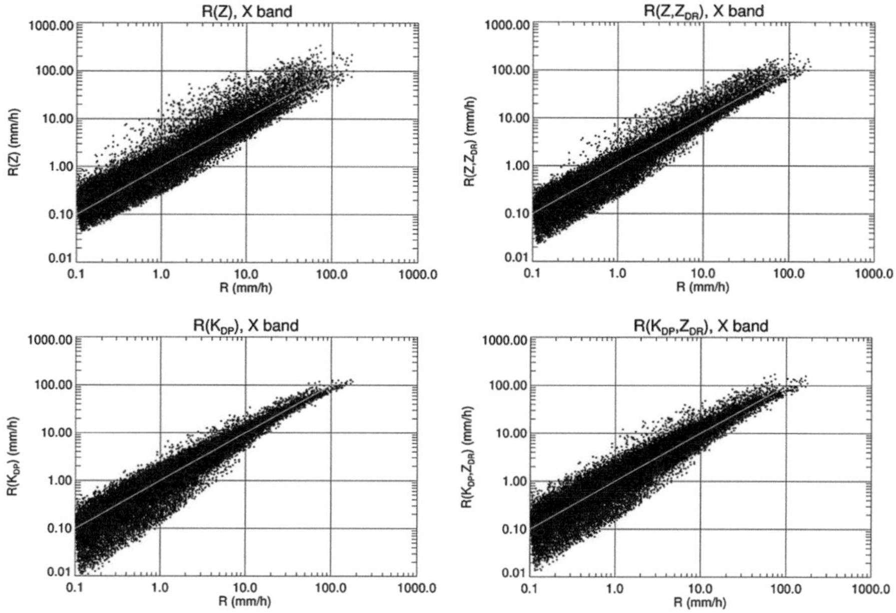

Fig. 10.14 Scatterplots of rain rates versus their estimates from relations $R(Z) = 3.06 \times 10^{-2} Z^{0.639}$, $R(Z, Z_{\mathrm{DR}}) = 1.50 \times 10^{-2} Z^{0.790} Z_{dr}^{-2.17}$, $R(K_{\mathrm{DP}}) = 16.9\, K_{\mathrm{DP}}^{0.801}$, and $R(K_{\mathrm{DP}}, Z_{\mathrm{DR}}) = 12.9\, K_{\mathrm{DP}}^{0.727} Z_{dr}^{0.831}$ optimized for the Oklahoma DSD dataset at X band ($\lambda = 3.2$ cm)

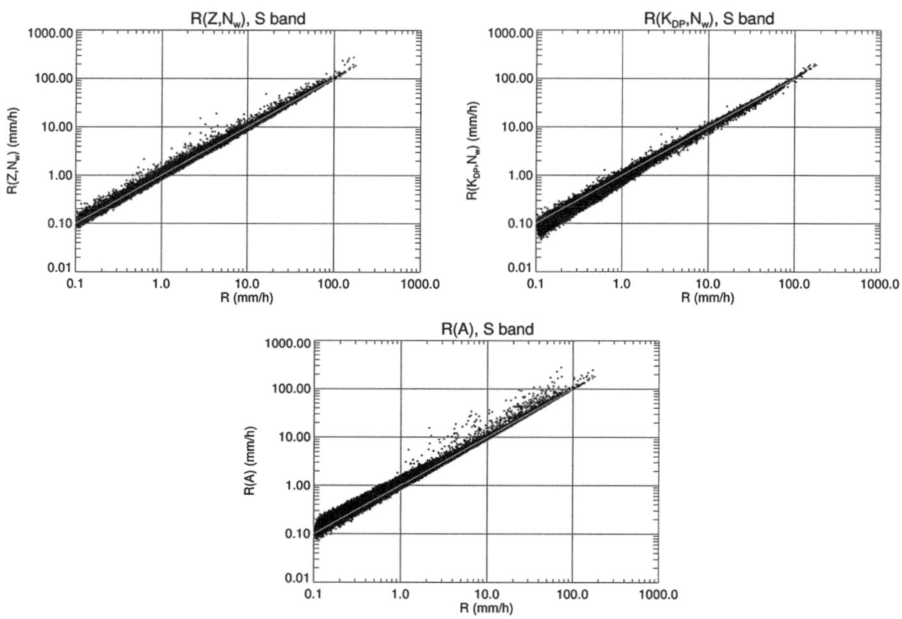

Fig. 10.15 Scatterplots of rain rates versus their estimates from relations $R(Z, N_w) = 1.57 \times 10^{-3} N_w^{0.365} Z^{0.635}$, $R(K_{\mathrm{DP}}, N_w) = 5.73\, N_w^{0.240} K_{\mathrm{DP}}^{0.760}$, and $R(A) = 4120\, R^{1.03}$ optimized for the Oklahoma DSD dataset at S band ($\lambda = 11.0$ cm)

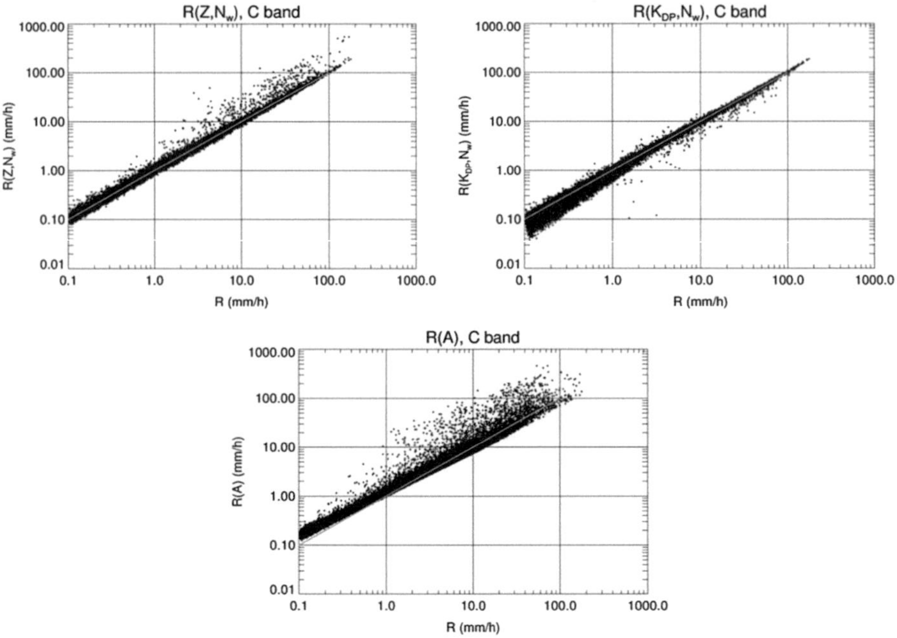

Fig. 10.16 Scatterplots of rain rates versus their estimates from relations $R(Z, N_w) = 1.73 \times 10^{-3} N_w^{0.348} Z^{0.645}$, $R(K_{\mathrm{DP}}, N_w) = 1.98\, N_w^{0.294} K_{\mathrm{DP}}^{0.707}$, and $R(A) = 294\, R^{0.89}$ optimized for the Oklahoma DSD dataset at C band ($\lambda = 5.3$ cm)

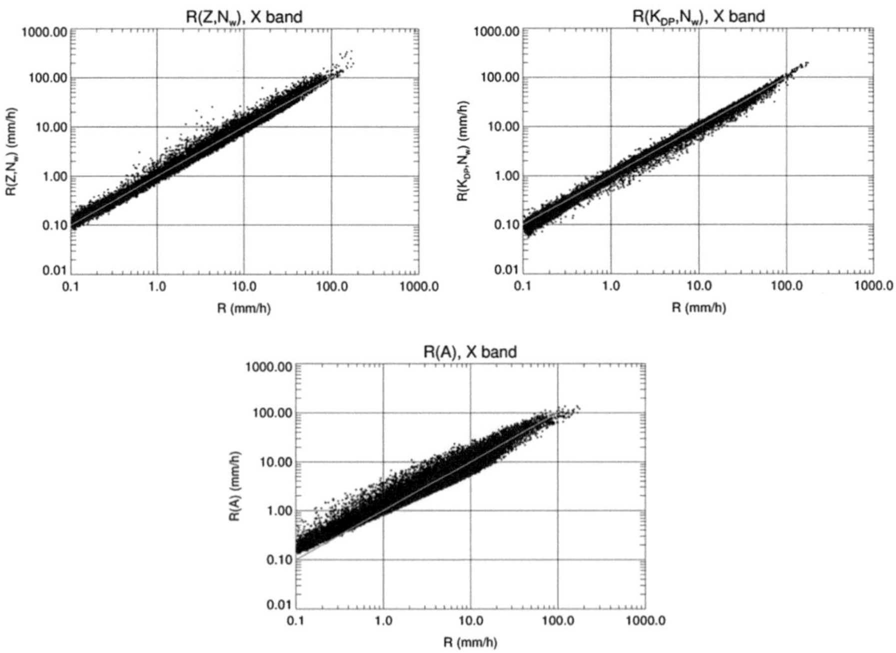

Fig. 10.17 Scatterplots of rain rates versus their estimates from relations $R(Z, N_w) = 2.05 \times 10^{-3} N_w^{0.356} Z^{0.606}$, $R(K_{\mathrm{DP}}, N_w) = 1.98\, N_w^{0.294} K_{\mathrm{DP}}^{0.707}$, and $R(A) = 43.5\, A^{0.79}$ optimized for the Oklahoma DSD dataset at X band ($\lambda = 3.2$ cm)

Table 10.2 Average bias ratios for different rainfall estimators of the "very tropical" and "very continental" rain types at *S* band ($\lambda = 11.0$ cm)

	Σ_r/Σ_g	
Rainfall relation at wavelength $\lambda = 11.0$ cm	Very tropical	Very continental
$R(Z) = 1.94 \times 10^{-2} Z^{0.694}$	0.67	2.27
$R(Z, Z_{\mathrm{DR}}) = 1.66 \times 10^{-2} Z^{0.738} Z_{dr}^{-0.979}$	0.72	1.88
$R(K_{\mathrm{DP}}) = 47.1 K_{\mathrm{DP}}^{0.790}$	0.75	1.66
$R(K_{\mathrm{DP}}, Z_{\mathrm{DR}}) = 39.4 K_{\mathrm{DP}}^{0.754} Z_{dr}^{0.537}$	0.74	1.91
$R(A) = 4120 \times A^{1.03}$	0.99	1.15
$R(Z, N_w) = 1.57 \times 10^{-3} N_w^{0.365} Z^{0.635}$	0.96	1.12
$R(K_{\mathrm{DP}}, N_w) = 5.73 N_w^{0.240} K_{\mathrm{DP}}^{0.760}$	0.95	1.03

Table 10.2 tend to underestimate very tropical rain and overestimate very continental rain, but the magnitude of the bias depends strongly on the algorithm. The conventional *R*(*Z*) underestimates very tropical rain on average by 33% and overestimates very continental rain by more than a factor of two which is consistent with most of observations reported in the literature. The magnitude of underestimation/overestimation is reduced for all polarimetric relations, but the *R*(*A*), $R(Z, N_w)$, and $R(K_{\mathrm{DP}}, N_w)$ relations yield the smallest negative and positive biases. The listed optimal relations have been computed using the disdrometer dataset from Oklahoma and were obtained with a particular type of regression. Nonetheless, the general conclusions regarding the performance of various radar algorithms and the sensitivity to the DSD variability (which can be drawn from Figs. 10.9, 10.10, 10.11, 10.10, 10.13, 10.14, 10.15, 10.16, 10.17 and Table 10.2) have a universal character.

The sensitivity to the DSD variability is the most important factor in choosing an optimal algorithm for rainfall estimation but not the only one. Next we overview various radar rainfall relations and their performance, starting with *R*(*Z*):

$$R(Z) = aZ^b. \tag{10.15}$$

Apart from being very sensitive to the DSD variability, the *R*(*Z*) estimate is prone to the reflectivity biases caused by radar miscalibration, attenuation, partial beam blockage, and wet radome. *R*(*Z*) generally overestimates rain rate if rain is mixed with hail. Attenuation effects are particularly strong at shorter wavelengths (*C* and *X* bands). Next we consider $R(Z, Z_{\mathrm{DR}})$ relation:

$$R(Z, Z_{\mathrm{DR}}) = aZ^b Z_{dr}^c. \tag{10.16}$$

In (10.16), both *Z* and Z_{dr} are expressed in linear scale. Historically, the $R(Z, Z_{\mathrm{DR}})$ was the first polarimetric relation suggested for the measurement of rain (Seliga et al. 1981). Equation (10.16) can be considered as the *R*(*Z*) relation (10.15) with the multiplier depending on Z_{dr}. This multiplier decreases with increasing Z_{dr} (if $c < 0$) and thus automatically adjusts multiplier *a* in (10.15) to match values expected from

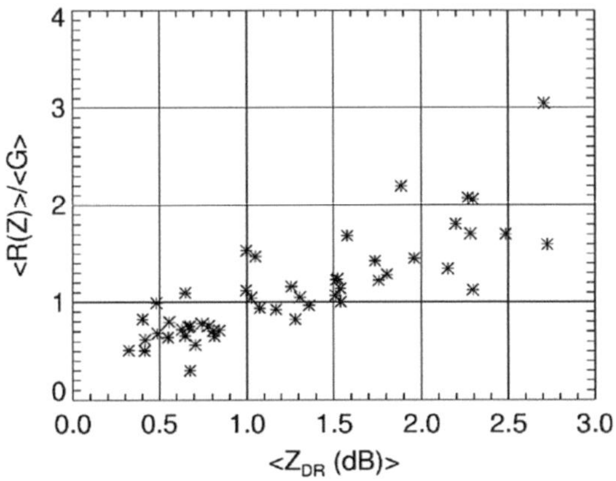

Fig. 10.18 Scatterplot of the ratio of mean areal rain rates obtained from radar using the standard WSR-88D relation $R(Z) = 1.70 \times 10^{-2} Z^{0.714}$ and gauge vs average values of differential reflectivity. From Ryzhkov et al. (2005). © American Meteorological Society. Used with permission

the DSD dominated by large raindrops (or with the deficit of small raindrops). This concept is illustrated in Fig. 10.18 where the scatterplot of the mean areal rain rates obtained from the radar ($<R>$) and gages ($<G>$) versus average values of differential reflectivity $<Z_{DR}>$ is shown.

The $<Z_{DR}>$ is defined as the weighted average Z_{DR} for a specific hour over a whole gauge network (Ryzhkov et al. 2005):

$$< Z_{DR} >= \frac{\sum\limits_{i,j} R^{(i,j)}(Z) Z_{DR}^{(i,j)}}{\sum\limits_{i,j} R^{(i,j)}(Z)}, \tag{10.17}$$

where superscript i characterizes the scan number within the 1-h time interval and superscript j stands for the gauge number. It is evident that $R(Z)$ tends to overestimate/underestimate rain at higher/lower net Z_{DR} proving the benefit of the $R(Z, Z_{DR})$ relation compared to the $R(Z)$. The improvement is tangible at rain rates >10 mm h^{-1} for all three frequency bands (Figs. 10.12, 10.13, and 10.14).

Various $R(Z, Z_{DR})$ relations proposed for S and C bands are listed in Table 10.3. Most have been obtained from simulations based on disdrometer data except the one by Ryzhkov et al. (2005) which was derived empirically via comparison with gauges. Two $R(Z, Z_{DR})$ relations from Table 10.3

$$R(Z, Z_{DR}) = 6.70 \times 10^{-3} Z^{0.927} Z_{dr}^{-3.43} \tag{10.18}$$

and

$$R(Z, Z_{DR}) = 1.42 \times 10^{-2} Z^{0.770} Z_{dr}^{-1.67} \tag{10.19}$$

are often used at S band (including WSR-88D). Equation (10.18) is more suitable for tropical rain type, whereas Eq. (10.19) is expected to work better for continental rain.

Table 10.3 The $R(K_{DP})$, $R(Z, Z_{DR})$, and $R(K_{DP}, Z_{DR})$ relations used in studies of rainfall estimation at *S*, *C*, and *X* bands

$R(K_{DP})$	$R(Z, Z_{DR})$	$R(K_{DP}, Z_{DR})$	Authors	Region
S band				
$41\,K_{DP}^{0.87}$	–	$52\,K_{DP}^{0.96}Z_{DR}^{-0.45}$	Ryzhkov and Zrnic (1995)	Oklahoma
$46\,K_{DP}^{0.87}\text{sign}(K_{DP})$	$7.910^{-3}Z^{0.97}Z_{dr}^{-4.8}$	$88\,K_{DP}^{0.98}Z_{dr}^{-1.9}$	Brandes et al. (2002)	Florida
$44\,K_{DP}^{0.82}\text{sign}(K_{DP})$	$\frac{1.710^{-2}Z^{0.714}}{0.4+5.0\lvert Z_{dr}-1\rvert^{1.3}}$	$\frac{44\lvert K_{DP}\rvert^{0.82}\text{sign}(K_{DP})}{0.4+3.5\lvert Z_{dr}-1\rvert^{1.7}}$	Ryzhkov et al. (2005)	Oklahoma
$44\,K_{DP}^{0.82}\text{sign}(K_{DP})$	$1.410^{-2}\,Z^{0.77}Z_{dr}^{-1.7}$		Giangrande and Ryzhkov (2008)	Oklahoma
$41\,K_{DP}^{0.85}$	$6.7\times10^{-3}Z^{0.93}Z_{dr}^{-3.4}$	$91\,K_{DP}^{0.93}Z_{dr}^{-1.7}$	Cifelli et al. (2011)	Colorado
$40\,K_{DP}^{0.85}$	$5.4\times10^{-3}Z^{0.94}Z_{dr}^{-3.59}$	$93\,K_{DP}^{0.95}Z_{dr}^{-1.75}$	Chen et al. (2017)	Iowa
$44\,K_{DP}^{0.80}$	$1.7\times10^{-2}\,Z^{0.84}Z_{dr}^{-4.5}$	$89\,K_{DP}^{0.88}Z_{dr}^{-2.5}$	Pepler et al. (2011)	Australia
$56.0\,K_{DP}^{0.80}$	$8.5\times10^{-3}\,Z^{0.92}Z_{dr}^{-5.24}$	$96.6\,K_{DP}^{0.93}Z_{dr}^{-2.11}$	Thompson et al. (2018)	Tropics
C band				
$35\,K_{DP}^{0.83}$			May et al. (1999)	Australia
$29\,K_{DP}^{0.85}$			Bringi et al. (2006)	Japan
$25\,K_{DP}^{0.81}$	$1.210^{-2}\,Z^{0.82}Z_{dr}^{-1.7}$		Bringi et al. (2011)	UK
$20\,K_{DP}$ sign (K_{DP})			Vulpiani et al. (2012)	Italy
$34\,K_{DP}^{0.79}$			Figueras et al. (2012)	France
$35\,K_{DP}^{0.80}$			Wang et al. (2013)	Taiwan
$30.6\,K_{DP}^{0.78}$	$8.6\ \ 10^{-3}Z^{0.91}Z_{dr}^{-4.21}$	$28.1\,K_{DP}^{0.88}Z_{dr}^{-1.67}$	Thompson et al. (2018)	Tropics
X band				
$14\,K_{DP}^{0.80}$			Matrosov et al. (2005)	California
$20\,K_{DP}^{0.82}$			Park et al. (2005)	Japan
$15\,K_{DP}^{0.76}$			Matrosov et al. (2006)	Colorado
$17\,K_{DP}^{0.73}$			Matrosov (2010)	California

(continued)

Table 10.3 (continued)

$R(K_{DP})$	$R(Z, Z_{DR})$	$R(K_{DP}, Z_{DR})$	Authors	Region
$18\,K_{DP}^{0.79}$			Wang and Chandrasekar (2010); Chen and Chandrasekar (2015)	Oklahoma Texas
$19\,K_{DP}^{0.85}$			Anagnostou et al. 2010	Greece
$17\,K_{DP}^{0.73}\,\text{sign}(K_{DP})$			Borowska et al. (2011)	Germany
$14\,K_{DP}^{0.77}$			Matrosov et al. (2013)	Colorado
$17\lvert K_{DP}\rvert^{0.80}$ sign (K_{DP})			Diederich et al. (2015)	Germany
$18.7\,K_{DP}^{0.77}$			Thompson et al. (2018)	Tropics

$R(Z,\ Z_{DR})$ relations require that Z_{DR} is calibrated with an accuracy of at least ±0.2 dB (Ryzhkov et al. 2005; Zrnic et al. 2010). This accuracy ensures that the biases in the estimates of rain rate from (10.18) to (10.19) are within 15–17% and 8–9%.

Note that the rain rate estimates obtained from the $R(Z,\ Z_{DR})$ relations with higher absolute value of the exponent c in the Z_{dr}^{c} factor are more vulnerable to the biases in Z_{DR}. At shorter wavelengths, both Z and Z_{DR} can be significantly biased by attenuation/differential attenuation; therefore careful correction for attenuation is required. Heavy differential attenuation hampers utilization of the $R(Z,\ Z_{DR})$ relations even after attenuation correction (Chap. 6) because the residual errors in Z_{DR} can easily exceed the required 0.2 dB limit. Therefore, the $R(Z,\ Z_{DR})$ algorithms are rarely used at X and C bands, and the $R(K_{DP})$ and $R(A)$ alternatives are the preferred choice which we examine next. We start with $R(K_{DP})$

$$R(K_{DP}) = aK_{DP}^{b}. \tag{10.20}$$

This type of power-law relation was proposed by Sachidananda and Zrnic (1987). The actual coefficients in Sachidananda and Zrnic (1987) are equivalent to the specific one listed in the first row of Table 10.3 with the following explanation. The coefficients listed by Sachidananda and Zrnic (1987) are computed for the Marshal-Palmer DSD at the wavelength $\lambda\ =\ 10$ cm and the two-way specific differential phase $K_{DP}^{\text{two way}} = 2\,K_{DP}$. Adjusting the multiplier a for the one-way $R(K_{DP})$ and the wavelength of $\lambda\ =\ 11$ cm as in Doviak and Zrnic (2006), Eq. (8.31a) produces the relation $R(K_{DP}) = 41\,K_{DP}^{0.87}$.

The specific differential phase K_{DP} normalized by the total concentration of raindrops is proportional to D^{α} where α varies between 4.3 and 4.9 for $D > 1$ mm, and it is somewhat higher for smaller raindrops (Chap. 5). Hence, K_{DP} is a lower-order moment of the DSD than Z, and the $R(K_{DP})$ estimate is less sensitive to the DSD variability than the $R(Z)$; this is illustrated in Figs. 10.12, 10.13, and 10.14. The $R(K_{DP})$ relation produces the narrowest scatter for higher rain rates among other relations in Figs. 10.12, 10.13, and 10.14. Such advantage of the $R(K_{DP})$ estimator gradually vanishes at lower rain rates. Therefore, the $R(K_{DP})$ estimate remains affected by the DSD variations but to a smaller extent than the $R(Z)$ and $R(Z, Z_{DR})$ algorithms. This is illustrated by Tables 10.2 and 10.3 where various $R(K_{DP})$ relations are listed. Indeed, some climatological adjustment is required for the parameters of the $R(K_{DP})$ relation. The intercept of the $R(K_{DP})$ relation optimized for tropics and maritime regions is generally higher than for a more continental rain.

Because K_{DP} is proportional to the radial derivative of differential phase Φ_{DP}, it is immune to radar miscalibration, attenuation, partial beam blockage, and presence of water or snow on the surface of radome. The immunity of K_{DP} to attenuation makes it particularly beneficial for rainfall measurements at shorter wavelengths. The $R(K_{DP})$ estimate is less affected by contamination from hail (Balakrishnan and Zrnic 1990). This is due to the fact that dry or slightly wet hailstones have chaotic orientation and do not contribute much to K_{DP}. This is not the case for melting hailstones at the advanced stage of melting, but even in this situation, K_{DP} is much less affected by the contribution from hail than Z and Z_{DR}. Ryzhkov et al. (2013b) demonstrate that the $R(K_{DP})$ relation is a good choice for estimating rainfall in the presence of hail at S, C, and X bands but it may produce positive bias at S band for larger-size hail. The $R(K_{DP})$ estimator is less affected by ground clutter canceling (Zrnic and Ryzhkov 1996) and has advantages for measuring rain in the areas impacted by ground clutter and anomalous propagation (Ryzhkov and Zrnic 1998a).

Significant limitations of the $R(K_{DP})$ algorithm are the degraded radial resolution and the noisiness of the K_{DP} estimate in light rain. Specific differential phase is defined as a half of the slope of the radial profile of Φ_{DP} ($K_{DP} = \frac{1}{2}\frac{d\Phi_{DP}}{dr}$) and estimated as a least-squares linear fit of the profile of Φ_{DP} in the range interval $N\Delta r$ where Δr is the spacing in range and N is the number of samples. The standard deviation of the K_{DP} estimate is (Chap. 6)

$$SD(K_{DP}) = \frac{\sqrt{3}SD(\Phi_{DP})}{N^{3/2}\Delta r} \tag{10.21}$$

where SD(Φ_{DP}) is the standard deviation of the Φ_{DP} estimate at a specific range location. It follows from (10.21) that SD(K_{DP}) decreases with increasing number of samples within the averaging window; hence the estimation of K_{DP} involves a compromise between the accuracy and radial resolution. A better radial resolution is required for convective rain causing larger statistical errors, whereas poorer radial resolution can be tolerated for stratiform rain. This motivated Ryzhkov and Zrnic (1996) to suggest obtaining two estimates of K_{DP}: "lightly filtered" with $N = 17$ and "heavily filtered" with $N = 49$, both for samples spaced 150 m apart. In the processing of differential phase currently implemented on WSR-88D, the pair of

$N = 9$ and $N = 25$ is used which corresponds to a 2.25 km and 6.25 km radial resolutions at the spacing $\Delta r = 0.25$ km. Then the "lightly filtered" estimate of K_{DP} is selected for the range locations where $Z > 40$ dBZ, and the "heavily filtered" estimate is chosen otherwise. More sophisticated techniques for the K_{DP} processing have been proposed by Wang and Chandrasekar (2009), Otto and Russchenberg (2011), and Maesaka et al. (2012). The algorithm of Wang and Chandrasekar (2009) uses adaptive smoothing to automatically optimize the estimate of K_{DP} depending on the noisiness of the Φ_{DP} data and the slope of the Φ_{DP} profiles which is somewhat equivalent to changing N adaptively.

The standard deviation of Φ_{DP}, SD(Φ_{DP}), is usually about 1°, and it would, according to (10.21), induce a SD(K_{DP}) = 0.26°km^{-1} for $N = 9$ and 0.055° km^{-1} for $N = 25$ if $\Delta r = 0.25$ km. Because the fractional standard error (FSE) of the rain rate estimate SD(R)/R is approximately equal to the product of the factor b in (10.20) and the ratio SD(K_{DP})/K_{DP}, the value of FSE is inversely proportional to the magnitude of K_{DP} and linearly decreases with radar wavelength. The FSE equal to 30% is achieved for R between 15 and 27 mm h^{-1} at S band if $N = 17$ and the $R(K_{DP})$ relation of Ryzhkov et al. (2005) is used (see Table 10.3). Similar FSE is achieved for R between 5 and 8 mm h^{-1} for $N = 17$ at X band and the $R(K_{DP})$ relation from Matrosov et al. (2013). The "heavily filtered" estimate of K_{DP} should be used for lower rain rates at both wavelengths. Therefore, the $R(K_{DP})$ algorithm is particularly beneficial at shorter wavelengths where reliable measurements of precipitation are possible in a wider range of rain intensities. Direct comparisons of the $R(K_{DP})$ estimates using simultaneous measurements at S and X bands by Matrosov et al. (2006) show that at X band the estimates are reliable for rain rates as low as 2.5–3 mm h^{-1} which correspond to reflectivities of about 27–28 dBZ, whereas at S band the low end of measurable rain rates is 8–10 mm h^{-1} corresponding to $Z \sim 37$–38 dBZ. For lower rain rates and reflectivities, either $R(Z)$ or $R(Z, Z_{DR})$ (at S band) is preferred.

The impact of inherent noisiness of the K_{DP} data in lighter rain can be reduced via integration over sufficiently large spatial/temporal domain. Hence, for estimation of areal rainfall or long-term rain total, the $R(K_{DP})$ algorithm is adequate for lower rain rates than specified in the Matrosov et al. (2006) study. This is demonstrated in the studies of Ryzhkov et al. (2000) at S band and Wang and Chandrasekar (2010) and Borowska et al. (2011) at X band. Note that in light rain Eq. (10.20) should be replaced with

$$R(K_{DP}) = a|K_{DP}|^b \operatorname{sign}(K_{DP}). \tag{10.22}$$

At first glance, retaining the sign of K_{DP} in the $R(K_{DP})$ relation produces negative instantaneous rain rates which seem unphysical. This is, however, justified by the need to average both positive and negative statistical fluctuations of the K_{DP} estimates to obtain unbiased rain average over a sufficiently large spatial/temporal interval (Fig. 10.19). Indeed, K_{DP} remains positive if statistical fluctuations of K_{DP} are small for lower intrinsic K_{DP} (left panel in Fig. 10.19) or if both fluctuations of

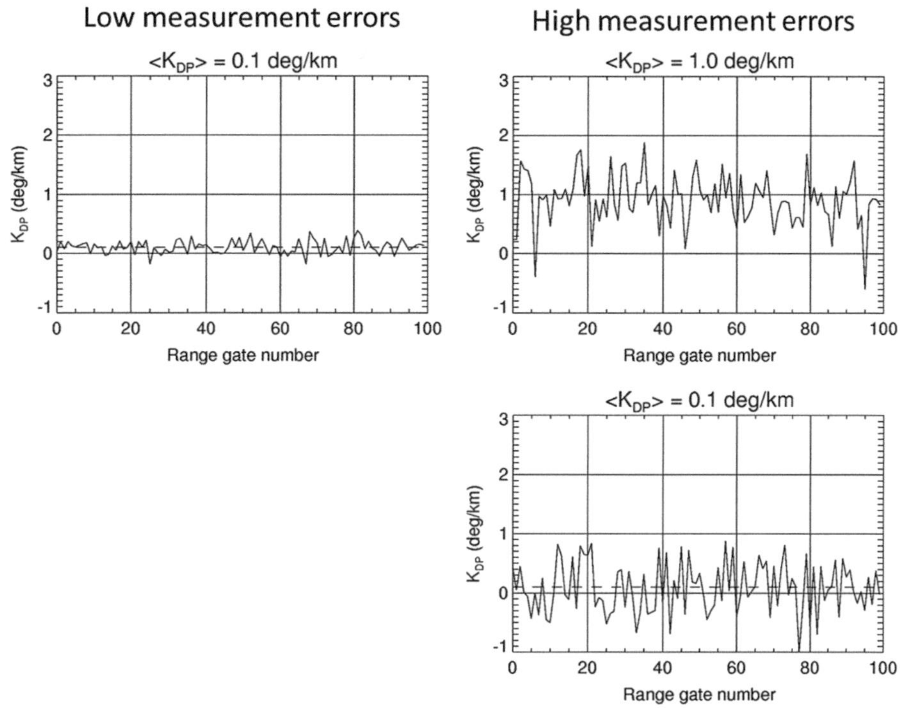

Fig. 10.19 Radial profiles of the K_{DP} estimate illustrating the need to retain its sign in the $R(K_{DP})$ relations. The spacing of range gates is 250 m. The dashed line indicates K_{DP} averaged over 100 range gates (i.e., $<K_{DP}>$)

K_{DP} and its mean value are large (right top panel). The K_{DP} estimate can become negative if K_{DP} fluctuations are large, but the mean value of K_{DP} is small (right bottom panel). Then, chopping negative K_{DP} or using absolute value of K_{DP} would produce positive bias of the mean rain rate.

Statistical fluctuations of K_{DP} appear as small-scale variations in range. This is contrary to larger negative and positive deviations caused by contribution from the backscatter differential phase δ which can be significant at C and X band (Fig. 5.17) or nonuniform beam filling (Chap. 6). In both cases, large negative excursion of K_{DP} is followed by large positive excursion (Ryzhkov and Zrnic 1998b). Again, eliminating only negative bias of K_{DP} while retaining its positive bias would create a positive bias in the rainfall estimate. Various techniques to separate δ and K_{DP} are discussed by Otto and Russchenberg (2011), Schneebeli and Berne (2012), and Troemel et al. (2013). It is also necessary to recognize the areas of enhanced nonuniform beam filling and avoid using the K_{DP}-based rainfall relation there.

In areas where the quality of the K_{DP} estimate is compromised, a K_{DP} substitute (K'_{DP}) can be obtained from the $K_{DP}(Z)$ power-law relation (Lim et al. 2013; Zhang et al. 2013):

$$K'_{DP}(r) = a[Z(r)]^b, \tag{10.23}$$

where the multiplier a can be estimated after integrating both sides of (10.23) over the part of the radial (r_1, r_2) where the K_{DP} quality is suspicious

$$\Delta\Phi_{DP} = \Phi_{DP}(r_2) - \Phi_{DP}(r_1) = 2a\int_{r_1}^{r_2} [Z(s)]^b ds, \tag{10.24}$$

so that

$$a = \frac{\Delta\Phi_{DP}}{2\int_{r_1}^{r_2} [Z(s)]^b ds}. \tag{10.25}$$

The ratio a should be determined separately for every radial or radial segment. It is assumed to be constant along the propagation path between ranges r_1 and r_2. In this formulation, K_{DP}' is not affected by a constant bias in Z along the radial segment (possibly caused by radar miscalibration, partial beam blockage, or wet radome). Note that attenuation-related bias of Z should be eliminated before applying (10.23) and (10.25).

The QPE algorithm utilizing K_{DP} and Z_{DR} was suggested by Jameson (1991) and tested by Ryzhkov and Zrnic (1995). It is expressed as

$$R(K_{DP}, Z_{DR}) = aK_{DP}^b Z_{dr}^c. \tag{10.26}$$

Later it was used in the studies at S band by Petersen et al. (1999), Brandes et al. (2002), Ryzhkov et al. (2005), Cifelli et al. (2011), Pepler et al. (2011), and Thompson et al. (2018) (Table 10.3). The $R(K_{DP}, Z_{DR})$ relation is slightly more sensitive to the DSD variations than $R(K_{DP})$ (Table 10.2 and Figs. 10.13, 10.14, and 10.15) as can be concluded from the simulations based on disdrometer measurements in Oklahoma. Maki et al. (2005) claim that it is the least sensitive to variations in the DSD measured in Japan. A similar conclusion is made by Thompson et al. (2018) with respect to rain in the tropics. The estimator is less affected by the uncertainty of raindrop shape because for a negative value of c in (10.23), the product ${K_{DP}}^b {Z_{dr}}^c$ does not change much as a function of the assumed raindrop shape vs diameter relations. Although some marginal improvement in rainfall estimates from $R(K_{DP}, Z_{DR})$ compared to $R(K_{DP})$ was reported by Ryzhkov and Zrnic (1995), Petersen et al. (1999), Ryzhkov et al. (2005, 2012), and Thompson et al. (2018), measurement errors of Z_{DR} (such as those attributed to miscalibration of Z_{DR} and differential attenuation) may compromise the $R(K_{DP}, Z_{DR})$ estimator. From practical considerations, the simpler $R(K_{DP})$ relation is more robust and reliable (Maki et al. 2005).

It has to be noted that all radar rainfall relations derived from the disdrometer data including the ones listed in Tables 10.2 and 10.3 are very sensitive to the choice of particular DSD dataset, the details of regression method, and the range of rain intensities (or reflectivities) for which they were optimized. For example, the $R(K_{DP}, Z_{DR})$ power-law relation (10.26) usually has negative exponent c if it is optimized for heavier rain rates (at which K_{DP} is less noisy and more practical). However, the exponent c has positive sign (see Table 10.2) if the optimization is performed for all rain intensities.

The radar rainfall relation

$$R(Z, N_w) = aN_w^b Z^c \tag{10.27}$$

based on the joint use of Z and normalized concentration of raindrops N_w is remarkably robust with respect to the DSD variability (Figs. 10.15, 10.16, and 10.17). This was demonstrated by Testud et al. (2000, 2001) and Illingworth and Blackman (2002). The $R(Z, N_w)$ relation at S band derived using simulations based on the Oklahoma DSD dataset has the following form (Ryzhkov et al. 2012):

$$R(Z, N_w) = 1.57 \times 10^{-3} N_w^{0.365} Z^{0.635}, \tag{10.28}$$

where Z is expressed in mm^{6}m^{-3} and N_w is in m^{-3} mm^{-1}. Two $R(Z, N_w)$ relations for C band have been used in the studies of Illingworth and Thompson (2005) and Tabary et al. (2011):

$$R(Z, N_w) = 1.87 \times 10^{-3} N_w^{0.333} Z^{0.667} \tag{10.29}$$

$$R(Z, N_w) = 1.07 \times 10^{-3} N_w^{0.381} Z^{0.619}. \tag{10.30}$$

The first is the inversion of the equation $Z = a(N_w)R^{1.5}$.

The problem with application of (10.27)–(10.30) is the need to accurately estimate the normalized drop concentration N_w. Illingworth and Thompson (2005) suggest using the Z–Z_{DR} pair if neither Z nor Z_{DR} is biased by attenuation and both are accurately measured ("ZZDR technique"). This can be easily done at S band and at C band (for relatively light precipitation) but is very problematic at X band (Ryzhkov et al. 2012; Tabary et al. 2011). If Z or Z_{DR} are significantly biased by attenuation, then the estimate of N_w can be obtained using the ZPHI method (Chap. 6, Testud et al. 2000; Tabary et al. 2011). According to the ZPHI method, first a radial profile of specific attenuation A is estimated from Z and Φ_{DP} along the beam, and then N_w is computed from A and Z (corrected for attenuation) using the relation $A = f(Z, N_w)$. Because N_w varies with range, the whole propagation path in rain is segmented into relatively small segments in which N_w is considered approximately constant. The deficiency of this technique is that the estimate of N_w is prone to the biases of Z caused by radar miscalibration, partial beam blockage, and wet radome. Nevertheless, Tabary et al. (2011) demonstrate that the $R(Z, N_w)$ algorithm performs relatively well on the data collected by the operational C-band radar near Paris, France. The "ZZDR" methodology (Tabary et al. 2011) with attenuation correction proved to be the best for estimating N_w and hourly rain totals less than 5 mm, whereas the ZPHI method is the best for heavier rain.

Theoretically, the $R(K_{DP}, N_w)$ relation

$$R(K_{\mathrm{DP}}, N_w) = aN_w^b K_{\mathrm{DP}}^c \tag{10.31}$$

is least sensitive to the DSD variability among all the relations reviewed in this chapter (Table 10.2 and Figs. 10.15, 10.16, and 10.17). Ryzhkov et al. (2012) suggest the following relation optimized for the Oklahoma DSD dataset at S band ($\lambda = 11.0$ cm):

$$R(K_{\mathrm{DP}}, N_w) = 5.73 N_w^{0.240} K_{\mathrm{DP}}^{0.760}. \tag{10.32}$$

The performance of this relation has not been validated by systematic comparisons with gauges. It is obvious that it has all advantages and limitations of the $R(K_{DP})$ estimator. Again, the normalized drop concentration N_w should be estimated using either ZZDR or ZPHI techniques as suggested for the $R(Z, N_w)$ algorithm. Note that the exponent on N_w in the $R(K_{DP}, N_w)$ formula is significantly lower than the similar exponent in the $R(Z, N_w)$ relations (10.29–10.31). This is consistent with the fact that the multiplier in the $R(Z)$ relation varies more with changes in DSD characterized by N_w than the multiplier in the $R(K_{DP})$ relation.

10.4 The *R*(*A*) Algorithm

Specific attenuation A is an attractive alternative to traditional variables such as Z, Z_{DR}, and K_{DP} utilized for rainfall estimation. Contrary to Z, A is more directly related to rain rate and liquid water content, and its estimates are affected very little by radar miscalibration and blockage of the radar beam. Low sensitivity of the $R(A)$ relation to the DSD variability was pointed out by Atlas and Ulbrich (1977) and Matrosov et al. (2005); it is illustrated in Table 10.2 and Figs. 10.11, 10.15, 10.16, and 10.17. The specific attenuation can be estimated from the radial profile of attenuated reflectivity Z_a and known two-way path-integrated attenuation PIA along the propagation path (r_1, r_2) using formulas (Sect. 6.2)

$$A(r) = \frac{[Z_a(r)]^b C(b, \mathrm{PIA})}{I(r_1, r_2) + C(b, \mathrm{PIA}) I(r, r_2)}, \tag{10.33}$$

where

$$I(r_1, r_2) = 0.46 b \int_{r_1}^{r_2} [Z_a(s)]^b ds, \tag{10.34}$$

$$I(r, r_2) = 0.46b \int_r^{r_2} [Z_a(s)]^b ds, \quad (10.35)$$

$$C(b, \text{PIA}) = \exp(0.23b, \text{PIA})^{-1}, \quad (10.36)$$

$$\text{PIA}(r_1, r_2) = \alpha[\Phi_{\text{DP}}(r_2) - \Phi_{\text{DP}}(r_1)] = \alpha\Delta\Phi_{\text{DP}}. \quad (10.37)$$

In (10.33)–(10.36), the factor b is a constant parameter (usually between 0.6 and 0.9 at microwave frequencies), and α is equal to the average ratio of A and K_{DP}.

Radial profiles of A have been used by Testud et al. (2000) as part of the ZPHI method for rainfall estimation. However, the method does not use specific attenuation directly to estimate rain rate. Instead, it uses it as an interim parameter for calculating the normalized concentration of raindrops N_w, and the rain rate is computed from the combination of Z (corrected for attenuation) and N_w via (10.27). The need for segmentation of the path and the vulnerability to reflectivity biases (by possible radar miscalibration, partial beam blockage, and wet radome) are the weaknesses of the ZPHI method. Ryzhkov et al. (2014) suggest using A directly for estimating rain rate as follows:

$$R(A) = \gamma A^{\delta}. \quad (10.38)$$

The advantages of this approach are simplicity and immunity to the biases of Z.

Because A is a function of the wavelength and raindrop temperature, the parameters of power-law $R(A)$ rainfall relations are also wavelength- and temperature-dependent. In Table 10.4, optimal power-law $R(A)$ relations are listed for all three frequency bands, different temperatures, and horizontal polarization.Specific attenuation depends strongly on the wavelength. Therefore, the parameters of the $R(A)$ relation can change significantly even within a particular microwave band (S, C, or X). For the WSR-88D radars operating at S band, the multiplier in the $R(A)$ relation increases by 35% as the wavelength changes from 10.0 to 11.0 cm. At S band, the dependencies of the multiplier on temperature and wavelength can be approximated by the formula (Ryzhkov et al. 2014):

$$R(A) = c_1(T)c_2(\lambda)A_h^{1.03}, \quad (10.39)$$

where

Table 10.4 $R(A)$ relations at different radar wavelengths and temperatures for horizontal polarization

Temperature, °C	S band ($\lambda = 11.0$ cm)	C band ($\lambda = 5.3$ cm)	X band ($\lambda = 3.2$ cm)
0	$R = 2.23\ 10^3\ A^{1.03}$	$R = 221\ A^{0.92}$	$R = 49.1\ A^{0.87}$
10	$R = 3.10\ 10^3\ A^{1.03}$	$R = 250\ A^{0.91}$	$R = 45.5\ A^{0.83}$
20	$R = 4.12\ 10^3\ A^{1.03}$	$R = 294\ A^{0.89}$	$R = 43.5\ A^{0.79}$
30	$R = 5.33\ 10^3\ A^{1.03}$	$R = 352\ A^{0.89}$	$R = 43.0\ A^{0.76}$

R is in mm h^{-1}, A is in dB km^{-1}

$$c_1(T) = \left(2.23 + 0.078\,T + 0.00085\,T^2\right), \tag{10.40}$$

and

$$c_2(\lambda) = 1 - 0.26(11.0 - \lambda). \tag{10.41}$$

In (10.40) and (10.41), T is the temperature (°C) and λ is the wavelength in cm. Strong dependence of the multiplier in the power-law *R*(*A*) relation on temperature and wavelength seems, at first glance, to complicate rainfall estimation. However, for relatively small *PIA*, the factor *C*(*b*, *PIA*) in (10.36) can be expanded as

$$\begin{aligned} C(b, \mathrm{PIA}) &= \exp(0.23\,b, \mathrm{PIA}) - 1 \approx 0.23\,b, \mathrm{PIA} \\ &= 0.23\,b\alpha\Delta\Phi_{\mathrm{DP}}(r_1, r_2), \end{aligned} \tag{10.42}$$

and the specific attenuation A retrieved from (10.33) depends approximately linearly on $\alpha = A/K_{\mathrm{DP}}$. Therefore, the rain rate in (10.38) is proportional to the product $\gamma\alpha^{\delta}$. Because the exponent δ is very close to 1, the wavelength/temperature dependences of γ and α almost perfectly cancel in their product. This is because γ increases with λ but α decreases in the same proportion; therefore their product remains almost unchanged. Similar conclusion is valid for the dependence of α on temperature if it does not vary much along the propagation path in rain. Thus, to first approximation, it is possible to use a single *R*(*A*) at a given microwave frequency band and ignore the dependencies of γ and α on the wavelength and temperature. This was proven in practice by testing the single *R*(*A*) relation:

$$R(A) = 4120A^{1.03}, \tag{10.43}$$

valid at $T = 20$ °C and $\lambda = 11.0$ cm (Table 10.3), on the network of WSR-88Ds. However, the temperature dependence of γ should be accounted for if the temperature varies significantly along the propagation path in rain. This is because a constant α is used for a whole path in which γ changes and, therefore, product $\gamma\alpha^{\delta}$ does too. The temperature impact on the performance of the *R*(*A*) method at *C* and *X* bands is investigated in the studies of Wang et al. (2014) and Diederich et al. (2015), and both conclude that it is relatively weak.

The multiplier γ and the exponent δ in the *R*(*A*) power-law relation (10.38) are practically unaffected by the DSD variability (at least at *S* band), but the estimate of *A* from (10.33) to (10.37) is affected through α, and determining the correct α poses the biggest challenge to the *R*(*A*) method. This is also a problem in the polarimetric technique for attenuation correction because it is based on the estimate of the path-integrated attenuation *PIA* using a total span of differential phase $\Delta\Phi_{\mathrm{DP}}$ along the propagation path (Chap. 6).

The factor α (i.e., the ratio A/K_{DP}) depends on differential reflectivity Z_{DR} (Fig. 6.16). It monotonically decreases with increasing Z_{DR} at *S* band and reaches a relatively stable value for $Z_{\mathrm{DR}} > 1$–1.5 dB. A nonmonotonic dependence of α on

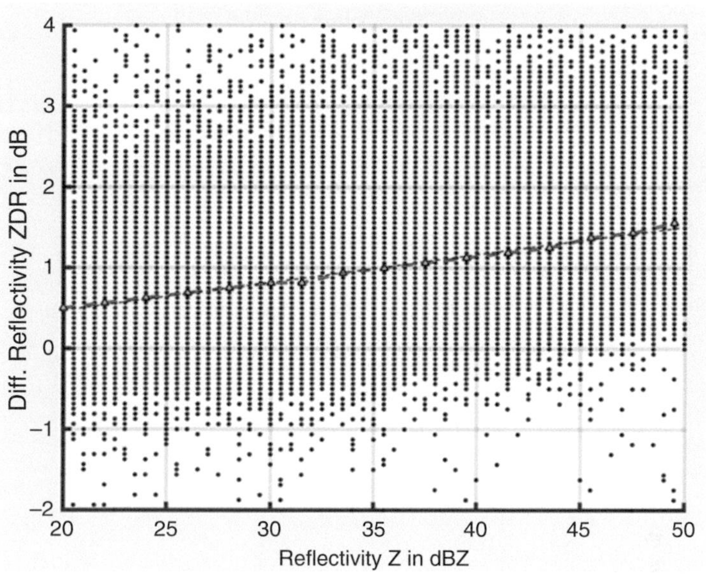

Fig. 10.20 Illustration of the Z_{DR} slope determination from the scatterplot of the measured Z–Z_{DR} pairs in rain over a whole volume scan

Z_{DR} occurs at C and X bands. A typical value of α for continental rain commonly characterized by larger Z_{DR} is 0.015 dB deg^{-1} at S band (Ryzhkov et al. 2014). Because the rain rate estimated from the $R(A)$ relation is roughly proportional to α, its estimate with fixed α tends to be negatively biased in tropical rain or light rain in general (which are characterized by lower values of Z_{DR}) if the value of α typical for continental rain is utilized.

To determine the optimal α, the following procedure is suggested. It uses the Z_{DR} slope $K = dZ_{DR}/dZ$ which is determined from a large number of Z–Z_{DR} pairs measured in rain as indicated in Fig. 10.20. Median Z_{DR} is computed for every 2 dB interval of Z, and the slope K is a linear fit to the resulting curve. The estimate of the Z_{DR} slope is updated every scan and converted into the value of α as follows:

$$\alpha = a_1 - a_2 K + a_3 K^2, \tag{10.44}$$

where the coefficients a_1, a_2, and a_3 are functions of Z_m, the maximal Z within the interval between 30 and 50 dBZ for which the median $Z_{DR}(Z_m)$ can be reliably estimated:

$$a_1 = 0.530 - 0.0188\,Z_m + 0.000194\,Z_m^2 \tag{10.45}$$

$$a_2 = 25.0 - 0.983\,Z_m + 0.0103\,Z_m^2 \tag{10.46}$$

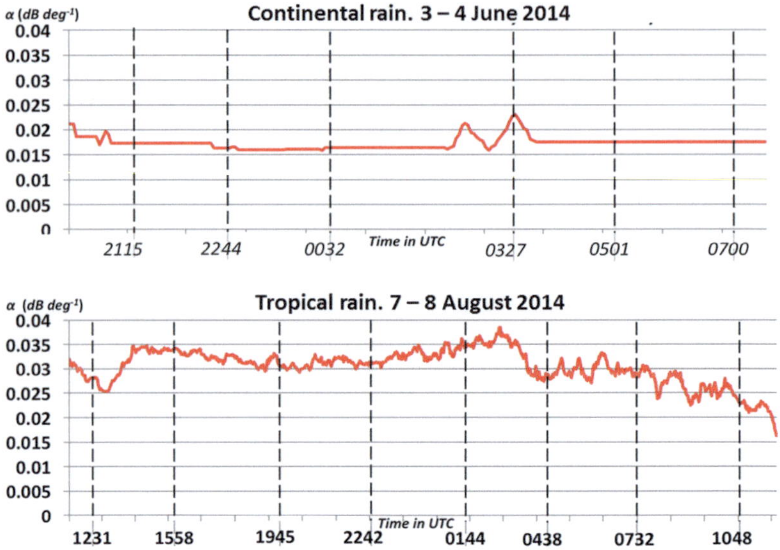

Fig. 10.21 Evolution of the parameter α in typical continental and tropical rain events observed by KOAX (Omaha, NE) and (KCLC (Charleston, SC) WSR-88Ds. Courtesy of Dr. S. Cocks

$$a_3 = 303 - 12.5 Z_m + 0.133 Z_m^2 \tag{10.47}$$

In (10.45)–(10.47), Z_m is in dBZ, and these relations have been obtained from simulations using disdrometer data and are valid at *S* band. It can be shown that α computed from (10.44) varies from 0.015 dB deg^{-1} to 0.065 dB deg^{-1}. More details about estimation of the Z_{DR} slope and α are in Wang et al. (2019) and Cocks et al. (2019). In the algorithm version described in these studies, a single value of the factor α is used in a whole radar coverage area containing rain, and this value is updated every scan.

Examples of the temporal variability of α during typical continental and tropical rain events observed by different WSR-88Ds are presented in Fig. 10.21. In the case of continental rain (top panel), the value of α hovers between 0.015 dB deg^{-1} and 0.025 dB deg^{-1}. In contrast, the tropical rain exhibits much higher and more variable α changing mostly between 0.025 dB deg^{-1} and 0.04 dB deg^{-1} (bottom panel).

The adjustment of α is also needed at *C* band; it varies from 0.06 to 0.07 dB deg^{-1} for continental rain (Giangrande et al. 2014) to about 0.09 dB deg^{-1} for tropical rain (Wang et al. 2013). Diederich et al. (2015) show that α between 0.27 and 0.30 dB deg^{-1} yields the best QPE results in Germany at *X* band. To produce robust *R*(*A*) estimates, the total span of differential phase $\Delta\Phi_{DP}$ along the beam should be at least few degrees (4–6° is suggested by Diederich et al. (2015) at *X* band and 3–4° at *S* band is suggested by Ryzhkov et al. (2014)). Otherwise, the *R*(*Z*) estimate is recommended to fill the gaps.

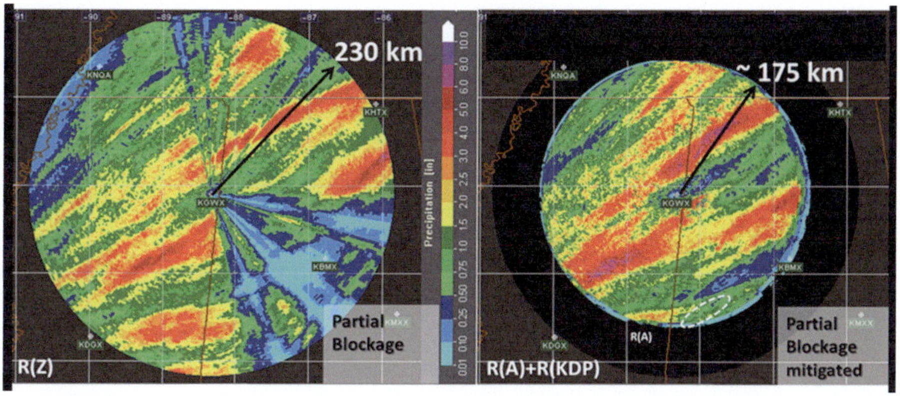

Fig. 10.22 The maps of 24-hour rain totals estimated using the $R(Z)$ relation (left panel) and the combination of the $R(A)$ and $R(K_{DP})$ relations (right panel) for the storm on 29 April 2014 observed with the KGWX (Columbus Air Force Base, MS) WSR-88D. The $R(K_{DP})$ relation is utilized in areas where rain is mixed with hail. Courtesy of Dr. S. Cocks

The immunity of $R(A)$ to partial beam blockage is demonstrated in Fig. 10.22. The maps of 24-hour rain totals estimated using the $R(Z)$ (left panel) and $R(A)$ (right panel) relations are compared for the storm on 29 April 2014 observed with the KGWX WSR-88D. The reflectivity is negatively biased in a wide SE sector due to the partial beam blockage caused by the trees in close proximity to the radar. Partial beam blockage of this type can't be compensated by using the digital elevation maps (DEM) as these do not contain tree heights. The $R(A)$ estimate of the storm rain total is not biased in the blocked sector.

Another important benefit of the $R(A)$ methodology is that it produces rain rate fields with high spatial resolution same as the one obtained from the $R(Z)$ relation. Figure 10.23 shows an example of the fields of rain rate derived from the $R(Z)$, $R(A)$, and $R(K_{DP})$ relations. The retrieved rain rates are plotted only at locations where $Z > 45$ dBZ and where the K_{DP}-based estimate of R can be potentially used at S band (assuming the statistical fluctuations of K_{DP} are not too high). The $R(A)$ estimate retains the same radial resolution as $R(Z)$ (i.e., 0.25 km). In the WSR-88D processing, the radial resolution of the $R(K_{DP})$ estimate is 2 km at $Z > 40$ dBZ which is eight times lower than the radial resolution of the $R(Z)$ and $R(A)$ estimates. Moreover, small-scale rain structures might be distorted, and negative rain rates do appear in the $R(K_{DP})$ map.

The $R(A)$ relation can be effectively utilized in a wide range of rain intensities, whereas the $R(K_{DP})$ is suitable only at sufficiently high rain rates. Because the $R(A)$ estimate is immune to the biases affecting Z and Z_{DR}, it is a good choice for networking and compositing of rainfall data collected by multiple radars (Diederich et al. 2015; Ryzhkov et al. 2014).

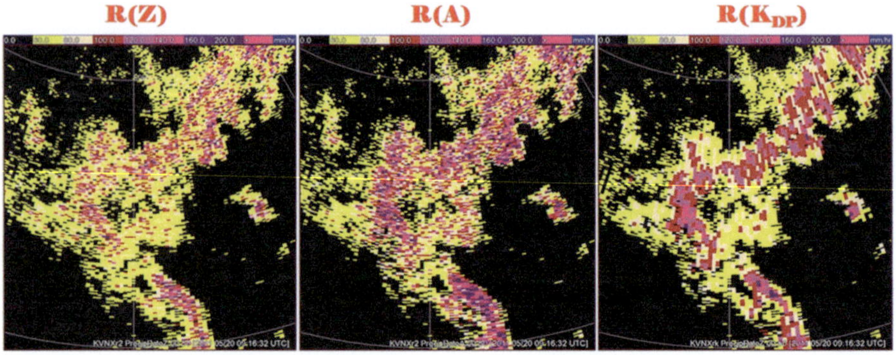

Fig. 10.23 Comparison of the texture of the fields of $R(Z)$, $R(A)$, and $R(K_{DP})$ measured by the KVNX (Vance Air Force Base, OK) WSR-88D on 20 May 2011 in northern Oklahoma. Only the data with $Z > 45$ dBZ are displayed. From Ryzhkov et al. (2014). © American Meteorological Society. Used with permission

Practical advantages of the $R(A)$ technique can be summarized as follows:

(1) Low sensitivity to the DSD variability
(2) Immunity to radar miscalibration, partial beam blockage, attenuation, and impact of wet radome
(3) High spatial resolution (same as of $R(Z)$)
(4) Uniformly good performance for light and heavy rain
(5) Seamless for networking and compositing (particularly for the radars operating at different wavelengths)

10.5 Composite Algorithms

All discussed radar rainfall relations have advantages and disadvantages at different rain rates and radar wavelengths. Hence it can be beneficial to use composite algorithms by combining various relations. This was suggested by Chandrasekar et al. (1993) and further explored by Ryzhkov et al. (2005), Matrosov et al. (2005), Giangrande and Ryzhkov (2008), Cifelli et al. (2011), Tabary et al. (2011), and Figueras i Ventura et al. (2013), among others. The most common approach in the past was to use the $R(K_{DP})$ or $R(K_{DP}, Z_{DR})$ relations at high rain rates and the $R(Z, Z_{DR})$ or $R(Z)$ relations at low rain rates. The criteria for switching between the different relations were based on the value of Z (Ryzhkov et al. 2005) with additional K_{DP} and Z_{DR} thresholds (Cifelli et al. 2011) or on the results of hydrometeor classification (Giangrande and Ryzhkov 2008). The $R(K_{DP})$ relation seems to be a sole solution for rain mixed with hail where Z is overestimated and Z_{DR} can be negatively biased by the presence of hail. The $R(Z, Z_{DR})$ relations are not practical at C band and are even less suited to X band as the impact of attenuation and differential attenuation are difficult to account for with a required precision. The $R(A)$ emerges as a primary choice for rain regardless of its intensity (at least at *S* band). It is complemented by the *R*(*Z*) relation if there is no sufficient span of Φ_{DP} along the propagation path in rain or by $R(K_{DP})$ for the rain-hail mixture.

The *R*(*Z*) relations overestimate rain rate if rain is mixed with hail. Capping *Z* at certain level (e.g., 53 dBZ on the WSR-88D network in the USA) generally reduces the bias. *R*(*Z*) also overestimates rain rate at the periphery of hail cells where $Z < 53$ dBZ because of the abundance of large drops originated from melting hail and small water-coated melting hailstones. The $R(Z, Z_{DR})$ relations are also not applicable in hail because of the significant increase of *Z* and decrease of Z_{DR} caused by hail mixed with rain. The *R*(*A*) relation is also not designed to estimate rain in the presence of hail.

The observational data provide ample evidence that the rainfall estimation algorithm based on the use of specific differential phase K_{DP} is the best choice in the situations of rain mixed with hail because K_{DP} is relatively insensitive to the presence of hail (e.g., Ryzhkov et al. 2005). The results of modeling study of melting hail mixed with rain (see Chap. 7) support this notion, although with some reservations.

Vertical profiles of "true" rain rates retrieved from the 1D model of melting hail (Chap. 7) and the corresponding $R(K_{DP})$ estimates for different hail sizes at the three wavelengths are displayed in Fig. 10.24. The $R(K_{DP})$ relation has the power-law form (10.20) in which the coefficients *a* and *b* were determined from simulations using a large set of DSD measurements in Oklahoma. The values are $a = 47.1$, 25.3, and 16.9 and $b = 0.790$, 0.776, and 0.801 at *S* ($\lambda = 11.0$ cm), *C* ($\lambda = 5.3$ cm), and *X* ($\lambda = 3.2$ cm) bands, respectively. In the cases of no hail and small hail, the vertical profiles of $R(K_{DP})$ (dashed curves in Fig. 10.24) follow the actual profiles of rain rate (solid curves in Fig. 10.24) quite well. For moderate and large hail sizes, $R(K_{DP})$ tends to overestimate rain rate close to the surface at *S* and *C* bands. Such overestimation is more pronounced at *S* band and is very mild at *C* band. This fully agrees with the results of theoretical simulations (Ryzhkov et al. 2013a)

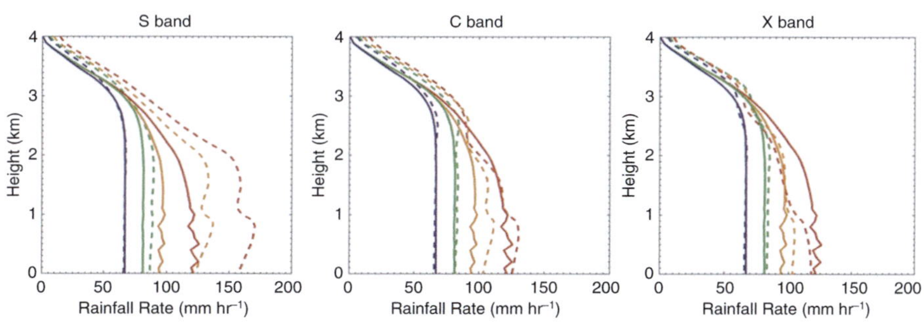

Fig. 10.24 Vertical profiles of simulated rain rates in the mixture of melting hail and rain (solid lines) and the estimates from the $R(K_{DP})$ relations (10.20) (dashed lines) at *S*, *C*, and *X* bands in the cases of no hail (blue curves), small hail (green curves), moderate hail (orange curves), and large hail (red curves). From Ryzhkov et al. (2013b). © American Meteorological Society. Used with permission

indicating that relative contribution of melting hail (i.e., hydrometeors with sizes exceeding 8 mm) to K_{DP} is more significant at *S* band than at *C* band. Measurements with *S*-band WSR-88Ds routinely exhibit very high K_{DP} in hailstorms often exceeding 7 km^{-1} corresponding to rain rates higher than 218 mm h^{-1}, obviously too high, even for the cases of severe hail. Another possible reason for overestimation is that rain mixed with hail is characterized by abundance of very large raindrops (some of them containing ice cores) and can be treated as a "very continental rain." For this rain type, the $R(K_{DP})$ relation with the multiplier $a = 44.0$ or 47.1 at *S* band may not be applicable. The $R(K_{DP})$ relation optimized for "very continental" rain (defined in Sect. 10.2) has much smaller a equal to 29.0. Therefore, the $R(K_{DP})$ relation with smaller a is the choice for areas of rain-hail mixture (Cocks et al. 2019).

With these modifications, the K_{DP}-based relations exhibit much better performance for hail-bearing storms than other options. The $R(Z)$ relation with Z capped at 53 dBZ (as on WSR-88D) overestimates significantly rain in the presence of small hail and underestimates it if larger hail is present. Anomalously high K_{DP} (6 deg km^{-1} or more at *S* band) indicates large or giant hail and can be used for hail detection and sizing (Chap. 9). The advantages of using K_{DP} for rainfall estimation in the presence of hail have been documented in observational studies at *S* band (Ryzhkov et al. 2013b) and *X* band (Matrosov et al. 2013) where the relation $R(K_{DP}) = 14\,K_{DP}^{0.77}$ was used.

10.6 Estimation of Surface Rain from Measurements in the Melting Layer and Snow

Once the radar beam intercepts the melting layer (ML) and snow/ice aloft, the relation between reflectivity (and other variables) and rain rate at the surface becomes complicated. This relation has to take into account the presence of nonliquid hydrometeors in the ML and the relative size of the beamwidth with respect to the ML thickness. To assess the impact of both factors on the measured vertical profiles of the polarimetric variables, a model of the intrinsic vertical profiles of Z, Z_{DR}, and ρ_{hv} within the melting layer and above in stratiform precipitation can be used (Fig. 10.25, upper panels).

The Z profile is characterized by a number of parameters: (1) maximal Z within the melting layer (Z_{peak}), (2) Z in rain below the melting layer (Z_{rain}), (3) Z in snow just above the melting layer (Z_{snow}), (4) vertical gradient of Z above the melting layer (β), (5) height of the top of the melting layer (H_t), and (6) the thickness of the melting layer (ΔH). Existing measurements of the vertical profile of Z through the melting layer with vertically pointing radars show that the difference Z_{peak}–Z_{rain} varies between 5 and 12 dB, vertical gradient β changes from 3 to 5 dB km^{-1}, and the thickness of the ML ΔH varies between 0.2 and 0.6 km (Fabry and Zawadzki 1995; Troemel et al. 2017; Wolfensberger et al. 2015).

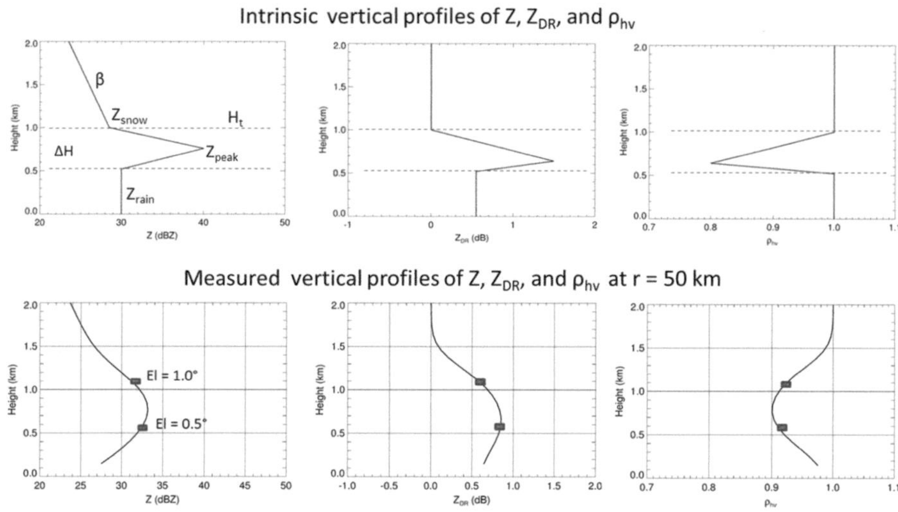

Fig. 10.25 Example of intrinsic (top panels) and measured (bottom panels) vertical profiles of Z, Z_{DR}, and ρ_{hv} at the distance 50 km from the radar for the antenna beamwidth of 1°. Black squares show the measurements available at the 0.5° and 1.0° antenna tilts

The true vertical profiles of Z, Z_{DR}, and ρ_{hv} can be transformed into "apparent" or "measured" profiles of these variables at any distance from the radar if the antenna pattern is known. As an example, the measured profiles at the distance of 50 km from the radar with the antenna beamwidth of 1° are displayed in Fig. 10.25 (bottom panels). These profiles would be measured if the beam is slowly scanning in the vertical plane. It is evident that the measured profiles of the polarimetric variables are smoothed. For typical antenna scanning patterns, the continuous vertical "apparent" profiles cannot be measured but only discrete points corresponding to specific beam elevations would be available. For example, the measurement points corresponding to beam elevations of 0.5° and 1.0° are marked by black squares in the bottom panels in Fig. 10.25. It is evident that the measured *Z* at the elevation 0.5° and $r = 50$ km is about 32.5 dBZ, i.e., 2.5 dB higher than the "true" one at the same height (left top panel in Fig. 10.25). Because *Z* below 0.5 km is constant in this model, 2.5 dB should be subtracted from the measured *Z* to obtain correct rain rate.

Assuming horizontally uniform rain, one can compute radial profiles of Z, Z_{DR}, and ρ_{hv} at the 0.5° elevation (Fig. 10.26). Evidently, the impact of the melting layer is manifested by the increase of *Z* and Z_{DR} and decrease of ρ_{hv} in the range interval affected by the bright band. The 30–130 km range interval is well marked by the drop of ρ_{hv} and the corresponding bump in Z_{DR}. Therefore, the radar rainfall relation for pure rain can be safely applied to distances less than 30 km, and the relation for pure snow can be applied at the distances exceeding 130 km. Once the radar beam passes through the ML, the reflectivity steadily decreases at the rate determined by the vertical gradient β of the reflectivity above the melting layer.

Fig. 10.26 Radial profiles of Z, Z_{DR}, and ρ_{hv} for $H_t = 1$ km, El = 0.5°, $Z_{peak} - Z_{rain} = 10$ dB, and $\Delta H = 0.4$ km

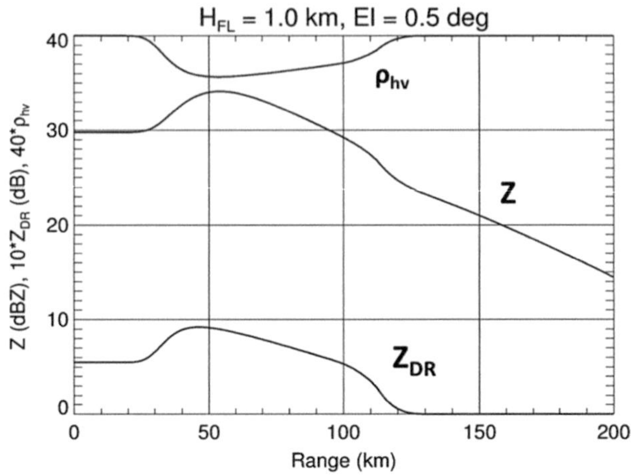

Fig. 10.27 Radial profiles of Z for different heights of the ML. Values of H_t (km) are shown next to each curve, Z_{peak}–$Z_{rain} = 8.5$ dB, $\Delta H = 0.4$ km, and $\beta = 5$ dB km^{-1}

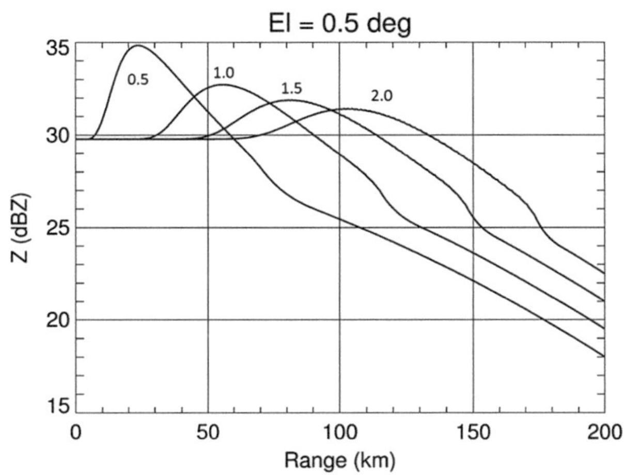

Overestimation of Z in the melting layer (ML) tends to decrease with increasing height of the ML (Fig. 10.27). Comparison of radial profiles of Z at El = 0.5° and 1.0° is illustrated in Fig. 10.28 where the difference in Z between the two elevations is approximately constant starting from about 60 km and dependent on the vertical gradient β. This makes possible the estimation of β by comparing radial profiles of Z at two close elevations.

Ideally, the method that eliminates the biases caused by the ML enhancement and beam overshooting should yield approximately range profile of Z close to Z_{rain} in Figs. 10.26, 10.27, and 10.28. Thus, the measured Z along the radials should be adjusted by subtracting the positive bias caused by the ML and adding the negative bias in the snow region above the ML.

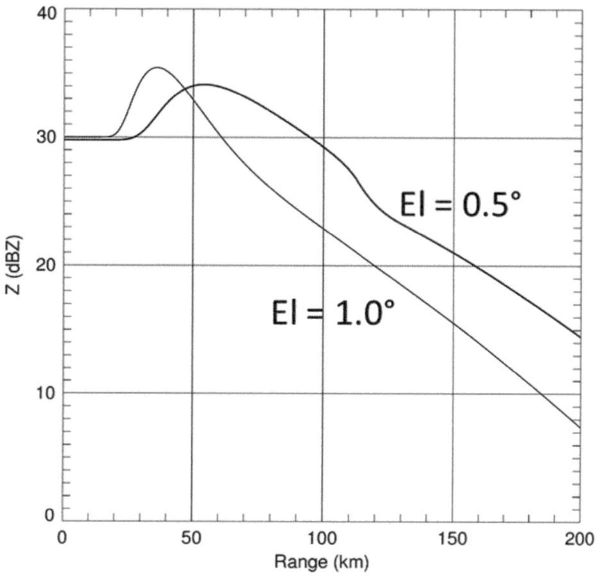

Fig. 10.28 Radial profiles of Z at El = 0.5° and El = 1.0°

Various methods have been devised to address the melting layer problem (e.g., Germann and Joss 2002; Kitchen et al. 1994; Matrosov et al. 2007; Seo et al. 2000). Common to these is the estimation of a vertical profile of reflectivity (VPR) and its use to retrieve rain rate near the ground. One of the ways to obtain the VPR is to scan the storm, reconstruct its structure near the radar, and extrapolate its effect to long range assuming horizontal homogeneity (Germann and Joss 2002; Koistinen 1991). Such volume scan VPR corrections are often part of operational algorithms for QPE. Andrieu and Creutin (1995) and Andrieu et al. (1995) added sophistication by applying discrete inverse theory based on hourly averaged VPRs at two elevation angles (1° and 3°). Local VPRs correction over small areas (20 km in range and 20° in azimuth) using an inverse filtering of the beam effects had been proposed by Vignal et al. (1999) and proven to be better in Switzerland (Vignal et al. 2000) and the USA (Vignal and Krajewski 2001) than "global" corrections.

Zhang and Qi (2010) introduced the concept of an "apparent" VPR or AVPR. It is the range (height) profile of azimuthally averaged reflectivities in the area affected by BB contamination. They use the lowest beams to obtain the AVPR's parameters. The AVPR procedure automatically includes the beam-smoothing effects. The idea of the method is illustrated in Fig. 10.29. The first step is discrimination between convective and stratiform parts of the storm based on the vertically integrated liquid (VIL) water. Then the "bright band affected area" (BBA) within the stratiform part of the storm is identified using a threshold of composite reflectivity. Next, the mean parametrized vertical profile of reflectivity (Z_a) within BBA is determined assuming horizontally uniform vertical structure within BBA. The VPR correction is done via subtraction of Z_a from the observed local profile of Z, and the corrected value of Z at the lowest level is used to estimate rain rate at the surface.

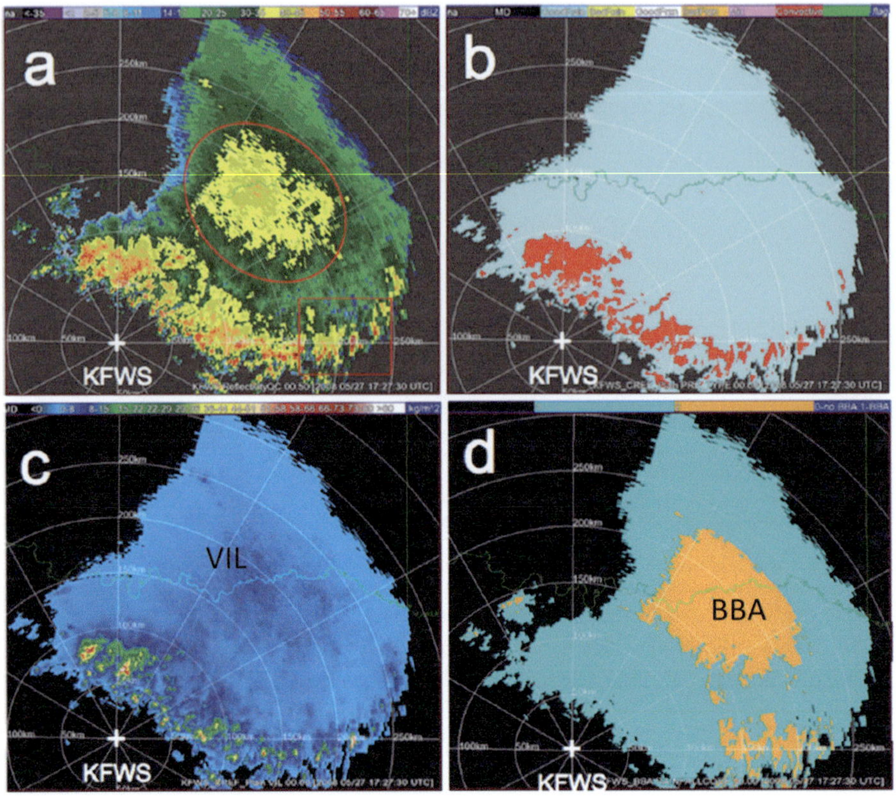

Fig. 10.29 (**a**) Reflectivity at elevation 0.5°, (**b**) precipitation type, (**c**) VIL, and (**d**) the bright band affected area (BBA) fields from the KFWS (Dallas Fr. Worth TX) WSR-88D on 27 May 2008, 1727 UTC. The red circle and squares in (**a**) indicate the area affected by the bright band. The colors in (**b**) represent convective (red) and stratiform (blue) precipitation regions. The colors in (**d**) represent BBA (orange) and non-BBA (blue). From Zhang and Qi (2010). © American Meteorological Society. Used with permission

Dual-polarization radars add a new dimension to identification of the BB and QPE because the cross-correlation coefficient ρ_{hv} provides much better delineation of the BBA than Z (Matrosov et al. 2007; Qi et al. 2013; Tabary et al. 2006; Zrnic et al. 1993, 1994). These studies report that inclusion of polarimetric information improves the VPR correction.

An alternative method to mitigate the bright band effect and beam overshooting uses the hydrometeor classification algorithm (HCA) to identify the affected range locations (Giangrande and Ryzhkov 2008). At these locations the following $R(Z)$ relations generate the QPEs:

$$R = 0.6R(Z) \tag{10.48}$$

in wet snow (WS),

$$R = R(Z) \tag{10.49}$$

in dry snow (DS) if the distance from the radar $r < R_{tt}$, and

$$R = 2.8R(Z) \tag{10.50}$$

in dry snow (DS) and crystals (CR) if $r > R_{tt}$, where

$$R(Z) = 1.69 \times 10^{-2} 10^{0.0714Z} \tag{10.51}$$

and R_{tt} is the distance where the bottom of the beam intercepts the top of the melting layer H_t (Fig. 9.6). Although (10.48)–(10.51) produce unbiased estimates if the same scatterer types fill the resolution volume, the QPE close to the boundary separating the hydrometeor classes is biased and exhibits a discontinuity at the boundary.

An alternative methodology to correct the effects of the BB is suggested by Troemel et al. (2017). It starts with modeling the intrinsic vertical profiles of Z, Z_{DR}, and ρ_{hv} by considering the physical and statistical correlations between these three variables and key parameters of the Z profile through the melting layer (ML) like the ML thickness, the difference between maximal Z in the ML and Z in rain below, vertical gradient of Z above the ML, etc. The second step involves generating massive lookup tables containing vertical profiles of the polarimetric variables smoothed by beam at various ranges. Finally, the appropriate intrinsic radial profile of Z or $R(Z)$ is selected from the lookup table depending on (1) the height of the ML and (2) the values of Z, Z_{DR}, and ρ_{hv} measured by the radar at the lowest elevation scan.

Figure 10.30 illustrates radial profiles of Z and ρ_{hv} for three strengths of the ML (characterized by the difference $\Delta Z = Z_{peak} - Z_{rain}$) and thicknesses ΔH of the ML. Because the shape of the radial profile of ρ_{hv} correlates well with ΔZ and ΔH, one can choose an appropriate Z-profile and estimate the bias of Z attributed to the bright band effect at every distance from the radar (i.e., the profile 3 of ρ_{hv} corresponds to the profile 3 of Z).

This methodology fully utilizes polarimetric data and makes no assumption that the vertical structure of a storm is horizontally uniform. Establishing statistical correlations between polarimetric variables in the melting layer is necessary before generating an optimal set of intrinsic vertical profiles through the ML and creation of lookup tables. At longer distances, where range dependence of Z is determined by the slope β, this slope and the corresponding negative bias of Z can be obtained from the difference of the measured Z at two successive antenna elevations (e.g., 0.5° and 1.0°) (Fig. 10.28).

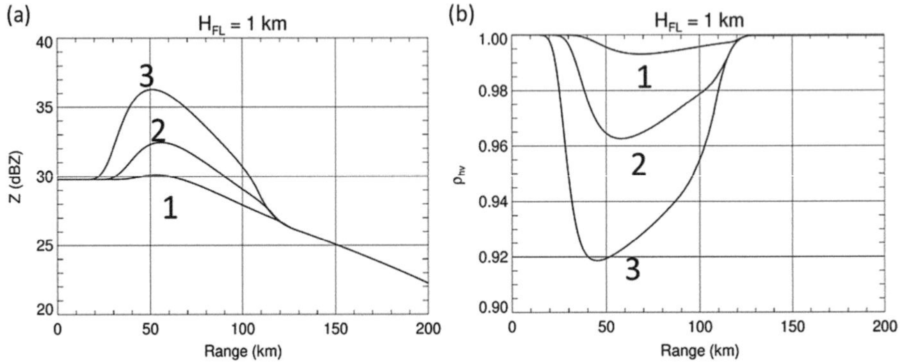

Fig. 10.30 Radial profiles of Z and ρ_{hv} for different ΔZ, i.e., the difference between maximal Z within the ML (Z_{peak}) and Z in rain below the ML (Z_{rain}), and the depth ΔH of the ML. Index "1" corresponds to $\Delta Z = 4$ dB and $\Delta H = 0.2$ km, index "2" to $\Delta Z = 8$ dB and $\Delta H = 0.4$ km, and index "3" to $\Delta Z = 12$ dB and $\Delta H = 0.6$ km. The freezing level (top of the ML) is at 1 km

10.7 Validation of Polarimetric Rainfall Estimators

Rain gauges are used to validate the performance of radar rainfall estimators by examining the following statistical measures: the bias *B*, the mean absolute error *MAE*, the root-mean-square error *RMSE*, the standard deviation *SD*, and the correlation coefficient *CC*. They are defined as

$$B = < T_R - T_G > \tag{10.52}$$

$$MAE = < | T_R - T_G | > \tag{10.53}$$

$$RMSE = < (T_R - T_G)^2 >^{1/2} \tag{10.54}$$

$$SD = \left(RMSE^2 - B^2\right)^{1/2} \tag{10.55}$$

$$CC = \frac{< (T_R - < T_R >)(T_G - < T_G >) >}{< (T_R - < T_R >)^2 >^{1/2} < (T_G - < T_G >)^2 >^{1/2}}, \tag{10.56}$$

where T_R and T_G are radar and gauge rain totals for a given radar-gauge pair and brackets indicate averaging over all pairs. Fractional errors defined via division of *B*, *MAE*, *RMSE*, and *SD* by $<T_G>$ are also common metrics of the QPE quality (e.g., Ryzhkov et al. 2005).

Massive statistical validations of different polarimetric QPE algorithms over a large number of rain events are relatively rare. First examples include the studies of Brandes et al. (2002) and Ryzhkov et al. (2005). Brandes et al. (2002) examine performance of four QPE relations: $R(Z)$, $R(K_{DP})$, $R(K_{DP}, Z_{DR})$, and $R(Z, Z_{DR})$ (Table 10.3) in subtropical Florida environment using the data from the S-Pol radar belonging to the National Center for Atmospheric Research. The study contains analysis of 17 rain events and data collected by 30 rain gauges located 60–90 km from the radar. The $R(Z, Z_{DR})$ estimator demonstrated the best performance in terms of *B*, *RMSE*, *SD*, and *CC*.

Ryzhkov et al. (2005) investigate the potency of the polarimetric prototype of the NEXRAD radar (KOUN) to estimate rainfall during the Joint Polarization Experiment (JPOLE) in central Oklahoma. A dense gauge network—the Agricultural Research Service (ARS) Micronet—was used to validate different polarimetric rainfall estimators. The Micronet contained 42 gauges located at the distances between 70 and 108 km from the KOUN radar. The dataset includes 24 rain events with 50 h of observations. In addition to the $R(Z)$, $R(K_{DP})$, $R(Z, Z_{DR})$, and $R(K_{DP}, Z_{DR})$ relations, a "synthetic" algorithm combining the three polarimetric variables has been tested (see Table 10.3). The synthetic polarimetric algorithm produces a significant reduction in the RMS errors of hourly rain estimates compared to the $R(Z)$ relation: 1.7 times for point measurements and 3.7 times for areal rainfall average (Fig. 10.31). The hourly rain total averaged over a small (1 km × 1°) area centered on a gauge is considered a point measurement. Areal mean hourly total or areal mean

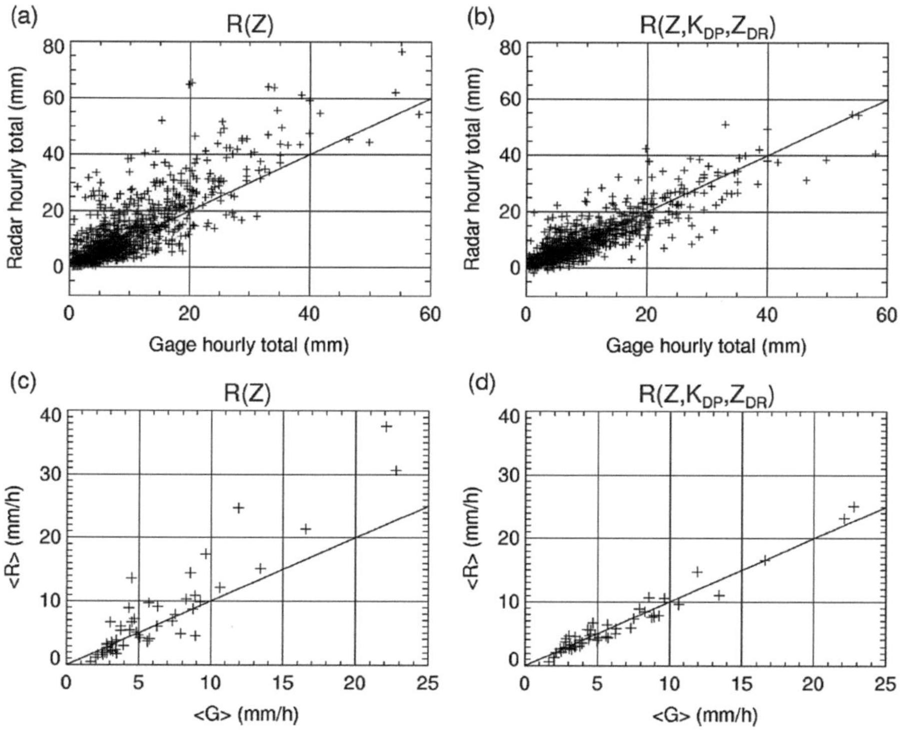

Fig. 10.31 One-hour (top) accumulations and mean areal rain rates (bottom) vs estimates from the $R(Z)$ relation and the "synthetic" $R(Z, K_{DP}, Z_{DR})$ relation of Ryzhkov et al. (2005) (24 rain events, 50 h of observations). Areal mean rain rate is estimated over the 50 × 40 km ARS Micronet area in central Oklahoma. From Ryzhkov et al. (2005). © American Meteorological Society. Used with permission

rain rate <*G*> is determined as a sum of hourly accumulations from all gauges that recorded rain divided by the number of these gauges. Better QPE improvement of areal rainfall estimates is obtained because the relations containing K_{DP} are particularly beneficial for large spatial/temporal domain (Borowska et al. 2011; Ryzhkov et al. 2000).

Another large QPE validation study was carried out by Giangrande and Ryzhkov (2008) for 43 rain events encompassing 179 h of observations with the KOUN WSR-88D in Oklahoma. They verified a number of radar QPE algorithms using the Oklahoma Mesonet gauges within 250 km of KOUN and demonstrate that the algorithm based on the polarimetric echo classification exhibits the best performance. It uses the $R(Z, Z_{DR})$ relation

$$R(Z, Z_{DR}) = 1.42 \times 10^{-2} Z^{0.770} Z_{dr}^{-1.67} \tag{10.57}$$

in the areas classified as pure rain (i.e., RA, HR, and BD according to the classification described in Chap. 9, Sect. 9.2.1) and the $R(K_{DP})$ relation

$$R(K_{DP}) = 44.0 |K_{DP}|^{0.822} \mathrm{sign}(K_{DP}) \tag{10.58}$$

if rain is mixed with hail (RH class). The $R(Z)$ relations with various "fudge" factors specified in (10.48) to (10.51) are applied at the distances where the dominant hydrometeors are classified as wet snow, dry snow, graupel, or ice crystals. From Fig. 10.32, it is evident that the polarimetric algorithm yields much smaller bias and RMS error than the conventional one, $R(Z)$, up to the ranges of 150 km. The most

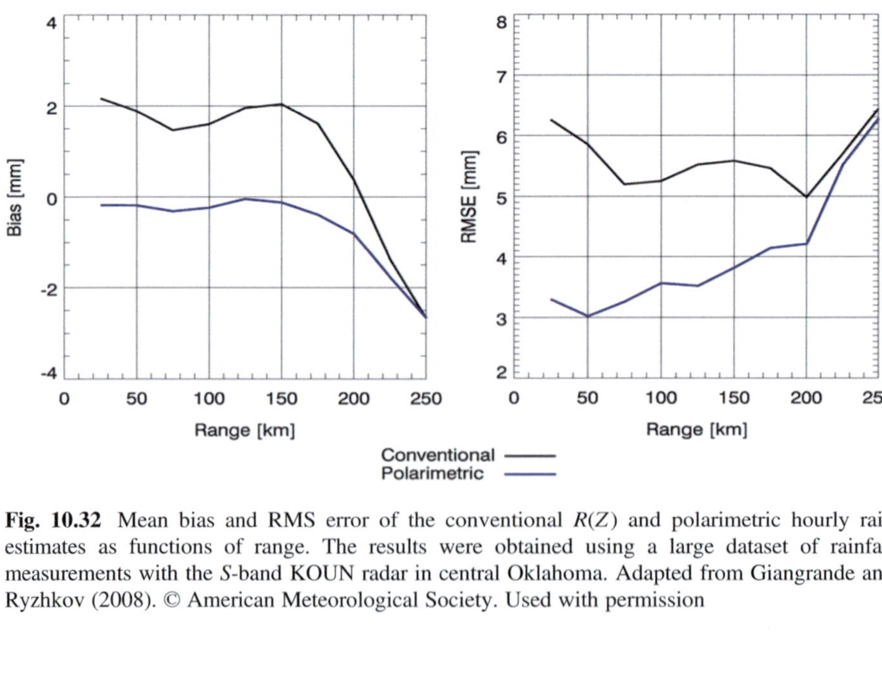

Fig. 10.32 Mean bias and RMS error of the conventional $R(Z)$ and polarimetric hourly rain estimates as functions of range. The results were obtained using a large dataset of rainfall measurements with the *S*-band KOUN radar in central Oklahoma. Adapted from Giangrande and Ryzhkov (2008). © American Meteorological Society. Used with permission

significant improvement is in convective storms where the RMS error of the hourly rain estimate is reduced by a factor of 2 at ranges shorter than 50 km. The improvement gradually decreases with range and becomes insignificant beyond 200 km.

In Europe, large QPE validation studies have been performed by Tabary et al. (2011), Figueras i Ventura et al. (2013), and Diederich et al. (2015) using polarimetric weather radars operating at *S*, *C*, and *X* bands. Tabary et al. (2011) studied the performance of the conventional *R*(*Z*) and two polarimetric methods, $R(Z, Z_{DR})$ and ZPHI (Testud et al. 2000), at C band using data collected during 12 rain events near Paris. They found that the $R(Z, Z_{DR})$ is the best for hourly rainfall accumulations smaller than 5 mm (if *Z* and Z_{DR} are corrected for attenuation using differential phase Φ_{DP}), whereas the ZPHI method outperforms others for larger rain totals. The Tabary et al.'s (2011) study was followed by Figueras i Ventura and Tabary (2013) on a larger dataset (35 events) where the conventional and polarimetric QPE methods are tested at *S*-, *C*-, and X-band radars comprising the Meteo France operational radar network. The results clearly show superior performance of the polarimetric methods, particularly at shorter wavelengths where polarimetric attenuation correction and the $R(K_{DP})$ relations demonstrate obvious benefits.

Rain rate retrievals from specific attenuation at *X* band have been thoroughly investigated by Diederich et al. (2015). They examine continuous time series of overlapping measurements by two polarimetric X-band radars in Germany during the summers of 2011–2013 and analyze various aspects of the *R*(*A*) method like miscalibration correction, ground clutter contamination, partial beam blockage, sensitivity to precipitation characteristics, and sensitivity to temperature. They conclude that *R*(*A*) complemented with *R*(*Z*) in segments where the total differential phase is small, correlates best with gauges, and has the lowest bias and RMSE. Next is $R(K_{DP})$ with *R*(*Z*) substituted at rain rates below 8 mm h^{-1}.

Several validation tests of the *R*(*A*) algorithm (complemented by $R(K_{DP})$ for $Z > 50$ dBZ) have been done using data from the network of WSR-88Ds. One of these includes measurements by 37 WSR-88Ds on 56 calendar days during the 2014–2016 warm seasons (Cocks et al. 2019). The performance of three QPE algorithms is analyzed in the Cocks et al. (2019) study: (1) *R*(*A*) with fixed value of $\alpha = 0.015$ dB deg^{-1}, (2) *R*(*A*) with variable α optimized using the Z_{DR} slope, and (3) QPE algorithm initially implemented on WSR-88Ds that utilizes $R(Z, Z_{DR})$ in pure rain and $R(K_{DP})$ in rain mixed with hail (Giangrande and Ryzhkov 2008) labeled as $R(Z, Z_{DR}, K_{DP})$. Scatterplots of the storm totals measured by the gauges versus their radar estimates from the three algorithms are displayed in Fig. 10.33. It is obvious that the *R*(*A*) algorithm with variable α outperforms the two others in all statistical metrics. As expected, *R*(*A*) with fixed $\alpha = 0.015$ dB deg^{-1} tends to underestimate tropical rain characterized by higher values of α.

The *R*(*A*) algorithm with variable α is particularly efficient for estimating heavy or extreme rain totals associated with potential flash floods. This is clearly indicated in a normalized Hit/Miss contingency table (Wilks 2006)—Table 10.5. The 24-hour gauge totals *G* were divided into five categories: very light (VL, $G < 12.7$ mm), light (L, $12.7 < G < 38.1$ mm), moderate (M, $38.1 < G < 101.6$ mm), heavy (H,

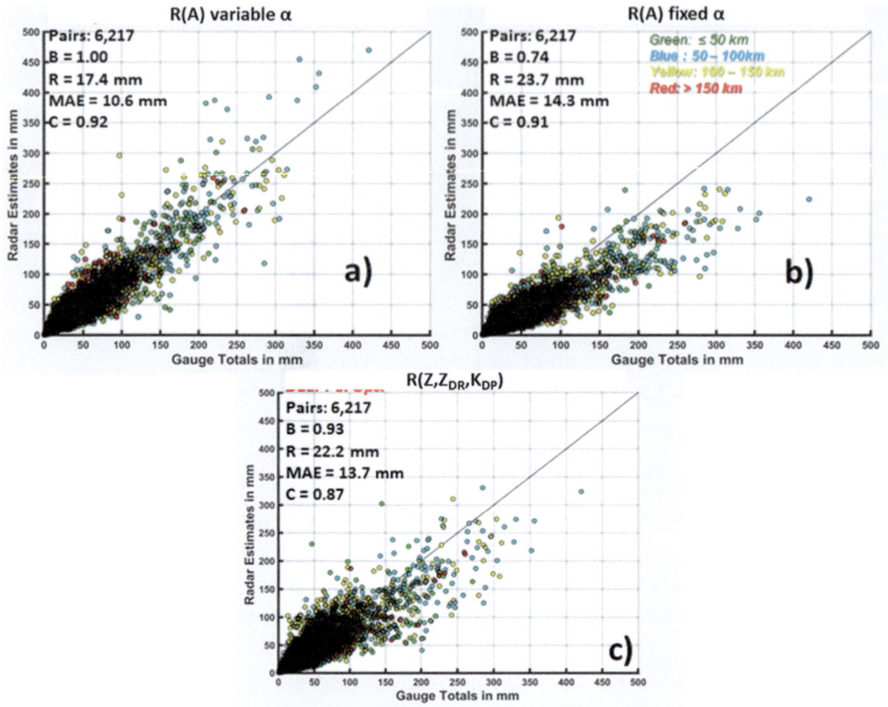

Fig. 10.33 Scatterplots of storm totals measured by gauges versus their radar estimates using three algorithms: (1) $R(A)$ with variable α, (2) $R(A)$ with fixed α, and (3) $R(Z, Z_{DR}, K_{DP})$ initially implemented on WSR-88Ds (Giangrande and Ryzhkov 2008). The data from 56 calendar days are accumulated in the scatterplot. Symbols of different colors indicate distance ranges from the radars. Symbols B, R, MAE, and C stand for radar-to-gauge bias ratio, root-mean-square error, mean absolute error, and correlation coefficient. Courtesy of Dr. S. Cocks

101.6 < G < 152.4 mm), and extreme (E, G > 152.4 mm). If the radar total category matches the gauge total category, then it is considered as a "Hit"; otherwise it is classified as a "Miss." The "Hit" scores are shown in bold on the diagonals of the three matrices in Table 10.5. The "Hit" score of the $R(A)$ algorithm with variable α is the highest with large margin for the categories of heavy and extreme rain totals (H and E).

In the statistics of rainfall measurements displayed in Fig. 10.33, the cases of poorly calibrated Z and Z_{DR} are excluded. However, the impact of miscalibration is devastating in some including the Louisiana flood event on 12 August 2016 (Fig. 10.34). During that event, Z_{DR} measured by the KLIX WSR-88D in New Orleans had negative bias of about 1.3 dB, and Z was positively biased by about 0.8 dB. The combination of these two biases causes large overestimation of rain by the $R(Z, Z_{DR}, K_{DP})$ algorithm (left panel in Fig. 10.34).

$R(A)$ with dynamically updated α produces much better results because the estimate of A is immune to radar miscalibration (right panel in Fig. 10.34).

Table 10.5 Hit/Miss contingency matrices for storm rain totals estimated by three QPE algorithms. "Hit" scores are shown in bold. Courtesy of Dr. S. Cocks

Algorithm	Rain total category	VL	L	M	H	E
$R(A)$ variable α	VL	**0.79**	*0.09*			
	L	*0.21*	**0.74**	*0.18*		
	M		*0.16*	**0.77**	*0.45*	*0.03*
	H			*0.05*	**0.48**	*0.19*
	E				*0.07*	**0.78**
		VL	**L**	**M**	**H**	**E**
$R(A)$ fixed α	VL	**0.90**	*0.25*	*0.01*		
	L	*0.10*	**0.69**	*0.41*	*0.02*	
	M		*0.06*	**0.57**	*0.87*	*0.16*
	H			*0.01*	**0.11**	*0.48*
	E					**0.36**
		VL	**L**	**M**	**H**	**E**
$R(Z, Z_{DR}, K_{DP})$	VL	**0.83**	*0.14*			
	L	*0.17*	**0.68**	*0.23*	*0.01*	
	M		*0.18*	**0.71**	*0.59*	*0.20*
	H			*0.05*	**0.38**	*0.31*
	E				*0.02*	**0.49**

Symbols VL, L, M, H, and E stand for very light, light, moderate, heavy, and extreme rain totals

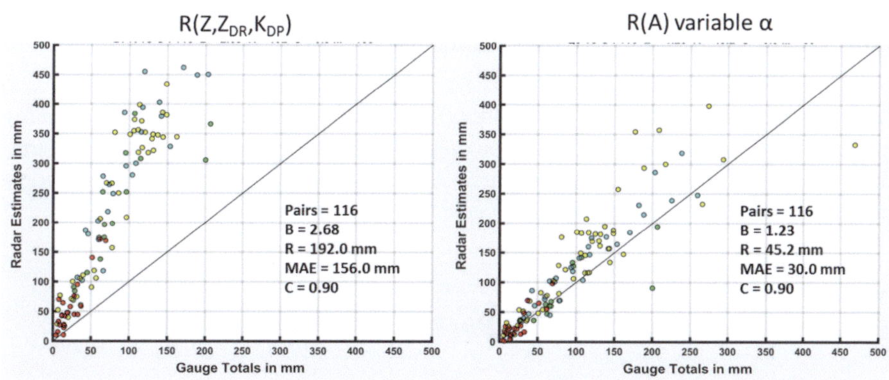

Fig. 10.34 Scatterplots of storm total measured by gauges versus their estimates from the $R(Z, Z_{DR}, K_{DP})$ algorithm (left panel) and the $R(A)$ algorithm with variable α (right panel) for the Louisiana flash flood event on 12 August 2016. Z_{DR} measured by the KLIX (New Orleans LA) WSR-88D was negatively biased by 1.3 dB, and Z was positively biased by 0.8 dB. Courtesy of Dr. S. Cocks

A larger-scale evaluation of the *R(A)* method was made for 269 events across the continental USA during May–September 2017. These consisted of mesoscale convective systems, hurricanes (Harvey and Irma), widespread stratiform rain in the east and northeast, scattered convective storms in the semiarid regions of the southwest, etc. The results summarized by Zhang et al. (2017) and presented in Fig. 10.35 are

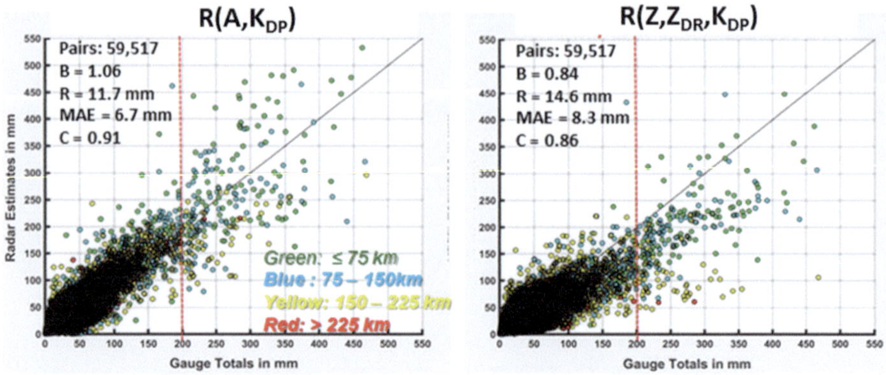

Fig. 10.35 Scatterplots of 24-hour rain totals measured by gauges versus their estimates from the $R(A)/R(K_{DP})$ algorithm and the $R(Z, Z_{DR}, K_{DP})$ algorithm for 269 rain events observed by 102 WSR-88Ds in May–September 2017. Courtesy of Dr. S. Cocks

the scatterplots of 24-h rain totals in gauges versus radar estimates from the two QPE methodologies. Again, the $R(A, K_{DP})$ algorithm exhibits better performance than the $R(Z, Z_{DR}, K_{DP})$ algorithm.

10.8 Polarimetric Measurements of Snow

Radar measurements of snow are challenging due to tremendous variability of snow particle size distributions (PSDs), their density, water content, shape, and orientation. Therefore, radar quantitative precipitation estimation (QPE) of snow is very difficult.

10.8.1 Snow Measurements Based on Reflectivity

Past studies of snowfall estimation with radar used power-law relations between the equivalent reflectivity (Z_e, herein Z) and snow water equivalent (SWE) snowfall rate S in the form $Z = aS^b$ (e.g., Fujiyoshi et al. 1990; Gunn and Marshall 1958; Heymsfield et al. 2016; Huang et al. 2010, 2015; Koistinen et al. 2003; Matrosov et al. 2007, 2009; Ohtake and Henmi 1970; Puhakka 1975; Rasmussen et al. 2003; Sekhon and Srivastava 1970; Szyrmer and Zawadzki 2010; Von Lerber et al. 2017; Wolfe and Snider 2012; Zhang et al. 2011). These relations (some are presented in Table 10.6) exhibit roughly an order of magnitude difference in the estimates of snowfall rate for the same reflectivity factor Z.

To explain this large variability, we derive analytical expressions for S and Z using simplified assumptions about snow size distributions and snow density as

Table 10.6 Summary of $Z(S)$ relations for dry snow listed in the literature and utilized by the WSR-88D network in the USA

Source	$Z(S)$ relation for dry snow
Gunn and Marshall (1958)	$Z = 448\ S^2$
Sekhon and Srivastava (1970)	$Z = 399\ S^{2.21}$
Ohtake and Henmi (1970)	$Z = (90\text{–}739)\ S^{(1.5\text{–}1.7)}$
Puhakka (1975)	$Z = 235\ S^2$
Koistinen et al. (2003)	$Z = 400\ S^2$
Matrosov et al. (2009)	$Z = (100\text{–}130)\ S^{(1.3\text{–}1.55)}$
Huang et al. (2010)	$Z = (106\text{–}305)\ S^{(1.11\text{–}1.92)}$
Saltikoff et al. (2010)	$Z = 100\ S^2$
Szyrmer and Zawadzki (2010)	$Z = 494\ S^{1.44}$
Wolfe and Snider (2012)	$Z = 110\ S^2$
Huang et al. (2015)	$Z = (130\text{–}209)\ S^{(1.44\text{–}1.81)}$
Von Lerber et al. (2017)	$Z = (53\text{–}782)\ S^{(1.19\text{–}1.61)}$
WSR-88D, Northeast	$Z = 120\ S^2$
WSR-88D, Great Lakes	$Z = 180\ S^2$
WSR-88D, North Plains/Upper Midwest	$Z = 180\ \mathrm{S}^2$
WSR-88D, High Plains	$Z = 130\ S^2$
WSR-88D, Inter-mountain West	$Z = 40\ S^2$
WSR-88D, Sierra Nevada	$Z = 222\ S^2$

a function of snowflake size and degree of riming. We start with the following basic equations for S and Z (see also Eqs. (10.4) and (5.14):

$$S = 6 \times 10^{-4}\pi \int_0^{D_{\max}} \frac{\rho_s(D)}{\rho_w} D^3 V_t^{(s)}(D) N(D) dD, \tag{10.59}$$

and

$$Z = \frac{|K_i|^2}{|K_w|^2} \int_0^{D_{\max}} \frac{\rho_s^2(D)}{\rho_i^2} D^6 N(D) dD. \tag{10.60}$$

We assume the density of dry snow decreases with equivolume diameter of snowflake D as specified by Eq. (4.26):

$$\rho_s(D) = \alpha D^{-\beta} = \alpha_u f_{\mathrm{rim}} D^{-\beta}. \tag{10.61}$$

According to Zawadzki et al. (2005, 2010), Zhang et al. (2011), and Moisseev et al. (2017), the terminal velocity of snowflakes can be approximated by

$$V_t^{(s)} = a_u f_{\text{rim}}^{\chi} D^{b_v}, \tag{10.62}$$

where a_u is a function of the temperature T_s at the location of snow and its distance from the cloud top H_{st},

$$a_u = 0.73 + 0.037 H_{st} + 0.011 T_s, \tag{10.63}$$

χ varies from 0.33 to 0.50, and b_v is between 0.17 and 0.22.

In (10.60), (10.61), (10.63), and (10.64), D is in mm, $V_{\text{t}}^{(s)}$ is in m s^{-1}, H_{st} is in km, and T_s is in °C.

We assume that size distributions of snowflakes are exponential:

$$N(D) = N_{0s}\exp(-\Lambda_s D), \tag{10.64}$$

where the slope Λ_{s} usually varies between 1 and 10 mm^{-1} (e.g., Heymsfield et al. 2002; Ryan 2000). The maximal size of dry snowflakes D_{max} commonly changes from 1 to 10 mm and is related to the slope Λ_{s} as (Heymsfield et al. 2002)

$$D_{\text{max}} = 11.6\,\Lambda_s^{-0.91}, \tag{10.65}$$

where D_{max} is in mm and Λ_s is in mm^{-1}. Integration over the size spectrum in (10.59) and (10.60) between 0 and ∞ produces the following expressions for S and Z:

$$S = \frac{6 \times 10^{-4} \pi \alpha_u a_u \Gamma(4 - \beta + b_v)}{\rho_w} \frac{f_{\text{rim}}^{1+\chi} N_{0s}}{\Lambda_s^{(4-\beta+b_v)}} \tag{10.66}$$

and

$$Z = \frac{|K_i^2|}{|K_w^2|} \frac{\alpha_u^2 \Gamma(7 - \beta)}{\rho_i^2} \frac{f_{\text{rim}}^2 N_{0s}}{\Lambda_s^{(7-\beta)}}. \tag{10.67}$$

Combining (10.66) and (10.67) and assuming $\beta = 1$, $\chi = 0.4$, and $b_{\text{v}} = 0.2$, one can obtain

$$S \sim f_{\text{rim}}^{0.12} N_{0s}^{0.36} Z^{0.64} \tag{10.68}$$

and

$$Z \sim f_{\text{rim}}^{-0.19} N_{0s}^{-0.56} S^{1.56}. \tag{10.69}$$

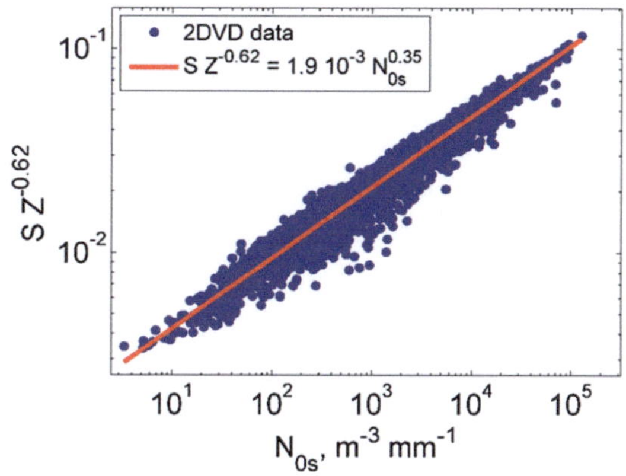

Fig. 10.36 Scatterplot of N_{0s} vs $S/Z^{0.62}$ from 2DVD retrievals (blue dots). The best fit to 2DVD data, $S\,Z^{-0.62} = 1.9\ 10^{-3}\ N_{0s}{}^{0.35}$, is the overlaid red line. From Bukovcic et al. (2018). © American Meteorological Society. Used with permission

In (10.68) and (10.69), the multipliers of Z and S are practically insensitive to the degree of snow riming f_{rim} but depend strongly on the intercept of the exponential size distribution N_{0s}.

The dependence of the prefactor a of the power-law relation $S = aZ^b$ on N_{0s} has been investigated experimentally by Bukovcic et al. (2018) using 2D-video disdrometer data from 16 snowstorms in Oklahoma. Disdrometer-based estimates of S, N_{0s}, and Z (computed for S band) have been used to determine the exponent b in the $S(Z)$ relation. Bukovcic et al. (2018) found the maximum correlation coefficient (0.978) between N_{0s} and S/Z^b occurs if $b = 0.62$, a value very close to 0.64 in (10.68) and (10.69) obtained from theoretical considerations.

The corresponding scatterplot of $S/Z^{0.62}$ (Fig. 10.36) clearly demonstrates S can vary more than an order of magnitude for a given Z and four orders of magnitude change in N_{0s}. The best linear fit to the scatterplot in Fig. 10.36 is

$$S = 1.9 \times 10^{-3} N_{0s}^{0.35} Z^{0.62}, \tag{10.70}$$

and the exponents on N_{0s} and Z are in excellent agreement with the theoretical prediction 0.36 and 0.64 specified in (10.68).The graph in Fig. 10.36 explains the large variability of the $Z(S)$ relations listed in Table 10.6.

Because N_{0s} can't be measured directly by radar, a "best guess" can be made using the dependence of N_{0s} on air temperature (Chap. 4, Sect. 4.1.3). N_{0s} and Λ_s depend on the snow habit and temperature (e.g., Heymsfield et al. 2002; Ryan 2000; Woods et al. 2008). Both increase with height and decreasing temperatures. The increase can be dramatic within a particular storm as indicated in the example taken from the study of Molthan et al. (2010) (Fig. 10.37). The primary reason for decrease of N_{0s}, Λ_s, and snow density ρ_s with increasing temperature and decreasing altitude is aggregation of ice crystals and snowflakes. An average climatological dependence of Λ_s on temperature as specified by Ryan (2000) is

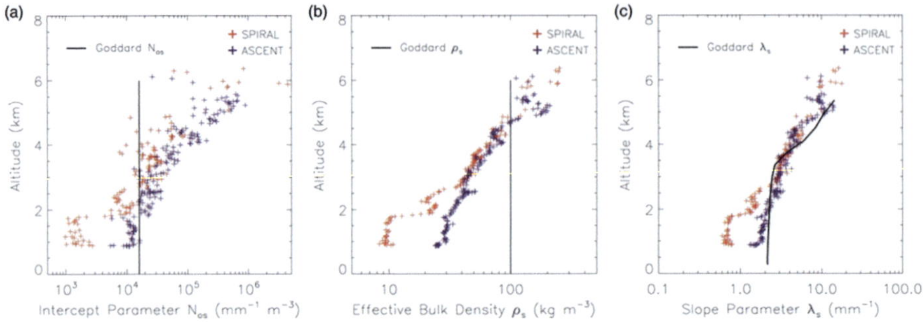

Fig. 10.37 Example of aircraft observations of the vertical profiles of N_{0s}, snow density ρ_s, and Λ_s. From Molthan et al. (2010). © American Meteorological Society. Used with permission

$$\Lambda_s\left(\text{mm}^{-1}\right) = 1.21\exp(-0.0564\,T). \tag{10.71}$$

Whereas Λ_s decreases with temperature, the temperature dependence of N_{0s} is much less certain (Fig. 4.10 in Chap. 4). However, for the snow size distributions of Gunn and Marshall (1958) and Sekhon and Srivastava (1970), the dependence can be obtained using the $N_{0s}(\Lambda_s)$ relations derived by Matrosov et al. (2009):

$$N_{0s} = 444\,\Lambda_s^{2.09} \tag{10.72}$$

for the Sekhon-Srivastava (SS) dataset and

$$N_{0s} = 700\,\Lambda_s^{1.81}, \tag{10.73}$$

for the Gunn-Marshall (GM) dataset. Note that Braham (1990) didn't find any dependence of N_{0s} on Λ_s for lake-effect snow; hence the dependencies (10.72) and (10.73) are not universal. Nevertheless, one can derive the $N_{0s}(T)$ relations using (10.71)–(10.73):

$$N_{0s} = 650\exp(-0.118\,T) \tag{10.74}$$

for the SS dataset and

$$N_{0s} = 988\exp(-0.102\,T) \tag{10.75}$$

for the GM dataset.

Substituting (10.74) and (10.75) into (10.68) and assuming $a_u = 0.9$ results in

$$S = 2.24 \times 10^{-2}\exp(-0.043\,T)Z^{0.64} \tag{10.76}$$

and

$$S = 2.59 \times 10^{-2} \exp(-0.037\,T) Z^{0.64}. \quad (10.77)$$

The corresponding $Z(S)$ relations are

$$Z = 375 \exp(0.067\,T) S^{1.56} \quad (10.78)$$

and

$$Z = 299 \exp(0.058\,T) S^{1.56}. \quad (10.79)$$

These indicate the multipliers in the $S(Z)$ and $Z(S)$ relations depend on temperature in agreement with Hunter et al. (2001) who suggest an exponential multiplier dependent on the height of radar beam above the freezing level in the $Z(S)$ relation. Equation (10.76) compares very well with the $S(Z)$ relation in Boudala et al. (2006):

$$S = 3.6\left(1.74 \times 10^{-5} T^2 + 0.0069\right) Z^{0.64} \quad (10.80)$$

derived using ice particle spectra from aircraft measurements in midlatitude and Arctic regions. Indeed, the temperature-dependent multipliers $a(T)$ in the $S = a(T)Z^b$ given by (10.76) and (10.80) agree remarkably well (Fig. 10.38). The multiplier in the $S(Z)$ relation increases almost four times if the temperature drops from 0 °C to −30 °C.

Assuming temperature lapse rate is 6.5 deg km^{-1} and the snow flux S does not depend on height (which is approximately valid if aggregation dominates deposition and riming), the vertical gradients of reflectivities above the freezing level computed from (10.73) and (10.74) are 4.4 dB km^{-1} and 3.8 dB km^{-1}. Larger vertical gradients of Z in dry snow are expected if S decreases with height (mainly due to deposition and riming). Numerous studies show that vertical gradients of Z derived from (10.78) and (10.79) are consistent with observed gradients ranging from 2.5 to

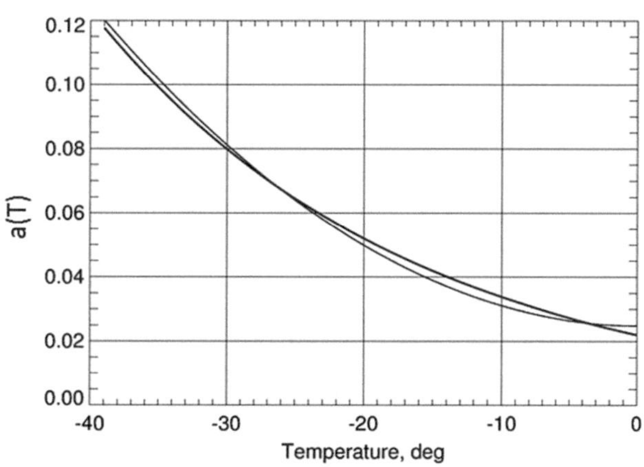

Fig. 10.38 Temperature dependencies of the multiplier a in $S = a(T)Z^{0.64}$ given by (10.76) (thick line) and $S = a(T)Z^{0.64}$ given by (10.80) (thin line)

5.1 dB km^{-1} (Bellon et al. 2005; Fabry and Zawadzki 1995; Matrosov et al. 2009; Steiner et al. 1995; Vignal and Krajewski 2001).

Although N_{0s} tends to decrease with temperature, it can also vary significantly at a fixed temperature depending on snow type and degree of riming (Fig. 4.10a). It usually increases with ice water content or snow flux S (Houze et al. 1979).

10.8.2 Snow Measurements Based on Z and K_{DP}

The underlying reason for poor performance of the $S(Z)$ relations is similar to the one compromising the $R(Z)$ relations in rain. In both relations, the reflectivity factor is disproportionally affected by few large raindrops or snowflakes, whereas the bulk of rain and snow is dominated by smaller hydrometeors. In snow, Z is roughly proportional to the 4th moment of the size distribution because of the approximately inverse dependence of snow density on the snow particle diameter (Eqs. (4.25), and (10.61). Meanwhile, the snow rate S is roughly proportional to the 2nd moment of the size distribution (10.59). The specific differential phase K_{DP} is proportional to the first moment of the snow size distribution (Chap. 5, Sect. 5.4.1).

Bukovcic et al. (2018) suggest estimating snow rate with the bivariate power-law relation:

$$S = \gamma_1 K_{\mathrm{DP}}^{\alpha_1} Z^{\beta_1}, \tag{10.81}$$

anticipating the product of the first and fourth moments of the size distribution may approximate well the second moment, i.e., snow rate S. Because S, Z, and K_{DP} are proportional to the concentration of snowflakes, the sum of exponents α_1 and β_1 should be close to 1.

In addition to S and Z, Bukovcic et al. (2018) simulated K_{DP} from the snow disdrometer data (obtained in Oklahoma) assuming the aspect ratio of snowflakes is 0.65 (consistent with the measurements by Korolev and Isaac (2003) of irregular or aggregated ice particles) and the width σ of the canting angle distribution is equal to zero. Simulations produced the relation

$$S = 1.48\, K_{\mathrm{DP}}^{0.61} Z^{0.33}. \tag{10.82}$$

Bukovcic et al. (2018) compare the performances of the $S(K_{\mathrm{DP}}, Z)$ relation with the ones of the $S(Z)$ relation:

$$S = 0.115\, Z^{0.5} \tag{10.83}$$

recommended by Vasiloff (1997) and the $S(K_{\mathrm{DP}})$ relation

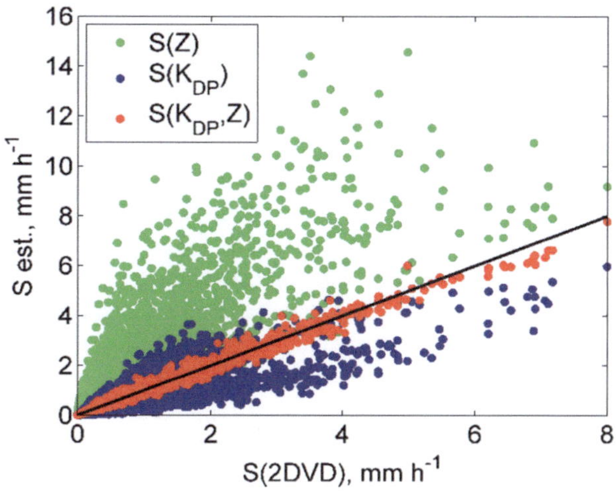

Fig. 10.39 Scatterplots of snow rate measured by disdrometer, S(2DVD), versus the estimates from the $S(Z)$ (green dots), $S(K_{\mathrm{DP}})$ (blue dots), and $S(K_{\mathrm{DP}}, Z)$ (red dots) relations. Courtesy of Dr. Bukovcic

$$S = 55.6 K_{\mathrm{DP}}^{1.08} \tag{10.84}$$

retrieved from the same disdrometer dataset.

The scatterplots of snow rates estimated by the disdrometer versus expected values computed from the three methods using the disdrometer data are displayed in Fig. 10.39. The width of the $S(Z)$ vs S(2DVD) scatterplot (green dots in Fig. 10.39) is very large because Z is proportional to the 4th moment of the particle size distribution for aggregated snow. Moreover, the $S(Z)$ tends to overestimate light and moderate snowfall. This is caused by the dependence of the multiplier in the $S(Z)$ relation on the parameters of particle size distributions. The $S(K_{\mathrm{DP}})$ estimates (blue dots in Fig. 10.39) are less spread about the one-to-one line compared to the $S(Z)$ estimates. The scatterplot of the $S(K_{\mathrm{DP}}, Z)$ vs S(2DVD) is the tightest, concentrated along the one-to-one line, and not biased.

The relation (10.82) is obtained from the disdrometer analysis of 16 snow events in Oklahoma. Besides very good performance in Oklahoma, this relation does reasonably well for snow disdrometer datasets in Colorado and very well for data in Canada, two distinct climate regions suggesting potentially universal character of this relation (Bukovcic et al. 2018).

The results of the Bukovcic et al.'s (2018) study demonstrate the $S(K_{\mathrm{DP}}, Z)$ estimate of snow rate dramatically reduces the adverse impact of the snow particle size distribution variability on the snow rate estimate compared to the Z-based relations. However, the relation (10.82) is derived with assumptions about snowflake shapes and orientations which may affect strongly the value of K_{DP}. Sensitivity tests indicate the exponents α_1 and β_1 in (10.81) are almost constant independent of the variability of snow density (i.e., degree of riming f_{rim}), the aspect ratio, and the width of the canting angle distribution. This simplifies the adjustment of this relation because only the multiplier γ_1 needs to be optimized.

Bukovcic et al. (2018) examine the sensitivity of γ_1 to the variability of the snowflake aspect ratio a/b, degree of riming f_{rim}, and the width σ of the canting angle distribution. They found the corresponding uncertainty in the magnitude of γ_1 is about 18–19% if the aspect ratio changes within the interval −0.1 to 0.1 about its mean. Because the width σ of the canting angle distribution generally varies between 10° and 40° (Chap. 4), the multiplier $\gamma_1 = 1.48$ in (10.82) should be increased by 4% for $\sigma = 10°$ and 79% for $\sigma = 40°$. It can be also shown that the magnitude of γ_1 is roughly proportional to $f_{\mathrm{rim}}^{-0.56}$, i.e., γ_1 is lower for rimed snow. Sensitivity of (10.82) to snowflake shapes, orientations, and degree of riming (density change) is a primary source of uncertainty in the polarimetric quantification of snow. This uncertainty could be evaluated using radar observations and snow gauge measurements at the surface or in situ aircraft probes. Thus, the suggested polarimetric relation should be "calibrated" experimentally using radar data. One possibility to mitigate the impact of the variability in snowflake shape and orientation is to combine measurements of K_{DP} with the measurements of differential reflectivity Z_{DR} or reflectivity difference $Z_{\mathrm{DP}} = Z_h - Z_v$ as discussed in Chap. 11.

Analysis of K_{DP} measurements in snow at *S* band indicates it is usually low and noisy in heavily aggregated dry snow and its reliable estimation may require spatial averaging over relatively large areas (Ryzhkov and Zrnic 1998c). Less averaging is needed at *C* and *X* bands because K_{DP} is inversely proportional to the radar wavelength. Because Eq. (10.82) is valid for *S* band (at $\lambda = 11.0$ cm), the corresponding relations at other wavelengths within this band and at shorter wavelengths can be obtained by appropriately scaling K_{DP} with wavelength.

The quasi-vertical profile technique (QVP, Ryzhkov et al. 2016; Griffin et al. 2018) or columnar vertical profiles (CVP, Ryzhkov et al. 2018) based on azimuthal averaging to reduce the statistical error of the K_{DP} estimate can improve significantly the quality of radar snowfall measurements. Oversampling of data in range at spacing considerably lower than the length of the radar pulse could additionally improve the K_{DP} accuracy (Chap. 6). Another possibility is to capitalize on the K_{DP} measurements in the dendritic growth layer (DGL) aloft in the temperature interval between −20 °C and −10 °C where K_{DP} is significantly higher than at warmer temperatures below the DGL (e.g., Bechini et al. 2013; Kennedy and Rutledge 2011) and make projections down to the surface along the snowflakes' trajectories assuming that the snow rate is conserved in the process of aggregation.

References

Anagnostou, M., Kalogiros, J., Anagnostou, E., Tarolli, M., Papadopoulos, A., & Borga, M. (2010). Performance evaluation of high-resolution estimation from X-band dual-polarization radar for flash flood applications in mountainous basins. *Journal of Hydrology, 394*, 4–16.

Andrieu, H., & Creutin, J. D. (1995). Identification of vertical profiles of radar reflectivity for hydrological applications using an inverse method. Part I: Formulation. *Journal of Applied Meteorology, 34*, 225–239.

Andrieu, H., Delrieu, G., & Creutin, J. D. (1995). Identification of vertical profiles of radar reflectivity for hydrological applications using an inverse method. Part II: Sensitivity analysis and case study. *Journal of Applied Meteorology, 34*, 240–259.

Atlas, D., Srivastava, R., & Sekhon, R. (1973). Doppler radar characteristics of precipitation at vertical incidence. *Reviews of Geophysics and Space Physics, 2*, 1–35.

Atlas, D., & Ulbrich, C. (1977). Path- and area-integrated rainfall measurements by microwave attenuation in the 1 – 3 cm band. *Journal of Applied Meteorology, 16*, 1322–1331.

Balakrishnan, N., & Zrnic, D. (1990). Estimation of rain and hail rates in mixed-phase precipitation. *Journal of the Atmospheric Sciences, 47*, 565–583.

Battan, L. J. (1973). *Radar observations of the atmosphere* (324 pp.). Chicago, IL: The University of Chicago Press.

Bechini, R., Baldini, L., & Chandrasekar, V. (2013). Polarimetric radar observations in the ice region of precipitating clouds at C-band and X-band radar frequencies. *Journal of Applied Meteorology and Climatology, 52*, 1147–1169.

Bellon, A., Lee, G., & Zawadzki, I. (2005). Error statistics of VPR corrections in stratiform precipitation. *Journal of Applied Meteorology, 44*, 998–1015.

Borowska, L., Zrnic, D., Ryzhkov, A., Zhang, P., & Simmer, C. (2011). Polarimetric estimates of a 1-month accumulation of light rain with a 3-cm wavelength radar. *Journal of Hydrometeorology, 12*, 1024–1039.

Boudala, F., Isaac, G., & Hudak, D. (2006). Ice water content and precipitation rate as a function of equivalent radar reflectivity and temperature based on in situ observations. *Journal of Geophysical Research, 111*, D11202. https://doi.org/10.1029/2005JD006499.

Braham, R. (1990). Snow particle size spectra in lake effect snows. *Journal of Applied Meteorology, 29*, 200–207.

Brandes, E., Ikeda, K., Zhang, G., Schoenhuber, M., & Rasmussen, R. (2007). A statistical and physical description of hydrometeor distributions in Colorado snowstorms using a video-disdrometer. *Journal of Applied Meteorology, 46*, 634–650.

Brandes, E., Zhang, G., & Vivekanandan, J. (2002). Experiments in rainfall estimation with a polarimetric radar in a subtropical environment. *Journal of Applied Meteorology, 41*, 674–685.

Bringi, V., Rico-Ramirez, M., & Thurai, M. (2011). Rainfall estimation with an operational polarimetric C-band radar in the United Kingdom: Comparison with a gage network and error analysis. *Journal of Hydrometeorology, 12*, 935–954.

Bringi, V., Thurai, M., Nakagawa, K., Huang, G., Kobayashi, T., Adachi, A., Hanado, H., & Sekizawa, S. (2006). Rainfall estimation from C-band polarimetric radar in Okinawa, Japan: Comparisons with 2D-video disdrometer and 400 MHz wind profiler. *Journal of the Meteorological Society of Japan, 84*, 705–724.

Bukovcic, P., Ryzhkov, A., Zrnic, D., & Zhang, G. (2018). Polarimetric radar relations for quantification of snow based on the disdrometer data. *Journal of Applied Meteorology and Climatology, 57*, 102–120.

Chandrasekar, V., Gorgucci, E., & Scarchilli, G. (1993). Optimization of multiparameter radar estimates of rainfall. *Journal of Applied Meteorology, 32*, 1288–1293.

Chen, H., & Chandrasekar, V. (2015). The quantitative precipitation estimation system for Dallas - Fort Worth (DFW) urban remote sensing network. *Journal of Hydrology, 531*, 259–271.

Chen, H., Chandrasekar, V., & Bechini, R. (2017). An improved dual-polarization radar rainfall algorithm (DROPS2.0): Application in NASA IFloodS field campaign. *Journal of Hydrometeorology, 18*, 917–937.

Cifelli, R., Chandrasekar, V., Lim, S., Kennedy, P., Wang, Y., & Rutledge, S. (2011). A new dual-polarization radar rainfall algorithm: Application in Colorado precipitation events. *Journal of Atmospheric and Oceanic Technology, 28*, 352–364.

Cocks, S., Y. Wang, L. Tang, P. Zhang, A. Ryzhkov, B. Kaney, et al. (2019). A prototype quantitative precipitation estimation algorithm for operational S-band polarimetric radar using specific attenuation and specific differential phase. Part II: Case study analysis and performance verification. *Journal of Hydrometeorology*. Conditionally accepted.

Diederich, M., Troemel, S., Ryzhkov, A., Zhang, P., & Simmer, C. (2015). Use of specific attenuation for rainfall measurements at X-band radar wavelengths. Part II: Rainfall estimates and comparison with rain gauges. *Journal of Hydrometeorology, 16*, 503–516.

Doviak, R. J., & Zrnic, D. S. (2006). *Doppler radar and weather observations.*, (p. 562). Reprint, Mineola, NY: Dover.

Fabry, F., & Zawadzki, I. (1995). Long-term radar observations of the melting layer of precipitation and their interpretation. *Journal of the Atmospheric Sciences, 52*, 838–851.

Figueras i Ventura, J., Baumahmoud, A.-A., Fradon, B., Dupuy, P., & Tabary, P. (2012). Long-term monitoring of French polarimetric radar data quality and evaluation of several polarimetric quantitative precipitation estimators in ideal conditions for operational implementation at C band. *Quarterly Journal of the Royal Meteorological Society, 138*, 2212–2228.

Figueras i Ventura, J., & Tabary, P. (2013). The new French operational polarimetric radar rainfall rate product. *Journal of Applied Meteorology and Climatology, 52*, 1817–1835.

Fujiyoshi, Y., Endoh, T., Yamada, T., Tsuboki, K., Tachibana, Y., & Wakahama, G. (1990). Determination of a Z–R relationship for snowfall using a radar and sensitive snow gauges. *Journal of Applied Meteorology, 29*, 147–152.

Germann, U., & Joss, J. (2002). Mesobeta profiles to extrapolate radar precipitation measurements above the Alps to the ground level. *Journal of Applied Meteorology, 41*, 542–557.

Giangrande, S., Collis, S., Theisen, A., & Tokay, A. (2014). Precipitation estimation from the ARM distributed radar network during the MC3E campaign. *Journal of Applied Meteorology and Climatology, 53*, 2130–2147.

Giangrande, S., & Ryzhkov, A. (2008). Estimation of rainfall based on the results of polarimetric echo classification. *Journal of Applied Meteorology, 47*, 2445–2462.

Griffin, E., Schuur, T., & Ryzhkov, A. (2018). A polarimetric analysis of ice microphysical processes in snow, using quasi-vertical profiles. *Journal of Applied Meteorology and Climatology, 57*, 31–50.

Gunn, K., & Marshall, J. (1958). The distribution with size of aggregated snowflakes. *Journal of Meteorology, 15*, 452–461.

Heymsfield, A., Bansemer, A., Field, P., Durden, S., Stith, J., Dye, J., Hall, W., & Grainger, C. (2002). Observations and parametrizations of particle size distributions in deep tropical cirrus and stratiform precipitating clouds: Results from in situ observations in TRMM field campaigns. *Journal of the Atmospheric Sciences, 59*, 3457–3490.

Heymsfield, A., Matrosov, S., & Wood, N. (2016). Toward improving ice water content and snow-rate retrievals from radars. Part I: X and W bands, emphasizing CloudSat. *Journal of Applied Meteorology and Climatology, 55*, 2063–2090.

Heymsfield, A., Schmitt, C., & Bansemer, A. (2013). Ice cloud particle size distributions and pressure – dependent terminal velocities from in situ observations at temperatures from 0°C to -86°C. *Journal of the Atmospheric Sciences, 70*, 4123–4154.

Houze, R., Hobbs, P., Herzegh, P., & Parsons, D. (1979). Size distributions of precipitation particles in frontal clouds. *Journal of the Atmospheric Sciences, 36*, 156–162.

Huang, G.-J., Bringi, V., Cifelli, R., Hudak, D., & Petersen, W. (2010). A methodology to derive radar reflectivity-liquid equivalent snow rate relations using C-band radar and a 2D video disdrometer. *Journal of Atmospheric and Oceanic Technology, 27*, 637–651.

Huang, G.-J., Bringi, V., Moisseev, D., Petersen, W., Bliven, L., & Hudak, D. (2015). Use of 2D-video disdrometer to derive mean density-size and Z_e – SR relations: Four snow cases from the light precipitation validation environment. *Atmospheric Research, 153*, 34–48.

Hunter, S., Holroyd, E., & Hartzell, C. (2001). Improvements to the WSR-88D snow accumulation algorithm. *Preprints, 30th International Conference on Radar Meteorology*. Munich, Germany: American Metrological Society (pp. 716–718). Retrieved from http://ams.confex.com/ams/pdfpapers/20852.pdf

Illingworth, A., & Blackman, T. (2002). The need to represent raindrop size spectra as normalized gamma distributions for the interpretation of polarization radar observations. *Journal of Applied Meteorology, 41*, 286–297.

Illingworth, A., & Thompson, A. (2005). *The estimation of moderate rain rates with operational polarization radar*. In 32nd International Conference on Radar Meteorology. Albuquerque, NM. P9R.1.

Jameson, A. (1991). A comparison of microwave techniques for measuring rainfall. *Journal of Applied Meteorology, 30*, 32–54.

Kennedy, P. C., & Rutledge, S. A. (2011). S-band dual-polarization radar observations of winter storms. *Journal of Applied Meteorology and Climatology, 50*, 844–858.

Kitchen, M., Brown, R., & Davies, A. (1994). Real-time correction of weather radar data for the effects of bright band, range and orographic growth in widespread precipitation. *Quarterly Journal of the Royal Meteorological Society, 120*, 1231–1254.

Koistinen, J. (1991). Operational correction of radar rainfall errors due to vertical reflectivity profile. In *Preprints, 25th International Conference on Radar Meteorology* (pp. 297–302). Paris, France: American Meteorological Society.

Koistinen, J., Michelson, D., Hohti, H., & Peura, M. (2003). Operational measurement of precipitation in cold climates. In P. Meischner (Ed.), *Weather radar, principles and advanced applications* (pp. 78–114). Berlin: Springer.

Korolev, A., & Isaac, G. (2003). Roundness and aspect ratio of particles in ice clouds. *Journal of the Atmospheric Sciences, 60*, 1795–1808.

Lim, S., Cifelli, R., Chandrasekar, V., & Matrosov, S. (2013). Precipitation classification and quantification using X-band dual-polarization weather radar: Application in the Hydrometeorology Testbed. *Journal of Atmospheric and Oceanic Technology, 30*, 2108–2120.

Maesaka, T., Iwanami, K., & Maki, M. (2012). *Non-negative KDP estimation by monotone increasing ΦDP assumption below melting layer*. In Extended abstracts. The 7th European conference on radar meteorology and hydrology.

Maki, M., Park, S.-G., & Bringi, V. (2005). Effect of natural variations in rain drop size distributions on rain rate estimators of 3 cm wavelength polarimetric radar. *Journal of the Meteorological Society of Japan, 83*, 871–893.

Marshall, J. S., & Palmer, W. (1948). The distribution of raindrops with size. *Journal of Meteorology, 5*, 165–166.

Matrosov, S. (2010). Evaluating polarimetric X-band radar rainfall estimators during HMT. *Journal of Atmospheric and Oceanic Technology, 27*, 122–134.

Matrosov, S., Campbell, C., Kingsmill, D., & Sukovich, E. (2009). Assessing snowfall rates from X-band radar reflectivity measurements. *Journal of Atmospheric and Oceanic Technology, 26*, 2324–2339.

Matrosov, S., Cifelli, R., & Gochis, D. (2013). Measurements of heavy convective rainfall in presence of hail in flood-prone areas using an X-band polarimetric radar. *Journal of Applied Meteorology and Climatology, 52*, 395–407.

Matrosov, S., Cifelli, R., Kennedy, P., Nesbitt, S., Rutledge, S., Bringi, V., & Martner, B. (2006). A comparative study of rainfall retrievals based on specific differential phase shifts at X- and S-band frequencies. *Journal of Atmospheric and Oceanic Technology, 23*, 952–963.

Matrosov, S., Kingsmill, D., & Ralph, F. (2005). The utility of X-band polarimetric radar for quantitative estimates of rainfall parameters. *Journal of Hydrometeorology, 6*, 248–262.

Matrosov, S. Y., Clark, K. A., & Kingsmill, D. E. (2007). A polarimetric radar approach to indentify rain, melting-layer, and snow regions for applying corrections to vertical profiles of reflectivity. *Journal of Applied Meteorology and Climatology, 46*, 154–166.

May, P., Keenan, T., Zrnic, D., Carey, L., & Rutledge, S. (1999). Polarimetric radar measurements of tropical rain at 5-cm wavelength. *Journal of Applied Meteorology, 38*, 750–765.

Mitchell, D., & Heymsfield, A. (2005). Refinements in the treatment of ice particle terminal velocities, highlighting aggregates. *Journal of the Atmospheric Sciences, 62*, 1637–1644.

Moisseev, D., von Lerber, A., & Tiira, J. (2017). Quantifying the effect of riming on snowfall using ground-based observations. *Journal of Geophysical Research – Atmospheres, 122*, 4019–4037.

Molthan, A., Petersen, W., Nesbitt, S., & Hudak, D. (2010). Evaluating the snow crystal size distribution and density assumptions within a single-moment microphysics scheme. *Monthly Weather Review, 138*, 4254–4267.

Ohtake, T., & Henmi, T. (1970). Radar reflectivity of aggregated snowflakes. In *Preprints, 14th Conference on Radar Meteorology* (pp. 332–335). Tucson, AZ: American Meteorological Society.

Otto, T., & Russchenberg, H. W. J. (2011). Estimation of specific differential phase and differential backscatter phase from polarimetric weather radar measurements of rain. *IEEE Geoscience and Remote Sensing Letters, 8*(5), 988–992.

Park, S.-G., Maki, M., Iwanami, K., Bringi, V., & Chandrasekar, V. (2005). Correction of radar reflectivity and differential reflectivity for rain attenuation at X band. Part II: Evaluation and application. *Journal of Atmospheric and Oceanic Technology, 22*, 1633–1655.

Pepler, A., May, P., & Thurai, M. (2011). A robust error-based rain estimation method for polarimetric radar. Part I: Development of a method. *Journal of Applied Meteorology and Climatology, 50*, 2092–2103.

Petersen, W., et al. (1999). Mesoscale and radar observations of the Fort Collins flash flood of 28 July 1997. *The Bulletin of the American Meteorological Society, 80*, 191–216.

Puhakka, T. (1975). On the dependence of the Z – R relation on the temperature in snowfall. In *Preprints, 18th conference on radar meteorology* (pp. 504–507). Houston, TX: American Meteorological Society.

Qi, Y., Zhang, J., Zhang, P., & Cao, Q. (2013). VPR correction of bright band effects in radar QPEs using polarimetric radar observations. *Journal of Geophysical Research-Atmospheres, 118*, 3627–3633.

Rasmussen, R., Dixon, M., Vasiloff, S., Hage, F., Knight, S., Vivekanandan, J., & Xue, M. (2003). Snow nowcasting using a real-time correlation of radar reflectivity with snow gauge accumulation. *Journal of Applied Meteorology, 42*, 20–36.

Ryan, B. (2000). A bulk parametrization of the ice particle size distribution and the optical properties in ice clouds. *Journal of the Atmospheric Sciences, 57*, 1436–1451.

Ryzhkov, A., Bukovcic, P., Murphy, A., Zhang, P., & McFarquhar, G. (2018). *Ice microphysical retrievals using polarimetric radar data.* In 10th European conference on radar in meteorology and hydrology, 1–6 July, The Netherlands, # 40. Retrieved from https://projects.knmi.nl/erad2018/ERAD2018_extended_abstract_040.pdf

Ryzhkov, A., Diederich, M., Zhang, P., & Simmer, C. (2014). Utilization of specific attenuation for rainfall estimation, mitigation of partial beam blockage, and radar networking. *Journal of Atmospheric and Oceanic Technology, 31*, 599–619.

Ryzhkov, A., Kumjian, M., Ganson, S., & Khain, A. (2013a). Polarimetric radar characteristics of melting hail. Pt I: Theoretical simulations using spectral microphysical modeling. *Journal of Applied Meteorology and Climatology, 52*, 2849–2870.

Ryzhkov, A., Kumjian, M., Ganson, S., & Zhang, P. (2013b). Polarimetric radar characteristics of melting hail. Pt II: Practical implications. *Journal of Applied Meteorology and Climatology, 52*, 2871–2886.

Ryzhkov, A., Zhang, P., Krause, J., Schuur, T., Palmer, R., & Zrnic, D. (2012). Simultaneous measurements of precipitation using S-band and C-band polarimetric radars. In *Weather radar and hydrology* (pp. 87–92) (Symposium held in Exeter, UK, April 2011) (IAHS Publ. 351, 2012).

Ryzhkov, A., Zhang, P., Reeves, H., Kumjian, M., Tschallener, T., Troemel, S., & Simmer, C. (2016). Quasi-vertical profiles – a new way to look at polarimetric radar data. *Journal of Atmospheric and Oceanic Technology, 33*, 551–562.

Ryzhkov, A., & Zrnic, D. (1995). Comparison of dual-polarization radar estimators of rain. *Journal of Atmospheric and Oceanic Technology, 12*, 249–256.

Ryzhkov, A., & Zrnic, D. (1996). Assessment of rainfall measurement that uses specific differential phase. *Journal of Applied Meteorology, 35*, 2080–2090.

Ryzhkov, A., & Zrnic, D. (1998a). Polarimetric rainfall estimation in the presence of anomalous propagation. *Journal of Atmospheric and Oceanic Technology, 15*, 1321–1331.
Ryzhkov, A., & Zrnic, D. (1998b). Beamwidth effects on the differential phase measurements of rain. *Journal of Atmospheric and Oceanic Technology, 15*, 624–634.
Ryzhkov, A., & Zrnic, D. (1998c). Discrimination between rain and snow with a polarimetric radar. *Journal of Applied Meteorology, 37*, 1228–1240.
Ryzhkov, A. V., Giangrande, S. E., & Schuur, T. J. (2005). Rainfall estimation with a polarimetric prototype of the WSR-88D radar. *Journal of Applied Meteorology, 44*, 502–515.
Ryzhkov, A. V., Zrnic, D. S., & Fulton, R. (2000). Areal rainfall estimates using differential phase. *Journal of Applied Meteorology, 39*, 263–268.
Sachidananda, M., & Zrnic, D. S. (1986). Differential propagation phase shift and rainfall rate estimation. *Radio Science, 21*, 235–247.
Sachidananda, M., & Zrnic, D. S. (1987). Rain rate estimates from differential polarization measurements. *Journal of Atmospheric and Oceanic Technology, 4*, 588–598.
Saltikoff, E., Huuskonen, A., Hohti, H., Koistinen, J., & Jarvinen, H. (2010). Quality assurance in the FMI Doppler weather radar network. *Boreal Environment Research, 15*, 579–594.
Schneebeli, M., & Berne, A. (2012). An extended Kalman filter framework for polarimetric X-band weather radar data processing. *Journal of Atmospheric and Oceanic Technology, 29*, 711–730.
Schuur, T. J., Ryzhkov, A. V., & Clabo, D. R. (2005). *Climatological analysis of DSDs in Oklahoma as revealed by 2D-video disdrometer and polarimetric WSR-88D*. In Preprints, 32nd Conference on Radar Meteorology, CD-ROM, 15R.4.
Sekhon, R., & Srivastava, R. (1970). Snow size spectra and radar reflectivity. *Journal of the Atmospheric Sciences, 27*, 299–307.
Seliga, T., Bringi, V., & Al-Khatib, H. (1981). A preliminary study of comparative measurement of rainfall rate using the differential reflectivity radar technique and a raingauge network. *Journal of Applied Meteorology, 20*, 1362–1368.
Seo, D.-J., Breidenbach, J., Fulton, R., Miller, D., & O'Bannon, T. (2000). Real-time adjustment of range-dependent biases in WSR-88D rainfall estimates due to nonuniform vertical profile of reflectivity. *Journal of Hydrometeorology, 1*, 222–240.
Steiner, M., Houze, R., & Yuter, S. (1995). Climatological characterization of three-dimensional storm structure from operational radar and rain gauge data. *Journal of Applied Meteorology, 34*, 1978–2007.
Szyrmer, W., & Zawadzki, I. (2010). Snow studies. Part II: Average relationship between mass of snowflakes and their terminal fall velocity. *Journal of the Atmospheric Sciences, 67*, 3319–3335.
Tabary, P., Boumahmoud, A., Andrieu, H., Thompson, R., Illingworth, A., LeBouar, E., & Testud, J. (2011). Evaluation of two "integrated" polarimetric quantitative precipitation estimation (QPE) algorithms at C band. *Journal of Hydrology, 405*(3), 248–260.
Tabary, P., Le Henaff, A., Vulpiani, G., Parent-du-Chatelet, J., & Gourley, J. (2006). Melting layer characterization and identification with a C-band dual-polarization radar: A long-term analysis. In *Preprints, Fourth European conference on radar in meteorology and hydrology* (ERAD 2006), Barcelona (pp. 17–20).
Testud, J., Le Bouar, E., Obligis, E., & Ali-Mehenni, M. (2000). The rain profiling algorithm applied to polarimetric weather radar. *Journal of Atmospheric and Oceanic Technology, 17*, 332–356.
Testud, J., Oury, S., Black, R., Amayenc, P., & Dou, X. (2001). The concept of "normalized" distribution to describe raindrop spectra: A tool for cloud physics and cloud remote sensing. *Journal of Applied Meteorology, 40*, 1118–1140.
Thompson, E., Rutledge, S., Dolan, B., Thurai, M., & Chandrasekar, V. (2018). Dual-polarization radar rainfall estimation over tropical oceans. *Journal of Applied Meteorology and Climatology, 57*, 755–775.
Troemel, S., Kumjian, M., Ryzhkov, A., & Simmer, C. (2013). Backscatter differential phase – estimation and variability. *Journal of Applied Meteorology and Climatology, 52*, 2529–2548.

Troemel, S., Ryzhkov, A., & Simmer, C. (2017). Climatology of vertical profiles of polarimetric radar variables at X band in stratiform clouds. In *38th Conference on Radar Meteorology* (p. 5). Chicago, IL: American Meteorological Society. Retrieved from https://ams.confex.com/ams/38RADAR/meetingapp.cgi/Paper/320485

Vasiloff, S. (1997) *Interpretation of radar data during snow events in mountainous terrain*. Western Region Tech. Attachment 97–35, 16 pp. Retrieved from https://www.weather.gov/media/wrh/online_publications/TAs/ta9735.pdf

Vignal, B., Andrieu, H., & Creutin, J. D. (1999). Identification of vertical profiles of reflectivity from volume-scan radar data. *Journal of Applied Meteorology, 38*, 1214–1228.

Vignal, B., Galli, G., Joss, J., & Germann, U. (2000). Three methods to determine profiles of reflectivity from volumetric radar data to correct precipitation estimates. *Journal of Applied Meteorology, 39*, 1715–1726.

Vignal, B., & Krajewski, W. (2001). Large-sample evaluation of two methods to correct range-dependent error for WSR-88D rainfall estimates. *Journal of Hydrometeorology, 2*, 490–504.

Von Lerber, A., Moisseev, D., Bliven, L., Petersen, W., Harri, A., & Chandrasekar, V. (2017). Microphysical properties of snow and their link to Z_e–S relations during BAECC 2014. *Journal of Applied Meteorology and Climatology, 56*, 1561–1582.

Vulpiani, G., Montopoli, M., Passeri, L., Giola, A., Giordano, P., & Marzano, F. (2012). On the use of dual-polarized C-band radar for operational rainfall retrieval in mountainous areas. *Journal of Applied Meteorology and Climatology, 51*, 405–425.

Wang, Y., & Chandrasekar, V. (2009). Algorithm for estimation of the specific differential phase. *Journal of Atmospheric and Oceanic Technology, 26*, 2565–2578.

Wang, Y., & Chandrasekar, V. (2010). Quantitative precipitation estimation in the CASA X-band dual-polarization radar network. *Journal of Atmospheric and Oceanic Technology, 27*, 1665–1676.

Wang, Y., Cocks, S., Tang, L., Ryzhkov, A., Zhang, P., Zhang, J. (2019). A prototype quantitative precipitation estimation algorithm for operational S-band polarimetric radar using specific attenuation and specific differential phase. Part I: Algorithm description and initial results. *Journal of Hydrometeorology*. Conditionally accepted.

Wang, Y., Zhang, J., Ryzhkov, A., & Tang, L. (2013). C-band polarimetric radar QPE based on specific differential propagation phase for extreme typhoon rainfall. *Journal of Atmospheric and Oceanic Technology, 30*, 1354–1370.

Wang, Y., Zhang, P., Ryzhkov, A., Zhang, J., & Chang, P.-L. (2014). The application of specific attenuation for tropical rainfall estimation in complex terrain. *Journal of Hydrometeorology, 15*, 2250–2266.

Wilks, D. (2006). *Statistical methods in the atmospheric sciences* (704 pp.). Amsterdam, The Netherlands: Elsevier.

Wolfe, J., & Snider, J. (2012). A relationship between reflectivity and snow rate for a high-altitude S-band radar. *Journal of Applied Meteorology and Climatology, 51*, 1111–1128.

Wolfensberger, D., Scipion, D., & Berne, A. (2015). Detection and characterization of the melting layer based on polarimetric radar scans. *Quarterly Journal of the Royal Meteorological Society, 142*, 108–124.

Woods, C., Stoelinga, T., & Locatelli, D. (2008). Size spectra of snow particles measured in wintertime precipitation in the Pacific Northwest. *Journal of the Atmospheric Sciences, 65*, 189–205.

Zawadzki, I., Jung, E., & Lee, G. (2010). Snow studies. Part I: A study of natural variability of snow terminal velocity. *Journal of the Atmospheric Sciences, 67*, 1591–1604.

Zawadzki, I., Szyrmer, W., Bell, C., & Fabry, F. (2005). Modeling of the melting layer. Part III: The density effect. *Journal of the Atmospheric Sciences, 62*, 3705–3723.

Zhang, G., Luchs, S., Ryzhkov, A., Xue, M., Ryzhkova, L., & Cao, Q. (2011). Winter precipitation microphysics characterized by polarimetric radar and video disdrometer observations in central Oklahoma. *Journal of Applied Meteorology and Climatology, 50*, 1558–1570.

Zhang, J., & Qi, Y. (2010). A real-time algorithm for the correction of brightband effects in radar-derived QPE. *Journal of Hydrometeorology, 11*, 1157–1171.

Zhang, J., Qi, Y., Tang, L., Cocks, S., Wang, Y., Zhang, P., Ryzhkov, A., Langston, C., & Kaney, B. (2017). MRMS dual-polarization radar synthetic QPE. In *38th Conference on Radar Meteorology* (p. 22B.3). Chicago, IL: American Meteorological Society. Retrieved from https://ams.confex.com/ams/38RADAR/meetingapp.cgi/Paper/320875

Zhang, P., Zrnic, D., & Ryzhkov, A. (2013). Partial beam blockage correction using polarimetric radar measurements. *Journal of Atmospheric and Oceanic Technology, 30*, 861–872.

Zrnic, D., & Ryzhkov, A. (1996). Advantages of rain measurements using specific differential phase. *Journal of Atmospheric and Oceanic Technology, 13*, 454–464.

Zrnic, D. S., Balakrishnan, N., Ziegler, C. L., Bringi, V. N., Aydin, K., & Matejka, T. (1993). Polarimetric signatures in the stratiform region of a mesoscale convective system. *Journal of Applied Meteorology, 32*, 678–693.

Zrnic, D. S., Doviak, R. J., Zhang, G., & Ryzhkov, R. V. (2010). Bias in differential reflectivity due to cross-coupling through the radiation patterns of polarimetric weather radars. *Journal of Atmospheric and Oceanic Technology, 27*, 1624–1637.

Zrnic, D. S., Raghavan, R., & Chandrasekar, V. (1994). Observation of copolar correlation coefficient through a bright band at vertical incidence. *Journal of Applied Meteorology, 33*, 45–52.

Chapter 11
Polarimetric Microphysical Retrievals

Assimilation of polarimetric radar data into storm-scale numerical weather prediction (NWP) models shows promise for improving their microphysical parameterization and subsequent performance (e.g., Jung et al. 2008a, b; Li and Mecikalski 2010; Posselt et al. 2015; Zhang 2016; Carlin et al. 2017; Putnam et al. 2017). This can be done either by comparing prognostic model variables with their retrieved radar estimates or by converting the model output to the fields of polarimetric variables and comparing these directly with radar observations. The first methodology implies radar retrieval of the mixing ratios q_i for different hydrometeor types (or water and ice contents), median or mean volume diameter, and particle number concentration. The second methodology requires use of the forward polarimetric radar operators (e.g., Pfeifer et al. 2008; Jung et al. 2008a; Ryzhkov et al. 2011). In this chapter, we discuss polarimetric microphysical retrievals essential for the first approach.

11.1 DSD Retrievals in Pure Rain

Knowledge of raindrop size distribution and its temporal/spatial evolution is a key for understanding the processes of precipitation formation and their adequate treatment in the cloud models. The radar retrieval techniques aim at separate estimation of the primary parameters of rain DSD: liquid water content (*LWC*), median or mean volume diameter (D_0 or D_m), concentration of raindrops characterized either by the intercept N_0 of the exponential or gamma size distribution or the intercept parameter N_w of the normalized DSD, and the shape parameter μ of the gamma DSD (Chap. 4). Multiparameter measurements available from dual-polarization radars provide ample opportunities to reach this goal.

A. V. Ryzhkov, D. S. Zrnic, *Radar Polarimetry for Weather Observations*,
Springer Atmospheric Sciences, https://doi.org/10.1007/978-3-030-05093-1_11

11.1.1 Estimation of Liquid Water Content (LWC)

Radar reflectivity was exclusively used for estimation of *LWC* before introduction of polarimetric radars. Among the first *LWC*(*Z*) relations is the one suggested by Greene and Clark (1972). It has been used for computing the vertically integrated liquid (VIL), still a popular parameter for characterizing storm intensity and its potential for hail production:

$$LWC = 3.44 \times 10^{-3} Z^{0.57}. \tag{11.1}$$

In (11.1), *LWC* is expressed in g m^{-3} and *Z* in mm^6 m^{-3}. Because *Z* is proportional to the sixth moment of the DSD and *LWC* to the third moment, the relation (11.1) is even less accurate than the *R*(*Z*) relations for estimating rain rate (proportional to a higher-order moment of DSD than *LWC*) (Chap. 10). Similar to the *R*(*Z*) relation, the multiplier in the *LWC*(*Z*) power-law relation is a strong function of the intercept parameter N_w (Chap. 4). Carlin et al. (2016) derived the following *LWC*(*Z*) relations parameterized by N_w using a large disdrometer dataset from Oklahoma:

$$LWC(Z, N_w) = 1.16 \times 10^{-4} N_w^{0.411} Z^{0.550} \tag{11.2}$$

at S band and

$$LWC(Z, N_w) = 1.51 \times 10^{-4} N_w^{0.363} Z^{0.572} \tag{11.3}$$

at C band. As in the case of QPE, a generic power-law *LWC*(*Z*) relation with a fixed multiplier tends to overestimate *LWC* in continental rain and underestimate it in tropical rain.

With the advent of the polarimetric radars, the K_{DP}-based relations for estimating *LWC* became available. The $R(K_{DP})$ relations are less sensitive to the DSD variability because K_{DP} is a lower-order moment of DSD than *Z* (at least for rain with larger-size drops). For example, one $LWC(K_{DP})$ relation is listed by Doviak and Zrnic (2006)

$$LWC = 0.34 (K_{DP} \lambda)^{0.702} \tag{11.4}$$

and another one by Bringi and Chandrasekar (2001)

$$LWC = 0.26 (K_{DP} \lambda)^{0.77}. \tag{11.5}$$

In (11.4) and (11.5), K_{DP} is in deg km^{-1} and the wavelength λ is in cm.

Successful use of the specific attenuation *A* for radar QPE motivates a similar approach for estimating *LWC*. Power-law relations *LWC*(*Z*), $LWC(K_{DP})$, and *LWC*(*A*) have been derived for S, C, and X bands by simulations on a large dataset

Table 11.1 The $LWC(Z)$, $LWC(K_{DP})$, and $LWC(A)$ relations optimized for S, C, and X bands

$LWC = a_1 Z^{b_1}$		
Frequency band	a_1	b_1
S ($\lambda = 11.0$ cm)	1.70×10^{-3}	0.644
C ($\lambda = 5.3$ cm)	1.58×10^{-3}	0.659
X ($\lambda = 3.2$ cm)	2.44×10^{-3}	0.601
$LWC = a_2 K_{DP}^{b_2}$		
Frequency band	a_2	b_2
S ($\lambda = 11.0$ cm)	2.25	0.723
C ($\lambda = 5.3$ cm)	1.28	0.713
X ($\lambda = 3.2$ cm)	0.878	0.730
$LWC = a_3 A^{b_3}$ ($T = 20°$)		
Frequency band	a_3	b_3
S ($\lambda = 11.0$ cm)	119	0.924
C ($\lambda = 5.3$ cm)	11.6	0.794
X ($\lambda = 3.2$ cm)	2.10	0.698

collected by a disdrometer in Oklahoma. The parameters of these relations are summarized in Table 11.1. The polarimetric variables at S ($\lambda = 11.0$ cm), C ($\lambda = 5.3$ cm), and X ($\lambda = 3.2$ cm) wavelengths were computed from each DSD at 20 °C assuming a 10° width of the canting angle distributions (with a mean of 0°) and the axis ratio relation specified by Brandes et al. (2002) (Eq. (4.29)). The retrieval equations were derived by weighted least-squares regressions on the median of each variable within bins of log(*LWC*) ranging from −2.2 to 1.0 in 0.1 intervals. Each bin was weighted by the center *LWC* within the bin multiplied by the number of points within that bin.

The scatterplots of *LWC* from disdrometer versus estimates from *Z*, K_{DP}, and *A* at S, C, and X bands using relations in Table 11.1 are shown in Fig. 11.1. As expected, these scatterplots are quite similar to the ones for rain rate (Figs. 10.12–10.17). Indeed, at all three frequency bands, the *A*-based relations demonstrate advantage for estimating low and moderate *LWC*s. At S band, the *LWC*(*A*) estimate exhibits a significantly smaller spread than the *LWC*(*Z*) and $LWC(K_{DP})$ estimates in the entire range of *LWC* values. The results at C and X band are different. The $R(K_{DP})$ relations are the best for *LWC* exceeding 1 g m^{-3}. However, for very low *LWC* (below 0.1 g m^{-3}), the K_{DP}-based relations are the most sensitive to the DSD variability. Their poor performance at low *LWC* is further exacerbated by the inherent noisiness of K_{DP} in light precipitation. Moreover, the radial resolution of the K_{DP}-based estimates of *LWC* (and rain rate *R*) is worse than the one of *LWC*(*A*).

Both K_{DP} and *A* are strong functions of the wavelength (Chap. 5). The $LWC(K_{DP})$ relations listed in Table 11.1 should be scaled according to the wavelength within a particular frequency band (S, C, or X). The specific attenuation also depends on the temperature of raindrops which causes temperature dependence of the parameters of the power-law *LWC*(*A*) relations (Table 11.2). The relation

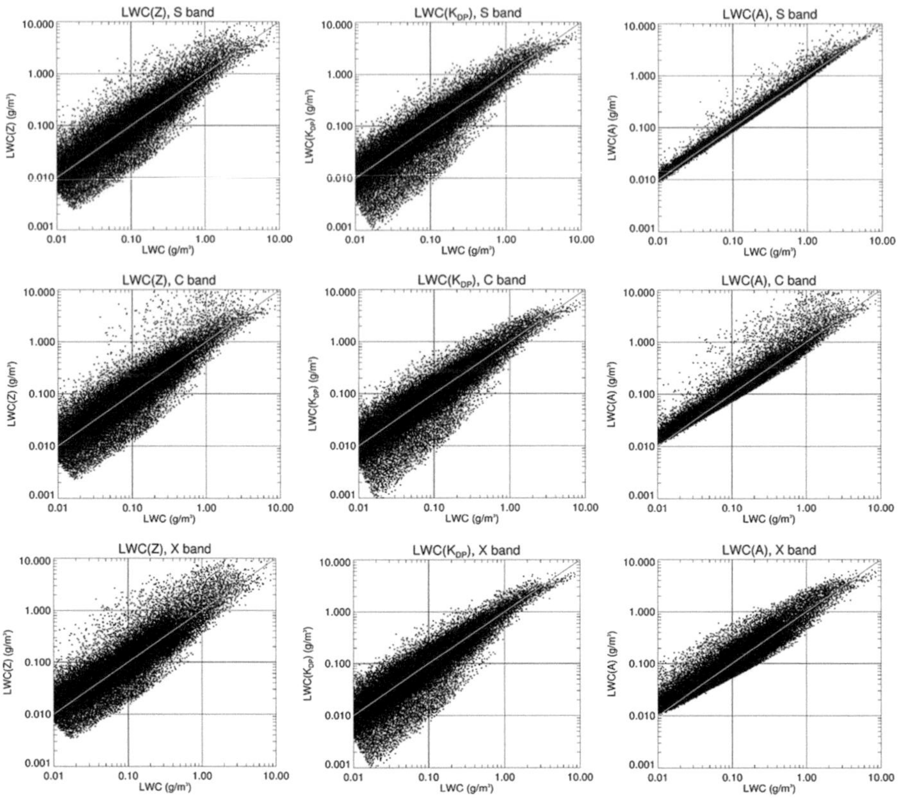

Fig. 11.1 Scatterplots of *LWC* versus its estimates from *Z*, K_{DP}, and *A* at S, C, and X bands using relations from Table 11.1. Large disdrometer dataset from Oklahoma is used for simulations

Table 11.2 The *LWC*(*A*) relations at different radar wavelengths and temperatures

Temperature (°C)	S band ($\lambda = 11.0$ cm)	C band ($\lambda = 5.3$ cm)	X band ($\lambda = 3.2$ cm)
0	$LWC = 70.3\, A^{0.921}$	$LWC = 8.57\, A^{0.814}$	$LWC = 2.25\, A^{0.767}$
10	$LWC = 92.4\, A^{0.922}$	$LWC = 9.79\, A^{0.800}$	$LWC = 2.14\, A^{0.731}$
20	$LWC = 119\, A^{0.924}$	$LWC = 11.6\, A^{0.794}$	$LWC = 2.10\, A^{0.698}$
30	$LWC = 150\, A^{0.926}$	$LWC = 14.1\, A^{0.794}$	$LWC = 2.10\, A^{0.669}$

$$LWC = \left(70.3 + 1.99T + 0.023T^2\right)A^{0.924} \tag{11.6}$$

parameterizes the temperature dependence of the *LWC*(*A*) relation at S band where the exponent in the power-law relation is almost insensitive to the temperature. In (11.6), *T* is the temperature in °C.

11.1.2 DSD Retrieval

Most existing radar techniques for DSD retrievals assume the DSDs can be approximated by the three-parameter gamma function

$$N(D) = N_0 D^{\mu} \exp(-\Lambda D), \tag{11.7}$$

where

$$\Lambda = \frac{4+\mu}{D_m} \tag{11.8}$$

if D_m is the mean volume or mass-weighted diameter of raindrop or

$$\Lambda = \frac{3.67+\mu}{D_0} \tag{11.9}$$

if D_0 is the median volume diameter (Chap. 4). Other forms of the gamma DSD are explicitly expressed via the normalized intercept parameter N_w:

$$N(D) = N_w f_0(\mu)\left(\frac{D}{D_0}\right)^{\mu} \exp\left(-\frac{3.67+\mu}{D_0} D\right) \tag{11.10}$$

or

$$N(D) = N_w f_m(\mu)\left(\frac{D}{D_m}\right)^{\mu} \exp\left(-\frac{4+\mu}{D_m} D\right), \tag{11.11}$$

where

$$f_0(\mu) = \frac{6(3.67+\mu)^{(4+\mu)}}{(3.67)^4 \Gamma(4+\mu)} \tag{11.12}$$

and

$$f_m(\mu) = \frac{6(4+\mu)^{(4+\mu)}}{4^4 \Gamma(4+\mu)}. \tag{11.13}$$

The mean volume diameter D_m is slightly larger than the median volume diameter D_0 for positive values of μ:

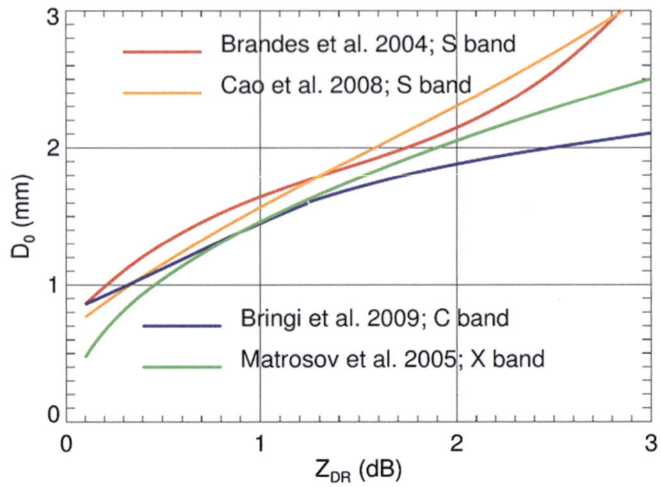

Fig. 11.2 Average dependencies of the median volume diameter D_0 on Z_{DR} reported in four different studies

$$D_m = \frac{4+\mu}{3.67+\mu} D_0. \tag{11.14}$$

Because the differential reflectivity Z_{DR} is related to the shape of raindrop and the shape depends on its size, Z_{DR} is commonly used to estimate the mean or median volume diameter (e.g., Zhang et al. 2001; Gorgucci et al. 2002; Bringi et al. 2002, 2003, 2009, 2012; Brandes et al. 2004; Matrosov et al. 2005b; Cao et al. 2008; Anagnostou et al. 2008; Kim et al. 2010; Thurai et al. 2012). The typical average dependencies of D_0 on Z_{DR} at S, C, and X bands reported in four different studies conducted in various parts of the world (Fig. 11.2) are

$$D_0 = 0.171\,Z_{\mathrm{DR}}^3 - 0.725\,Z_{\mathrm{DR}}^2 + 1.48\,Z_{\mathrm{DR}} + 0.717 \tag{11.15}$$

from Brandes et al. (2004) (Florida, S band),

$$D_0 = 0.0436\,Z_{\mathrm{DR}}^3 - 0.216\,Z_{\mathrm{DR}}^2 + 1.08\,Z_{\mathrm{DR}} + 0.659 \tag{11.16}$$

from Cao et al. (2008) (Oklahoma, S band),

$$\begin{cases} D_0 = 0.0203Z_{\mathrm{DR}}^4 - 0.149Z_{\mathrm{DR}}^3 + 0.221Z_{\mathrm{DR}}^2 + 0.557Z_{\mathrm{DR}} + 0.801 & \text{if } Z_{\mathrm{DR}} < 1.25\,\mathrm{dB} \\ D_0 = 0.0355Z_{\mathrm{DR}}^3 - 0.302Z_{\mathrm{DR}}^2 + 1.06Z_{\mathrm{DR}} + 0.684 & \text{if } Z_{\mathrm{DR}} > 1.25\,\mathrm{dB} \end{cases} \tag{11.17}$$

from Bringi et al. (2009) (Darwin, Australia, C band), and

$$D_0 = 1.46 Z_{\mathrm{DR}}^{0.49}, \tag{11.18}$$

from Matrosov et al. (2005b) (California, X band). In (11.15)–(11.18), D_0 is in mm and Z_{DR} is in dB. All four curves in Fig. 11.2 are similar for $Z_{\mathrm{DR}} < 2$ dB and $D_0 < 2$ mm. Therefore, any of these four D_0–Z_{DR} relations can be used if Z_{DR} is less than 2 dB. For larger Z_{DR}, the curves diverge due to the impact of resonance scattering at C and X bands on Z_{DR} for larger-size raindrops. The biggest increase in Z_{DR} of resonance-size raindrops is at C band (see Fig. 5.11).

Therefore, the D_0–Z_{DR} relation "flattens" for $Z_{\mathrm{DR}} > 2$ dB signaling that Z_{DR} increases significantly with relatively small increase of D_0.

Different methodologies exist for estimating the DSD shape parameter μ and the parameters characterizing total concentration of raindrops (N_0 and N_w). Analyzing the DSD data in Florida, Zhang et al. (2001) found that there was a strong correlation between the shape (μ) and slope (Λ) of the DSD approximated by the gamma function (11.6). Zhang et al. (2001) introduced a "constrained-gamma" DSD model in which μ depends on Λ (or D_0) as

$$\mu = -0.0201\Lambda^2 + 0.902\Lambda - 1.718 \tag{11.19}$$

where Λ is expressed in mm^{-1}. Then the intercept N_0 in (11.7) or the normalized intercept N_w in (11.10) and (11.11) can be obtained from the reflectivity using equations (Zhang 2016)

$$Z = N_0 I(\mu, D_0) \tag{11.20}$$

or

$$Z = N_w \frac{f_0(\mu)}{D_0^{\mu}} I(\mu, D_0), \tag{11.21}$$

where

$$I(\mu, D_0) = \frac{\lambda^4}{\pi^5 |K_w|^2} \int \sigma(D) D^{\mu} \exp\left(-\frac{3.67+\mu}{D_0} D\right) dD. \tag{11.22}$$

In (11.22), $\sigma(D)$ is the radar cross section of the raindrop with diameter D (Chap. 5). In the Rayleigh approximation,

$$\begin{aligned} I(\mu, D_0) &= \int D^{6+\mu} \exp\left(-\frac{3.67+\mu}{D_0} D\right) dD \\ &= \left(\frac{D_0}{3.67+\mu}\right)^{(7+\mu)} \Gamma(7+\mu), \end{aligned} \tag{11.23}$$

and (11.21) is transformed into

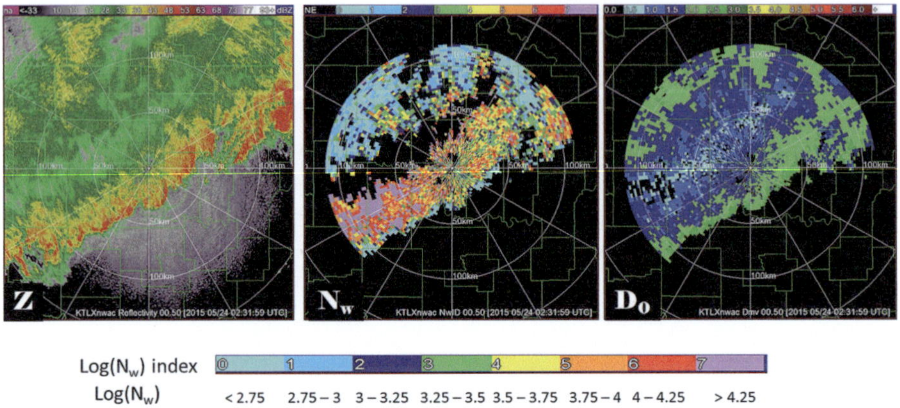

Fig. 11.3 Composite PPI of Z, N_w, and D_0, retrieved from the measurements by the KTLX WSR-88D radar in Oklahoma on 24 May 2015. N_w is in m^{-3} mm^{-1}; D_0 is in mm. Courtesy of Dr. P. Zhang

$$Z = f_0(\mu)\frac{\Gamma(7+\mu)}{(3.67+\mu)^{7+\mu}} N_w D_0^7. \tag{11.24}$$

Alternatively, N_w can be determined from the combination of Z and Z_{DR} using the lines of constant log(N_w) in the Z–Z_{DR} plane (Fig. 10.7) (Illingworth and Thompson 2005; Tabary et al. 2011; Ryzhkov et al. 2014).

An example of the fields of Z, N_w, and D_0 retrieved from the measurements by the KTLX WSR-88D radar in Oklahoma on 24 May 2015 is displayed in Fig. 11.3. The composite plot illustrates how the normalized concentration N_w gradually increases southward along the squall line of the mesoscale convective system, whereas the corresponding D_0 tends to decrease. This indicates the transition from a more continental type of the DSD in the NE part of a squall line toward a more tropical one in the humid environment at the SW flank of a squall line. Another example of a spatial distribution of N_w within the storms of different types is presented in Fig. 10.8.

The described DSD retrieval is deterministic and does not consider the uncertainty caused by the measurement/model errors. A statistical approach based on the Bayesian theory can quantify the DSD retrieval errors and uses the a priori information about the natural variability of DSDs. The parameters of the DSD (e.g., D_0, log(N_w), and others) are grouped in the state vector **x**, and the measurement vector **y** consists of the polarimetric variables. Both are random variables characterized by the probability density functions (PDF). The Bayesian theorem expresses the posterior conditional PDF of **x** for a given measurement **y** as

$$p_{\mathrm{pc}}(\mathbf{x}/\mathbf{y}) = \frac{p_{\mathrm{fo}}(\mathbf{y}/\mathbf{x})p_{\mathrm{pr}}(\mathbf{x})}{\int p_{\mathrm{fo}}(\mathbf{y}/\mathbf{x})p_{\mathrm{pr}}(\mathbf{x})d\mathbf{x}}, \tag{11.25}$$

where $p_{\text{fo}}(\mathbf{y}/\mathbf{x})$ is a conditional PDF of $\mathbf{y}$ (for a given state $\mathbf{x}$) and $p_{\text{pr}}(\mathbf{x})$ is the a priori PDF of $\mathbf{x}$. The challenge of this approach is to determine this a priori probability; the conditional probability can be obtained from data or simulations. If these probabilities are found, the expected value $\langle \mathbf{x} \rangle$ and the standard deviation σ_{x} can be calculated as (Papoulis 1991)

$$\langle \mathbf{x} \rangle = \frac{\int \mathbf{x} p_{\text{fo}}(\mathbf{y}/\mathbf{x}) p_{\text{pr}}(\mathbf{x})\, d\mathbf{x}}{\int p_{\text{fo}}(\mathbf{y}/\mathbf{x}) p_{\text{pr}}(\mathbf{x})\, d\mathbf{x}} \tag{11.26}$$

and

$$\sigma_x = \left(\frac{\int (\mathbf{x} - \langle \mathbf{x} \rangle)^2 p_{\text{fo}}(\mathbf{y}/\mathbf{x}) p_{\text{pr}}(\mathbf{x})\, d\mathbf{x}}{\int p_{\text{fo}}(\mathbf{y}/\mathbf{x}) p_{\text{pr}}(\mathbf{x})\, d\mathbf{x}} \right)^{1/2}. \tag{11.27}$$

The conditional PDF $p_{\text{fo}}(\mathbf{y}/\mathbf{x})$ is usually represented by a bivariate Gaussian distribution function. The dispersions and mean values of the polarimetric variables (Z or Z_{DR}) are obtained from the physical relation between the polarimetric variables and drops weighted by a specific DSD (i.e., prescribed D_0 and N_w). Details about the Bayesian DSD retrieval are described by Cao et al. (2010) and Zhang (2016).

Precise DSD radar retrievals require the measurements of Z and Z_{DR} to be free of attenuation-related and other biases. Attenuation correction of Z and Z_{DR} is based on assumptions about the DSD along the propagation path. Hence, the DSD retrieval and attenuation correction can't be done independently. This issue is particularly challenging at shorter wavelengths (C and X bands). A variational approach is one possible solution (Cao et al. 2013; Yoshikawa et al. 2014, 2016; Zhang 2016). According to this methodology, the forward radar operator simultaneously accounts for the effects of backscattering and forward propagation, i.e., attenuation/differential attenuation.

11.2 Microphysical Retrievals in Ice and Snow

11.2.1 Estimation of Ice Water Content (IWC)

Similar to liquid water content *LWC*, the reflectivity Z was the primary radar measurement for estimating ice water content *IWC*. There have been a number of studies suggesting various *IWC*(Z) relations in ice/snow clouds (e.g., Sassen 1987; Atlas et al. 1995; Liu and Illingworth 2000; Hogan et al. 2006; Delanoe et al. 2014; Heymsfield et al. 2005, 2016; Protat et al. 2016). The inherent uncertainty in the *IWC*(Z) estimate is caused by the variability of the size distribution (SD) of ice/snow particles and (to a lesser extent) by the variations of density.

Simple expressions for Z and IWC can be obtained assuming the exponential size distribution

$$N(D) = N_{0s}\exp(-\Lambda_s D) \tag{11.28}$$

and a power-law dependence of the snow density ρ_s on the equivolume diameter D

$$\rho_s(D) = \alpha D^{-\beta}, \tag{11.29}$$

where the exponent β is close to 1 (Brown and Francis 1995; Heymsfield et al. 2004; Matrosov 2007; Brandes et al. 2007; Erfani and Mitchell 2017; Tiira et al. 2016) and α is much more variable and proportional to the degree of riming f_{rim} ($\alpha = \alpha_u f_{\mathrm{rim}}$) (Zawadzki et al. 2005).

If $\beta = 1$ then

$$Z = \frac{|K_i|^2}{|K_w|^2}\frac{1}{\rho_i^2}\int \rho_s^2(D)D^6N(D)dD = 5.42\frac{\alpha^2 N_{0s}}{\Lambda_s^5} = 5.30\times 10^{-3}\alpha^2 N_{0s}D_m^5 \tag{11.30}$$

and

$$IWC = \frac{\pi}{6}\int \rho_s(D)D^3N(D)dD = 1.05\times 10^{-3}\frac{\alpha N_{0s}}{\Lambda_s^3} = 1.64\times 10^{-5}\alpha N_{0s}D_m^3. \tag{11.31}$$

From (11.30) and (11.31), one can obtain analytical expressions for $IWC(Z)$ parameterized either by the intercept N_{0s} of the size distributions

$$IWC = 3.81\times 10^{-4}\alpha^{-0.2}N_{0s}^{0.4}Z^{0.6} \tag{11.32}$$

or by the mean volume diameter $D_m = 4/\Lambda_s$ of ice particles

$$IWC = 3.09\times 10^{-3}\frac{Z}{\alpha D_m^2}. \tag{11.33}$$

In (11.32) and (11.33), IWC is in g m^{-3}, Λ_s is in mm^{-1}, N_{0s} is in m^{-3} mm^{-1}, D_m is in mm, and Z is in mm^6 m^{-3}.

It follows from (11.32) and (11.33) that the multipliers in the power-law $IWC(Z)$ relations are almost entirely dependent on N_{0s} (because the dependence on α is very weak in (11.32)) or the product αD_m^2. Theoretical relation (11.32) is similar to the one derived by Hogan et al. (2006) and consistent with the relation empirically derived by Delanoe et al. (2014) from a very large dataset of in situ aircraft measurements

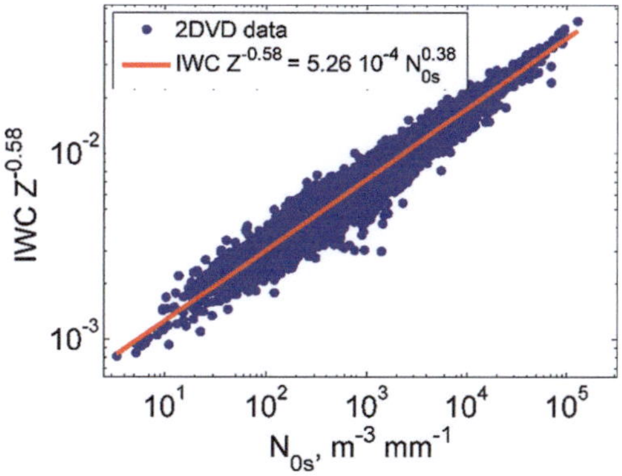

Fig. 11.4 Scatterplot of $IWC/Z^{0.58}$ vs N_{0s} from 2DVD retrievals (blue dots). The best fit to 2DVD data, $IWC\,Z^{-0.58} = 5.26 \times 10^{-4} N_{0s}^{0.38}$, is the overlaid red line. From Bukovcic et al. (2018). © American Meteorological Society. Used with permission

$$IWC = 2.36 \times 10^{-4} N_{0s}^{0.42} Z^{0.58}. \tag{11.34}$$

Bukovcic et al. (2018) analyzed a 2D-video disdrometer dataset of snow measurements in Oklahoma containing 16 events and obtained a relation

$$IWC = 5.26 \times 10^{-4} N_{0s}^{0.38} Z^{0.58} \tag{11.35}$$

with the same exponent 0.58 on Z. The scatterplot of the ratio $IWC/Z^{0.58}$ vs N_{0s} (Fig. 11.4) from Bukovcic et al. (2018) spans a broad range of N_{0s} and the corresponding prefactor in (11.35). It is evident that the intercept N_{0s} varies 4 orders of magnitude in typical snow near the ground and the corresponding multiplier in the $IWC(Z)$ relation changes by a factor of 15. This explains the tremendous variability of the $IWC(Z)$ relations found in the literature.

Because N_{0s} usually increases with height and exhibits strong dependence on temperature, Hogan et al. (2006) parameterize the dependence of the prefactor in the $IWC(Z)$ relation by temperature as

$$IWC = 0.02 \times 10^{-0.02T} Z^{0.6} \tag{11.36}$$

where T is the temperature in °C.

Polarimetric retrievals of ice remain relatively unexplored. Vivekanandan et al. (1994) suggested using specific differential phase K_{DP} to estimate IWC. Because K_{DP} depends on the particle aspect ratio and orientation, whereas IWC does not, the use of K_{DP} requires the knowledge of the particles' shapes and orientations. Ryzhkov et al. (1998) found the polarimetric relation

$$IWC = 4.46 \times 10^{-3} \frac{K_{\mathrm{DP}}\lambda}{Z_{\mathrm{DP}}} Z = 4.46 \times 10^{-3} \frac{K_{\mathrm{DP}}\lambda}{1 - Z_{\mathrm{dr}}^{-1}} \quad (11.37)$$

is practically insensitive to the shape and orientations of ice particles. In (11.37), *IWC* is in g m^{-3}, K_{DP} is in deg km^{-1}, wavelength λ is in mm, and

$$Z_{\mathrm{DP}} = Z_h - Z_v \quad (11.38)$$

is the difference between reflectivity factors at horizontal and vertical polarizations expressed in linear scale. Therefore, the ratio Z/Z_{DP} equals $1/(1 - Z_{\mathrm{dr}}^{-1})$, where Z_{dr} is the differential reflectivity expressed in linear scale ($Z_{\mathrm{dr}} = 10^{0.1Z_{\mathrm{DR}}}$). Testing (11.37) by comparison with in situ *IWC* measurements on board of the T-28 aircraft showed that it exhibits a noticeably better performance than the *IWC*(*Z*) relation in a stratiform part of the thunderstorm in Oklahoma (Ryzhkov et al. 1998). Nguyen et al. (2017) made a similar conclusion after analyzing data collected with the airborne X-band dual-polarization radar on board the Convair-580 aircraft. The plane flew through areas of high ice water content during the High-Altitude Ice Crystal-High Ice Water Content (HAIC-HIWC) international field campaign in Cayenne, French Guiana, in May 2015. They demonstrate that the polarimetric estimator (11.37) provides significant improvement over the conventional IWC(Z) estimator in clouds containing mixture of small ice particles and large aggregates, although some additional tuning of the polarimetric relation might be needed.

In the Bukovcic et al. (2018) study, the bivariate power-law relation

$$IWC = 0.71 K_{\mathrm{DP}}^{0.65} Z^{0.28} \quad (11.39)$$

is derived using disdrometer snow data from Oklahoma and assuming the snowflakes' aspect ratio of 0.65 and no canting. This relation is much less sensitive to the variability of snow size distributions compared to the Hogan et al. (2006) *IWC*(*Z*) relation (red and green dots in Fig. 11.5). There is also improvement with respect to the *IWC*(K_{DP}) relation

$$IWC(K_{\mathrm{DP}}) = 14.4 K_{\mathrm{DP}}^{1.05} \quad (11.40)$$

optimized for the Oklahoma disdrometer dataset (blue dots in Fig. 11.5).

The relation (11.39) was tested using snow disdrometer measurements in Colorado and Ontario, Canada, and demonstrated robust performance in these distinct climate regions, which attests to its potentially universal character. Note that (11.39) is derived assuming no canting of snowflakes. Bukovcic et al. (2018) show the multiplier in (11.39) depends on the width of the canting angle distribution. This necessitates calibration of the *IWC*(K_{DP}, *Z*) relation by systematic comparisons of the *IWC* estimates from radar data and in situ measurements on the ground or on board the aircraft.

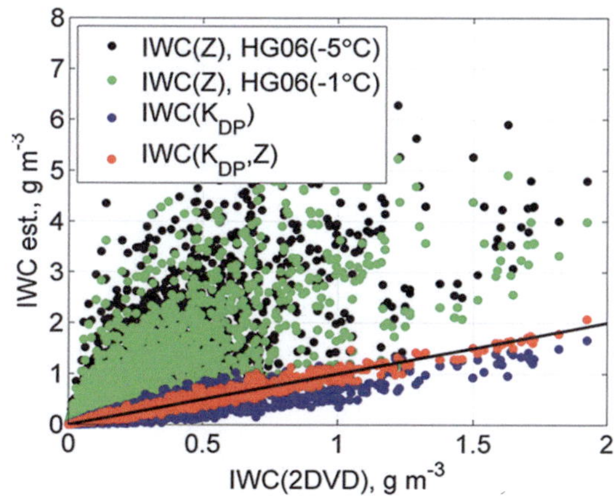

Fig. 11.5 Scatterplots of ice water content measured by disdrometer, *IWC* (2DVD), versus the estimates *IWC*(*Z*) (black and green dots), *IWC*(K_{DP}) (blue dots), and *IWC*(K_{DP}, *Z*) (red dots) relations. Two *IWC*(*Z*) relations corresponding to different temperatures are taken from Hogan et al. (2006). Courtesy of Dr. Bukovcic

The accuracy of the polarimetric relations for snow measurements depends on the accuracy of the K_{DP} estimates. The latter are notoriously noisy in aggregated snow where K_{DP} is small. The effects of these statistical fluctuations can be reduced by spatial averaging. Measurements in regions of high K_{DP} are spared of relatively intense fluctuations and are therefore suitable for application of the polarimetric *IWC* relation (11.39). High K_{DP} is typically observed in the dendritic growth layer centered at the -15 °C isotherm.

11.2.2 Snow Size Distribution Retrieval

Estimating the mean size and concentration of snow is a challenging task. These two key parameters of the snow size distribution are almost impossible to retrieve from single-polarization radar measurements due to the large variety of ice/snow habits and their microphysical properties. Some correlation between the median volume diameter of ice particles D_0 and the reflectivity factor is reported by Matrosov and Heymsfield (2017) (see Fig. 11.6). They quantify the correlation using two datasets from two-dimensional microphysical probes during aircraft flights in subtropical and midlatitude regions. Scatterplots of D_0 (tagged as D_{mv}) vs Ka-band reflectivity Z are displayed in Fig. 11.6 representing the data obtained at warmer (left panel) and colder (right panel) temperatures during the GCPEX and CRYSTAL-FACE field campaigns. Average ambient air temperatures were approximately -14 and -33 °C for these datasets. Although the correlation between D_0 and Z is high for both cases, the multipliers in the power-law $D_0(Z)$ relations are quite different because smaller-size ice particles in higher concentration prevail at higher altitudes and lower temperatures.

A more accurate estimate of the ice particle size can be obtained from the polarimetric measurements. This is because the radar reflectivity factor Z is proportional to the fourth moment of snow size distribution and the specific differential

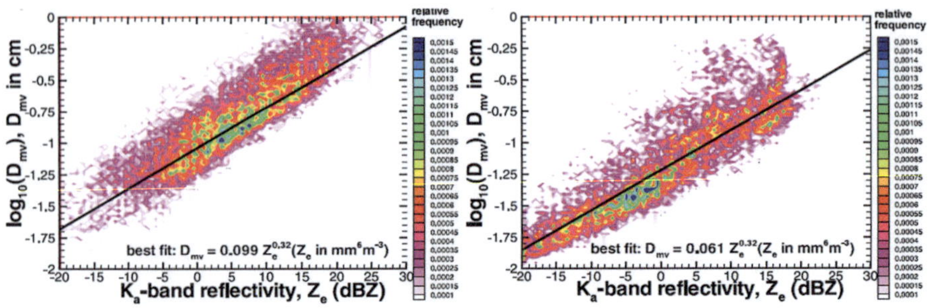

Fig. 11.6 Scatterplots of the median volume diameter D_{mv} of ice particles derived from the (left) GCPEX and (right) CRYSTAL-FACE microphysical data vs Ka-band radar reflectivity. From Matrosov and Heymsfield (2017). © American Meteorological Society. Used with permission

phase K_{DP} is proportional to the first moment (see Chap. 5). Therefore, their ratio is approximately proportional to the cube of D_m (or D_0). Indeed, according to Eq. (5.27) in Chap. 5, the expression for K_{DP} in the Rayleigh approximation is

$$K_{\mathrm{DP}} = \frac{0.27\pi}{\lambda\rho_i^2}\left(\frac{\varepsilon_i - 1}{\varepsilon_i + 2}\right)^2 \int F_{\mathrm{shape}} F_{\mathrm{orient}} \rho_s^2(D) D^3 N(D) dD, \tag{11.41}$$

where ε_i is the dielectric constant of solid ice, $\rho_i = 0.92$ g cm^{-3} is the density of solid ice, λ is the wavelength in mm, F_{shape} is the "shape factor" determined by the aspect ratio of the particle, and F_{orient} is the "orientation factor" depending on the width of the canting angle distribution σ (Chap. 3):

$$F_{\mathrm{orient}} = \frac{1}{2}\exp\left(-2\sigma^2\right)\left[1 + \exp\left(-2\sigma^2\right)\right]. \tag{11.42}$$

The dependence of F_{shape} on the aspect ratio of oblate spheroids is illustrated in Fig. 11.7, and the dependence of F_{orient} on σ is in Fig. 3.5. Equation (11.41) can be simplified using (11.28) and (11.29) and assuming that $\beta = 1$ and the shape and orientation of the particles do not change across the size spectrum

$$K_{\mathrm{DP}} = 1.02 \times 10^{-2} F_{\mathrm{shape}} F_{\mathrm{orient}} \frac{\alpha^2 N_{0s}}{\lambda} D_m^2. \tag{11.43}$$

Therefore, taking into account Eq. (11.30) for Z,

$$\frac{Z}{K_{\mathrm{DP}}\lambda} = 0.520 \frac{D_m^3}{F_{\mathrm{shape}} F_{\mathrm{orient}}} \tag{11.44}$$

and the mean volume diameter D_m can be estimated from a cubic root of the ratio Z/K_{DP} (or $Z/(K_{\mathrm{DP}}\lambda)$) if the shapes and orientations of ice particles are roughly known:

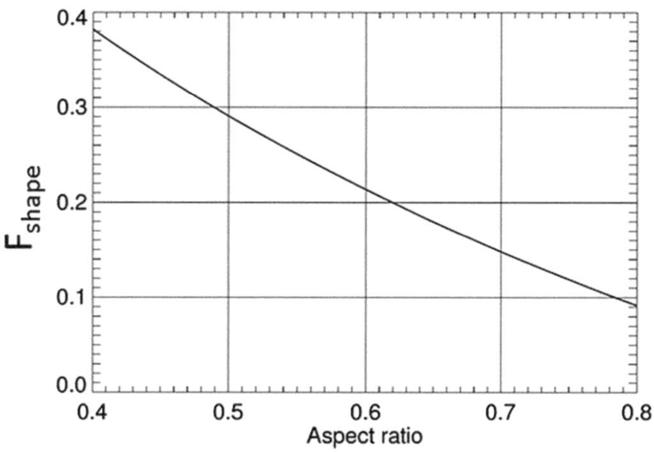

Fig. 11.7 Shape factor F_{shape} as a function of the aspect ratio of oblate spheroid

$$D_m = 1.24 \ \left[F_{\text{shape}} F_{\text{orient}}\right]^{1/3} \left(\frac{Z}{K_{\text{DP}}\lambda}\right)^{1/3}, \tag{11.45}$$

where D_m and λ are expressed in mm, Z is in mm^6 m^{-3}, and K_{DP} is in deg km^{-1}.

Aspect ratios of ice crystals and snowflakes change in a wide range. However, the aspect ratio of dry aggregated snowflakes and ice crystals with irregular shape varies in a relatively narrow interval between 0.5 and 0.7 (Straka et al. 2000; Korolev and Isaac 2003) with a median value close to 0.6. Irregularly shaped ice particles comprise the majority of ice hydrometeors in the atmosphere (Korolev et al. 2000). Therefore, a shape-dependent factor $F_{\text{shape}}{}^{1/3}$ in (11.45) of these particles does not change much (within 10–15%). The width of the canting angle distribution σ of ice hydrometeors is poorly known, but a few polarimetric radar studies indicate it varies from about 10° for pristine dendrites (Matrosov et al. 2005a; Melnikov and Straka 2013) to 40° in moderate-to-heavy snow consisting of dry aggregates (Hendry et al. 1987). This corresponds to the change in the $[F_{\text{orient}}]^{1/3}$ factor in (11.45) from 0.97 to 0.69. The dependencies of D_m on

$$\xi = \left(\frac{Z}{K_{\text{DP}}\lambda}\right)^{1/3} \tag{11.46}$$

for various aspect ratios $a/b < 1$, different values of σ, and exponential size distribution of ice particles are illustrated in Fig. 11.8.

For the gamma size distribution, the $D_m(\xi)$ is somewhat sensitive to the size distribution shape factor μ in the range between -1 and 1 (Fig. 11.9). This range contains the majority of μ estimates from in situ observations (Brandes et al. 2007; McFarquhar et al. 2007; Delanoe et al. 2014; Tiira et al. 2016; Matrosov and Heymsfield 2017).

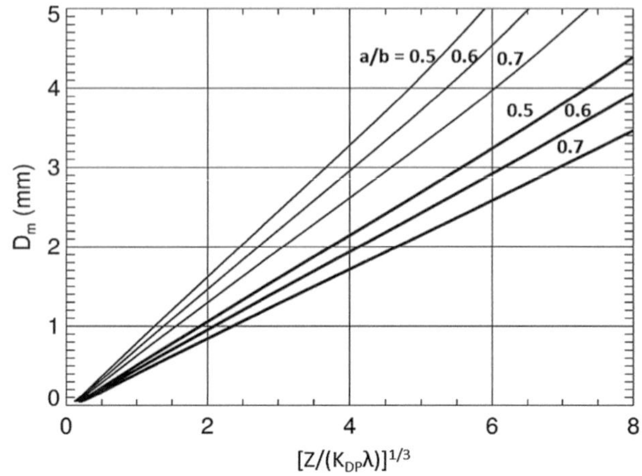

Fig. 11.8 Dependence of the mean volume diameter D_m of ice particles on the parameter $\xi = [Z/(K_{DP}\lambda)]^{1/3}$ for irregularly shaped/aggregated snow for different aspect ratios *a*/*b* and two widths of the canting angle distributions: $\sigma = 10°$ (thin curves) and $\sigma = 40°$ (thick curves). Exponential size distribution of snowflakes is assumed

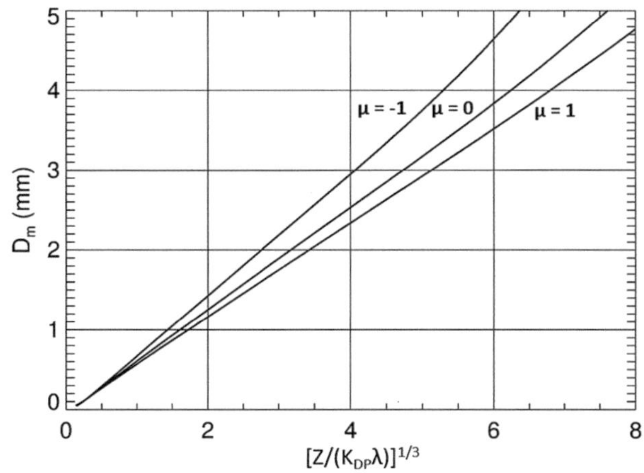

Fig. 11.9 Dependence of the mean volume diameter D_m of ice particles on the parameter $\xi = [Z/(K_{DP}\lambda)]^{1/3}$ for irregularly shaped/aggregated snow and different values of the shape parameter μ of the gamma size distribution. The assumed aspect ratio $a/b = 0.6$ and the width of the canting angle distribution $\sigma = 25°$

The advantage of the D_m estimator based on the Z/K_{DP} ratio is its low sensitivity to the size distribution variability and changes of snow density and degree of riming (the latter because both Z and K_{DP} are proportional to α^2 or f_{rim}^2). However, this estimator does not account for the diversity of ice particle shapes and orientations both of which strongly affect K_{DP}. Therefore, the estimator is better suited to work for irregularly shaped and aggregated snow (for which the aspect ratio does not vary much around the value of 0.6) assuming that the width σ of the canting angle distribution of ice is roughly known.

Information about mean particles' shapes and orientations can be retrieved by estimating the circular depolarization ratio (*CDR*) from the combination of Z_{DR} and ρ_{hv} measured by dual-polarization radars with simultaneous transmission/reception of orthogonally polarized waves (Ryzhkov et al. 2017). Matrosov et al. (2017) report promising results regarding the *CDR*-based retrieval of the aspect ratio of ice particles. The results are in good agreement with measurements by a video ice particle samplers on a tethered balloon system (TBS) and also by a ground-based multi-angle snowflake camera.

Secondary ice generated in the form of pristine dendrites, plates, or needles/columns usually has aspect ratios very different from 0.6. Moreover, the shape of pristine ice particles varies with size as they grow by vapor deposition. This diversity of ice habits and shapes is not accounted for in the estimator (11.45).

An alternative estimator of D_m immune to the variations of particle shapes and orientations can be constructed. It uses the ratio Z_{DP}/K_{DP} where $Z_{DP} = Z_h - Z_v$ is the reflectivity difference. The ratio Z_{DP}/K_{DP} is very robust with respect to the variability of the particles' aspect ratios and orientations because these two factors affect similarly Z_{DP} and K_{DP}, and, therefore, it remains almost intact (Ryzhkov et al. 1998, 2018). It can be shown that Z_{DP} is roughly proportional to the third moment of the size distribution of ice/snow. Therefore, the ratio Z_{DP}/K_{DP} is expected to be linearly dependent on D_m^2.

To further substantiate this idea, we present simulated D_m and the ratio

$$\eta = \left(\frac{Z_{DP}}{K_{DP}\lambda}\right)^{1/2} \tag{11.47}$$

for 12 snow habits consisting of irregular/aggregated snow (with the aspect ratio 0.6) and 11 types of pristine ice crystals listed in Table 4.1 (Chap. 4). All particles are modeled as oblate or prolate spheroids with aspect ratios dependent on the size approximated by the power-law relation

$$h = cL^d, \tag{11.48}$$

where h is the smaller and L is the larger dimension of a crystal, both expressed in mm.

The hydrometeors with oblate shapes (irregular/aggregates, dendrites, solid thick plates, and hexagonal plates) are assumed to have a Gaussian distribution of orientation of their rotation axis with zero mean canting angle and the width σ ranging from 10° to 40°. The particles with prolate shape (columns, bullets, needles) are supposed to have random orientation in the horizontal plane. The computations use the gamma size distribution with the parameter μ ranging from −1 to 1 and the degree of riming f_{rim} equal to 1 (unrimed snow) and 2.

The results of simulations are presented in Fig. 11.10 as D_m vs η dependencies. It is remarkable that the curves for 12 ice habits corresponding to exponential SD and $\alpha = 0.2$ are practically indistinguishable. The average dependence of D_m on η characterizing the 12 curves can be approximated by

$$D_m = -0.1 + 2.0\eta \tag{11.49}$$

where D_m is in mm and the η is computed assuming Z_{DP} is in mm^6 m^{-3}, K_{DP} is in deg km^{-1}, and the wavelength λ is in mm. Changing the angular orientation spread σ from 10° to 40° has no visible impact on $D_m(\eta)$ dependence.

Fig. 11.10 Dependence of the mean volume diameter D_m of ice particles on $[Z_{DP}/(K_{DP}\lambda)]^{1/2}$ for 12 various ice habits for exponential size distribution and $\alpha = 0.2$

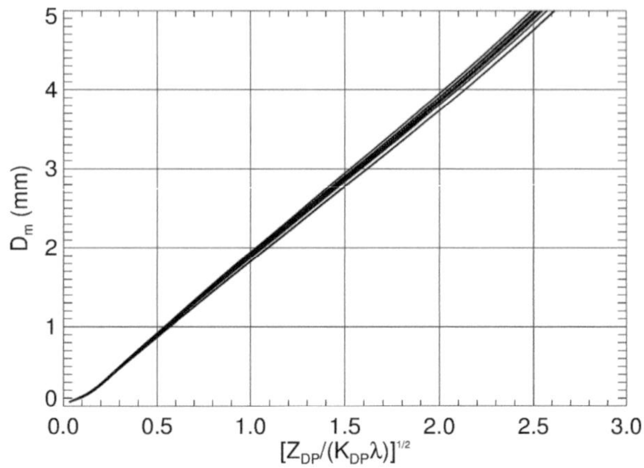

As opposed to the estimate of $D_m(\xi)$ from (11.45), the $D_m(\eta)$ estimate is affected by the degree of riming. Hence, it is more suitable for regions of lower temperatures where riming is less likely.

Once D_m is determined, the ice water content *IWC* can be estimated from the combination of *Z* and D_m using (11.33). In fact, it is the product

$$\gamma = \alpha D_{\mathrm{m}}^2 \tag{11.50}$$

that is needed to compute *IWC* from (11.33) rather than D_m. It can be shown that γ is highly correlated with η^2,

$$\gamma \approx 0.78\eta^2 = 0.78\frac{Z_{\mathrm{DP}}}{K_{\mathrm{DP}}\lambda}, \tag{11.51}$$

and the dependence of γ on η is almost insensitive to the ice habit, aspect ratio, orientation, and density (or degree of riming) of the ice particles for an exponential size distribution. Combining (11.51) with (11.33) yields

$$IWC \approx 4.0 \times 10^{-3}\frac{K_{\mathrm{DP}}\lambda}{1 - Z_{\mathrm{dr}}^{-1}}, \tag{11.52}$$

which is very close to the relation (11.37) derived by Ryzhkov et al. (1998). Note that the radar wavelength λ is expressed in mm in (11.52).

The total concentration of ice particles N_t can be estimated from the combination of *Z* and γ. Indeed, for exponential size distribution, $N_t = N_{0s}D_m/4$ and (11.30) can be modified to

$$Z = 2.12 \times 10^{-2}\gamma^2 N_t, \tag{11.53}$$

where N_t has the units of m^{-3}. Therefore, $\log(N_t)$ can be estimated as

$$\log(N_t) = 0.1Z(\text{dBZ}) - 2\log(\gamma) - 1.33. \qquad (11.54)$$

In (11.54), N_t is expressed in (1/L) to facilitate comparisons of the retrieved N_t with in situ probe observations which commonly use the units (1/L) for N_t.

11.2.3 Measurement Errors

The accurate and reliable measurements of K_{DP} are critical for implementing the polarimetric techniques for estimation of the mean volume diameter D_m. The big advantage of K_{DP} is that it is not biased by noise, attenuation, and radar miscalibration. On the other hand, the magnitude of K_{DP} in ice and snow is relatively small, particularly at longer radar wavelengths. At S band, K_{DP} in dry ice and snow is usually smaller than 0.3 deg km^{-1}, and typical values are below 0.05 deg km^{-1} in aggregated snow at warmer temperatures (Ryzhkov and Zrnic 1998; Griffin et al. 2018). Because the mean volume diameter D_m determined from (11.45) is inversely proportional to the cubic root of K_{DP}, the fractional standard deviation of the D_m estimate is proportional to the ratio 1/3 SD(K_{DP})/K_{DP}. The value of the standard deviation of the K_{DP} estimate SD(K_{DP}) is equal to 0.11 deg km^{-1} for typical parameters of the WSR-88D processing at low reflectivity (<40 dBZ) signals and radial resolution of 6 km. Therefore, the fractional standard deviation of the D_m estimate is about 70% if $K_{DP} = 0.05$ deg km^{-1} and can be much higher if K_{DP} is lower. Additional azimuthal averaging helps to reduce the statistical error of the D_m estimate. The quasi-vertical profile (QVP) methodology (Ryzhkov et al. 2016) prescribes azimuthal averaging of K_{DP} over a full 360° circle, and this decreases the SD(K_{DP}) by the factor $(360)^{1/2} \approx 19$, so D_m can be estimated with the accuracy better than 4% (for $K_{DP} = 0.05$ deg km^{-1}). The corresponding accuracy is about 20% if $K_{DP} = 0.01$ deg km^{-1}. Averaging in a limited azimuthal sector may be needed if a substantial horizontal nonuniformity of the radar echo occurs within a full circle of the QVP averaging.

Calibration biases of Z and Z_{DR} may affect the $D_m(\xi)$ and $D_m(\eta)$ estimates. Because D_m is proportional to the cubic root of Z, the impact of the calibration bias of Z is not very significant. The 1 dB and 2 dB biases of Z would cause 8% and 16% fractional bias of D_m. The adverse effect of miscalibrated Z_{DR} on the quality of the $D_m(\eta)$ estimate depends on the actual value of Z_{DR}. It rapidly increases for lower Z_{DR}. It can be shown that a Z_{DR} bias of 0.2 dB would produce a 5.5% fractional bias in D_m if $Z_{DR} = 1.0$ dB and the bias exceeds 25% if $Z_{DR} < 0.2$ dB. Therefore, the estimator $D_m(\eta)$ is expected to be much better in areas of high Z_{DR} (like the dendritic growth layer).

Given the enhanced sensitivity of the $D_m(\eta)$ estimator to the degree of riming and Z_{DR} biases (at small Z_{DR}), the method is expected to be more accurate at relatively low temperatures (say, lower than −10 to −15 °C) in stratiform clouds where significant riming is less likely and in areas of higher Z_{DR} (exceeding 0.2 dB). The

$D_m(\xi)$ estimator might be the preferred choice at higher temperatures (>−10 °C) in mixed-phase clouds where riming can be significant.

11.2.4 Validation of the Algorithm

The algorithm for snow size distribution retrieval described in a previous section may require further refinement and is a subject of extensive testing and validation. Herein, we provide initial validation results using snow disdrometer measurements at the surface and in situ measurements on board of research airplane.

The disdrometer measurements of snow during 16 winter events in Oklahoma (Bukovcic et al. 2018) have been used to test the procedure for estimating the mean volume diameter of snowflakes D_m from the ratio Z/K_{DP} in (11.45). A scatterplot of D_m measured by the disdrometer versus the retrieved value is in Fig. 11.11. Note that the values of Z and K_{DP} in this example are calculated from the disdrometer data assuming fixed aspect ratio a/b and the width of the canting angle distribution σ and using the factors F_{shape} and F_{orient} in (11.45) matched with the selected a/b and σ. Therefore, the difference between the actual and retrieved values of D_m is attributed to the variability of snow density and size distributions which are not considered in the theoretical simulations and the disdrometer measurement errors. The agreement between actual and retrieved D_m (Fig. 11.13) is quite good although the retrieved values are slightly higher.

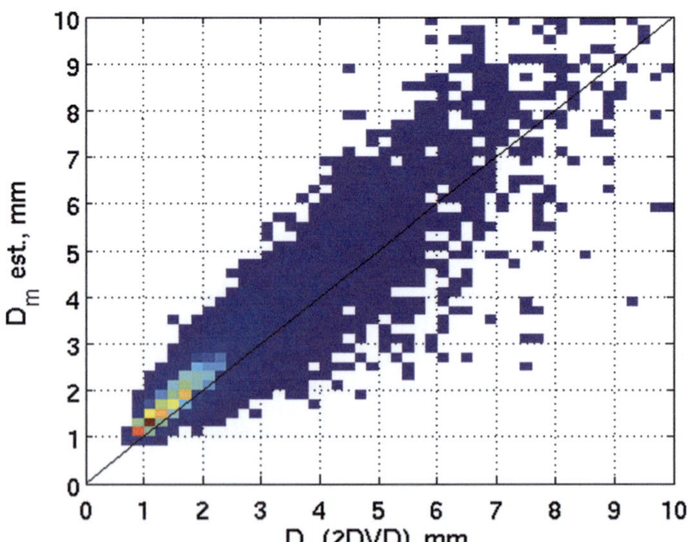

Fig. 11.11 Scatterplot of the mean volume diameter D_m estimated from the ratio Z/K_{DP} using (11.45) vs measurement with the 2D-video disdrometer. Data are from 16 snow events in Oklahoma

Because K_{DP} in snow is low and noisy, aggressive spatial averaging is required to ensure suitable quality of the K_{DP} estimates. This is a primary reason for introducing the QVP methodology. However, the QVP product is radar-centric and has relatively coarse horizontal resolution degrading with altitude because the azimuthal averaging is performed on a conical surface (Ryzhkov et al. 2016). For direct comparison of the ground-based radar and aircraft data, the columnar vertical profiles (CVP) methodology for processing and displaying the radar data might be more appropriate (Murphy et al. 2018). As opposed to QVP, the CVP technique prescribes azimuthal averaging in a limited azimuthal sector using all available tilts of radar data instead of one. As a result, vertical profiles of polarimetric radar variables are generated in a column with much smaller horizontal dimension. This column can follow the aircraft track to facilitate direct comparison of the radar and in situ aircraft data.

An example of the CVP product generated from the KVNX (Vance Air Force Base, OK) WSR-88D radar data for the case of mesoscale convective system observed during the Midlatitude Continental Convective Clouds Experiment (MC3E) on 20 May 2011 is shown in Fig. 11.12a–d. The horizontal size of the vertical column that follows research aircraft is 20 km $\times$ 20° (shown at the top of the plot). The aircraft altitude at any given time is depicted by a black line in the plot. It is important to contrast the CVP product with a 4D-grid WSR-88D product suggested by Homeyer (2014) and Homeyer and Kumjian (2015) in a column following aircraft for the same storm ((e)) (Tian et al. 2016). The latter has horizontal and vertical resolution of 2 km and 1 km, respectively. It is obvious that the CVP product has a better vertical resolution (at the expense of degraded horizontal resolution) compared to the 4D-grid product.

Temporal dependencies of Z, Z_{DR}, ρ_{hv}, and K_{DP} measured by the radar along the aircraft track are displayed in Fig. 11.13. Shaded areas in the plot indicate time intervals when the aircraft penetrated the melting layer in the stratiform part of the MCS. These are marked by a strong depression of ρ_{hv} and increase in Z and Z_{DR}. The retrieved D_m, N_t, and IWC using Eqs. (11.49), (11.54), and (11.52) from polarimetric data in ice regions outside of the shaded intervals are presented in Fig. 11.14. The best agreement of the retrieved microphysical parameters (dots) and in situ measurements (solid curves) is achieved for the mean volume diameter and ice water content in the dendritic growth layer at the temperature between −10 and −20 °C (top panel in Fig. 11.14). The initial validation results are promising and call for more comprehensive and systematic studies.

11.2.5 Retrievals in a Tropical Cyclone

Polarimetric ice retrievals made in tropical cyclones and hurricanes reveal common microphysical features characterized by very high concentrations of small ice particles aloft. As an example, a composite QVP of Z, Z_{DR}, ρ_{hv}, and K_{DP} retrieved from the KMHX (Morehead, NC) WSR-88D data collected during the hurricane Irene on 27 August 2011 is presented in Fig. 11.15.

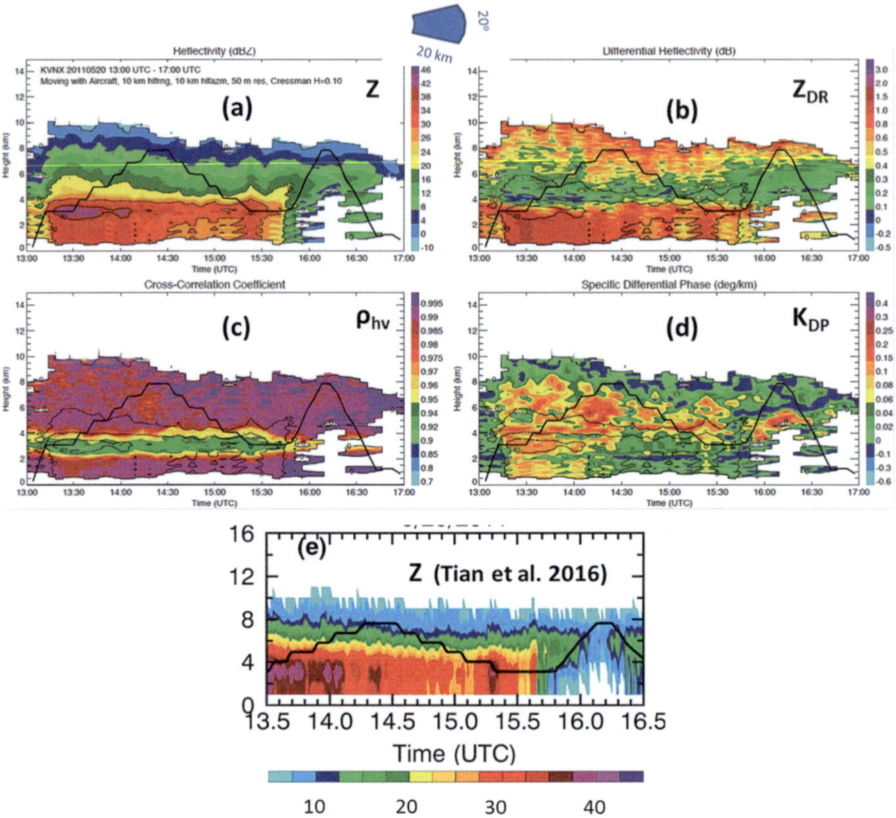

Fig. 11.12 Vertical profiles of polarimetric variables in a vertical column following the aircraft represented in a height vs time format for the MCS on 20 May 2011. (**a**)–(**d**) Z, Z_{DR}, ρ_{hv}, and K_{DP} estimated from the KVNX WSR-88D (Vance Air Force Base) data using the CVP methodology. (**e**) 4D-grid Z product from WSR-88D (Tian et al. 2016). Aircraft height vs time is depicted with the black curve

A striking feature of this storm is the evidence of extended regions of enhanced K_{DP} associated with very low Z above the freezing level. Reflectivity is below 20 dBZ above the melting layer, while the maximal K_{DP} is above 0.3 deg km^{-1}. This is a strong indication of high concentrations of very small ice in a whole depth of the cold part of the storm. Another evidence of the dominance of small hydrometeors is a uniformly high ρ_{hv} above and below the melting layer typical for tropical storms. The top of the storm is highest at the time when the eyewall of the hurricane passed in the proximity of the radar between 12 and 15 UTC.

This time period is also marked by the ascension of the melting layer clearly visible in the Z_{DR} and ρ_{hv} panels.

The corresponding plot illustrating the time evolution of D_m, log(N_t), and log(IWC) in the vertical cross section of the storm (Fig. 11.16) shows the mean volume diameter of ice does not exceed 0.1–0.2 mm in the large vertically

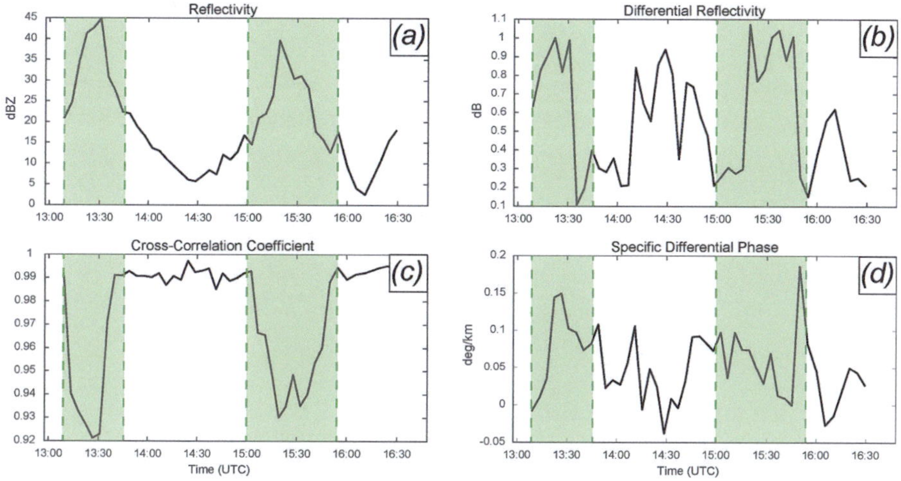

Fig. 11.13 Variability of (**a**) Z, (**b**) Z_{DR}, (**c**) ρ_{hv}, and (**d**) K_{DP} measured by the KVNX (Vance Air Force Base, OK) WSR-88D radar along the aircraft track depicted in Fig. 11.14. Shaded areas indicate periods when the aircraft was within the melting layer

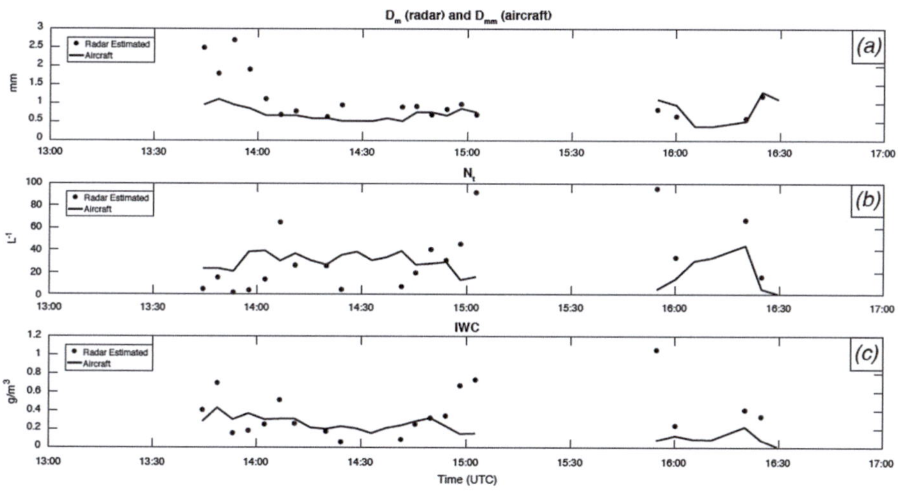

Fig. 11.14 Comparison of (**a**) D_m, (**b**) N_t, and (**c**) IWC retrieved from the polarimetric radar measurements (dots) and in situ aircraft probes (solid lines) for the MCS on 20 May 2011

extended areas of the storm where the total ice concentration is higher than 1000 1/L. The resulting *IWC* in these areas is significant and may be as high as 5 g m^{-3}. These retrieval results are consistent with the documented evidence of high ice water content at the periphery of tropical cyclones which may dramatically impact the performance of airplane engines and therefore pose significant aviation hazard (Fridlind et al. 2015; Strapp et al. 2016; Leroy et al. 2017; Stanford et al. 2017).

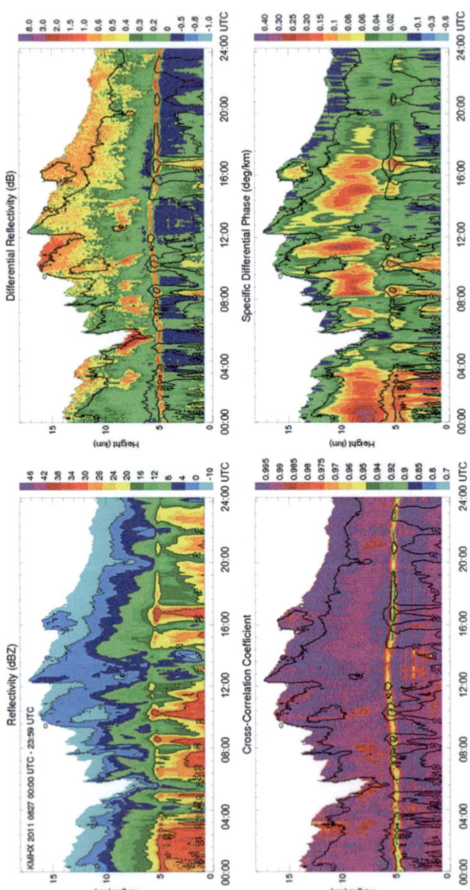

Fig. 11.15 Composite QVP of Z, Z_{DR}, ρ_{hv}, and K_{DP} retrieved from the KMHX (Morehead, NC) WSR-88D data collected during the hurricane Irene on 27 August 2011. Overlaid are contours of $Z = 10$, 20, and 30 dBZ. Courtesy of Dr. P. Zhang

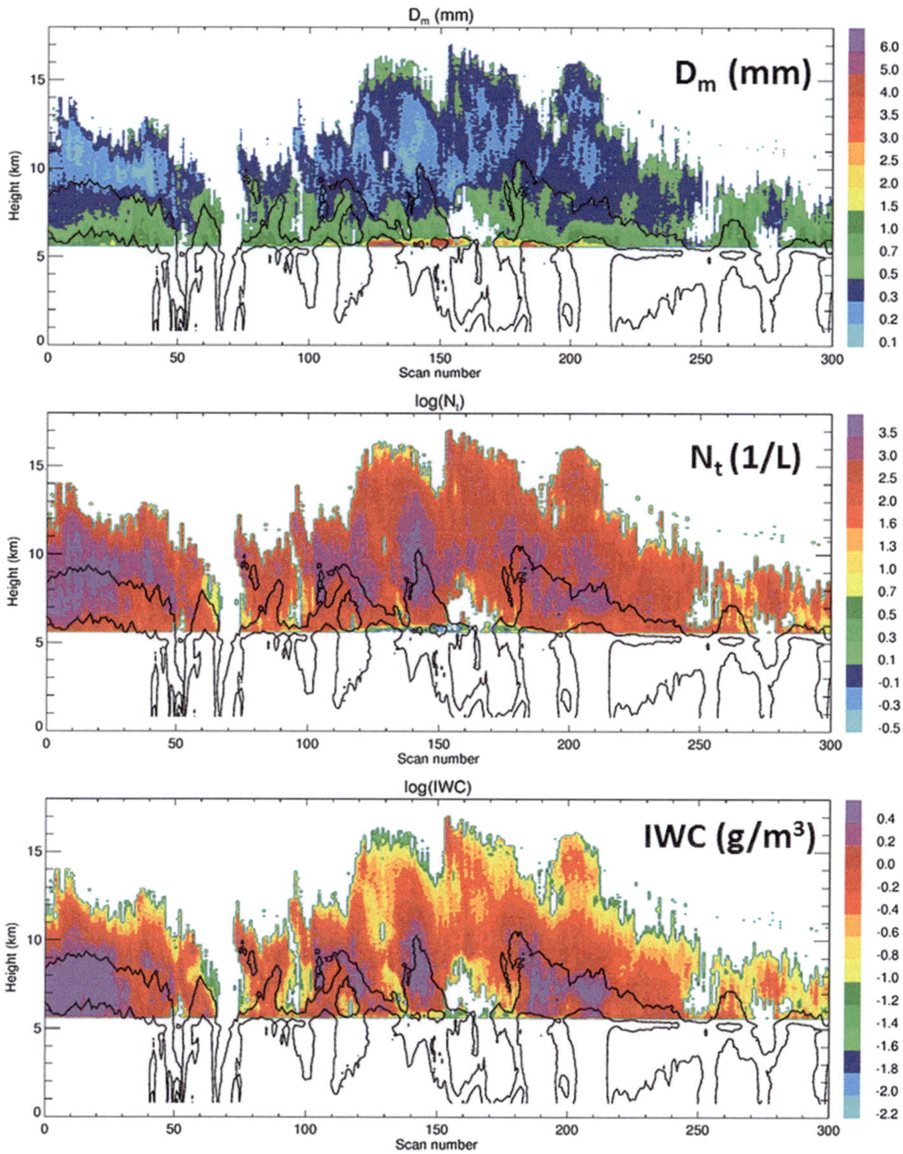

Fig. 11.16 Vertical cross sections of the mean volume diameter D_m, total number concentration N_t, and ice water content *IWC* above the freezing level retrieved from the QVP displayed in Fig. 11.15 for the hurricane Irene on 27 August 2011. The N_t and *IWC* data are in logarithmic scale. Overlaid are contours of $Z = 10$, 20, and 30 dBZ (increasing from top to bottom). Courtesy of Dr. P. Zhang

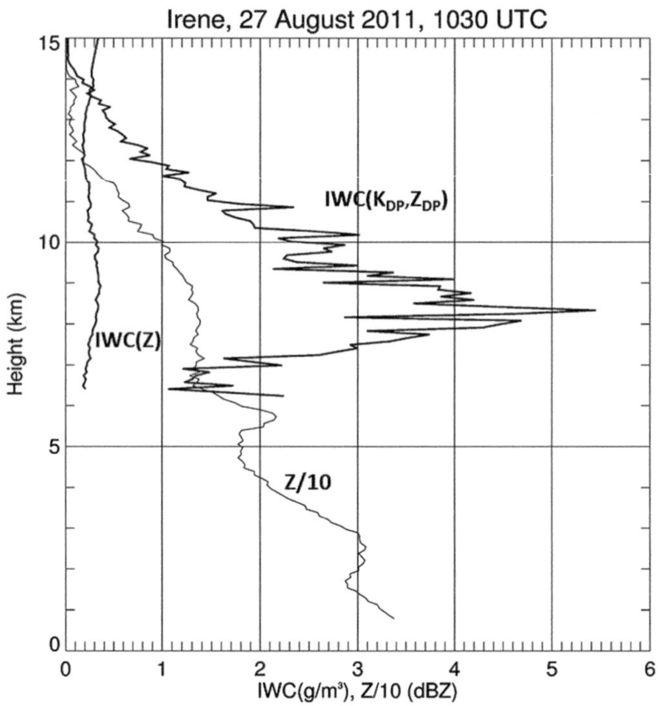

Fig. 11.17 Vertical profiles of *Z*/10 and *IWC* retrieved from the *IWC*(*Z*) and *IWC*(K_{DP},Z_{DP}) relations applied to data from the hurricane Irene observed by the KMHX (Morehead, NC) WSR-88D on 27 August 2011 at 1030 UTC

Ice water content computed from the *IWC*(*Z*) relation (11.36) is almost an order of magnitude lower than the one obtained from the polarimetric retrieval algorithm using *Z*, K_{DP}, and Z_{DP}. This is seen in the vertical profiles of *Z*, *IWC*(*Z*), and *IWC*(K_{DP}, Z_{DP}) taken at 1030 UTC (Fig. 11.17). The reflectivity is generally below 15 dBZ above the freezing level in this storm due to the very small dominant size of ice.

References

Anagnostou, M., Anagnostou, E., Vivekanandan, J., & Ogden, F. (2008). Comparison of two raindrop size distribution retrieval algorithms for X-band dual polarization observations. *Journal of Hydrometeorology, 9*, 589–600.

Atlas, D., Matrosov, S., Heymsfield, A., Chou, M., & Wolf, D. (1995). Radar and radiation properties of ice clouds. *Journal of Applied Meteorology, 34*, 2329–2345.

Brandes, E., Zhang, G., & Vivekanandan, J. (2002). Experiments in rainfall estimation with polarimetric radar in a subtropical environment. *Journal of Applied Meteorology, 41*, 674–685.

Brandes, E., Zhang, G., & Vivekanandan, J. (2004). Drop size distribution retrieval with polarimetric radar: Model and application. *Journal of Applied Meteorology, 43*, 461–475.

Brandes, E., Ikeda, K., Zhang, G., Schoenhuber, M., & Rasmussen, R. (2007). A statistical and physical description of hydrometeor distributions in Colorado snowstorms using a video-disdrometer. *Journal of Applied Meteorology, 46*, 634–650.

Bringi, V. N., & Chandrasekar, V. (2001). *Polarimetric Doppler weather radar. Principles and applications*. Cambridge, UK: Cambridge University Press. 636pp.

Bringi, V., Huang, G., Chandrasekar, V., & Gorgucci, E. (2002). A methodology for estimating the parameters of a gamma raindrop size distribution model from polarimetric radar data: Application to a squall-event from the TRMM/Brazil campaign. *Journal of Atmospheric and Oceanic Technology, 19*, 633–645.

Bringi, V., Carey, L., Gatlin, P., Schultz, E., Petersen, W. (2012). Estimating the accuracy of polarimetric radar-based retrievals of drop-size distribution parameters and rain rate: an application of error variance separation using radar-derived spatial correlations. *J. Hydrometerology, 13*, 1066–1079.

Bringi, V., Chandrasekar, V., Hubbert, J., Gorgucci, E., Randeu, W., & Schoenhuber, M. (2003). Raindrop size distribution in different climatic regimes from disdrometer and dual-polarized radar analysis. *Journal of the Atmospheric Sciences, 60*, 354–365.

Bringi, V., Williams, C., Thurai, M., & May, P. (2009). Using dual-polarized radar and dual-frequency profiler for DSD characterization: A case study from Darwin, Australia. *Journal of Atmospheric and Oceanic Technology, 26*, 2107–2122.

Brown, P. R. A., & Francis, P. N. (1995). Improved measurements of ice water content in cirrus using a total-water probe. *Journal of Atmospheric and Oceanic Technology, 12*, 410–414.

Bukovcic, P., Ryzhkov, A., Zrnic, D., & Zhang, G. (2018). Polarimetric radar relations for quantification of snow based on the disdrometer data. *Journal of Applied Meteorology and Climatology, 57*, 103–120.

Cao, Q., Zhang, G., Brandes, E., Schuur, T., Ryzhkov, A., & Ikeda, K. (2008). Analysis of video disdrometer and polarimetric radar data to characterize rain microphysics in Oklahoma. *Journal of Applied Meteorology and Climatology, 47*, 2238–2255.

Cao, Q., Zhang, G., Brandes, E., & Schuur, T. (2010). Polarimetric radar rain estimation through retrieval of drop size distribution using a Bayesian approach. *Journal of Applied Meteorology and Climatology, 49*, 973–990.

Cao, Q., Zhang, G., & Xue, M. (2013). A variational approach for retrieving raindrop size distribution from polarimetric radar measurements in the presence of attenuation. *Journal of Applied Meteorology and Climatology, 52*, 169–185.

Carlin, J., Ryzhkov, A., Snyder, J., & Khain, A. (2016). Hydrometeor mixing ratio retrievals for storm-scale radar data assimilation: Utility of current equations and potential benefits of polarimetry. *Monthly Weather Review, 144*, 2981–3001.

Carlin, J., Gao, J., Snyder, J., & Ryzhkov, A. (2017). Assimilation of Z_{DR} columns for improving the spin-up and forecast of convective storms in storm-scale models: Proof-of-concept experiments. *Monthly Weather Review, 145*, 5033–5057.

Delanoe, J., Heymsfield, A., Protat, A., Bansemer, A., & Hogan, R. (2014). Normalized particle distribution for remote sensing application. *Journal of Geophysical Research – Atmospheres, 119*, 4204–4227.

Doviak, R., & Zrnic, D. (2006). *Doppler radar and weather observations* (2nd ed.). Reprint, Mineola, NY: Dover, 562 pp.

Erfani, E., & Mitchell, D. (2017). Growth of ice particle mass and projected area during riming. *Atmospheric Chemistry and Physics, 17*, 1241–1257.

Fridlind, A., Ackerman, A., Grandin, A., Dezitter, F., Weber, M., Strapp, J., Korolev, A., & Williams, C. (2015). High ice water content at low radar reflectivity near deep convection – Part 1: Consistency of in situ and remote-sensing observations with stratiform rain column simulations. *Atmospheric Chemistry and Physics, 15*, 11713–11728.

Gorgucci, E., Chandrasekar, V., Bringi, V., & Scarchilli, G. (2002). Estimation of raindrop size distribution parameters from polarimetric radar measurements. *Journal of the Atmospheric Sciences, 59*, 2373–2384.

Greene, D., & Clark, R. (1972). Vertically integrated liquid water – A new analysis tool. *Monthly Weather Review, 100*, 548–552.

Griffin, E., Schuur, T., & Ryzhkov, A. (2018). A polarimetric analysis of ice microphysical processes in snow, using quasi-vertical profiles. *Journal of Applied Meteorology and Climatology, 57*, 31–50.

Hendry, A., Antar, Y., & McCormick, G. (1987). On the relationship between the degree of preferred orientation in precipitation and dual-polarization radar echo characteristics. *Radio Science, 22*, 37–50.

Heymsfield, A., Bansemer, A., Schmitt, C., Twohy, C., & Poellot, M. (2004). Effective ice particle densities derived from aircraft data. *Journal of the Atmospheric Sciences, 61*, 982–1003.

Heymsfield, A., Wang, Z., & Matrosov, S. (2005). Improved radar ice water content retrieval algorithms using coincident microphysical and radar measurements. *Journal of Applied Meteorology, 44*, 1391–1412.

Heymsfield, A., Matrosov, S., & Wood, N. (2016). Toward improving ice water content and snowrate retrievals from radars. Part I: X and W bands, emphasizing CloudSat. *Journal of Applied Meteorology and Climatology, 55*, 2063–2090.

Hogan, R., Mittermaier, M., & Illingworth, A. (2006). The retrievals of ice water content from radar reflectivity factor and temperature and its use in evaluating a mesoscale model. *Journal of Applied Meteorology and Climatology, 45*, 301–317.

Homeyer, C. (2014). Formation of the enhanced-V infrared cloud top feature from high-resolution three-dimensional radar observations. *Journal of the Atmospheric Sciences, 71*, 332–348.

Homeyer, C., & Kumjian, M. (2015). Microphysical characteristics of overshooting convection from polarimetric radar observations. *Journal of the Atmospheric Sciences, 72*, 870–891.

Illingworth, A., & Thompson, A. (2005). The estimation of moderate rain rates with operational polarization radar. In *32nd Conference on Radar Meteorology*, Albuquerque, NM, P9R.1.

Jung, Y., Zhang, G., & Xue, M. (2008a). Assimilation of simulated polarimetric radar data for a convective storm using the ensemble Kalman filter. Part I: Observation operators for reflectivity and polarimetric variables. *Monthly Weather Review, 136*, 2228–2245.

Jung, Y., Zhang, G., Xue, M., & Straka, J. (2008b). Assimilation of simulated polarimetric radar data for a convective storm using the ensemble Kalman filter. Part II: Impact of polarimetric radar data on storm analysis. *Monthly Weather Review, 136*, 2246–2260.

Kim, D.-S., Maki, M., & Lee, D.-I. (2010). Retrieval of three-dimensional raindrop size distribution using X-band polarimetric radar data. *Journal of Atmospheric and Oceanic Technology, 27*, 1265–1285.

Korolev, A., & Isaac, G. (2003). Roundness and aspect ratios of particles in ice clouds. *Journal of the Atmospheric Sciences, 60*, 1795–1808.

Korolev, A., Isaac, G., & Hallett, J. (2000). Ice particle habits in stratiform clouds. *Quarterly Journal of the Royal Meteorological Society, 126*, 2873–2902.

Leroy, D., Fontane, E., Schwarzenboeck, A., Strapp, J., Korolev, A., McFarquhar, G., Dupuy, R., Courbeyre, C., Lilie, L., Protat, A., Delanoe, J., Dezitter, F., & Grandin, A. (2017). Ice crystal sizes in high ice water content clouds. Part II: Statistics of mass diameter percentiles in tropical convection observed during the HAIC/HIWC project. *Journal of Atmospheric and Oceanic Technology, 34*, 117–136.

Li, X., & Mecikalski, J. (2010). Assimilation of the dual-polarization Doppler radar data for a convective storm with a warm-rain radar forward operator. *Journal of Geophysical Research, 115*, 1–16.

Liu, C., & Illingworth, A. (2000). Toward more accurate retrievals of ice water content from radar measurements of clouds. *Journal of Applied Meteorology, 39*, 1130–1146.

Matrosov, S. (2007). Modeling backscatter properties of snowfall at millimeter wavelengths. *Journal of the Atmospheric Sciences, 64*, 1727–1736.

Matrosov, S., & Heymsfield, A. (2017). Empirical relations between size parameters of ice hydrometeor populations and radar reflectivity. *Journal of Applied Meteorology and Climatology, 56*, 2479–2488.

Matrosov, S., Reinking, R., & Djalalova, I. (2005a). Inferring fall attitudes of pristine dendritic crystals from polarimetric radar data. *Journal of the Atmospheric Sciences, 62*, 241–250.

Matrosov, S., Kingsmill, D., Martner, B., & Ralph, F. (2005b). The utility of X-band polarimetric radar for quantitative estimates of rainfall parameters. *Journal of Hydrometeorology, 6*, 248–262.

Matrosov, S., Schmitt, C., Maahn, M., & de Boer, G. (2017). Atmospheric ice particle shape estimates from polarimetric radar measurements and in situ observations. *Journal of Atmospheric and Oceanic Technology, 34*, 2569–2586.

McFarquhar, G., Timlin, M., Rauber, R., Jewitt, B., & Grim, J. (2007). Vertical variability of cloud hydrometeors in the stratiform region of mesoscale convective systems and bow echoes. *Monthly Weather Review, 135*, 3405–3428.

Melnikov, V., & Straka, J. (2013). Axis ratios and flutter angles of cloud ice particles: Retrievals from radar data. *Journal of Atmospheric and Oceanic Technology, 30*, 1691–1703.

Murphy, A., Ryzhkov, A., Zhang, P., McFarquhar, G., Wu, W., & Stechman, D. (2018). *A polarimetric and microphysical analysis of the stratiform rain region of MCSs*. In Annual American Meteorological Society Meeting, January 8–11, Austin, TX.

Nguyen, C., Wolde, M., Baibakov, K., & Korolev, A. (2017). Detection and estimation of high ice water content using X-band and W-band dual-polarization airborne radar data. In *38th Conference on Radar Meteorology, Chicago, IL* (p. 89). American Meteorological Society.

Papoulis, A. (1991). *Probability, random variables and stochastic processes*. Boston, MA: McGraw-Hill.

Pfeifer, M., Craig, C., Hagen, M., & Keil, C. (2008). A polarimetric radar forward operator for model evaluation. *Journal of Applied Meteorology and Climatology, 47*, 3202–3220.

Posselt, D., Li, X., Tushaus, A., & Mecikalski, J. (2015). Assimilation of dual-polarization radar observations in mixed- and ice-phase regions of convective storms: Information content and forward model errors. *Monthly Weather Review, 143*, 2611–2636.

Protat, A., Delanoe, J., Strapp, J., Fontaine, E., Leroy, D., Schwarzenboeck, A. et al. (2016). The measured relationship between ice water content and cloud radar reflectivity in tropical convective clouds. *Journal of Applied Meteorology and Climatology, 55*, 1707–1729.

Putnam, B., Xue, M., Jung, Y., Snook, N., & Zhang, G. (2017). Ensemble probabilistic prediction of a mesoscale convective system and associated polarimetric radar variables using single-moment and double-moment microphysics schemes and EnKF radar data assimilation. *Monthly Weather Review, 145*, 2257–2279.

Ryzhkov, A., & Zrnic, D. (1998). Discrimination between rain and snow with a polarimetric radar. *Journal of Applied Meteorology, 37*, 1228–1240.

Ryzhkov, A., Zrnic, D., & Gordon, B. (1998). Polarimetric method for ice water content determination. *Journal of Applied Meteorology, 37*, 125–134.

Ryzhkov, A., Pinsky, M., Pokrovsky, A., & Khain, A. (2011). Polarimetric radar observation operator for a cloud model with spectral microphysics. *Journal of Applied Meteorology and Climatology, 50*, 873–894.

Ryzhkov, A., Diederich, M., Zhang, P., & Simmer, C. (2014). Utilization of specific attenuation for rainfall estimation, mitigation of partial beam blockage, and radar networking. *Journal of Atmospheric and Oceanic Technology, 31*, 599–619.

Ryzhkov, A., Zhang, P., Reeves, H., Kumjian, M., Tschallener, T., Troemel, S., & Simmer, C. (2016). Quasi-vertical profiles – A new way to look at polarimetric radar data. *Journal of Atmospheric and Oceanic Technology, 33*, 551–562.

Ryzhkov, A., Matrosov, S., Melnikov, V., Zrnic, D., Zhang, P., Cao, Q., Knight, M., Simmer, C., & Troemel, S. (2017). Estimation of depolarization ratio using radars with simultaneous transmission/reception. *Journal of Applied Meteorology and Climatology, 56*, 1797–1816.

Ryzhkov, A., Bukovcic, P., Murphy, A., Zhang, P., & McFarquhar, G. (2018). Ice microphysical retrievals using polarimetric radar data. In *10th European Conference on Radar in Meteorology and Hydrology, 1–6 July, The Netherlands*, # 40. Retrieved from https://projects.knmi.nl/erad2018/ERAD2018_extended_abstract_040.pdf

Sassen, K. (1987). Ice cloud content from radar reflectivity. *Journal of Climate and Applied Meteorology, 26*, 1050–1053.

Stanford, M., Warble, A., Zipser, E., Strapp, J., Leroy, D., Schwarzenboeck, A., Potts, R., & Protat, A. (2017). A ubiquitous ice size bias in simulations of tropical deep convection. *Atmospheric Chemistry and Physics, 17*, 9599–9621.

Straka, J. M., Zrnic, D. S., & Ryzhkov, A. V. (2000). Bulk hydrometeor classification and quantification using multiparameter radar data. Synthesis of relations. *Journal of Applied Meteorology, 39*, 1341–1372.

Strapp, J., Korolev, A., Ratvasky, T., Potts, R., Protat, A., May, P., et al. (2016). The High Ice Water Content (HIWC) study of deep convective clouds: Report on science and technical plan. (FAA Rep. No. DOT/FAA/TC-14/31), 105pp. Retrieved from https://www.tc.faa.gov/its/worldpac/techrpt/tc14-31.pdf

Tabary, P., Boumahmoud, A., Andrieu, H., Thompson, R., Illingworth, A., LeBouar, E., & Testud, J. (2011). Evaluation of two "integrated" polarimetric quantitative precipitation estimation (QPE) algorithms at C band. *Journal of Hydrology, 405*(3), 248–260.

Thurai, M., Bringi, V. N., Carey, L. D., Gatlin, P., Schultz, E., & Petersen, W. A. (2012). Estimating the accuracy of polarimetric radar-based retrievals of drop-size distribution parameters and rain rate: An application of error variance separation using radar-derived spatial correlations. *Journal of Hydrometeorology, 13*, 1066–1079.

Tian, J., Dong, X., Xi, B., Wang, J., Homeyer, C., McFarquhar, G., & Fan, J. (2016). Retrievals of ice cloud microphysical properties of deep convective systems using radar measurements. *Journal of Geophysical Research – Atmospheres, 121*, 10820–10839.

Tiira, J., Moisseev, D., Lerber, A., Ori, D., Tokay, A., Bliven, L., & Petersen, W. (2016). Ensemble mean density and its connection to other microphysical properties of falling snow as observed in Southern Finland. *Atmospheric Measurement Techniques, 9*, 4825–4841.

Vivekanandan, J., Bringi, V., Hagen, M., & Meischner, P. (1994). Polarimetric radar studies of atmospheric ice particles. *IEEE Transactions on Geoscience and Remote Sensing, 32*, 1–10.

Yoshikawa, E., Chandrasekar, V., & Ushio, T. (2014). Raindrop size distribution (DSD) retrieval for X-band dual-polarization radar. *Journal of Atmospheric and Oceanic Technology, 31*, 387–403.

Yoshikawa, E., Chandrasekar, V., Ushio, T., & Matsuda, T. (2016). A Bayesian approach for integrated raindrop size distribution (DSD) retrieval on an X-band dual-polarization radar network. *Journal of Atmospheric and Oceanic Technology, 33*, 377–389.

Zawadzki, I., Szyrmer, W., Bell, C., & Fabry, F. (2005). Modeling of the melting layer. Part III: The density effect. *Journal of the Atmospheric Sciences, 62*, 3705–3723.

Zhang, G. (2016). *Weather radar polarimetry*. Boca Raton, FL: CRC Press. 304pp.

Zhang, G., Vivekanandan, J., & Brandes, E. (2001). A method for estimating rain rate and drop size distribution from polarimetric radar measurements. *IEEE Transactions on Geoscience and Remote Sensing, 39*, 830–841.

Appendix A

Attenuation of Microwaves in Atmospheric Gases

Two atmospheric gases, oxygen and water vapor, contribute to attenuation of microwave radiation in clear atmosphere. Specific attenuation caused by oxygen A_{O_2} depends on the microwave frequency f, atmospheric pressure$P$, and temperature T (Ulaby et al. 1981):

$$A_{O_2} = 1.110^{-2} f^2 \frac{P}{1013}\left(\frac{300}{T}\right)^2 \gamma \left[\frac{1}{\left(f - f_{O_2}\right)^2 + \gamma^2} + \frac{1}{f^2 + \gamma^2}\right]. \tag{A.1}$$

In (A.1), $f_{O_2} = 60$ GHz,

$$\gamma = \gamma_0 \frac{P}{1013}\left(\frac{300}{T}\right)^{0.85} \quad \text{(GHz)},$$

$$\gamma_0 = \left\{ \begin{array}{c} 0.59 \quad P > 333 \text{ mb} \\ 0.59\left[1 + 3.1 \times 10^{-3}(333 - P)\right] \quad 25 < P < 333 \text{ mb}. \\ 1.18 \quad P < 25 \text{ mb} \end{array} \right.$$

Specific attenuation caused by water vapor depends on the same parameters and absolute humidityρ_v:

$$A_{H_2O} = 2f^2 \rho_v \left(\frac{300}{T}\right)^{3/2} \gamma_1 \left[\left(\frac{300}{T}\right) e^{-644/T} \frac{1}{\left(f^2 - f_{H_2O}^2\right)^2} + 1.210^{-6}\right], \tag{A.2}$$

where $f_{H_2O} = 22.2$ GHz and
$\gamma_1 = 2.85 \frac{P}{1013} \left(\frac{300}{T}\right)^{0.626} \left(1 + 0.018 \frac{\rho_v T}{P}\right)$ (GHz).

A. V. Ryzhkov, D. S. Zrnic, *Radar Polarimetry for Weather Observations*,
Springer Atmospheric Sciences, https://doi.org/10.1007/978-3-030-05093-1

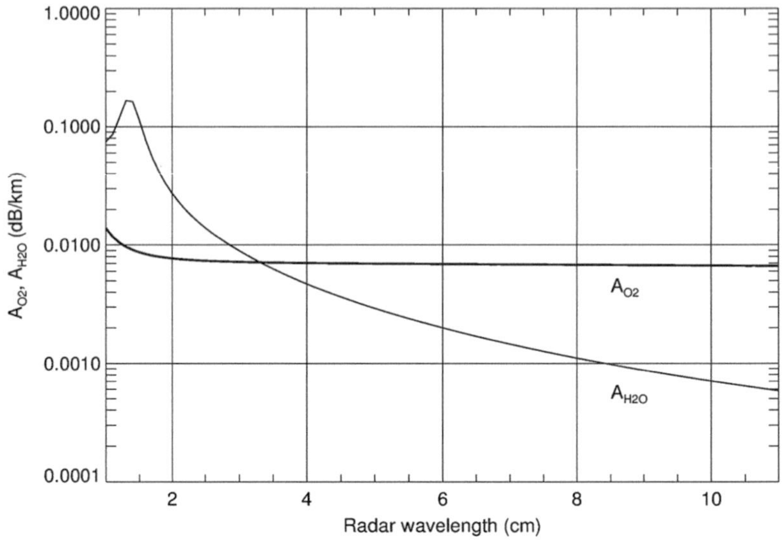

Fig. A.1 Wavelength dependencies of specific attenuations of the microwave radiation in oxygen and water vapor ($\rho_v = 7.5$ g m^{-3})

In (A.1) and (A.2), A_{O_2} and A_{H_2O} are expressed in dB km^{-1}, f is in GHz, P is in mb, T is in K°, and ρ_v is in g m^{-3}. Approximations (A.1) and (A.2) are valid for $f < 45$ GHz and $f < 100$ GHz.

Wavelength dependencies of A_{O_2} and A_{H_2O} are plotted in Fig. A.1 for $P = 1013$ mb, $T = 20$°C, and $\rho_v = 7.5$ g m^{-3}. It is evident that A_{H_2O} is a strong function of wavelength and reaches its maximum at $\lambda = 1.35$ cm corresponding to the water vapor absorption line, whereas A_{O_2} does not vary much for λ changing between 3 and 11 cm. The equations (A.1) and (A.2) indicate the dependence on temperature and pressure. These variables vary systematically with height above ground and cause predictable decrease in attenuation with increase in altitude.

Reference

Ulaby, F., Moore, R., & Fung, A. (1981). *Microwave remote sensing. Active and passive* (Vol. 1). Boston: Artech House. 456pp.

Appendix B

Angular Moments

Here we present analytical formulas for the angular moments defined by (3.45)

$$
\begin{aligned}
&A_1 =< \sin^2\psi \cos^2\alpha > , \quad A_2 =< \sin^2\psi \sin^2\alpha > , \quad A_3 =< \sin^4\psi \cos^4\alpha > , \\
&A_4 =< \sin^4\psi \sin^4\alpha > , \quad A_5 =< \sin^4\psi \cos^2\alpha \sin^2\alpha >
\end{aligned}
\tag{B.1}
$$

for the following three special cases.

Random Orientation of Hydrometeors

Assume that the scatterers are randomly oriented in the horizontal plane (i.e., azimuth angle ϕ is between 0 and 2π), the beam elevation is 0° (a good approximation for surveillance radars), and the angle between axis of scatterer and horizontal plane is uniformly distributed from 0 to $\pi/2 - \theta_v$ (angle θ_v is measured with respect to the true vertical). The maximum deviation of ($\pi/2 - \theta_v$) is referred as a "flutter angle" and this model is well suited for representing prolate shaped ice crystals like needles or chaff (Zrnic and Ryzhkov 2004; Melnikov and Straka 2013). The probability density function that represents a uniform distribution of orientation within these prescribed limits is

$$
p(\theta, \phi) = \sin\theta / (2\pi \cos\theta_v). \tag{B.2}
$$

The relations between the α, ψ, θ, and ϕ angles

A. V. Ryzhkov, D. S. Zrnic, *Radar Polarimetry for Weather Observations*,
Springer Atmospheric Sciences, https://doi.org/10.1007/978-3-030-05093-1

$$\begin{aligned}\sin\theta\cos\phi &= \sin\alpha\sin\psi\\ \cos\theta &= \sin\psi\cos\alpha\\ \cos\psi &= \sin\theta\sin\phi\end{aligned} \tag{B.3}$$

are needed to integrate various terms in (B.1). Two of the equations in (B.3) are independent, but three are listed for convenience (these are substituted into various integrands). Then integration with the prescribed distribution produces the following closed-form solutions for the angular moments $<A_i>$:

$$\begin{aligned}&A_1 = \cos^2\theta_v/3; \quad A_2 = \sin^2\theta_v/6 + 1/3; \quad A_3 = \cos^4\theta_v/5\\ &A_4 = 3(\sin^4\theta_v - 4\cos^2\theta_v/3 + 4)/40; \quad A_5 = (\cos^2\theta_v/3 - \cos^4\theta_v/5)/2.\end{aligned} \tag{B.4}$$

For a fully random orientation $\theta_v = 0$ and (B.4) reduces to

$$A_1 = A_2 = 1/3; A_3 = A_4 = 1/5; A_5 = 1/15. \tag{B.5}$$

This result is independent of the elevation angle.

Random Orientation in the Horizontal Plane

This orientation is typical for columnar crystals; these tend to be oriented with their major dimensions in the horizontal plane.

$$A_1 = \frac{1}{2}\sin^2\beta, \quad A_2 = \frac{1}{2}, \quad A_3 = \frac{3}{8}\sin^4\beta, \quad A_4 = \frac{3}{8}, \quad A_5 = \frac{1}{8}\sin^2\beta \tag{B.6}$$

In (B.6), β is the elevation angle of the radar beam (Fig. 1.5).

Two-Dimensional Axisymmetric Gaussian Distribution of Orientation

This distribution can be approximated by (3.46) and it represents a wide class of atmospheric scatterers that can be modeled as oblate spheroids, including raindrops, snowflakes, graupel, and hail. The approximation (3.46) makes possible averaging over angles ψ and α independently, and the angular moments A_1—A_5 can be expressed as the products of the following factors:

$$
\begin{aligned}
&< \sin^2\psi >= \frac{1}{2}\left(1 - r_\sigma \cos 2\overline{\psi}\right),\\
&< \sin^4\psi >= \frac{3}{8} - \frac{1}{2} r_\sigma \cos 2\overline{\psi} + \frac{1}{8} r_\sigma^4 \cos 4\overline{\psi},\\
&< \cos^2\alpha >= \frac{1}{2}\left(1 + r_{\sigma\alpha} \cos 2\overline{\alpha}\right),\\
&< \sin^2\alpha >= \frac{1}{2}\left(1 - r_{\sigma\alpha} \cos 2\overline{\alpha}\right),\\
&< \cos^4\alpha >= \frac{3}{8} + \frac{1}{2} r_{\sigma\alpha} \cos 2\overline{\alpha} + \frac{1}{8} r_{\sigma\alpha}^4 \cos 4\overline{\alpha},\\
&< \sin^4\alpha >= \frac{3}{8} - \frac{1}{2} r_{\sigma\alpha} \cos 2\overline{\alpha} + \frac{1}{8} r_{\sigma\alpha}^4 \cos 4\overline{\alpha},\\
&< \cos^2\alpha \sin^2\alpha >= \frac{1}{8}\left(1 - r_{\sigma\alpha}^4 \cos 4\overline{\alpha}\right)
\end{aligned}
\tag{B.7}
$$

where

$$
r_\sigma = \exp\left(-2\sigma^2\right), \quad r_{\sigma\alpha} = \exp\left(-2\sigma_\alpha^2\right) \tag{B.8}
$$

If the mean canting angle is zero (a reasonable assumption for a majority of hydrometeors (Ryzhkov et al. 2002)), then at low antenna elevation angle $\overline{\psi} \approx \pi/2$ and the angular moments A_1—A_5 can be approximated by (Ryzhkov 2001)

$$
\begin{aligned}
&A_1 = \frac{1}{4}(1 + r_\sigma)^2, \quad A_2 = \frac{1}{4}\left(1 - r_\sigma^2\right), \quad A_3 = \left(\frac{3}{8} + \frac{1}{2} r_\sigma + \frac{1}{8} r_\sigma^4\right)^2,\\
&A_4 = \left(\frac{3}{8} - \frac{1}{2} r_\sigma + \frac{1}{8} r_\sigma^4\right)\left(\frac{3}{8} + \frac{1}{2} r_\sigma + \frac{1}{8} r_\sigma^4\right), \quad A_5 = \frac{1}{8}\left(1 - r_\sigma^4\right)\left(\frac{3}{8} + \frac{1}{2} r_\sigma + \frac{1}{8} r_\sigma^4\right)
\end{aligned}
\tag{B.9}
$$

References

Melnikov, V., & Straka, J. (2013). Axis ratios and flutter angles of cloud ice particles: Retrievals from radar data. *Journal of Atmospheric and Oceanic Technology, 30*, 1691–1703.

Ryzhkov, A. (2001). Interpretation of polarimetric radar covariance matrix for meteorological scatterers: Theoretical analysis. *Journal of Atmospheric and Oceanic Technology, 18*, 315–328.

Ryzhkov, A., Zrnic, D., Hubbert, J., Bringi, V., Vivekanandan, J., & Brandes, E. (2002). Polarimetric radar observations and interpretation of co-cross-polar correlation coefficients. *Journal of Atmospheric and Oceanic Technology, 19*, 340–354.

Zrnic, D., & Ryzhkov, A. (2004). Polarimetric properties of chaff. *Journal of Atmospheric and Oceanic Technology, 21*, 1017–1024.

Appendix C

Dielectric Constants of Water and Solid Ice

Formulas for the dielectric constants of water and solid ice as functions of temperature and wavelength are summarized here. The real and imaginary parts of ε for water and ice can be expressed as (Ray 1972)

$$\mathrm{Re}(\varepsilon) = \varepsilon_\infty + \frac{(\varepsilon_0 - \varepsilon_\infty)\left[1 + (\lambda_0/\lambda)^{1-\alpha}\sin(\alpha\pi/2)\right]}{1 + 2(\lambda_0/\lambda)^{1-\alpha}\sin(\alpha\pi/2) + (\lambda_0/\lambda)^{2(1-\alpha)}}, \tag{C.1}$$

$$\mathrm{Im}(\varepsilon) = \pm\left(\frac{(\varepsilon_0 - \varepsilon_\infty)(\lambda_0/\lambda)^{1-\alpha}\cos(\alpha\pi/2)}{1 + 2(\lambda_0/\lambda)^{1-\alpha}\sin(\alpha\pi/2) + (\lambda_0/\lambda)^{2(1-\alpha)}} + \frac{\sigma\lambda}{18.8496 \times 10^{10}}\right), \tag{C.2}$$

where in the case of water

$$\begin{aligned}
\varepsilon_0 &= 78.54[1.0 - 4.579 \times 10^{-3}(T - 25.0) + 1.19 \times 10^{-5}(T - 25)^2 \\
&\qquad -2.8 \times 10^{-8}(T - 25)^3] \\
\varepsilon_\infty &= 5.27137 + 0.021647\ T - 0.00131198\ T^2 \\
\alpha &= -16.8129/(T + 273) + 0.0609265 \\
\lambda_0 &= 0.00033836 \exp[2513.98/(T + 273)] \\
\sigma &= 12.5664 \times 10^8
\end{aligned} \tag{C.3}$$

and in the case of ice

A. V. Ryzhkov, D. S. Zrnic, *Radar Polarimetry for Weather Observations*, Springer Atmospheric Sciences, https://doi.org/10.1007/978-3-030-05093-1

$$
\begin{aligned}
\varepsilon_0 &= 203.168 + 2.5\,T + 0.15\,T^2 \\
\varepsilon_\infty &= 3.168 \\
\alpha &= 0.288 + 0.0052\,T + 0.00023\,T^2 \\
\lambda_0 &= 0.0009990288\,\exp\{13200/[(T + 273)\,1.9869]\} \\
\sigma &= 1.26\,\exp\{-12500/[(T + 273)\,1.9869]\}.
\end{aligned}
\tag{C.4}
$$

In Formulas (C.1)–(C.4), λ is the wavelength expressed in cm and T is the temperature in $^\circ$C. Positive sign in Eq. (C.2) should be selected if the $\exp(-j\omega t)$ convention is assumed, and negative sign is appropriate in the case of the $\exp(j\omega t)$ convention.

Reference

Ray, P. (1972). Broadband complex refractive indices of ice and water. *Applied Optics, 11*, 1836–1844.

Appendix D

The Impact of Different Microphysical Factors on the Cross-Correlation Coefficient

To clarify how physical properties of atmospheric particles affect ρ_{hv}, let's consider a simple model with two scatterers, so that

$$\begin{aligned} | \rho_{hv} | e^{j\bar{\delta}} &= \frac{s_{hh}^{(1)*} s_{vv}^{(1)} + s_{hh}^{(2)*} s_{vv}^{(2)}}{\left(\left| s_{hh}^{(1)} \right|^2 + \left| s_{hh}^{(2)} \right|^2 \right)^{1/2} \left(\left| s_{vv}^{(1)} \right|^2 + \left| s_{vv}^{(2)} \right|^2 \right)^{1/2}} \\ &= \frac{| s_{hh}^{(1)} || s_{vv}^{(1)} | e^{j\delta_1} + | s_{hh}^{(2)} || s_{vv}^{(2)} | e^{j\delta_2}}{\left(\left| s_{hh}^{(1)} \right|^2 + \left| s_{hh}^{(2)} \right|^2 \right)^{1/2} \left(\left| s_{vv}^{(1)} \right|^2 + \left| s_{vv}^{(2)} \right|^2 \right)^{1/2}}, \end{aligned} \tag{D.1}$$

where superscripts 1 and 2 denote two different scatterers and δ_1 and δ_2 are their backscatter differential phases defined in (3.38). If $\delta_1 = \delta_2 = 0$ (which is a valid assumption for smaller size raindrops (Fig. 5.17)) then ρ_{hv} is a real number and

$$\rho_{hv} = \frac{| s_{hh}^{(1)} || s_{vv}^{(1)} | + | s_{hh}^{(2)} || s_{vv}^{(2)} |}{\left(\left| s_{hh}^{(1)} \right|^2 + \left| s_{hh}^{(2)} \right|^2 \right)^{1/2} \left(\left| s_{vv}^{(1)} \right|^2 + \left| s_{vv}^{(2)} \right|^2 \right)^{1/2}}. \tag{D.2}$$

Next consider several special cases to illustrate the impact of different microphysical factors on ρ_{hv}.

(a) Particles are spherical and have different sizes (Fig. D.1a), so that

A. V. Ryzhkov, D. S. Zrnic, *Radar Polarimetry for Weather Observations*, Springer Atmospheric Sciences, https://doi.org/10.1007/978-3-030-05093-1

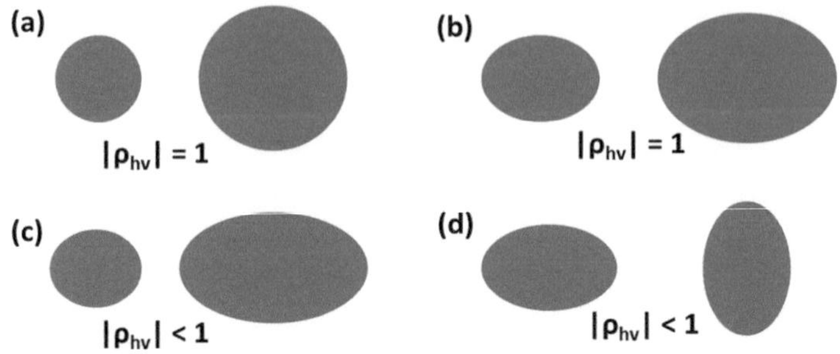

Fig. D.1 Magnitudes of the cross-correlation coefficient for different combinations of two particles in the radar resolution volume. (**a**) Spherical particles, (**b**) Nonspherical particles with equal axis ratios, (**c**) Nonspherical particles with different axis ratios, (**d**) Identical particles with different orientations

$$| s_{\mathrm{hh}}^{(1)} |=| s_{\mathrm{vv}}^{(1)} |= s_1, \quad | s_{\mathrm{hh}}^{(2)} |=| s_{\mathrm{vv}}^{(2)} |= s_2. \tag{D.3}$$

Then it follows from (D.2) that $\rho_{\mathrm{hv}} = 1$ for arbitrary number of spherical particles of various sizes.

(b) Particles are nonspherical, have different sizes, similar shape, and same orientation (Fig. D.1b), so that

$$| s_{\mathrm{hh}}^{(1)} |= p\, | s_{\mathrm{vv}}^{(1)} |, \quad | s_{\mathrm{hh}}^{(2)} |= p\, | s_{\mathrm{vv}}^{(2)} |, \quad | s_{\mathrm{vv}}^{(1)} |\neq| s_{\mathrm{vv}}^{(2)} |. \tag{D.4}$$

Again, $\rho_{\mathrm{hv}} = 1$. This means that if the axis ratio (which determines the parameter p) and canting angle do not change across the particle size distributions, then the magnitude of the cross-correlation coefficient is identically one. Note that the symmetry axis of both scatterers is vertically oriented in this example. Nonetheless, the result is invariant to the orientation as long as both (all) scatterers have the same orientation.

(c) Particles are nonspherical and have different sizes and same orientation, and their axis ratio changes as a function of size which is typical for rain because larger raindrops are more oblate than smaller ones (Fig. D.1c). Therefore

$$| s_{\mathrm{hh}}^{(1)} |= p_1\, | s_{\mathrm{vv}}^{(1)} |, \quad | s_{\mathrm{hh}}^{(2)} |= p_2\, | s_{\mathrm{vv}}^{(2)} |, \quad | s_{\mathrm{vv}}^{(1)} |\neq| s_{\mathrm{vv}}^{(2)} |, \quad p_1 \neq p_2, \tag{D.5}$$

and $\rho_{\mathrm{hv}} < 1$. This leads to the important conclusion that variability of particle shapes across particle size spectra causes reduction of ρ_{hv}.

(d) Particles are nonspherical and have same size but different orientation. As an example, one scatterer has its rotation axis directed vertically and the other horizontally (Fig. D.1d). In this case,

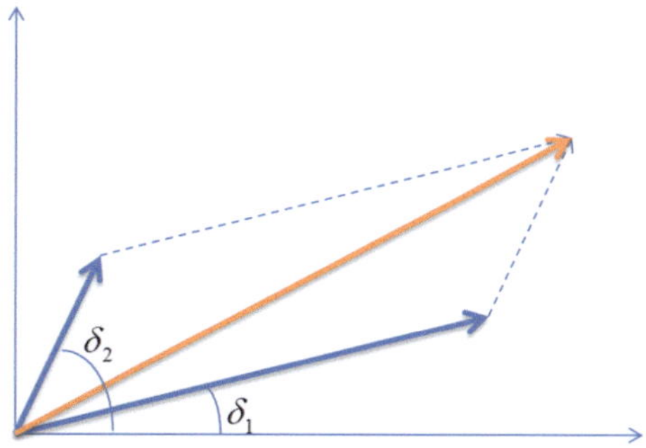

Fig. D.2 The impact of the difference between backscatter differential phases of two particles δ_2-δ_1 on the magnitude of the cross-correlation coefficient which is proportional to the length of the sum (orange) vector

$$|s_{hh}^{(1)}| = p\,|s_{vv}^{(1)}|, \quad |s_{hh}^{(2)}| = \frac{1}{p}|s_{vv}^{(2)}|, \quad |s_{vv}^{(1)}| = |s_{hh}^{(2)}|, \tag{D.6}$$

and it can be shown that $\rho_{hv} < 1$. Hence, variability of orientation of particles (identical or not identical) causes the drop in the cross-correlation coefficient.

(e) Two particles of the same size, shape, and orientation but with different phase composition (e.g., raindrop and dry/wet graupel, or two melting graupel particles with different fractional mass water contents).

The situation with such two particles is described by conditions (D.4) and (D.5) because neither their reflectivities nor differential reflectivities (i.e., p_1^2 and p_2^2) are equal. Therefore, $\rho_{hv} < 1$. Larger difference in the effective dielectric constants of the two hydrometeors lowers more their ρ_{hv}.

So far we assumed that $\delta_1 = \delta_2 = 0$. If $\delta_1 \neq \delta_2$, then ρ_{hv} depends on the difference between δ_1 and δ_2 because two terms in the numerator of (D.1) are summed up as vectors (Fig. D.2). The value of ρ_{hv} can be substantially reduced with increase in $|\delta_1 - \delta_2|$. Therefore, if the backscatter differential phase changes significantly across the particle size spectrum, ρ_{hv} tends to decrease. This happens, for example, in rain at C band if raindrop size distribution contains drops with sizes between 5 and 7 mm for which δ changes from $-15°$ to $45°$ (Fig. 5.17). The cross-correlation coefficient also decreases if hydrometeors with small intrinsic δ (such as raindrops) are mixed with hydrometeors having large δ (such as wet snowflakes or hailstones).

Index

A. V. Ryzhkov, D. S. Zrnic, *Radar Polarimetry for Weather Observations*, Springer Atmospheric Sciences, https://doi.org/10.1007/978-3-030-05093-1

Q

R

S